LIFE SCIENCES

(Continued on inside of back cover)

EPIC: EXPLORATION PROGRAMS IN CALCULUS, 2/E

by James W. Burgmeier and Larry L. Kost

This interactive software package for the IBM Personal Computer is a collection of programs that allows users to explore a variety of calculus concepts, including: equations of lines, intersections of curves, maxima and minima, differentiation, and anti-derivatives. The hardware requirements include a 256K memory, one disk drive, and a graphics card. The software also offers the following features:

1. *Users may superimpose graphs of functions with parameters.*
2. *Users have complete control over the range used for the x and y variables on graphs.*
3. *Functions may include parameters that users can alter (e.g., users may enter a sin bx and specify a,b later).*
4. *Users can invoke options or gain information by a single keystroke (e.g., they can change the range along the x axis, get the max and min of the function, or trace polar curves).*

A departmental version of this software package is available and offers the same features listed above, as well as:

1. *A built-in math calculator with 25 memory addresses and 10 formula saving registers*
2. *Help screens and on-screen reminders*
3. *Extended options for: function plotting, integration, and parametric/polar equations*

ORDER YOUR COPY TODAY!

Please send me:

☐ **EPIC: EXPLORATION PROGAMS IN CALCULUS, 2/E**
(student version) (Title Code #28325–9) . . . . . $ 13.00

☐ **EPIC: EXPLORATION PROGRAMS IN CALCULUS, 2/E**
(departmental version) (Title Code #28323–4) $225.00

QUANTITY	TITLE/AUTHOR	TITLE CODE	PRICE	TOTAL
(Sample Entry) 2	Burgmeier/EPIC (Student)	28325–9	$13.00	$26.00

TOTAL PRICE: _____

SAVE!
If payment accompanies order, plus your state's sales tax where applicable, Prentice Hall pays postage and handling charges. Same return privilege refund guaranteed.

☐ PAYMENT ENCLOSED—shipping and handling to be paid by publisher (please include your state's tax where applicable).

MAIL TO:
PRENTICE HALL
Book Distribution Center
Route 59 at Brook Hill Drive
West Nyack, NY 10995

NAME _____

ORGANIZATION _____

STREET ADDRESS _____

CITY _____ STATE _____ ZIP _____

I Prefer to charge my ☐ VISA ☐ MasterCard
Card Number _____ Expiration Date _____

Signature _____

All prices in this catalog are subject to change without notice. Prices not valid outside the U.S.

Dept. 1   D–EPSD–FS(0)

Calculus

2ND EDITION

Calculus

For the Managerial, Life, and Social Sciences

Ernest F. Haeussler, Jr.

Richard S. Paul

The Pennsylvania State University

PRENTICE HALL, Englewood Cliffs, N.J. 07632

Library of Congress Cataloging-in-Publication Data

Haeussler, Ernest F.
  Calculus for the managerial, life, and social
sciences.

  Bibliography: p.
  Includes index.
  1. Calculus.   I. Paul, Richard S.   II.   Title.
QA303.H15   1988   515   87-18834
ISBN 0-13-111147-7

Interior design: Janet Schmid
Cover design: Janet Schmid
Cover photograph: © Paul Silverman
Manufacturing buyer: Paula Benevento

 © 1988, 1980 by Prentice Hall
A Division of Simon & Schuster
Englewood Cliffs, New Jersey 07632

Printed in the United States of America

10 9 8 7 6 5 4 3 2 1

ISBN 0-13-111147-7

Prentice-Hall International (UK) Limited, *London*
Prentice-Hall of Australia Pty. Limited, *Sydney*
Prentice-Hall Canada Inc., *Toronto*
Prentice-Hall Hispanoamericana, S.A., *Mexico*
Prentice-Hall of India Private Limited, *New Delhi*
Prentice-Hall of Japan, Inc., *Tokyo*
Simon & Schuster Asia Pte. Ltd., *Singapore*
Editora Prentice-Hall do Brasil, Ltda., *Rio de Janeiro*

Contents

Preface

This second edition provides a foundation in those topics of calculus that are relevant to students in the managerial, life, and social sciences. It begins with functions and progresses through both single-variable and multivariable calculus. Technical proofs, conditions, and so on, are sufficiently described but not overdone. At times, informal intuitive arguments are given to preserve clarity. An algebra review is included in an appendix and provides a convenient reference for students.

An abundance and variety of applications for the intended audience appear throughout the book so that students continually see ways in which the mathematics they are learning can be used. These applications cover such diverse areas as business, economics, biology, medicine, sociology, psychology, ecology, statistics, and archaeology. An index of applications is provided on the endpapers. Many of these real-world situations are drawn from literature and are documented by references. Such documentation is indicated by a boldface number enclosed in brackets, such as [1], and the source appears in the reference list at the back of the book. In many of the applications drawn from source material, the background and context are given in order to stimulate interest. However, the text is virtually self-contained in the sense that it assumes no prior exposure to the concepts on which the applications are based.

For this edition several changes have been made. Considerable effort has been devoted to improving the exposition and the exercise sets. Material has been rewritten and reorganized to afford greater clarity. Chapter summaries are now included. An optional chapter on trigonometry, including the calculus of trigonometric functions, has been added (Chapter 9). New topics are Newton's method (Sec. 4.2), elasticity of demand (Sec. 4.3), and consumers' and producers' surplus (Sec. 5.7). Functions are now considered in two sections (Secs. 1.1 and 1.2). Also, lines and applications of linear functions now occur in separate sections (Secs. 1.5 and 1.6). Continuity is treated more intuitively than before. Curve sketching is now a separate chapter. The approach to the definite integral has been revised so as to be more accessible to today's students. Although the definite integral is defined as a limit of a sum, the actual computation of such limits is not emphasized.

Interspersed throughout the text are many warnings to the student that point out commonly made errors. These warnings are indicated with the heading "Pitfall."

Each chapter has a review section that contains a list of important terms and symbols, a chapter summary, and numerous review problems.

More than 2800 exercises are included. In each exercise set, the problems are given in increasing order of difficulty. However, there is always a sufficient number that the student can do without difficulty so that confidence may be gained and the basic concepts studied may be fixed firmly in mind. Answers to odd-numbered problems appear at the end of the book. For many of the differentiation problems in Chapter 2, the answers appear in both unsimplified and simplified forms. This allows students to check their procedures and work.

As an aid in planning a course outline, we remark that, should time not permit, the following sections may be deleted: 4.2, 4.4, 5.6, 5.7, 7.2, 7.3, 7.4, 7.6, 8.6, 8.7, and 8.8. Section 6.5 (logarithmic differentiation) may be omitted if the derivative of u^v is of no concern.

Available from the publisher is an extensive instructor's manual that contains solutions to all exercises. Also available is a test bank and a student's solution manual.

We express our appreciation to the following colleagues who contributed comments and suggestions that were valuable to us in developing the manuscript:

Ron Atkinson (University of Tennessee at Nashville),

Wilson Banks (Illinois State University),

John Bishir (North Carolina State University at Raleigh),

Micheal Dyer (University of Oregon),

Anne Grams (University of Tennessee at Nashville),

Robert Hathway (Illinois State University),

Burt C. Horne, Jr. (Virginia Polytechnic Institute and State University),

John Laible (Eastern Illinois University),

Ralph McWilliams (Florida State University),

Glenda K. Owens (Central State University),

Franklin E. Schroeck, Jr. (Florida Atlantic University),

and Donald R. Sherbert (University of Illinois at Urbana-Champaign).

We are indebted to Jerry Covert of the Department of Biology of The Pennsylvania State University for his review and comments relative to biological applications.

In addition, we especially wish to acknowledge and sincerely thank those colleagues who reviewed the entire manuscript for this edition and offered many helpful suggestions for its improvement:

Stephen Andrielli (LaSalle College),

Dennis Bertholf (Oklahoma State University),

Charles Clever (South Dakota State University),

Harvey Fox (University of Wisconsin),

and John Gilbert (Mississippi State University).

Finally, we express our sincere appreciation to John Morgan, our production editor, for his patience, expert assistance, and enthusiastic cooperation.

Ernest F. Haeussler, Jr.
Richard S. Paul

Functions and Limits

1.1 FUNCTIONS

In the seventeenth century, Gottfried Wilhelm Leibniz, one of the inventors of calculus, introduced the term *function* into the mathematical vocabulary. The concept of a function is one of the most basic in all of mathematics, and it is essential to the study of calculus. As you will see, calculus involves two basic processes: *differentiation* and *integration*. Because both processes are performed on functions, a logical starting point for us is a discussion of functions.

Briefly, a function is a special type of input-output relation that expresses how one quantity (the *output*) depends on another quantity (the *input*). For example, when money is invested at some interest rate, the interest I (output) depends on the length of time t (input) that the money is invested. To express this dependence, we say that "I is a function of t." Functional relations like this are usually specified by a formula that shows what must be done to the input to find the output.

To illustrate, suppose that \$100 earns simple interest at an annual rate of 6 percent. Then it can be shown that interest and time are related by the formula

$$I = 100(0.06)t, \tag{1}$$

where I is in dollars and t is in years. For example,

$$\text{if } t = \tfrac{1}{2}, \quad \text{then} \quad I = 100(0.06)(\tfrac{1}{2}) = 3. \tag{2}$$

Thus Formula (1) assigns to the input $\tfrac{1}{2}$ the output 3. We can think of Formula (1) as defining a *rule*: Multiply t by $100(0.06)$. The rule assigns to each input number t exactly one output number I, which we symbolize by the following arrow notation:

$$t \rightarrow I \quad \text{or} \quad t \rightarrow 100(0.06)t.$$

This rule is an example of a *function* in the following sense:

Definition

*A **function** is a rule that assigns to each input number exactly one output number. The set of all input numbers to which the rule applies is called the **domain** of the function. The set of all output numbers is called the **range**.*

For the interest function defined by Formula (1), the input number t cannot be negative because negative time makes no sense. Thus the domain consists of all nonnegative numbers; that is, all $t \geq 0$. From (2) we see that when the input is $\frac{1}{2}$, the output is 3. Thus 3 is in the range.

We have been using the term *function* in a restricted sense because, in general, the inputs or outputs do not have to be numbers. For example, a list of states and their capitals assigns to each state its capital (exactly one output). Thus a function is implied. However, for the time being we shall consider only functions whose domains and ranges consist of real numbers.

A variable that represents input numbers for a function is called an **independent variable**. A variable that represents output numbers is called a **dependent variable** because its value *depends* on the value of the independent variable. We say that the dependent variable is a *function of* the independent (input) variable. That is, output is a function of input. Thus for the interest formula $I = 100(0.06)t$, the independent variable is t, the dependent variable is I, and I is a function of t.

As another example, the equation (or formula)

$$y = x + 2 \tag{3}$$

defines y as a function of x. It gives the rule: Add 2 to x. This rule assigns to each input x exactly one output $x + 2$, which is y. Thus if $x = 1$, then $y = 3$; if $x = -4$, then $y = -2$. The independent variable is x and the dependent variable is y.

Not all equations in x and y define y as a function of x. For example, let $y^2 = x$. If x is 9, then $y^2 = 9$, so $y = \pm 3$. Thus to the input 9 there are assigned not one but *two* output numbers, 3 and -3. This violates the definition of a function, so y is **not** a function of x.

On the other hand, some equations in two variables define either variable as a function of the other variable. For example, if $y = 2x$, then for each input x there is exactly one output, $2x$. Thus y is a function of x. However, solving the equation for x gives $x = y/2$. For each input y there is exactly one output, $y/2$. Thus x is a function of y.

Usually, the letters f, g, h, F, G, and so on are used to represent function rules. For example, Eq. (3) above ($y = x + 2$) defines y as a function of x, where the rule is add 2 to the input. Suppose we let f represent this rule.

Then we say that f is the function. To indicate that f assigns to the input 1 the output 3, we write $f(1) = 3$, which is read "f of 1 equals 3." Similarly, $f(-4) = -2$. More generally, if x is any input, we have the following notation:

$f(x)$, which is read "f of x," means the output number in the range of f that corresponds to the input number x in the domain.

$$\text{input} \\ \downarrow \\ f(x) \\ \uparrow \\ \text{output}$$

Thus the output $f(x)$ is the same as y. But since $y = x + 2$, we may write $y = f(x) = x + 2$ or simply

Pitfall

$f(x)$ does **not** mean f times x. $f(x)$ is the output that corresponds to the input x.

$$f(x) = x + 2.$$

For example, to find $f(3)$, which is the output corresponding to the input 3, we replace each x in $f(x) = x + 2$ by 3:

$$f(3) = 3 + 2 = 5.$$

Likewise,

$$f(8) = 8 + 2 = 10,$$
$$f(-4) = -4 + 2 = -2.$$

Output numbers such as $f(-4)$ are called **function values** (or functional values). Keep in mind that they are in the range of f.

Quite often, functions are defined by "functional notation." For example, the equation $g(x) = x^3 + x^2$ defines the function g that assigns to an input number x the output number $x^3 + x^2$.

$$g: \quad x \to x^3 + x^2.$$

Thus g adds the cube and the square of an input number. Some function values are

$$g(2) = 2^3 + 2^2 = 12,$$
$$g(-1) = (-1)^3 + (-1)^2 = -1 + 1 = 0,$$
$$g(t) = t^3 + t^2,$$
$$g(x + 1) = (x + 1)^3 + (x + 1)^2.$$

Note that $g(x + 1)$ was found by replacing each x in $x^3 + x^2$ by $x + 1$.

When we refer to the function g defined by $g(x) = x^3 + x^2$, we shall feel free to call the equation itself a function. Thus we speak of "the function $g(x) = x^3 + x^2$," and similarly, "the function $y = x + 2$."

Let's be specific about the domain of a function that is given by an equation. Unless otherwise stated, *the domain consists of all real numbers for which that equation makes sense and gives function values that are real numbers.* For example, suppose

$$h(x) = \frac{1}{x - 6}.$$

Here any real number can be used for x except 6, because the denominator is 0 when x is 6 (we cannot divide by zero). Thus the domain of h is understood to be all real numbers except 6.

EXAMPLE 1   *Find the domain of each function.*

a. $f(x) = \dfrac{x}{x^2 - x - 2}.$

We cannot divide by zero, so we must find any values of x that make the denominator 0. These *cannot* be input numbers. Thus we set the denominator equal to 0 and solve for x.

$$x^2 - x - 2 = 0 \qquad \text{(quadratic equation)},$$
$$(x - 2)(x + 1) = 0 \qquad \text{(factoring)},$$
$$x = 2, -1.$$

Therefore, the domain of f is all real numbers *except* 2 and -1.

b. $g(t) = \sqrt{2t - 1}.$

$\sqrt{2t - 1}$ is a real number if $2t - 1$ is greater than or equal to 0. If $2t - 1$ is negative, then $\sqrt{2t - 1}$ is not a real number (it is an *imaginary number*). Since function values must be real numbers, we must assume that

$$2t - 1 \geq 0,$$
$$2t \geq 1 \qquad \text{(adding 1 to both sides)},$$
$$t \geq \frac{1}{2} \qquad \text{(dividing both sides by 2)}.$$

Thus the domain is all real numbers t such that $t \geq \frac{1}{2}$.

EXAMPLE 2   *Domain and functional notation.*

Let $g(x) = 3x^2 - x + 5$. Any real number can be used for x, so the domain of g is all real numbers.

$$g(x) = 3x^2 - x + 5.$$

Find $g(z)$: $\quad g(z) = 3(z)^2 - z + 5 = 3z^2 - z + 5.$

Find $g(r^2)$: $\quad g(r^2) = 3(r^2)^2 - r^2 + 5 = 3r^4 - r^2 + 5.$

Find $g(x + h)$: $\quad g(x + h) = 3(x + h)^2 - (x + h) + 5$
$$= 3(x^2 + 2hx + h^2) - x - h + 5$$
$$= 3x^2 + 6hx + 3h^2 - x - h + 5.$$

Pitfall

Don't be confused by notation. In Example 2 we found $g(x + h)$ by replacing each x in $g(x) = 3x^2 - x + 5$ by the input $x + h$. **Don't** write the function and then add h. That is, $g(x + h) \neq g(x) + h$.

$$g(x + h) \neq 3x^2 - x + 5 + h.$$

Also, **don't** use the distributive law on $g(x + h)$. It does **not** stand for multiplication.

$$g(x + h) \neq g(x) + g(h).$$

EXAMPLE 3   If $f(x) = x^2$, find $\dfrac{f(x + h) - f(x)}{h}$ for $h \neq 0$.

Here the numerator is a difference of function values.

$$\frac{f(x + h) - f(x)}{h} = \frac{(x + h)^2 - x^2}{h}$$

$$= \frac{x^2 + 2hx + h^2 - x^2}{h} = \frac{2hx + h^2}{h}$$

$$= \frac{h(2x + h)}{h} = 2x + h.$$

In some cases the domain of a function is restricted for physical or economic reasons. For example, the previous interest function $I = 100(0.06)t$ has $t \geq 0$ because t represents time. Example 4 will give another illustration.

EXAMPLE 4

Suppose that the equation $p = 100/q$ describes the relationship between the price per unit, p, of a certain product and the number of units, q, of that product that consumers will buy (that is, demand) per week at that price. This equation is called a *demand equation* for the product. If q is an input number, then to each value of q there is assigned exactly one output number p:

$$q \to \frac{100}{q} = p.$$

For example,

$$20 \to \frac{100}{20} = 5;$$

that is, when q is 20, then p is 5. Thus price p is a function of quantity demanded, q. This function is called a **demand function**. The independent variable is q, and p is the dependent variable. Since q cannot be 0 (division by 0 is not defined) and cannot be negative (q represents quantity), the domain is all values of q such that $q > 0$.

We have seen that a function is essentially a *correspondence* whereby to each input number in the domain there is assigned exactly one output number in the range. For the correspondence given by $f(x) = x^2$, some sample assignments are shown by the arrows in Fig. 1.1. The next example shows a functional correspondence that is not given by an algebraic formula.

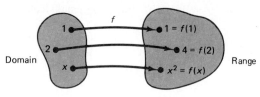

FIGURE 1.1

EXAMPLE 5

Supply schedule

p	q
Price per unit in dollars	Quantity supplied per week
500	11
600	14
700	17
800	20

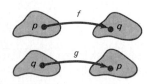

FIGURE 1.2

The table in Fig. 1.2 is a *supply schedule*. It gives a correspondence between the price p of a certain product and the quantity q that producers will supply per week at that price. For each price there corresponds exactly one quantity, and vice versa.

If p is the independent variable, then q is a function of p, say $q = f(p)$, and

$$f(500) = 11, \quad f(600) = 14, \quad f(700) = 17, \quad \text{and} \quad f(800) = 20.$$

Similarly, if q is the independent variable, then p is a function of q, say $p = g(q)$, and

$$g(11) = 500, \quad g(14) = 600, \quad g(17) = 700, \quad \text{and} \quad g(20) = 800.$$

We speak of f and g as **supply functions**. Notice from the supply schedule that as price per unit increases, the producers are willing to supply more units per week.

EXERCISE 1.1

In Problems **1–12**, *give the domain of each function.*

1. $f(x) = \dfrac{3}{x}$.

2. $g(x) = \dfrac{x}{3}$.

3. $h(x) = \sqrt{x - 5}$.

4. $H(z) = \dfrac{1}{\sqrt{z}}$.

5. $F(t) = 3t^2 + 5$.

6. $H(x) = \dfrac{x}{x + 2}$.

7. $f(x) = \dfrac{3x - 1}{2x + 5}$.

8. $g(x) = \sqrt{4x + 3}$.

9. $G(y) = \dfrac{4}{y^2 - y}$.

10. $f(x) = \dfrac{x + 1}{x^2 + 6x + 5}$.

11. $h(s) = \dfrac{4 - s^2}{2s^2 - 7s - 4}$.

12. $G(r) = \dfrac{2}{r^2 + 1}$.

In Problems **13–24**, *find the function values for each function.*

13. $f(x) = 5x$; $f(0)$, $f(3)$, $f(-4)$.

14. $H(s) = s^2 - 3$; $H(4)$, $H(\sqrt{2})$, $H(\tfrac{2}{3})$.

15. $G(x) = 2 - x^2$; $G(-8)$, $G(u)$, $G(u^2)$.

16. $f(x) = 7x$; $f(s)$, $f(t + 1)$, $f(x + 3)$.

17. $g(u) = u^2 + u$; $g(-2)$, $g(2v)$, $g(-x^2)$.

18. $h(v) = \dfrac{1}{\sqrt{v}}$; $h(16)$, $h\left(\dfrac{1}{4}\right)$, $h(1 - x)$.

19. $f(x) = x^2 + 2x + 1$; $f(1)$, $f(-1)$, $f(x + h)$.

20. $H(x) = (x + 4)^2$; $H(0)$, $H(2)$, $H(t - 4)$.

21. $g(x) = \dfrac{x - 5}{x^2 + 4}$; $g(5)$, $g(3x)$, $g(x + h)$.

22. $H(x) = \sqrt{4 + x}$; $H(-4)$, $H(-3)$, $H(x + 1) - H(x)$.

23. $f(x) = x^{4/3}$; $f(0)$, $f(64)$, $f(\tfrac{1}{8})$.

24. $g(x) = x^{2/5}$; $g(32)$, $g(-64)$, $g(t^{10})$.

In Problems 25–28, find the values of (a) $f(x + h)$ and (b) $\dfrac{f(x + h) - f(x)}{h}$; simplify your answers. Assume $h \neq 0$.

25. $f(x) = 3x - 4$.   26. $f(x) = \dfrac{x}{2}$.

27. $f(x) = x^2 + 2x$.   28. $f(x) = 2x^2 - 3x - 5$.

In Problems 29–32, is y a function of x? Is x a function of y?

29. $y - 3x - 4 = 0$.   30. $x^2 + y = 0$.

31. $y = 7x^2$.   32. $x^2 + y^2 = 1$.

33. The formula for the area A of a circle of radius r is $A = \pi r^2$. Is the area a function of the radius?

34. Suppose that $f(b) = ab^2 + a^2 b$. Find (a) $f(a)$ and (b) $f(ab)$.

35. A business with an original capital of \$10,000 has income and expenses each week of \$2000 and \$1600, respectively. If all profits are retained in the business, express the value V of the business at the end of t weeks as a function of t.

36. If a \$30,000 machine depreciates 2 percent of its original value each year, find a function f that expresses its value V after t years have elapsed.

37. If q units of a certain product are sold (q is nonnegative), the profit P is given by the equation $P = 1.25q$. Is P a function of q? What is the dependent variable; the independent variable?

38. An insurance company examined the records of a group of individuals hospitalized for a particular illness. It was found that the total proportion who had been discharged at the end of t days of hospitalization is given by $f(t)$, where

$$f(t) = 1 - \left(\frac{300}{300 + t}\right)^3.$$

a. Evaluate $f(0)$.

b. Evaluate $f(100)$.

c. Evaluate $f(300)$.

d. At the end of how many days was 0.999 of the group discharged?

39. A psychophysical experiment (adapted from [1]) was conducted to analyze human response to electrical shocks. The subjects received a shock of a certain intensity. They were told to assign a magnitude of 10 to this particular shock, called the standard stimulus. Then other shocks (stimuli) of various intensities were given. For each one the response R was to be a number that indicated the perceived magnitude of the shock relative to that of the standard stimulus. It was found that R was a function of the intensity I of the shock (I in microamperes) and was estimated by

$$R = f(I) = \frac{I^{4/3}}{2500}, \qquad 500 \le I \le 3500.$$

a. Evaluate $f(1000)$.

b. Evaluate $f(2000)$.

c. Suppose that I_0 and $2I_0$ are in the domain of f. Express $f(2I_0)$ in terms of $f(I_0)$. What effect does the doubling of intensity have on response?

40. In a paired-associate learning experiment [2], the probability of a correct response as a function of the number n of trials has the form

$$P(n) = 1 - \frac{1}{2}(1 - c)^{n-1}, \qquad n \ge 1,$$

where the estimated value of c is 0.344. Find $P(1)$ and $P(2)$ by using this value of c.

41. Table 1.1 is called a *demand schedule*. It gives a correspondence between the price p of a product and the quantity q that consumers will demand (that is, purchase) at that price.

a. If $p = f(q)$, list the numbers in the domain of f. Find $f(2900)$ and $f(3000)$.

b. If $q = g(p)$, list the numbers in the domain of g. Find $g(10)$ and $g(17)$.

TABLE 1.1
Demand Schedule

PRICE PER UNIT, p (DOLLARS)	QUANTITY DEMANDED PER WEEK, q
10	3000
12	2900
17	2300
20	2000

1.2 SPECIAL FUNCTIONS

In this section we look at functions having special forms and representations. We begin with perhaps the simplest type of function there is: a *constant function*.

EXAMPLE 1

Let $h(x) = 2$. The domain of h is all real numbers. All function values are 2. For example,

$$h(10) = 2, \quad h(-387) = 2, \quad h(x + 3) = 2.$$

We call h a *constant function*. More generally, we have this definition.

Any function of the form $h(x) = c$, where c is a *constant*, is called a **constant function**.

A constant function belongs to a broader class of functions, called *polynomial functions*. In general, a function of the form

$$f(x) = c_n x^n + c_{n-1} x^{n-1} + \cdots + c_1 x + c_0,$$

where n is a nonnegative integer and $c_n, c_{n-1}, \ldots, c_0$ are constants with $c_n \neq 0$, is called a **polynomial function** (in x). The number n is called the **degree** of the function, and c_n is the **leading coefficient**. Thus $f(x) = 3x^2 - 8x + 9$ is a polynomial function of degree 2 with leading coefficient 3. Likewise, $g(x) = 4 - 2x$ has degree 1 and leading coefficient -2. Polynomial functions of degree 1 or 2 are called **linear** or **quadratic functions,** respectively. Hence $g(x) = 4 - 2x$ is linear and $f(x) = 3x^2 - 8x + 9$ is quadratic. Note that a nonzero constant function, such as $f(x) = 5$ [which can be written as $f(x) = 5x^0$], is a polynomial function of degree 0. The constant function $f(x) = 0$ is also considered a polynomial function but has no degree assigned to it. The domain of any polynomial function is all real numbers.

EXAMPLE 2

a. $f(x) = x^3 - 6x^2 + 7$ is a polynomial (function) of degree 3 with leading coefficient 1.

b. $g(x) = \dfrac{2x}{3}$ is a linear function with leading coefficient $\dfrac{2}{3}$.

c. $f(x) = \dfrac{2}{x^3}$ is *not* a polynomial function. Because $f(x) = 2x^{-3}$ and the exponent for x is not a nonnegative integer, this function does not have the proper form for a polynomial. Similarly, $g(x) = \sqrt{x}$ is not a polynomial because $g(x) = x^{1/2}$.

Another type of function is a *rational function*, which involves polynomials.

> A function that is a quotient of polynomial functions is called a **rational function**.

EXAMPLE 3

a. $f(x) = \dfrac{x^2 - 6x}{x + 5}$ is a rational function since the numerator and denominator are each polynomials. Note that this rational function is not defined for $x = -5$.

b. $g(x) = 2x + 3$ is a rational function since $2x + 3 = \dfrac{2x + 3}{1}$. In fact, every polynomial function is also a rational function.

Sometimes more than one equation is needed to define a function, as Example 4 shows.

EXAMPLE 4

Let

$$F(s) = \begin{cases} 1, & \text{if } -1 \le s < 1, \\ 0, & \text{if } 1 \le s \le 2, \\ s - 3, & \text{if } 2 < s \le 3. \end{cases}$$

This is called a **compound function** because it is defined by more than one equation. Here s is the independent variable and the domain of F is all s such that $-1 \le s \le 3$. The value of s determines which equation to use.

Find $F(0)$:   Since $-1 \le 0 < 1$, we have $F(0) = 1$.

Find $F(2)$:   Since $1 \le 2 \le 2$, we have $F(2) = 0$.

Find $F(\frac{9}{4})$:   Since $2 < \frac{9}{4} \le 3$, we substitute $\frac{9}{4}$ for s in $s - 3$.

$$F(\tfrac{9}{4}) = \tfrac{9}{4} - 3 = -\tfrac{3}{4}.$$

EXAMPLE 5

The function $f(x) = |x|$ is called the *absolute value function*. Recall that the **absolute value**, or **magnitude**, of a real number x is denoted $|x|$ and is defined by

$$|x| = \begin{cases} x, & \text{if } x \ge 0, \\ -x, & \text{if } x < 0. \end{cases}$$

Thus the domain of f is all real numbers. Some function values are

$$f(16) = |16| = 16,$$
$$f(-\tfrac{4}{3}) = |-\tfrac{4}{3}| = -(-\tfrac{4}{3}) = \tfrac{4}{3},$$
$$f(0) = |0| = 0.$$

In our next examples we make use of *factorial notation*.

The symbol $r!$, where r is a positive integer, is read "**r factorial**." It represents the product of the first r positive integers:
$$r! = 1 \cdot 2 \cdot 3 \cdots r.$$
We define $0!$ to be 1.

EXAMPLE 6

a. $5! = 1 \cdot 2 \cdot 3 \cdot 4 \cdot 5 = 120.$

b. $3!(6 - 5)! = 3! \cdot 1! = (3 \cdot 2 \cdot 1)(1) = (6)(1) = 6.$

c. $\dfrac{4!}{0!} = \dfrac{4 \cdot 3 \cdot 2 \cdot 1}{1} = \dfrac{24}{1} = 24.$

EXAMPLE 7   *Suppose that two black guinea pigs are bred and produce exactly five offspring. Under certain conditions it can be shown that the probability P that exactly r of the offspring will be brown and the others black is a function of r, say P = P(r), where*

$$P(r) = \frac{5! \left(\dfrac{1}{4}\right)^{r} \left(\dfrac{3}{4}\right)^{5-r}}{r! \, (5 - r)!}, \qquad r = 0, 1, 2, \ldots, 5.$$

The letter P in P = P(r) is used in two ways. On the right side, P represents the function rule. On the left side, P represents the dependent variable. The domain of P is all integers from 0 to 5 inclusive. Find the probability that exactly three guinea pigs will be brown.

We want to find $P(3)$.

$$P(3) = \frac{5! \left(\frac{1}{4}\right)^3 \left(\frac{3}{4}\right)^2}{3!2!} = \frac{120\left(\frac{1}{64}\right)\left(\frac{9}{16}\right)}{6(2)} = \frac{45}{512}.$$

EXERCISE 1.2

In Problems 1–4, give the domain of each function.

1. $H(z) = 10.$ **2.** $f(t) = \pi.$

3. $f(x) = \begin{cases} 2x, & \text{if } x > 1, \\ 3, & \text{if } x \le 1. \end{cases}$

4. $f(x) = \begin{cases} 4, & \text{if } x = 3, \\ x^2, & \text{if } 1 \le x < 3. \end{cases}$

In Problems 5–8, give (a) the degree and (b) the leading coefficient of the given polynomial function.

5. $F(x) = 2x^3 - 5x^2 + 6.$ **6.** $f(x) = x.$

7. $f(x) = 2 - 3x^4 + 2x.$ **8.** $f(x) = 9.$

In Problems 9–14, find the function values for each function.

9. $f(x) = 12;$ $f(2),$ $f(t + 8),$ $f(-\sqrt{17}).$

10. $g(x) = |x - 3|;$ $g(10),$ $g(3),$ $g(-3).$

11. $F(t) =$
$\begin{cases} 1, & \text{if } t > 0 \\ 0, & \text{if } t = 0; \\ -1, & \text{if } t < 0 \end{cases}$ $F(10),$ $F(-\sqrt{3}),$ $F(0),$ $F(-\frac{18}{5}).$

12. $f(x) = \begin{cases} 4, & \text{if } x \ge 0 \\ 3, & \text{if } x < 0 \end{cases};$ $f(3),$ $f(-4),$ $f(0).$

13. $G(x) =$
$\begin{cases} x, & \text{if } x \ge 3 \\ 2 - x, & \text{if } x < 3 \end{cases};$ $G(8),$ $G(3),$ $G(-1),$ $G(1).$

14. $h(r) = \begin{cases} 3r - 1, & \text{if } r > 2 \\ r^2 - 4r + 7, & \text{if } r < 2 \end{cases};$ $h(3),$ $h(-3),$ $h(2).$

15. In manufacturing a component for a machine, the initial cost of a die is $850 and all other additional costs are $3 per unit produced.

a. Express the total cost C (in dollars) as a linear function of the number q of units produced.

b. How many units are produced if the total cost is $1600?

16. If a principal of P dollars is invested at a simple annual interest rate of r for t years, express the total accumulated amount of the principal and interest as a function of t. Is your result a linear function of t?

17. Under certain conditions, if two brown-eyed parents have exactly three children, the probability P that there will be exactly r blue-eyed children is given by the function $P = P(r)$, where

$$P(r) = \frac{3! \left(\frac{1}{4}\right)^r \left(\frac{3}{4}\right)^{3-r}}{r! \, (3 - r)!}, \qquad r = 0, 1, 2, 3.$$

Find the probability that exactly two of the children will be blue-eyed.

18. In Example 7 find the probability that all five offspring will be brown-eyed.

19. Bacteria are growing in a culture. The time t (in hours) for the number of bacteria to double in number (generation time) is a function of the temperature T (in °C) of the culture. This function is given by (adapted from [3])

$$t = f(T) = \begin{cases} \dfrac{1}{24}T + \dfrac{11}{4}, & \text{if } 30 \le T \le 36, \\ \dfrac{4}{3}T - \dfrac{175}{4}, & \text{if } 36 < T \le 39. \end{cases}$$

a. Determine the domain of f.

b. Find $f(30)$, $f(36)$, and $f(39)$.

1.3 COMBINATIONS OF FUNCTIONS _____

There are different ways of combining two functions to create a new function. Suppose that f and g are the functions given by

$$f(x) = x^2 \quad \text{and} \quad g(x) = 3x.$$

Adding $f(x)$ and $g(x)$ gives

$$f(x) + g(x) = x^2 + 3x.$$

This operation defines a new function called the *sum* of f and g, denoted by $f + g$. Its function value at x is $f(x) + g(x)$. That is,

$$(f + g)(x) = f(x) + g(x) = x^2 + 3x.$$

For example,

$$(f + g)(2) = 2^2 + 3(2) = 10.$$

In general, for any functions f and g, we define the **sum** $f + g$, the **difference** $f - g$, the **product** fg, and the **quotient** $\dfrac{f}{g}$ as follows:*

$$
\begin{aligned}
(f + g)(x) &= f(x) + g(x), \\
(f - g)(x) &= f(x) - g(x), \\
(fg)(x) &= f(x) \cdot g(x), \\
\frac{f}{g}(x) &= \frac{f(x)}{g(x)}.
\end{aligned}
$$

Thus for $f(x) = x^2$ and $g(x) = 3x$, we have

$$
\begin{aligned}
(f + g)(x) &= f(x) + g(x) = x^2 + 3x, \\
(f - g)(x) &= f(x) - g(x) = x^2 - 3x, \\
(fg)(x) &= f(x) \cdot g(x) = x^2(3x) = 3x^3, \\
\frac{f}{g}(x) &= \frac{f(x)}{g(x)} = \frac{x^2}{3x} = \frac{x}{3}, \qquad \text{for } x \neq 0.
\end{aligned}
$$

EXAMPLE 1   *If $f(x) = 3x - 1$ and $g(x) = x^2 + 2x$, find $(f + g)(x)$, $(f - g)(x)$, $(fg)(x)$, and $\dfrac{f}{g}(x)$.*

a. $(f + g)(x) = f(x) + g(x) = (3x - 1) + (x^2 + 2x) = x^2 + 5x - 1.$

b. $(f - g)(x) = f(x) - g(x) = (3x - 1) - (x^2 + 2x) = x - 1 - x^2.$

---

* In each of the four combinations, we assume that x is in the domains of both f and g. In the quotient we also do not allow any value of x for which $g(x)$ is 0.

c. $(fg)(x) = f(x)g(x) = (3x - 1)(x^2 + 2x) = 3x^3 + 5x^2 - 2x.$

d. $\dfrac{f}{g}(x) = \dfrac{f(x)}{g(x)} = \dfrac{3x - 1}{x^2 + 2x}.$

We may also combine two functions by first applying one function to a number and then applying the other function to the result. For example, suppose that $g(x) = 3x$, $f(x) = x^2$, and $x = 2$. Then $g(2) = 3(2) = 6$. Thus g sends the input 2 into the output 6:

$$2 \xrightarrow{g} 6.$$

Next, we let the output 6 become input to f.

$$f(6) = 6^2 = 36,$$

so f sends 6 into 36.

$$6 \xrightarrow{f} 36.$$

By first applying g and then f, we send 2 into 36.

$$2 \xrightarrow{g} 6 \xrightarrow{f} 36.$$

To be more general, let's replace the 2 by x, where x is in the domain of g (see Fig. 1.3). Applying g to x, we get the number $g(x)$, which we shall assume is in the domain of f. By applying f to $g(x)$, we get $f(g(x))$, read "f of g of x," which is in the range of f. This operation of applying g and then f defines a so-called "composite" function, denoted $f \circ g$. This function assigns to the input number x the output number $f(g(x))$ (see the bottom arrow in Fig. 1.3). Thus $(f \circ g)(x) = f(g(x))$. We can think of $f(g(x))$ as a function of a function.

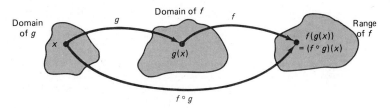

FIGURE 1.3

Definition

*If f and g are functions, the **composition of f with g** is the function $f \circ g$ defined by*

$$(f \circ g)(x) = f(g(x)),$$

where the domain of $f \circ g$ is the set of all x in the domain of g such that $g(x)$ is in the domain of f.

For $f(x) = x^2$ and $g(x) = 3x$, we can get a simple form for $f \circ g$:

$$(f \circ g)(x) = f(g(x)) = f(3x) = (3x)^2 = 9x^2.$$

For example, $(f \circ g)(2) = 9(2)^2 = 36$, as we saw before.

EXAMPLE 2   *Let* $f(x) = \sqrt{x}$ *and* $g(x) = x + 1$. *Find (a)* $(f \circ g)(x)$ *and (b)* $(g \circ f)(x)$.

a. $(f \circ g)(x)$ is $f(g(x))$ and f takes the square root of an input number, which is $g(x)$ or $x + 1$. Thus

$$(f \circ g)(x) = f(g(x)) = f(x + 1) = \sqrt{x + 1}.$$

The domain of g is all real numbers x, and the domain of f is all nonnegative reals. Hence the domain of the composition is all x for which $g(x) = x + 1$ is nonnegative. That is, the domain is all $x \geq -1$.

b. $(g \circ f)(x)$ is $g(f(x))$ and g adds 1 to an input number, which is $f(x)$ or $\sqrt{x}$. Thus g adds 1 to $\sqrt{x}$.

$$(g \circ f)(x) = g(f(x)) = g(\sqrt{x}) = \sqrt{x} + 1.$$

The domain of f is all $x \geq 0$ and the domain of g is all reals. Hence the domain of the composition is all $x \geq 0$ for which $f(x) = \sqrt{x}$ is real, namely all $x \geq 0$.

Pitfall

Generally, $f \circ g \neq g \circ f$. In Example 2, $(f \circ g)(x) = \sqrt{x + 1}$, whereas $(g \circ f)(x) = \sqrt{x} + 1$. Also, do not confuse $f(g(x))$ with $(fg)(x)$, which is the product $f(x)g(x)$. Here

$$f(g(x)) = \sqrt{x + 1}$$
$$\text{but} \quad f(x)g(x) = \sqrt{x}\,(x + 1).$$

EXAMPLE 3   *If* $F(p) = p^2 + 4p - 3$ *and* $G(p) = 2p + 1$, *find (a)* $F(G(p))$ *and (b)* $G(F(1))$.

a. $F(G(p)) = F(2p + 1) = (2p + 1)^2 + 4(2p + 1) - 3 = 4p^2 + 12p + 2.$

b. $G(F(1)) = G(1^2 + 4 \cdot 1 - 3) = G(2) = 2 \cdot 2 + 1 = 5.$

In calculus it is necessary at times to think of a particular function as a composition of two simpler functions, as the next example shows.

EXAMPLE 4

The function $h(x) = (2x - 1)^3$ can be considered to be a composition. We note that $h(x)$ is obtained by finding $2x - 1$ and cubing the result. Suppose that we let $g(x) = 2x - 1$ and $f(x) = x^3$. Then

$$h(x) = (2x - 1)^3 = [g(x)]^3 = f(g(x)) = (f \circ g)(x),$$

which gives h as a composition of two functions.

EXERCISE 1.3

1. If $f(x) = x + 1$ and $g(x) = x + 4$, find the following.
   a. $(f + g)(x)$.
   b. $(f + g)(0)$.
   c. $(f - g)(x)$.
   d. $(fg)(x)$.
   e. $(fg)(-2)$.
   f. $\dfrac{f}{g}(x)$.
   g. $(f \circ g)(x)$.
   h. $(f \circ g)(3)$.
   i. $(g \circ f)(x)$.

2. If $f(x) = 8x$ and $g(x) = 8 + x$, find the following.
   a. $(f + g)(x)$.
   b. $(f - g)(x)$.
   c. $(f - g)(4)$.
   d. $(fg)(x)$.
   e. $\dfrac{f}{g}(x)$.
   f. $\dfrac{f}{g}(2)$.
   g. $(f \circ g)(x)$.
   h. $(g \circ f)(x)$.
   i. $(g \circ f)(2)$.

3. If $f(x) = x^2$ and $g(x) = x^2 + x$, find the following.
   a. $(f + g)(x)$.
   b. $(f - g)(x)$.
   c. $(f - g)(-\frac{1}{2})$.
   d. $(fg)(x)$.
   e. $\dfrac{f}{g}(x)$.
   f. $\dfrac{f}{g}(-\frac{1}{2})$.
   g. $(f \circ g)(x)$.
   h. $(g \circ f)(x)$.
   i. $(g \circ f)(-3)$.

4. If $f(x) = x^2 - 1$ and $g(x) = 4$, find the following.
   a. $(f + g)(x)$.
   b. $(f + g)(\frac{1}{2})$.
   c. $(f - g)(x)$.
   d. $(fg)(x)$.
   e. $(fg)(4)$.
   f. $\dfrac{f}{g}(x)$.
   g. $(f \circ g)(x)$.
   h. $(f \circ g)(100)$.
   i. $(g \circ f)(x)$.

5. If $f(x) = 2x^2 + 3$ and $g(x) = 1 - 3x$, find $f(g(2))$ and $g(f(2))$.

6. If $f(p) = \dfrac{4}{p}$ and $g(p) = \dfrac{p - 2}{3}$, find $(f \circ g)(p)$ and $(g \circ f)(p)$.

7. If $F(t) = t^2 + 3t + 1$ and $G(t) = \dfrac{2}{t - 1}$, find $(F \circ G)(t)$ and $(G \circ F)(t)$.

8. If $F(s) = \sqrt{s}$ and $G(t) = 3t^2 + 4t + 2$, find $(F \circ G)(t)$ and $(G \circ F)(s)$.

9. If $f(w) = \dfrac{1}{w^2 + 1}$ and $g(v) = \sqrt{v + 2}$, find $(f \circ g)(v)$ and $(g \circ f)(w)$.

10. If $f(x) = x^2 + 3$, find $(f \circ f)(x)$.

In Problems **11–16**, *find functions* f *and* g *such that* $h(x) = f(g(x))$.

11. $h(x) = (3x + 1)^4$.
12. $h(x) = \sqrt{x^2 - 2}$.

13. $h(x) = \dfrac{1}{x^2 - 2}$.

14. $h(x) = (3x^3 - 2x)^3 - (3x^3 - 2x)^2 + 7$.

15. $h(x) = \sqrt[5]{\dfrac{x + 1}{3}}$.
16. $h(x) = \dfrac{x + 1}{(x + 1)^2 + 2}$.

17. A manufacturer determines that the total number of units of output per day, q, is a function of the number of employees, m, where $q = f(m) = (40m - m^2)/4$. The total revenue, r, that is received for selling q units is given by the function g, where $r = g(q) = 40q$. Determine $(g \circ f)(m)$. What does this composition function describe?

18. Studies have been conducted [4] concerning the statistical relations between a person's status, education, and income. Let S denote a numerical value of status based on annual income I. For a certain population, suppose that

$$S = f(I) = 0.45(I - 1000)^{0.53}.$$

Furthermore, suppose that a person's income I is a function of the number of years of education E, where

$$I = g(E) = 7202 + 0.29E^{3.68}.$$

Find $(f \circ g)(E)$. What does this function describe?

1.4 GRAPHS OF FUNCTIONS

A function with independent variable x and dependent variable y can be represented geometrically on an x, y-plane by its *graph*. The **graph of a function** f is simply the graph of the equation $y = f(x)$. It consists of all points (x, y), or $(x, f(x))$, where x is in the domain of f.

For example, to graph the function

$$f(x) = x^2 + 2x - 3,$$

we let $y = f(x)$ and graph

$$y = x^2 + 2x - 3.$$

To find points, we choose arbitrary values for x and find the corresponding values of y. For example, if $x = -4$, then $y = (-4)^2 + 2(-4) - 3 = 5$. Thus $(-4, 5)$ is a point of the graph. Other points are given in the table in Fig. 1.4(a) and have been plotted in Fig. 1.4(b). The vertical axis can be labeled either y or $f(x)$ and is referred to as the **function-value axis**. *We always label the horizontal axis with the independent variable.*

Since the graph has infinitely many points, it seems impossible to determine the graph precisely. However, we are concerned only with the graph's general shape. For this reason we plot enough points so that we may intelligently guess its proper shape. Then we join these points by a smooth curve wherever conditions permit. We start with the point having the least x-coordinate, namely $(-4, 5)$, and progress through the points having increasingly larger x-coordinates. We finish with the point having the greatest x-coordinate, namely $(2, 5)$ [see Fig. 1.4(c)]. Of course, the more points we plot, the better our graph is. Here we assume that the graph extends indefinitely upward, which is indicated by arrows. The point $(0, -3)$ where the curve intersects the y-axis is called the **y-intercept**. It is found by solving the equation for y when $x = 0$. The y-intercept is a convenient point to plot because it is usually easy to find. The

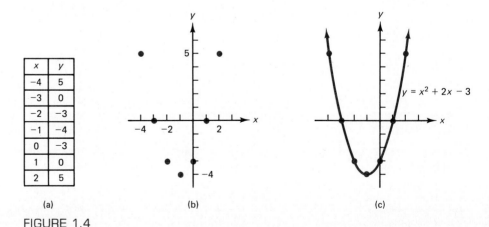

x	y
-4	5
-3	0
-2	-3
-1	-4
0	-3
1	0
2	5

(a)

(b)

(c)

FIGURE 1.4

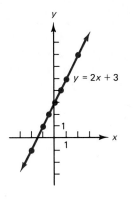

$y = 2x + 3$

x	0	$\frac{1}{2}$	$-\frac{1}{2}$	1	-1	2	-2
y	3	4	2	5	1	7	-1

FIGURE 1.5

points $(-3, 0)$ and $(1, 0)$ where the curve intersects the x-axis are called the **x-intercepts**. They are found by solving the equation for x when $y = 0$.

In a later chapter you will see that calculus is a *great* aid in graphing because it helps us analyze the shape of a graph. It provides powerful techniques for determining whether or not a curve "wiggles" between points.

EXAMPLE 1   *Graph $f(x) = 2x + 3$.*

The graph of f is the graph of $y = 2x + 3$, where $y = f(x)$ (see Fig. 1.5). From the table, note that the y-intercept is $(0, 3)$. The x-intercept can be found by solving the equation $2x + 3 = 0$:

$$2x + 3 = 0,$$
$$2x = -3,$$
$$x = -\frac{3}{2}.$$

Thus the x-intercept is $(-\frac{3}{2}, 0)$.

Figure 1.6 shows the graph of some function $y = f(x)$. Corresponding to the input number x on the horizontal axis is the output number $f(x)$ on the vertical axis. For example, corresponding to the input 4 is the output 3, so $f(4) = 3$. From the shape of the graph, it seems reasonable to assume that for any value of x, there is an output number, so the domain of f is all real numbers. Notice that the set of all y-coordinates of points on the graph is the set of all nonnegative numbers. Thus the range of f is all $y \geq 0$. This shows that we can make an "educated" guess about the domain and range of a function by looking at its graph. In general, *the domain consists of all x-values that are included in the graph, and the range is all y-values that are included*. For example, from Fig. 1.4 it is clear that the domain of $f(x) = x^2 + 2x - 3$ is all real numbers, and the range is all reals ≥ -4. From Fig. 1.5 we see that both the domain and range of $f(x) = 2x + 3$ are all real numbers.

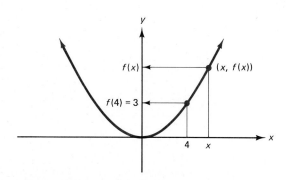

$f(x)$

$(x, f(x))$

$f(4) = 3$

4   x

FIGURE 1.6

EXAMPLE 2   *Graph $p = G(q) = |q|$ (absolute value function).*

We use the independent variable q to label the horizontal axis. The vertical axis can be labeled either $G(q)$ or p. The domain of G is all real numbers. From Fig. 1.7 it is clear that the range is all nonnegative numbers ($p \geq 0$). Note that the p-intercept and q-intercept are the same, namely (0, 0).

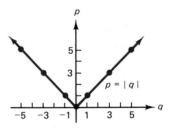

q	0	1	−1	3	−3	5	−5
p	0	1	1	3	3	5	5

FIGURE 1.7

EXAMPLE 3   *Graph $s = f(t) = \dfrac{100}{t}$.*

Here t is the independent variable, so the horizontal axis is labeled t. See Fig. 1.8. The function-value axis can be labeled either $f(t)$ or s. The domain of f is all real numbers except 0, so the graph has *no* point corresponding to $t = 0$. Thus there is no s-intercept. Moreover, there is no t-intercept since $\dfrac{100}{t}$ never has a value of 0. In general, the graph of $s = k/t$, where k is a nonzero constant, is called a *hyperbola*.

t	5	−5	10	−10	20	−20	25	−25	50	−50
s	20	−20	10	−10	5	−5	4	−4	2	−2

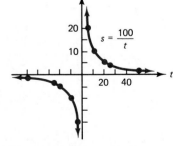

FIGURE 1.8

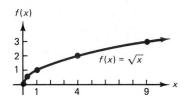

x	f(x)
0	0
$\frac{1}{4}$	$\frac{1}{2}$
1	1
4	2
9	3

FIGURE 1.9

EXAMPLE 4   *Graph $f(x) = \sqrt{x}$.*

See Fig. 1.9. Since $\sqrt{x}$ denotes the *principal* square root of x (see Appendix A, Sec. A.2), $f(9) = \sqrt{9} = 3$, not ± 3. Also, we cannot choose negative values for x because we don't want imaginary numbers for $\sqrt{x}$. That is, we must have $x \geq 0$, so the domain of f is all nonnegative numbers. From the graph, the range of f is clearly all nonnegative numbers.

EXAMPLE 5   *Graph the compound function*

$$f(x) = \begin{cases} x, & \text{if } 0 \leq x < 3, \\ x - 1, & \text{if } 3 \leq x \leq 5, \\ 4, & \text{if } 5 < x \leq 7. \end{cases}$$

The domain of f is $0 \leq x \leq 7$. The graph is given in Fig. 1.10, where the *hollow dot* means that the point is *not* included in the graph. Notice that the range of f is all real numbers y such that $0 \leq y \leq 4$.

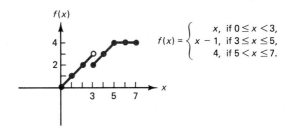

x	0	1	2	3	4	5	6	7
f(x)	0	1	2	2	3	4	4	4

FIGURE 1.10

There is an easy way to tell whether or not a curve is the graph of a function. In the leftmost diagram in Fig. 1.11, notice that with the given x there are associated *two* values of y—namely y_1 and y_2. Thus the curve is *not* the graph of a function of x. Looking at it another way, we have the

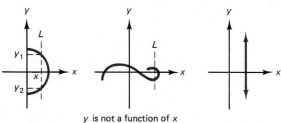

y is not a function of x

FIGURE 1.11

following general rule, called the **vertical-line test**. If a *vertical* line *L* can be drawn that intersects a curve in at least two points, the curve is *not* the graph of a function of *x*. When no such vertical line can be drawn, the curve *is* the graph of a function of *x*. Thus the curves in Fig. 1.11 do not represent functions of *x*, but those in Fig. 1.12 do.

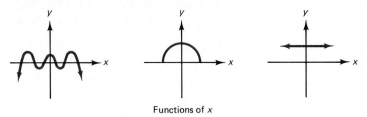

Functions of *x*

FIGURE 1.12

EXERCISE 1.4

1. Figure 1.13(a) shows the graph of $y = f(x)$.

   a. Estimate $f(0)$, $f(2)$, $f(4)$, and $f(-2)$.

   b. What is the domain of f?

   c. What is the range of f?

2. Figure 1.13(b) shows the graph of $y = f(x)$.

   a. Estimate $f(0)$ and $f(2)$.

   b. What is the domain of f?

   c. What is the range of f?

3. Figure 1.14(a) shows the graph of $y = f(x)$.

   a. Estimate $f(0)$, $f(1)$, and $f(-1)$.

   b. What is the domain of f?

   c. What is the range of f?

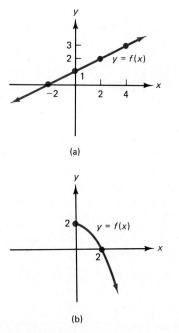

FIGURE 1.13

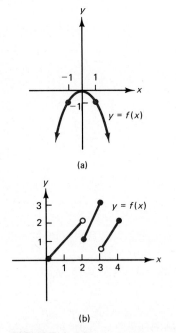

FIGURE 1.14

4. Figure 1.14(b) shows the graph of $y = f(x)$.

a. Estimate $f(0)$, $f(2)$, $f(3)$, and $f(4)$.

b. What is the domain of f?

c. What is the range of f?

In Problems 5–24, graph each function and give the domain and range.

5. $f(x) = 3x - 5$.

6. $g(x) = 3 - 2x$.

7. $s = f(t) = 4 - t^2$.

8. $f(x) = 5 - 2x^2$.

9. $y = g(x) = 2$.

10. $G(s) = -8$.

11. $y = h(x) = x^2 - 4x + 1$.

12. $y = f(x) = x^2 + 2x - 8$.

13. $f(t) = -t^3$.

14. $p = h(q) = q(2 - q)$.

15. $s = F(r) = \sqrt{r - 5}$.

16. $F(r) = -\dfrac{1}{r}$.

17. $f(x) = |2x - 1|$.

18. $v = H(u) = |u - 3|$.

19. $F(t) = \dfrac{16}{t^2}$.

20. $y = f(x) = \dfrac{2}{x - 4}$.

21. $c = g(p) = \begin{cases} p, & \text{if } 0 \le p < 2, \\ 2, & \text{if } p \ge 2. \end{cases}$

22. $f(x) = \begin{cases} 2x + 1, & \text{if } -1 \le x < 2, \\ 9 - x^2, & \text{if } x \ge 2. \end{cases}$

23. $g(x) = \begin{cases} x + 6, & \text{if } x \ge 3, \\ x^2, & \text{if } x < 3. \end{cases}$

24. $f(x) = \begin{cases} x + 1, & \text{if } 0 < x \le 3, \\ 4, & \text{if } 3 < x \le 5, \\ x - 1, & \text{if } x > 5. \end{cases}$

25. Which graphs in Fig. 1.15 represent functions of x?

26. Given the supply schedule (see Example 5 in Sec. 1.1) in Table 1.2, plot each quantity-price pair by choosing the horizontal axis for the possible quantities. Approximate the points in between the data by connecting the data points with a smooth curve. Thus you get a *supply curve*. From the graph, determine the relationship between price and supply. (That is, as price increases, what happens to the quantity supplied?) Is price per unit a function of quantity supplied?

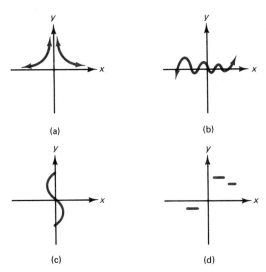

(a)   (b)

(c)   (d)

FIGURE 1.15

TABLE 1.2
Supply Schedule

QUANTITY SUPPLIED PER WEEK, q	PRICE PER UNIT, p (DOLLARS)
30	10
100	20
150	30
190	40
210	50

27. Table 1.3 is called a *demand schedule*. It indicates the quantities of Brand X that consumers will demand (that is, purchase) each week at certain prices per unit (in dollars). Plot each quantity-price pair by choosing the vertical axis for the possible prices. Connect the points

TABLE 1.3
Demand Schedule

QUANTITY DEMANDED, q	PRICE PER UNIT, p
5	20
10	10
20	5
25	4

with a smooth curve. In this way we approximate points in between the given data. The result is called a *demand curve*. From the graph, determine the relationship between the price of Brand X and the amount that will be demanded. (That is, as price decreases, what happens to the quantity demanded?) Is price per unit a function of quantity demanded?

28. In a psychological experiment on visual information, a subject briefly viewed an array of letters and was then asked to recall as many letters from the array as possible. The procedure was repeated several times. Suppose that y is the average number of letters recalled from arrays with x letters. The graph of the results approximately fits the graph of (adapted from [5])

$$y = f(x) = \begin{cases} x, & \text{if } 0 \le x \le 4, \\ \frac{1}{2}x + 2, & \text{if } 4 < x \le 5, \\ 4.5, & \text{if } 5 < x \le 12. \end{cases}$$

Graph this function.

29. Unicellular protein coagulates (denatures) at temperatures T aboye 60°C. This is indicated by noticing a decrease in solubility of the protein in a certain solution for $T > 60$. Suppose that y is the percentage of protein that remains soluble at time t. If y is estimated by

$$y = f(T) = \begin{cases} 100, & \text{if } 20 \le T \le 60, \\ 160 - T, & \text{if } 60 < T \le 90, \end{cases}$$

graph f and give the domain and range. (Adapted from [6].)

30. Sketch the graph of

$$y = f(x) = \begin{cases} -100x + 600, & \text{if } 0 \le x < 5, \\ -100x + 1100, & \text{if } 5 \le x < 10, \\ -100x + 1600, & \text{if } 10 \le x < 15. \end{cases}$$

A function such as this might describe the inventory y of a company at time x.

1.5   LINES

Many relationships between quantities can be represented conveniently by straight lines. One feature of a straight line is its "steepness." For example, in Fig. 1.16 line L_1 rises faster as it goes from left to right than does line L_2. In this sense L_1 is steeper.

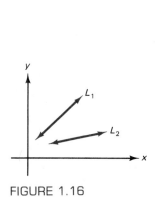

FIGURE 1.16

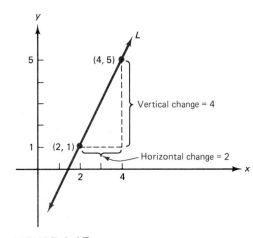

FIGURE 1.17

To measure the steepness of a line, we use the notion of *slope*. In Fig. 1.17, as we move along line L from (2, 1) to (4, 5), the x-coordinate increases from 2 to 4 and the y-coordinate increases from 1 to 5. The average rate of change of y with respect to x is the ratio

$$\frac{\text{change in } y}{\text{change in } x} = \frac{\text{vertical change}}{\text{horizontal change}} = \frac{5 - 1}{4 - 2} = \frac{4}{2} = 2.$$

This means that for each 1-unit increase in x, there is a 2-unit increase in y. Thus the line must *rise* from left to right. We say that the *slope* of the line is 2. If (x_1, y_1) and (x_2, y_2) are two other different points on L, then it can be shown that the ratio $\dfrac{\text{vertical change}}{\text{horizontal change}}$, or $\dfrac{y_2 - y_1}{x_2 - x_1}$, is also 2. In general we have the following.

Definition

*Let (x_1, y_1) and (x_2, y_2) be two points on a line, where $x_1 \neq x_2$. The **slope** of the line is the number m given by*

$$m = \frac{y_2 - y_1}{x_2 - x_1} \quad \left(= \frac{\text{vertical change}}{\text{horizontal change}}\right). \tag{1}$$

Slope is not defined for a vertical line because any two points on such a line must have $x_1 = x_2$ (see Fig. 1.18(a)). Thus the denominator in (1) would be 0. A horizontal line, on the other hand, does have slope. For such a line any two points must have $y_1 = y_2$ (see Fig. 1.18(b)). Thus the numerator in (1) is zero, so $m = 0$.

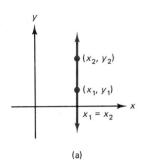

(a)

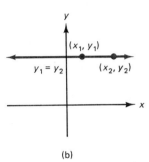

(b)

FIGURE 1.18

EXAMPLE 1   *The line in Fig. 1.19 shows the relationship between the price p of a widget (in dollars) and the quantity q of widgets (in thousands) that consumers will buy at that price. Find and interpret the slope.*

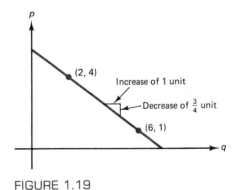

FIGURE 1.19

In the slope formula (1) we replace the x's by q's and the y's by p's. Either point in Fig. 1.19 may be chosen as (q_1, p_1). Letting $(2, 4) = (q_1, p_1)$ and $(6, 1) = (q_2, p_2)$, we have

$$m = \frac{p_2 - p_1}{q_2 - q_1} = \frac{1 - 4}{6 - 2} = \frac{-3}{4} = -\frac{3}{4}.$$

The slope is negative, $-\frac{3}{4}$. This means that for each 1-unit increase in quantity (one thousand widgets), there corresponds a **decrease** in price of $\frac{3}{4}$ (dollars per widget). Due to this decrease, the line **falls** from left to right.

In summary, we can characterize the orientation of a line by its slope:

Zero slope:	horizontal line,
Undefined slope:	vertical line,
Positive slope:	line rises from left to right,
Negative slope:	line falls from left to right.

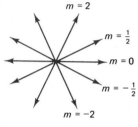

FIGURE 1.20

Lines with different slopes are shown in Fig. 1.20. Notice that *the closer the slope is to* 0, *the more nearly horizontal is the line. The greater the absolute value of the slope, the more nearly vertical is the line.* We point out that **two lines are parallel if and only if they have the same slope or are both vertical**.

If we are given a point on a line and also the slope of that line, we can find an equation whose graph is that line. Suppose that line L has slope m and passes through the point (x_1, y_1). If (x, y) is *any* other point on L (see Fig. 1.21), we can find an algebraic relationship between x and y. Using the

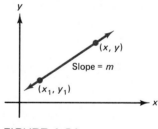

FIGURE 1.21

slope formula on the points (x_1, y_1) and (x, y), we must have

$$\frac{y - y_1}{x - x_1} = m,$$

$$y - y_1 = m(x - x_1). \tag{2}$$

That is, every point on L satisfies Eq. (2). It is also true that every point satisfying Eq. (2) must lie on L. Thus Eq. (2) is an equation for L and is given a special name:

$$y - y_1 = m(x - x_1)$$

is the **point-slope form** of an equation of the line through (x_1, y_1) with slope m.

EXAMPLE 2   *Find an equation of the line that has slope 2 and passes through* $(1, -3)$.

Here $m = 2$ and $(x_1, y_1) = (1, -3)$. Using the point-slope form, we have

$$y - (-3) = 2(x - 1),$$

which simplifies to

$$y + 3 = 2x - 2.$$

We can rewrite our answer as

$$2x - y - 5 = 0.$$

An equation of the line passing through two given points can be found easily, as Example 3 shows.

EXAMPLE 3   *Find an equation of the line passing through* $(-3, 8)$ *and* $(4, -2)$.

The line has slope

$$m = \frac{-2 - 8}{4 - (-3)} = -\frac{10}{7}.$$

Choosing $(-3, 8)$ as (x_1, y_1) in a point-slope form gives

$$y - 8 = -\tfrac{10}{7}[x - (-3)].$$
$$y - 8 = -\tfrac{10}{7}(x + 3),$$
$$7y - 56 = -10x - 30,$$

or

$$10x + 7y - 26 = 0.$$

Choosing $(4, -2)$ as (x_1, y_1) would give an equivalent result.

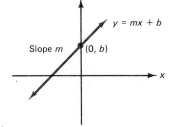

Recall that a point $(0, b)$ where a graph intersects the y-axis is called a **y-intercept** (Fig. 1.22). Sometimes we simply say that the number b is the y-intercept. If the slope m and y-intercept b of a line are known, an equation for the line is (by using the point-slope form)

$$y - b = m(x - 0).$$

Solving for y gives $y = mx + b$, called the *slope-intercept form* of an equation of the line.

FIGURE 1.22

$$y = mx + b$$

is the **slope-intercept form** of an equation of the line with slope m and y-intercept b.

EXAMPLE 4

a. An equation of the line with slope 3 and y-intercept -4 is

$$y = mx + b,$$
$$y = 3x + (-4),$$
$$y = 3x - 4.$$

b. The equation $y = 5(x - 3)$ can be written $y = 5x - 15$, which has the form $y = mx + b$ with $m = 5$ and $b = -15$. Thus its graph is a line with slope 5 and y-intercept -15.

If a *vertical* line passes through (a, b) (see Fig. 1.23), then any other point (x, y) lies on the line if and only if $x = a$. The y-coordinate can have any value. Hence an equation of the line is $x = a$. Similarly, an equation of the *horizontal* line passing through (a, b) is $y = b$ (see Fig. 1.24). Here the x-coordinate can have any value.

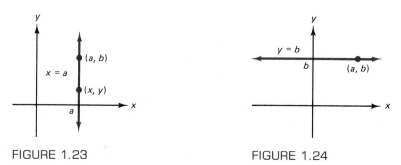

FIGURE 1.23   FIGURE 1.24

EXAMPLE 5

a. An equation of the vertical line through $(-2, 3)$ is $x = -2$. An equation of the horizontal line through $(-2, 3)$ is $y = 3$.

b. The x- and y-axes are horizontal and vertical lines, respectively. Since $(0, 0)$ lies on both axes, an equation of the x-axis is $y = 0$ and an equation of the y-axis is $x = 0$.

From our discussions we can show that every straight line is the graph of an equation of the form $Ax + By + C = 0$, where A, B, and C are constants and A and B are not both zero. We call this a **general linear equation** (or *an equation of the first degree*) **in the variables x and y**, and x and y are said to be **linearly related**. For example, a general linear equation for the line $y = 7x - 2$ is $(-7)x + (1)y + (2) = 0$. Conversely, the graph of a general linear equation is a straight line. For example, $3x + 4y + 5 = 0$ is equivalent to $y = (-\frac{3}{4})x + (-\frac{5}{4})$, so its graph is a straight line with slope $-\frac{3}{4}$ and y-intercept $-\frac{5}{4}$.

EXAMPLE 6   *Sketch the graph of $2x - 3y + 6 = 0$.*

Since this is a general linear equation, its graph is a straight line. Thus we need only determine two different points on the graph in order to sketch it. If $x = 0$, then $y = 2$. If $y = 0$, then $x = -3$. We now draw the line passing through $(0, 2)$ and $(-3, 0)$ (see Fig. 1.25). The point $(-3, 0)$ is an x-intercept, and $(0, 2)$ is a y-intercept.

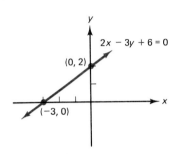

FIGURE 1.25

Table 1.4 gives the various forms of equations of straight lines.

TABLE 1.4
Forms of Equations of Straight Lines

Point-slope form:	$y - y_1 = m(x - x_1)$.
Slope-intercept form:	$y = mx + b$.
General linear form:	$Ax + By + C = 0$.
Vertical line:	$x = a$.
Horizontal line:	$y = b$.

EXERCISE 1.5

*In Problems **1–8**, find the slope of the straight line that passes through the given points.*

1. $(1, 2)$, $(4, 8)$.

2. $(-1, 9)$, $(1, 5)$.

3. $(6, -3)$, $(-7, 5)$.

4. $(2, -4)$, $(3, -4)$.

5. $(-2, 4)$, $(-2, 8)$.

6. $(0, -6)$, $(3, 0)$.

7. $(5, -2)$, $(4, -2)$.

8. $(1, -6)$, $(1, 0)$.

*In Problems **9–24**, find a general linear equation $(Ax + By + C = 0)$ of the straight line that has the indicated properties and sketch each line.*

9. Passes through $(1, 2)$ and has slope 6.

10. Passes through origin and has slope -5.

11. Passes through $(-2, 5)$ and has slope $-\frac{1}{4}$.

12. Passes through $(\frac{1}{2}, 6)$ and has slope $\frac{1}{3}$.

13. Passes through $(1, 4)$ and $(8, 7)$.

14. Passes through $(7, 1)$ and $(7, -5)$.

15. Passes through $(3, -1)$ and $(-2, -9)$.

16. Passes through $(0, 0)$ and $(2, 3)$.

17. Has slope 2 and y-intercept 4.

18. Has slope 7 and y-intercept -5.

19. Has slope $-\frac{1}{2}$ and y-intercept -3.

20. Has slope 0 and y-intercept $-\frac{1}{2}$.

21. Is horizontal and passes through $(-3, -2)$.

22. Is vertical and passes through $(-1, 4)$.

23. Passes through $(2, -3)$ and is vertical.

24. Passes through the origin and is horizontal.

In Problems **25–34**, *find, if possible, the slope and y-intercept of the straight line determined by the equation and sketch the graph.*

25. $y = 2x - 1$.

26. $x - 1 = 5$.

27. $x + 2y - 3 = 0$.

28. $y + 4 = 7$.

29. $x = -5$.

30. $x - 1 = 5y + 3$.

31. $y = 3x$.

32. $y - 7 = 3(x - 4)$.

33. $y = 1$.

34. $2y - 3 = 0$.

In Problems **35–40**, *find a general linear form and the slope-intercept form of the given equation.*

35. $x = -2y + 4$.

36. $3x + 2y = 6$.

37. $4x + 9y - 5 = 0$.

38. $2(x - 3) - 4(y + 2) = 8$.

39. $\dfrac{x}{2} - \dfrac{y}{3} = -4$.

40. $y = \dfrac{1}{300}x + 8$.

41. A straight line passes through $(1, 2)$ and $(-3, 8)$. Find the point on it that has a first coordinate of 5.

42. A straight line has slope 2 and y-intercept $(0, 1)$. Does the point $(-1, -1)$ lie on the line?

1.6 APPLICATIONS AND LINEAR FUNCTIONS

Straight lines are sometimes used in economics to describe situations. To illustrate, for each price level of a product there is a corresponding quantity of that product that consumers will demand (that is, purchase) during some time period. If the price per unit of the product is given by p and the corresponding quantity (in units) is given by q, then an equation relating p and q is called a **demand equation**. Its graph is called a **demand curve**. Figure 1.26 shows a demand curve which is a straight line; it is called a *linear demand curve*. There, p and q are linearly related. In keeping with the practice of most economists, the horizontal axis is the q-axis and the vertical axis is the p-axis. We shall assume that the price per unit is given in dollars and the time period is one week. Thus the point (a, b) in Fig. 1.26 indicates that, at a price of b dollars per unit, consumers will demand a units per week. Since negative prices or quantities are not meaningful, both a and b must be nonnegative. For most products, an increase in the quantity demanded corresponds to a decrease in price. Thus a linear demand curve typically has negative slope, as in Fig. 1.26.

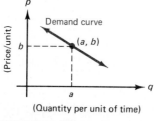

FIGURE 1.26

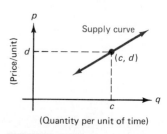

FIGURE 1.27

In response to various prices, there is a corresponding quantity of a product that *producers* are willing to supply to the market during some time period. If p denotes the price per unit and q denotes the corresponding quantity, then an equation relating p and q is called a **supply equation** and its graph is called a **supply curve**. Figure 1.27 shows a *linear* supply curve. If p is in dollars and the time period is one week, then the point (c, d) indicates that, at a price of d dollars each, producers will supply c units per week. Typically, a linear supply curve has positive slope because producers are willing to produce more when the price per unit is higher.

EXAMPLE 1 *Suppose that the demand per week for a product is 100 units when the price is $58 per unit, and 200 units at $51 each. Determine the demand equation, assuming that it is linear.*

Since the demand equation is linear, the demand curve must be a straight line. We are given that quantity q and price p are linearly related such that $p = 58$ when $q = 100$, and $p = 51$ when $q = 200$. Thus the given data can be represented in a q, p-coordinate plane by the points $(100, 58)$ and $(200, 51)$, which lie on a line whose slope is given by

$$m = \frac{51 - 58}{200 - 100} = -\frac{7}{100}.$$

An equation of the line (point-slope form) is

$$p - p_1 = m(q - q_1),$$

$$p - 58 = -\frac{7}{100}(q - 100).$$

Simplifying gives the demand equation

$$p = -\frac{7}{100}q + 65. \tag{1}$$

Customarily, a demand equation (as well as a supply equation) expresses p in terms of q and actually defines a function of q. For example, Eq. (1) defines p as a function of q and is called the *demand function* for the product.

In Sec. 1.2 a *linear function* was described. More formally we have the following definition.

Definition
*A function f is a **linear function** if and only if $f(x)$ can be written in the form $f(x) = ax + b$, where a and b are constants and $a \neq 0$.*

Suppose that $f(x) = ax + b$ is a linear function and we let $y = f(x)$. Then $y = ax + b$, which is an equation of a straight line with slope a and y-intercept b. Thus **the graph of a linear function is a straight line**. We say that $f(x) = ax + b$ has slope a.

EXAMPLE 2

a. *Graph* $f(x) = 2x - 1$.

Here f is a linear function (with slope 2), so its graph is a straight line. Since two points determine a straight line, we need only plot two points and then draw a line through them [see Fig. 1.28(a)]. Note that one of the points plotted is the vertical-axis intercept -1 that occurs when $x = 0$.

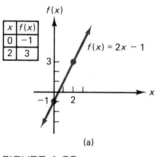

x	$f(x)$
0	-1
2	3

t	$g(t)$
0	5
6	1

$f(x) = 2x - 1$

$g(t) = \dfrac{15 - 2t}{3}$

(a)   (b)

FIGURE 1.28

b. *Graph* $g(t) = \dfrac{15 - 2t}{3}$.

Notice that

$$g(t) = \frac{15 - 2t}{3} = \frac{15}{3} - \frac{2t}{3} = -\frac{2}{3}t + 5.$$

Thus g is a linear function [see Fig. 1.28(b)]. Observe that since the slope is $-\frac{2}{3}$, then as t increases by 3 units, $g(t)$ decreases by 2.

EXAMPLE 3   *Suppose f is a linear function with slope 2 and $f(4) = 8$. Find $f(x)$.*

Since f is linear it has the form $f(x) = ax + b$. The slope is 2, so $a = 2$:

$$f(x) = 2x + b. \tag{2}$$

Now we determine b. Since $f(4) = 8$, in Eq. (2) we replace x by 4 and solve for b.

$$f(4) = 2(4) + b,$$
$$8 = 8 + b,$$
$$0 = b.$$

Hence $f(x) = 2x$.

EXAMPLE 4   *Suppose $y = f(x)$ is a linear function such that $f(-2) = 6$ and $f(1) = -3$. Find $f(x)$.*

The condition that $f(-2) = 6$ means that when $x = -2$, then $y = 6$. Thus $(-2, 6)$ lies on the graph of f, which is a straight line. Similarly, $f(1) = -3$ implies that $(1, -3)$ also lies on the line. If $(x_1, y_1) = (-2, 6)$ and $(x_2, y_2) = (1, -3)$, the slope of the line is given by

$$m = \frac{y_2 - y_1}{x_2 - x_1} = \frac{-3 - 6}{1 - (-2)} = \frac{-9}{3} = -3.$$

We can find an equation of the line by using a point-slope form.

$$y - y_1 = m(x - x_1),$$
$$y - 6 = -3[x - (-2)],$$
$$y - 6 = -3x - 6,$$
$$y = -3x.$$

Because $y = f(x)$, $f(x) = -3x$. Of course, the same result is obtained if we set $(x_1, y_1) = (1, -3)$.

In many studies data are collected and plotted on a coordinate system. An analysis of the results may indicate a functional relationship between the variables involved. For example, the data points may be approximated by points on a straight line. This would indicate a linear functional relationship, such as the one in Example 5.

EXAMPLE 5   *In testing an experimental diet for hens, it was determined that the average live weight w (in grams) of a hen was statistically a linear function of the number of days d after the diet began, where $0 \le d \le 50$. Suppose the average weight of a hen beginning the diet was 40 g and 25 days later it was 675 g.*
a. Determine w as a linear function of d.
b. Find the average weight of a hen when $d = 10$.

a. Since w is a linear function of d, its graph is a straight line. When $d = 0$ (the beginning of the diet) then $w = 40$. Thus $(0, 40)$ lies on the graph (see Fig. 1.29). Similarly, $(25, 675)$ lies on the graph. If $(d_1, w_1) = (0, 40)$ and

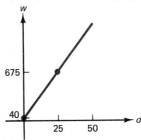

FIGURE 1.29

$(d_2, w_2) = (25, 675)$, the slope of the line is given by

$$m = \frac{w_2 - w_1}{d_2 - d_1} = \frac{675 - 40}{25 - 0} = \frac{635}{25} = \frac{127}{5}.$$

Using the point-slope form, we have

$$w - w_1 = m(d - d_1),$$

$$w - 40 = \frac{127}{5}(d - 0),$$

$$w - 40 = \frac{127}{5}d,$$

$$w = \frac{127}{5}d + 40,$$

which expresses w as a linear function of d.

b. When $d = 10$, then $w = \frac{127}{5}(10) + 40 = 254 + 40 = 294$. Thus the average weight of a hen 10 days after the beginning of the diet is 294 g.

EXERCISE 1.6

In Problems 1–6, find the slope and vertical-axis intercept of the linear function and sketch the graph.

1. $y = f(x) = -4x.$

2. $y = f(x) = x + 1.$

3. $g(t) = 2t - 4.$

4. $g(t) = 2(4 - t).$

5. $h(q) = \dfrac{7 - q}{2}.$

6. $h(q) = 0.5q + 0.25.$

In Problems 7–14, determine $f(x)$ if f is a linear function that has the given properties.

7. slope $= 5$, $f(3) = 1$.

8. $f(0) = 4$, $f(2) = -6$.

9. $f(2) = 3$, $f(-1) = 12$.

10. slope $= -6$, $f(\frac{1}{2}) = -2$.

11. slope $= -\frac{1}{2}$, $f(-\frac{1}{2}) = 4$.

12. $f(1) = 1$, $f(2) = 2$.

13. $f(-1) = -2$, $f(-3) = -4$.

14. slope $= 0.01$, $f(0.1) = 0.01$.

15. Suppose that consumers will demand 40 units of a product when the price is $12 per unit and 25 units when the price is $18 each. Find the demand equation assuming that it is linear. Find the price per unit when 30 units are demanded.

16. Suppose that a manufacturer of shoes will place on the market 50 (thousand pairs) when the price is 35 (dollars per pair) and 35 when the price is 30. Find the supply equation, assuming that it is linear.

17. Suppose that the cost to produce 10 units of a product is $40 and the cost of 20 units is $70. If cost c is linearly related to output q, find a linear equation relating c and q. Find the cost to produce 35 units.

18. A cancer patient is to receive drug and radiation therapies. Each cubic centimeter of the drug to be used contains 200 curative units, and each minute of radiation exposure gives 300 curative units. The patient requires 2400 curative units. If d cubic centimeters of the drug and r minutes of radiation are administered, determine an equation relating d and r. Graph the equation for $d \geq 0$ and $r \geq 0$; label the horoizontal axis as d.

19. Suppose that the value of a piece of machinery decreases each year by 10 percent of its original value. If the original value is $8000, find an equation that expresses the value v of the machinery after t years of purchase where $0 \leq t \leq 10$. Sketch the equation, choosing t as the horizontal axis and v as the vertical axis. What is the slope of the resulting line? This method of considering the value of equipment is called *straight-line depreciation*.

20. In production analysis, an *isocost line* is a line whose points represent all combinations of two factors of production that can be purchased for the same amount. Suppose that a farmer has allocated $20,000 for the purchase of x tons of fertilizer (costing $200 per ton) and y acres of land (costing $2000 per acre). Find an equation of the isocost line that describes the various combinations that can be purchased for $20,000. Observe that neither x nor y can be negative.

21. For reasons of comparison, a professor wants to rescale the scores on a set of test papers so that the maximum score is still 100 but the mean (average) is 80 instead of 56.

    a. Find a linear equation that will do this. [*Hint:* You want 56 to become an 80 and 100 to remain 100. Consider the points (56, 80) and (100, 100) and, more generally, (x, y), where x is the old score and y is the new score. Find the slope and use a point-slope form. Express y in terms of x.]

    b. If 60 on the new scale is the lowest passing score, what was the lowest passing score on the original scale?

22. A manufacturer produces products X and Y for which the profits per unit are $4 and $6, respectively. If x units of X and y units of Y are sold, then the total profit P is given by $P = 4x + 6y$, where $x, y \geq 0$.

    a. Sketch the graph of this equation for $P = 240$. The result is called an *isoprofit line* and its points represent all combinations of sales that produce a profit of $240.

    b. Determine the slope for $P = 240$.

    c. If $P = 600$, determine the slope.

    d. Are isoprofit lines for products X and Y parallel?

23. For sheep maintained at high environmental temperatures, respiratory rate r (per minute) increases as wool length l (in centimeters) decreases. Suppose that sheep with a wool length of 2 cm have a (average) respiratory rate of 160, and those with a wool length of 4 cm have a respiratory rate of 125, and that r is a linear function of l. (Adapted from [7].)

    a. Determine this function.

    b. Determine the respiratory rate of sheep with a wool length of 1 cm.

24. The result of Sternberg's psychological experiment [5] on information retrieval is that a person's reaction time R, in milliseconds, is statistically a linear function of memory set size N as follows:

$$R = 38N + 397.$$

Sketch the graph for $1 \leq N \leq 5$. What is the slope?

25. In a certain learning experiment involving repetition and memory [8], the proportion p of items recalled was estimated to be a linear function of effective study time t (in seconds), where t is between 5 and 9 inclusive. For an effective study time of 5 seconds, the proportion of items recalled was 0.32. For each 1-second increase in study time, the proportion recalled increased by 0.059.

    a. Find an equation that gives p in terms of t.

    b. What proportion of items was recalled with 9 seconds of effective study time?

26. In testing an experimental diet for pigs, it was determined that the (average) live weight w (in kilograms) of a pig was statistically a linear function of the number of days d after the diet was initiated, where $0 \leq d \leq 100$. If the weight of a pig beginning the diet was 20 kg and thereafter the pig gained 6.6 kg every 10 days, determine w as a function of d and find the weight of a pig 50 days after the beginning of the diet.

27. Biologists have found that the number of chirps made per minute by crickets of a certain species is related to the temperature. The relationship is very close to being linear. At 68°F, those crickets chirp about 124 times a minute. At 80°F, they chirp about 172 times a minute.

    a. Find an equation that gives Fahrenheit temperature t in terms of the number of chirps c per minute.

    b. If you count chirps for only 15 seconds, how can you quickly estimate the temperature?

1.7 QUADRATIC FUNCTIONS

In Sec. 1.2 a *quadratic function* was described as a polynomial function of degree 2. Here is a formal definition.

Definition

A function f is a **quadratic function** *if and only if f(x) can be written in the form f(x) = ax² + bx + c, where a, b, and c are constants and a ≠ 0.*

For example, the functions $f(x) = x^2 - 3x + 2$ and $F(t) = -3t^2$ are quadratic. However, $g(x) = \dfrac{1}{x^2}$ is *not* quadratic because it cannot be written in the form $g(x) = ax^2 + bx + c$.

The graph of the quadratic function $y = f(x) = ax^2 + bx + c$ is called a **parabola** and has a shape like the curves in Fig. 1.30. If $a > 0$, the graph extends upward indefinitely and we say that the parabola *opens upward* [Fig. 1.30(a)]. If $a < 0$, the parabola *opens downward* [see Fig. 1.30(b)].

Each parabola in Fig. 1.30 is *symmetric* about a vertical line, called the **axis of symmetry** of the parabola. That is, if the page were folded on one of these lines, the two halves of the corresponding parabola would coincide. The axis (of symmetry) is *not* part of the parabola, but is a useful aid in sketching the parabola.

Figure 1.30 also shows points each labeled **vertex**, where the axis cuts the parabola. If $a > 0$, the vertex is the "lowest" point on the parabola. This means that $f(x)$ has a minimum value at this point. By performing algebraic manipulations on $ax^2 + bx + c$ (referred to as *completing the square*), we can determine not only this minimum value but also where it occurs.

$$f(x) = ax^2 + bx + c$$
$$= (ax^2 + bx) + c.$$

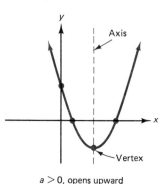

Parabola: $y = f(x) = ax^2 + bx + c$

Axis

Vertex

$a > 0$, opens upward

(a)

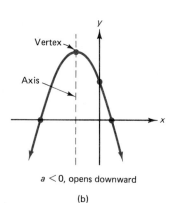

Vertex

Axis

$a < 0$, opens downward

(b)

FIGURE 1.30

Adding and subtracting $\dfrac{b^2}{4a}$ give

$$f(x) = \left(ax^2 + bx + \frac{b^2}{4a}\right) + c - \frac{b^2}{4a}$$
$$= a\left(x^2 + \frac{b}{a}x + \frac{b^2}{4a^2}\right) + c - \frac{b^2}{4a}$$
$$= a\left(x + \frac{b}{2a}\right)^2 + c - \frac{b^2}{4a}.$$

Since $\left(x + \dfrac{b}{2a}\right)^2 \geq 0$ and $a > 0$, it follows that $f(x)$ has a minimum value when $x + \dfrac{b}{2a} = 0$, that is, when $x = -\dfrac{b}{2a}$. The y-coordinate corresponding to this value of x is $f\left(-\dfrac{b}{2a}\right)$. Thus the vertex is given by

$$\text{vertex} = \left(-\frac{b}{2a}, f\left(-\frac{b}{2a}\right)\right).$$

This is also the vertex of a parabola that opens downward ($a < 0$), but in this case $f\left(-\dfrac{b}{2a}\right)$ is the *maximum* value of $f(x)$ [see Fig. 1.30(b)].

The point where the parabola $y = f(x) = ax^2 + bx + c$ intersects the y-axis (that is, the *y-intercept*) occurs when $x = 0$. The y-coordinate of this point is $f(0) = c$, so the y-intercept is $(0, c)$. More simply, we say that the y-intercept is c. In summary, we have the following.

The graph of the quadratic function

$$y = f(x) = ax^2 + bx + c$$

is a parabola.

1. If $a > 0$, the parabola opens upward.
   If $a < 0$, it opens downward.

2. The vertex is $\left(-\dfrac{b}{2a}, f\left(-\dfrac{b}{2a}\right)\right)$.

3. The y-intercept is c.

We can quickly sketch the graph of a quadratic function by first locating the vertex, the y-intercept, and a few other points like those where the parabola intersects the x-axis. These *x-intercepts* are obtained by setting $y = 0$ and solving for x. Once the intercepts and vertex are found, it is then relatively easy to pass the appropriate parabola through these points. In the event that the x-intercepts are very close to the vertex, or that no x-intercepts exist, we find a point on each side of the vertex so that we can give a reasonable sketch of the parabola. Keep in mind that passing a (dashed) vertical line through the vertex gives the axis of symmetry. By plotting points to one side of the axis, we can use symmetry and obtain corresponding points on the other side.

EXAMPLE 1   *Graph the quadratic function*

$$y = f(x) = -x^2 - 4x + 12.$$

Here $a = -1$, $b = -4$, and $c = 12$. Since $a < 0$, the parabola opens downward. The x-coordinate of the vertex is

$$-\frac{b}{2a} = -\frac{-4}{2(-1)} = -2.$$

The y-coordinate is $f(-2) = -(-2)^2 - 4(-2) + 12 = 16$. Thus the vertex (highest point) is $(-2, 16)$. Since $c = 12$, the y-intercept is 12. To find the x-intercepts, we let y be 0 in $y = -x^2 - 4x + 12$ and solve for x.

$$0 = -x^2 - 4x + 12,$$
$$0 = -(x^2 + 4x - 12),$$
$$0 = -(x + 6)(x - 2).$$

Thus $x = -6$ or $x = 2$, so the x-intercepts are -6 and 2. Now we plot the vertex, axis of symmetry, and intercepts [see Fig. 1.31(a)]. Since $(0, 12)$ is *two* units to the *right* of the axis, there is a corresponding point *two* units to the *left* of the axis with the same y-coordinate. Thus we get the point $(-4, 12)$. Through all points we draw a parabola opening downward [see Fig. 1.31(b)].

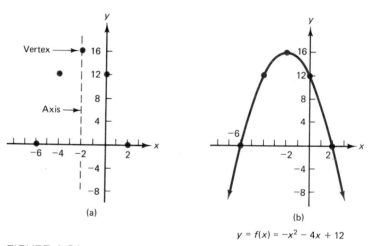

(a)

(b)

$y = f(x) = -x^2 - 4x + 12$

FIGURE 1.31

EXAMPLE 2   *Sketch the following quadratic functions.*

a. $p = 2q^2$.

Here p is a quadratic function of q, where $a = 2$, $b = 0$, and $c = 0$. Since $a > 0$, the parabola opens upward. The q-coordinate of the vertex is

$$-\frac{b}{2a} = -\frac{0}{2(2)} = 0,$$

and the p-coordinate is $2(0)^2 = 0$. Thus the vertex is $(0, 0)$. In this case the p-axis is the axis of symmetry. A parabola opening upward with vertex at $(0, 0)$ cannot have any other intercepts. Hence to draw a reasonable graph we plot a point on each side of the vertex. If $q = 2$, then $p = 8$. This gives the point $(2, 8)$ and, by symmetry, the point $(-2, 8)$ [see Fig. 1.32(a)].

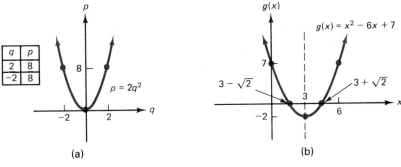

FIGURE 1.32

b. $g(x) = x^2 - 6x + 7$.

Here g is a quadratic function, where $a = 1$, $b = -6$, and $c = 7$. The parabola opens upward since $a > 0$. The x-coordinate of the vertex is

$$-\frac{b}{2a} = -\frac{-6}{2(1)} = 3,$$

and $g(3) = 3^2 - 6(3) + 7 = -2$. Thus the vertex is $(3, -2)$. Since $c = 7$, the vertical-axis intercept is 7. To find x-intercepts, we set $g(x) = 0$.

$$0 = x^2 - 6x + 7.$$

The right side does not factor easily, so we shall use the quadratic formula to solve for x.

$$x = \frac{-b \pm \sqrt{b^2 - 4ac}}{2a} = \frac{-(-6) \pm \sqrt{(-6)^2 - 4(1)(7)}}{2(1)}$$

$$= \frac{6 \pm \sqrt{8}}{2} = \frac{6 \pm \sqrt{4 \cdot 2}}{2} = \frac{6 \pm 2\sqrt{2}}{2}$$

$$= \frac{6}{2} \pm \frac{2\sqrt{2}}{2} = 3 \pm \sqrt{2}.$$

Thus the x-intercepts are $3 + \sqrt{2}$ and $3 - \sqrt{2}$. After plotting the vertex, intercepts, and (by symmetry) the point $(6, 7)$, we draw a parabola opening upward [see Fig. 1.32(b)].

EXAMPLE 3   *The demand function for a manufacturer's product is* $p = 1000 - 2q$, *where p is the price (in dollars) per unit when q units are demanded (per week) by consumers. Find the level of production that will maximize the manufacturer's total revenue, and determine this revenue.*

Total revenue r is given by

total revenue = (price)(quantity).

$$r = pq$$
$$= (1000 - 2q)q.$$
$$r = 1000q - 2q^2.$$

Note that r is a quadratic function of q, with $a = -2$, $b = 1000$, and $c = 0$. Since $a < 0$ (parabola opens downward), r is maximum when

$$q = -\frac{b}{2a} = -\frac{1000}{2(-2)} = 250.$$

The maximum value of r is given by

$$r = 1000(250) - 2(250)^2$$
$$= 250,000 - 125,000 = 125,000.$$

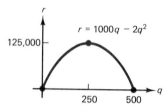

FIGURE 1.33

Thus the maximum revenue that the manufacturer can receive is $125,000, which occurs at a production level of 250 units. Figure 1.33 shows the graph of the revenue function. Only that portion for which $q \geq 0$ and $r \geq 0$ is drawn, since quantity and revenue cannot be negative.

EXERCISE 1.7

In Problems 1–8, state whether or not the function is quadratic.

1. $f(x) = 26 - 3x.$

2. $g(x) = (7 - x)^2.$

3. $g(x) = 4x^2.$

4. $h(s) = 6(4s + 1).$

5. $h(q) = \dfrac{1}{2q - 4}.$

6. $f(t) = 2t(3 - t) + 4t.$

7. $f(s) = \dfrac{s^2 - 4}{2}.$

8. $g(t) = (t^2 - 1)^2.$

In Problems 9–12, do not include a graph.

9. a. For the parabola $y = f(x) = -4x^2 + 8x + 7$, find the vertex.

   b. Does the vertex correspond to the highest point, or the lowest point, on the graph?

10. Repeat Problem 9 if $y = f(x) = 8x^2 + 4x - 1.$

11. For the parabola $y = f(x) = x^2 + 2x - 8$, find (a) the y-intercept, (b) the x-intercepts, and (c) the vertex.

12. Repeat Problem 11 if $y = f(x) = 3 + x - 2x^2.$

In Problems 13–22, graph each function. Give the intercepts and vertex.

13. $y = f(x) = x^2 - 6x + 5.$

14. $y = f(x) = -3x^2.$

15. $y = g(x) = -2x^2 - 6x.$

16. $y = f(x) = x^2 - 1.$

17. $s = h(t) = t^2 + 2t + 1.$

18. $s = h(t) = 2t^2 + 3t - 2.$

19. $y = f(x) = -9 + 8x - 2x^2.$

20. $y = H(x) = 1 - x - x^2.$

21. $t = f(s) = s^2 - 8s + 13.$

22. $t = f(s) = s^2 + 6s + 11.$

In Problems 23–26, state whether $f(x)$ has a maximum value or a minimum value, and find that value.

23. $f(x) = 100x^2 - 20x + 25.$

24. $f(x) = -2x^2 - 16x + 3.$

25. $f(x) = 4x - 50 - 0.1x^2.$

26. $f(x) = x(x + 3) - 12$.

27. The demand function for a manufacturer's product is $p = f(q) = 1200 - 3q$, where p is the price (in dollars) per unit when q units are demanded (per week). Find the level of production that maximizes the manufacturer's total revenue and determine this revenue.

28. A sociologist is hired by a city to study various programs that aid the education of preschool-age children. The sociologist estimates that n years after the beginning of a particular program, $f(n)$ thousand preschoolers will be enrolled, where

$$f(n) = \frac{10}{9}n(12 - n), \qquad 0 \le n \le 12.$$

Estimate the maximum number of preschoolers that will be enrolled in the program.

29. Biologists studied the nutritional effects on rats that were fed a diet containing 10 percent protein (adapted from [9]). The protein consisted of yeast and corn flour. By varying the percentage P of yeast in the protein mix, the group estimated that the average weight gain (in grams) of a rat over a period of time was $f(P)$, where

$$f(P) = -\frac{1}{50}P^2 + 2P + 20, \qquad 0 \le P \le 100.$$

Find the maximum weight gain.

30. Suppose that the height s of a ball thrown vertically upward from the ground is given by

$$s = -4.9t^2 + 58.8t,$$

where s is in meters and t is elapsed time in seconds. After how many seconds will the ball reach its maximum height? What is the maximum height?

1.8 LIMITS

Our study of calculus will begin in the next chapter. Because the notion of a *limit* lies at the foundation of calculus, we must develop some understanding of that concept. We shall first give you a "feeling" for limits by some examples.

Suppose that we examine the function

$$f(x) = 2x - 1$$

when x is "near" 2 but not equal to 2. Some values of $f(x)$ for x less than 2 and then greater than 2 are given in Table 1.5. It is apparent that as x takes

TABLE 1.5

$x < 2$	$x > 2$
$f(1.7) = 2.4$	$f(2.3) = 3.6$
$f(1.8) = 2.6$	$f(2.2) = 3.4$
$f(1.9) = 2.8$	$f(2.1) = 3.2$
$f(1.99) = 2.98$	$f(2.01) = 3.02$
$f(1.999) = 2.998$	$f(2.001) = 3.002$

on values closer and closer to 2, regardless of whether x approaches 2 *from the left* ($x < 2$) or *from the right* ($x > 2$), the corresponding values of $f(x)$ become closer and closer to one number, 3. This is also clear from the graph of f in Fig. 1.34. To express our conclusion we say that 3 is the **limit** of $f(x)$ as x approaches 2, which is written

$$\lim_{x \to 2} (2x - 1) = 3.$$

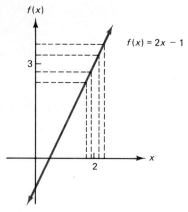

FIGURE 1.34

We can make the number $2x - 1$ as close to 3 as we wish by taking x sufficiently close to 2.

You may think that you can find the limit of a function as x approaches some number a by just evaluating the function when x is a. For the above function this is true: $f(2) = 2(2) - 1 = 3$, which is also the limit. But simple substitution does not always work. For example, consider the function

$$g(x) = \begin{cases} 2x - 1, & \text{if } x \neq 2, \\ 1, & \text{if } x = 2. \end{cases}$$

Notice that $g(2) = 1$. Let us find the limit of $g(x)$ as x approaches 2 (that is, as $x \to 2$). From the graph of g in Fig. 1.35, you can see that as x gets closer to 2 (but *not equal* to 2), then $g(x)$ gets closer to 3. Thus

$$\lim_{x \to 2} g(x) = 3,$$

which is *not* the same as $g(2)$.

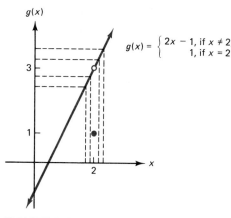

FIGURE 1.35

In general, for any function f we have the following definition of a limit.

Definition
*The **limit** of $f(x)$ as x approaches a is the number L, written*

$$\lim_{x \to a} f(x) = L,$$

provided that $f(x)$ is arbitrarily close to L for all x sufficiently close to, but not equal to, a.

We remind you that when finding a limit, we are concerned not with what happens to $f(x)$ when x *equals* a but only with what happens to it when x is *close to* a. We emphasize that a limit is independent of the way in which x

approaches a. The limit must be the same whether x approaches a from the left or from the right (for $x < a$ or $x > a$, respectively).

Here are some properties of limits which may seem reasonable to you.

1. **If $f(x) = c$ is a constant function, then $\lim_{x \to a} f(x) = \lim_{x \to a} c = c$.**

2. **$\lim_{x \to a} x^n = a^n$, for any positive integer n.**

EXAMPLE 1

a. $\lim_{x \to 2} 7 = 7$; $\lim_{x \to -5} 7 = 7$.   **b.** $\lim_{x \to -6} x^2 = (-6)^2 = 36$.

Some other properties of limits are as follows.

If $\lim_{x \to a} f(x) = L_1$ and $\lim_{x \to a} g(x) = L_2$, where L_1 and L_2 are real numbers, then

3. **$\lim_{x \to a} [f(x) \pm g(x)] = \lim_{x \to a} f(x) \pm \lim_{x \to a} g(x) = L_1 \pm L_2$.**

   That is, the limit of a sum or difference is the sum or difference, respectively, of the limits.

4. **$\lim_{x \to a} [f(x) \cdot g(x)] = \lim_{x \to a} f(x) \cdot \lim_{x \to a} g(x) = L_1 \cdot L_2$.**

   That is, the limit of a product is the product of the limits.

5. **$\lim_{x \to a} [cf(x)] = c \cdot \lim_{x \to a} f(x) = cL_1$, where c is a constant.**

   That is, the limit of a constant times a function is the constant times the limit of the function.

EXAMPLE 2

a. $\lim_{x \to 2} (x^2 + x) = \lim_{x \to 2} x^2 + \lim_{x \to 2} x$ (Property 3)

$$= 2^2 + 2 = 6 \qquad \text{(Property 2).}$$

b. Property 3 can be extended to the limit of a finite number of sums and differences. For example,

$$\lim_{q \to -1} (q^3 - q + 1) = \lim_{q \to -1} q^3 - \lim_{q \to -1} q + \lim_{q \to -1} 1$$
$$= (-1)^3 - (-1) + 1 = 1.$$

c. $\lim_{x \to 2} [(x + 1)(x - 3)] = \lim_{x \to 2} (x + 1) \cdot \lim_{x \to 2} (x - 3)$ (Property 4)

$$= [\lim_{x \to 2} x + \lim_{x \to 2} 1] \cdot [\lim_{x \to 2} x - \lim_{x \to 2} 3]$$
$$= [2 + 1] \cdot [2 - 3] = 3[-1] = -3.$$

d. $\lim\limits_{x \to -2} 3x^3 = 3 \cdot \lim\limits_{x \to -2} x^3$ (Property 5)

$$= 3(-2)^3 = -24.$$

EXAMPLE 3

Let $f(x) = c_n x^n + c_{n-1} x^{n-1} + \cdots + c_1 x + c_0$ be a polynomial function. Then

$$\lim_{x \to a} f(x) = \lim_{x \to a} (c_n x^n + c_{n-1} x^{n-1} + \cdots + c_1 x + c_0)$$

$$= c_n \cdot \lim_{x \to a} x^n + c_{n-1} \cdot \lim_{x \to a} x^{n-1} + \cdots + c_1 \cdot \lim_{x \to a} x + \lim_{x \to a} c_0$$

$$= c_n a^n + c_{n-1} a^{n-1} + \cdots + c_1 a + c_0 = f(a).$$

Thus **if f is a polynomial function, then**

$$\lim_{x \to a} f(x) = f(a).$$

That is, the limit of a polynomial function as x approaches a is merely the function value at a.

The result of Example 3 allows us to find many limits as $x \to a$ by just substituting a for x. For example,

$$\lim_{x \to -3} (x^3 + 4x^2 - 7) = (-3)^3 + 4(-3)^2 - 7 = 2,$$

$$\lim_{h \to 3} [2(h - 1)] = 2(3 - 1) = 4.$$

Our final two properties will concern limits involving quotients and roots.

If $\lim\limits_{x \to a} f(x) = L_1$ and $\lim\limits_{x \to a} g(x) = L_2$, where L_1 and L_2 are real numbers, then

6. $\lim\limits_{x \to a} \dfrac{f(x)}{g(x)} = \dfrac{\lim\limits_{x \to a} f(x)}{\lim\limits_{x \to a} g(x)} = \dfrac{L_1}{L_2}$, **if $L_2 \neq 0$.**

That is, the limit of a quotient is the quotient of the limits, provided that the denominator does not have a limit of 0.

7. $\lim\limits_{x \to a} \sqrt[n]{f(x)} = \sqrt[n]{\lim\limits_{x \to a} f(x)} = \sqrt[n]{L_1}.$*

---

* If n is even, we require that L_1 be positive.

EXAMPLE 4

a. $\lim\limits_{x \to 1} \dfrac{2x^2 + x - 3}{x^3 + 4} = \dfrac{\lim\limits_{x \to 1} (2x^2 + x - 3)}{\lim\limits_{x \to 1} (x^3 + 4)} = \dfrac{2 + 1 - 3}{1 + 4} = \dfrac{0}{5} = 0.$

b. $\lim\limits_{t \to 4} \sqrt{t^2 + 1} = \sqrt{\lim\limits_{t \to 4} (t^2 + 1)} = \sqrt{17}.$

c. $\lim\limits_{x \to 3} \sqrt[3]{x^2 + 7} = \sqrt[3]{\lim\limits_{x \to 3} (x^2 + 7)} = \sqrt[3]{16} = \sqrt[3]{8 \cdot 2} = 2\sqrt[3]{2}.$

EXAMPLE 5   *Find* $\lim\limits_{x \to -1} \dfrac{x^2 - 1}{x + 1}.$

As $x \to -1$, both numerator and denominator approach zero. Because the limit of the denominator is 0, we cannot use Property 6. However, since what happens to the quotient when x equals -1 is of no concern, we can assume $x \neq -1$ and write

$$\frac{x^2 - 1}{x + 1} = \frac{(x + 1)(x - 1)}{x + 1} = x - 1.$$

This algebraic manipulation on the original function $\dfrac{x^2 - 1}{x + 1}$ yields a new function $x - 1$, which is the same as the original function for $x \neq -1$. Thus

$$\lim_{x \to -1} \frac{x^2 - 1}{x + 1} = \lim_{x \to -1} \frac{(x + 1)(x - 1)}{x + 1} = \lim_{x \to -1} (x - 1) = -2.$$

Notice that although the original function is not defined at -1, it *does* have a limit as $x \to -1$.

In Example 5 the method of finding a limit by direct substitution does not work. Replacing x by -1 gives $0/0$, which has no meaning. When the meaningless form $0/0$ arises, algebraic manipulation (as in Example 5) may result in a form for which the limit *can* be determined. In fact, many important limits cannot be evaluated by substitution.

EXAMPLE 6   *Find* $\lim\limits_{h \to 0} \dfrac{(2 + h)^2 - 4}{h}.$

As $h \to 0$, both the numerator and denominator approach 0. Therefore, we shall try to express the function in a different form for $h \neq 0$.

$$\lim_{h \to 0} \frac{(2 + h)^2 - 4}{h} = \lim_{h \to 0} \frac{4 + 4h + h^2 - 4}{h}$$

$$= \lim_{h \to 0} \frac{4h + h^2}{h} = \lim_{h \to 0} \frac{h(4 + h)}{h}$$

$$= \lim_{h \to 0} (4 + h) = 4.$$

EXERCISE 1.8

In Problems **1–28**, *find the limits.*

1. $\lim\limits_{x \to 2} 16.$

2. $\lim\limits_{x \to 3} 2x.$

3. $\lim\limits_{x \to 4} (x + 3).$

4. $\lim\limits_{s \to 1} 2.$

5. $\lim\limits_{t \to -5} (t^2 - 5).$

6. $\lim\limits_{t \to 1/2} (3t - 5).$

7. $\lim\limits_{x \to -1} (x^3 - 3x^2 - 2x + 1).$

8. $\lim\limits_{r \to 9} \dfrac{4r - 3}{11}.$

9. $\lim\limits_{t \to -3} \dfrac{t - 2}{t + 5}.$

10. $\lim\limits_{x \to -6} \dfrac{x^2 + 6}{x - 6}.$

11. $\lim\limits_{h \to 0} \dfrac{h}{h^2 - 7h + 1}.$

12. $\lim\limits_{h \to 0} \dfrac{h^2 - 2h - 4}{h^3 - 1}.$

13. $\lim\limits_{p \to 4} \sqrt{p^2 + p + 5}.$

14. $\lim\limits_{y \to 9} \sqrt{y + 3}.$

15. $\lim\limits_{x \to -2} \dfrac{x^2 + 2x}{x + 2}.$

16. $\lim\limits_{x \to -1} \dfrac{x + 1}{x + 1}.$

17. $\lim\limits_{x \to 2} \dfrac{x^2 - x - 2}{x - 2}.$

18. $\lim\limits_{t \to 0} \dfrac{t^2 + 2t}{t^2 - 2t}.$

19. $\lim\limits_{x \to -1} \dfrac{x^2 + 2x + 1}{x + 1}.$

20. $\lim\limits_{t \to 1} \dfrac{t^2 - 1}{t - 1}.$

21. $\lim\limits_{x \to 3} \dfrac{x - 3}{x^2 - 9}.$

22. $\lim\limits_{x \to 0} \dfrac{x^2 - 2x}{x}.$

23. $\lim\limits_{x \to 4} \dfrac{x^2 - 9x + 20}{x^2 - 3x - 4}.$

24. $\lim\limits_{x \to 2} \dfrac{x^2 - 2x}{x - 2}.$

25. $\lim\limits_{x \to 2} \dfrac{3x^2 - x - 10}{x^2 + 5x - 14}.$

26. $\lim\limits_{x \to -4} \dfrac{x^2 + 2x - 8}{x^2 + 5x + 4}.$

27. $\lim\limits_{h \to 0} \dfrac{(2 + h)^2 - 2^2}{h}.$

28. $\lim\limits_{x \to 0} \dfrac{(x + 2)^2 - 4}{x}.$

29. Find $\lim\limits_{h \to 0} \dfrac{(x + h)^2 - x^2}{h}$ by treating x as a constant.

30. Find $\lim\limits_{h \to 0} \dfrac{2(x + h)^2 + 5(x + h) - 2x^2 - 5x}{h}$ by treating x as a constant.

31. If $f(x) = x + 5$, show that $\lim\limits_{h \to 0} \dfrac{f(x + h) - f(x)}{h} = 1$ by treating x as a constant.

32. If $f(x) = x^2$, show that $\lim\limits_{h \to 0} \dfrac{f(x + h) - f(x)}{h} = 2x$ by treating x as a constant.

33. The maximum theoretical efficiency E of a power plant is given in **[10]** by

$$E = \frac{T_h - T_c}{T_h},$$

where T_c and T_h are the respective absolute temperatures of the hotter and colder reservoirs. Find (a) $\lim\limits_{T_c \to 0} E$ and (b) $\lim\limits_{T_c \to T_h} E.$

1.9 LIMITS AT INFINITY

We now consider the behavior of the function $f(x) = \dfrac{1}{x}$ when x is positive and gets larger and larger. Table 1.6(a) shows, for example, that $f(10{,}000) = 0.0001$ and $f(100{,}000) = 0.00001$. It is clear that the larger the value of x, the

TABLE 1.6

x	$f(x)$	x	$f(x)$
1,000	0.001	−1,000	−0.001
10,000	0.0001	−10,000	−0.0001
100,000	0.00001	−100,000	−0.00001
1,000,000	0.000001	−1,000,000	−0.000001
(a)		(b)	

closer its reciprocal $1/x$ is to 0. To describe this, we say that as x **becomes positively infinite** or **increases without bound**, the limit of $1/x$ is 0. We denote this by using the infinity symbol ∞ and writing

$$\lim_{x\to\infty} \frac{1}{x} = 0.$$

We emphasize that the symbol ∞ does not represent a number. Similarly, as x becomes very negative, that is, x becomes **negatively infinite** or **decreases without bound** ($x \to -\infty$), Table 1.6(b) shows that the values of $1/x$ again approach 0. This is written

$$\lim_{x\to -\infty} \frac{1}{x} = 0.$$

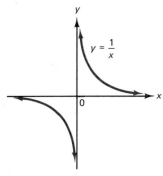

FIGURE 1.36

Both of these limits are clear from the graph of $y = f(x) = 1/x$ in Fig. 1.36. As $x \to \infty$ or $x \to -\infty$, the graph gets closer and closer to the line $y = 0$ (the x-axis) because the function values are getting closer to 0. In general,

$$\lim_{x\to\infty} \frac{1}{x^n} = 0 \quad \text{and} \quad \lim_{x\to -\infty} \frac{1}{x^n} = 0,$$

where n is positive.* For example,

$$\lim_{x\to\infty} \frac{1}{\sqrt[3]{x}} = \lim_{x\to\infty} \frac{1}{x^{1/3}} = 0.$$

We can easily determine limits at ∞ or $-\infty$ for a polynomial function. For example, consider

$$\lim_{x\to\infty} (8x^2 + 2x).$$

For *large* values of x, the term $8x^2$ has a much greater value than does the term $2x$, because $8x^2$ involves the greater power of x. Thus the term $8x^2$ dominates the sum $8x^2 + 2x$ and $8x^2 + 2x \approx 8x^2$ when x is large. The symbol $\approx$ means "approximately equals." That is, as $x \to \infty$, $8x^2 + 2x$ will behave like $8x^2$. Hence, to determine the limit of $8x^2 + 2x$, it suffices to determine the limit of $8x^2$:

$$\lim_{x\to\infty} (8x^2 + 2x) = \lim_{x\to\infty} 8x^2.$$

Now, as x becomes larger and larger, so does x^2. Moreover, 8 times a very large number is also very large. Thus the values of $8x^2$ are increasing without bound and are *not* approaching a specific number. Therefore, no limit exists and we write

$$\lim_{x\to\infty} (8x^2 + 2x) = \lim_{x\to\infty} 8x^2 = \infty.$$

* For $\lim\limits_{x\to -\infty} 1/x^n$, we assume that n is such that $1/x^n$ is defined for $x < 0$.

The use of the ∞ symbol here is a way of saying *why* there is no limit. In general,

> As $x \to \infty$ (or $x \to -\infty$), the limit of a polynomial function is the same as the limit of its term that involves the greatest power of x.

EXAMPLE 1

a. $\lim\limits_{x \to -\infty} (x^3 - x^2 + x - 2) = \lim\limits_{x \to -\infty} x^3$. As x becomes very negative, so does x^3. Thus

$$\lim_{x \to -\infty} (x^3 - x^2 + x - 2) = \lim_{x \to -\infty} x^3 = -\infty.$$

b. $\lim\limits_{x \to -\infty} (-2x^3 + 9x) = \lim\limits_{x \to -\infty} -2x^3 = \infty$, because -2 times a very negative number is very positive.

Let us now find the limit of the function

$$f(x) = \frac{4x^3 + x}{2x^3 + 3}$$

as $x \to \infty$. This function is a quotient of polynomials, which you recall is a rational function. As x gets larger and larger, both the numerator and denominator of $f(x)$ become infinite. However, the form of the quotient can be changed so that we can draw a conclusion as to whether or not it has a limit. To do this, we divide both the numerator and denominator by the greatest power of x that occurs in the denominator. Here it is x^3. This gives

$$\lim_{x \to \infty} \frac{4x^3 + x}{2x^3 + 3} = \lim_{x \to \infty} \frac{\dfrac{4x^3 + x}{x^3}}{\dfrac{2x^3 + 3}{x^3}}$$

$$= \lim_{x \to \infty} \frac{\dfrac{4x^3}{x^3} + \dfrac{x}{x^3}}{\dfrac{2x^3}{x^3} + \dfrac{3}{x^3}}$$

$$= \lim_{x \to \infty} \frac{4 + \dfrac{1}{x^2}}{2 + \dfrac{3}{x^3}} = \frac{\lim\limits_{x \to \infty} 4 + \lim\limits_{x \to \infty} \dfrac{1}{x^2}}{\lim\limits_{x \to \infty} 2 + 3 \cdot \lim\limits_{x \to \infty} \dfrac{1}{x^3}}.$$

Since $\lim\limits_{x\to\infty} 1/x^n = 0$, where n is positive,

$$\lim_{x\to\infty} \frac{4x^3 + x}{2x^3 + 3} = \frac{4 + 0}{2 + 3(0)} = \frac{4}{2} = 2.$$

Similarly, the limit as $x \to -\infty$ is 2.

For the preceding function, there is an easier way to find $\lim\limits_{x\to\infty} f(x)$. For *large* values of x, in the numerator the term $4x^3$ dominates the sum $4x^3 + x$, and the dominant term in the denominator, $2x^3 + 3$, is $2x^3$. Hence, to determine the limit of $f(x)$, it suffices to determine the limit of $(4x^3)/(2x^3)$. That is,

$$\lim_{x\to\infty} \frac{4x^3 + x}{2x^3 + 3} = \lim_{x\to\infty} \frac{4x^3}{2x^3} = \lim_{x\to\infty} 2 = 2,$$

as we saw before. In general, we have the following.

If $f(x)$ is a rational function and $a_n x^n$ and $b_m x^m$ are the terms in the numerator and denominator, respectively, with the greatest powers of x, then

$$\lim_{x\to\infty} f(x) = \lim_{x\to\infty} \frac{a_n x^n}{b_m x^m}$$

and

$$\lim_{x\to-\infty} f(x) = \lim_{x\to-\infty} \frac{a_n x^n}{b_m x^m}.$$

For example,

$$\lim_{x\to-\infty} \frac{x^4 - 3x}{5 - 2x} = \lim_{x\to-\infty} \frac{x^4}{-2x} = \lim_{x\to-\infty} \left(-\frac{1}{2}x^3\right) = \infty.$$

(Note that in the next-to-last step, as x becomes very negative, so does x^3; moreover, $-\frac{1}{2}$ times a very negative number is very positive.) Similarly,

$$\lim_{x\to\infty} \frac{x^4 - 3x}{5 - 2x} = \lim_{x\to\infty} \left(-\frac{1}{2}x^3\right) = -\infty.$$

From this illustration we conclude that *whenever the degree of the numerator of a rational function is greater than the degree of the denominator, the function has no limit as $x \to \infty$ or as $x \to -\infty$.*

EXAMPLE 2   *Find the indicated limits.*

a. $\lim\limits_{x\to\infty} \dfrac{x^2 - 1}{7 - 2x + 8x^2}.$

$$\lim_{x\to\infty} \frac{x^2 - 1}{7 - 2x + 8x^2} = \lim_{x\to\infty} \frac{x^2}{8x^2} = \lim_{x\to\infty} \frac{1}{8} = \frac{1}{8}.$$

b. $\lim\limits_{x \to -\infty} \dfrac{x}{(3x - 1)^2}.$

$$\lim_{x \to -\infty} \frac{x}{(3x - 1)^2} = \lim_{x \to -\infty} \frac{x}{9x^2 - 6x + 1} = \lim_{x \to -\infty} \frac{x}{9x^2}$$

$$= \lim_{x \to -\infty} \frac{1}{9x} = \frac{1}{9} \cdot \lim_{x \to -\infty} \frac{1}{x} = \frac{1}{9}(0) = 0.$$

c. $\lim\limits_{x \to \infty} \dfrac{x^5 - x^4}{x^4 - x^3 + 2}.$

Since the degree of the numerator is greater than that of the denominator, there is no limit. More precisely,

$$\lim_{x \to \infty} \frac{x^5 - x^4}{x^4 - x^3 + 2} = \lim_{x \to \infty} \frac{x^5}{x^4} = \lim_{x \to \infty} x = \infty.$$

EXAMPLE 3   *Find* $\lim\limits_{x \to -\infty} \sqrt{4 - x}.$

As x gets negatively infinite, $4 - x$ becomes positively infinite. Because square roots of large numbers are large numbers, we conclude that

$$\lim_{x \to -\infty} \sqrt{4 - x} = \infty.$$

EXERCISE 1.9

In Problems 1–28, find the indicated limits.

1. $\lim\limits_{x \to \infty} (x^2 - 2x + 1).$

2. $\lim\limits_{x \to -\infty} (3x - 4).$

3. $\lim\limits_{x \to -\infty} (7 - x^2 - x^3).$

4. $\lim\limits_{x \to \infty} (-3x^5 + 4x + 2).$

5. $\lim\limits_{x \to \infty} \dfrac{x^3 - 3x^2 - 2x}{4x^3 + 2x - 3}.$

6. $\lim\limits_{x \to -\infty} \dfrac{x^2 - x}{2 + 3x}.$

7. $\lim\limits_{x \to -\infty} \dfrac{3x + 5}{5x^2 + 7x - 2}.$

8. $\lim\limits_{x \to \infty} \dfrac{6x^4 - 3x^3 + x^2}{7 - x^2 - 3x^4}.$

9. $\lim\limits_{x \to \infty} \dfrac{x^3 - 6x + 2}{x^2 + 4}.$

10. $\lim\limits_{x \to \infty} \dfrac{1 - x^6}{x^7 - 1}.$

11. $\lim\limits_{x \to \infty} \dfrac{7}{2x + 1}.$

12. $\lim\limits_{x \to -\infty} \dfrac{1}{(4x - 1)^3}.$

13. $\lim\limits_{x \to \infty} \dfrac{x + 2}{x + 3}.$

14. $\lim\limits_{x \to \infty} \dfrac{2x - 4}{3 - 2x}.$

15. $\lim\limits_{x \to -\infty} \dfrac{x^2 - 1}{x^3 + 4x - 3}.$

16. $\lim\limits_{r \to \infty} \dfrac{r^3}{r^2 + 1}.$

17. $\lim\limits_{t \to \infty} \dfrac{5t^2 + 2t + 1}{7 - 4t}.$

18. $\lim\limits_{x \to -\infty} \dfrac{2x}{3x^6 - x + 4}.$

19. $\lim\limits_{x \to -\infty} \dfrac{3 - 4x - 2x^3}{5x^3 - 8x + 1}.$

20. $\lim\limits_{x \to \infty} \dfrac{7 - 2x - x^4}{9 - 3x^4 + 2x^2}.$

21. $\lim\limits_{w \to \infty} \dfrac{2w^2 - 3w + 4}{5w^2 + 7w - 1}.$

22. $\lim\limits_{x \to \infty} \dfrac{4 - 3x^3}{x^3 - 1}.$

23. $\lim\limits_{x \to \infty} \dfrac{2x^2}{(2x - 1)^2}.$

24. $\lim\limits_{x \to \infty} \dfrac{(x - 1)^2}{x^3 + 2}.$

25. $\lim\limits_{x \to \infty} \dfrac{3}{\sqrt{x}}.$

26. $\lim\limits_{x \to \infty} \dfrac{3}{2x\sqrt{x}}.$

27. $\lim\limits_{x \to \infty} \sqrt{x + 10}.$

28. $\lim\limits_{x \to -\infty} \sqrt{2 - 3x}.$

29. For a particular host-parasite relationship, it was determined that when host density (number of hosts per unit of area) is x, then the number of hosts parasitized over a period of time is y, where

$$y = \frac{900x}{10 + 45x}.$$

If the host density were to increase without bound, what value would y approach?

30. The population N of a certain small city t years from now is predicted to be

$$N = 20,000 + \frac{10,000}{(t + 2)^2}.$$

Determine the population in the long run; that is, find $\lim_{t \to \infty} N$.

31. If c is the total cost in dollars to produce q units of a product, then the average cost per unit $\bar{c}$ for an output of q units is given by $\bar{c} = c/q$. Thus if the total cost equation is given by $c = 5000 + 6q$, then $\bar{c} = (5000/q) + 6$. For example, the total cost of an output of 5 units is $5030, and the average cost per unit at this level of production is $1006. By finding $\lim_{q \to \infty} \bar{c}$, show that the average cost approaches a level of stability if the producer continually increases output. What is the limiting value of the average cost?

1.10 CONTINUITY

Many functions have the property that there is no "break" in their graphs. For example, compare Figs. 1.37 and 1.38. When $x = 1$ the graph of the

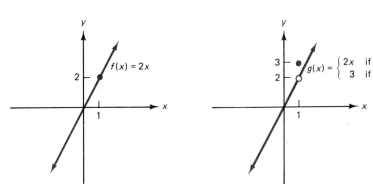

FIGURE 1.37   FIGURE 1.38

function f is unbroken, but the graph of g has a break. Stated another way, if you were to trace both graphs with a pencil, you would have to lift the pencil on the graph of g when $x = 1$, but you would not have to lift it on the graph of f. We characterize these situations by saying that f is *continuous* at $x = 1$ and g is *discontinuous* at $x = 1$.

Definition
*A function f is **continuous** at the point $x = a$ (or simply, continuous at a) if and only if there is no break in its graph at $x = a$. If f is not continuous at a point, then it is said to be **discontinuous** there.*

Our definition of continuity is geometrical in nature and is sufficient for our purposes. Alternatively, continuity can be defined in terms of limits. If the graph of a function f has no break at $x = a$, then for values of x near a, the values of $f(x)$ must be near $f(a)$. For example, the function in Fig. 1.37 is continuous at $x = 1$ and

$$\lim_{x \to 1} f(x) = 2 = f(1).$$

But in Fig. 1.38, g is discontinuous at $x = 1$ and

$$\lim_{x \to 1} g(x) = 2 \neq g(1).$$

In general, a function f is continuous at $x = a$ if and only if

$$\lim_{x \to a} f(x) = f(a).$$

EXAMPLE 1

a. In Fig. 1.39(a) the graph of f has no break, so f is continuous for all values of x. To describe this, we say that f is **continuous everywhere**, or simply f is **continuous**.

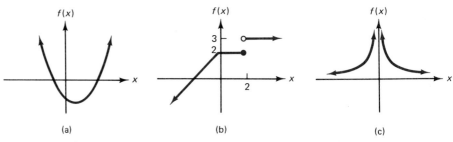

FIGURE 1.39

b. In Fig. 1.39(b) the graph has a break, or *discontinuity*, at $x = 2$ only. Thus f is discontinuous at 2 and is continuous otherwise.

c. In Fig. 1.39(c) the graph has a break at $x = 0$ only. Thus f is discontinuous at 0 and is continuous otherwise.

EXAMPLE 2   *Determine whether* $f(x) = \dfrac{x^2 - 1}{x - 1}$ *is continuous at* 1.

Because this function is not defined at $x = 1$, its graph has no point there. This causes a break in the graph, so f is discontinuous at 1. In general, *if a* *function is not defined at $x = a$, it is discontinuous there.* It is interesting to note that

$$f(x) = \frac{x^2 - 1}{x - 1} = \frac{(x + 1)(x - 1)}{x - 1} = x + 1, \qquad x \neq 1.$$

Thus the graph of f is the same as the graph of $y = x + 1$ (which is a straight line) except that it has a hole in it when $x = 1$ (see Fig. 1.40).

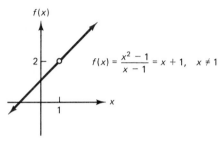

FIGURE 1.40

If a function has a certain form, we can easily determine whether or not it is continuous. It can be shown that:

1. A polynomial function is continuous everywhere.

2. A rational function is discontinuous at points where the denominator is 0 and is continuous otherwise.

EXAMPLE 3   *For each of the following functions, determine all points of discontinuity.*

a. $f(x) = 2x^2 + 5x + 4.$

Since f is a polynomial, it is continuous everywhere and hence has no points of discontinuity.

b. $F(x) = \dfrac{x^2 - 3}{x^2 + 2x - 8}.$

This rational function has denominator

$$x^2 + 2x - 8 = (x + 4)(x - 2),$$

which is 0 when $x = -4$ or $x = 2$. Thus F is discontinuous only at -4 and 2.

c. $h(x) = \dfrac{x + 4}{x^2 + 4}.$

The denominator of this rational function is never 0 (it is always positive). Thus h has no discontinuity.

EXAMPLE 4

The "post-office function"

$$c = f(x) = \begin{cases} 22, & \text{if } 0 < x \leq 1, \\ 39, & \text{if } 1 < x \leq 2, \\ 56, & \text{if } 2 < x \leq 3, \\ 73, & \text{if } 3 < x \leq 4 \end{cases}$$

gives the cost c (in cents) of mailing a parcel of weight x (ounces), where $0 < x \leq 4$, in January 1987. It is clear from its graph in Fig. 1.41 that f has discontinuities at 1, 2, and 3 and is constant for values of x between successive discontinuities. Such a function is called a *step function* because of the appearance of its graph.

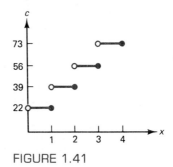

FIGURE 1.41

Often it is helpful to describe a situation by a continuous function. For example, the demand schedule in Table 1.7 indicates the number of units of a particular product that consumers will demand per week at various prices. This information can be given graphically as in Fig. 1.42(a) by plotting each quantity-price pair as a point. Clearly this graph does not represent a continuous

TABLE 1.7
Demand Schedule

PRICE/UNIT, p (DOLLARS)	QUANTITY PER WEEK, q
20	0
10	5
5	15
4	20
2	45
1	95

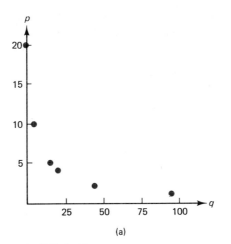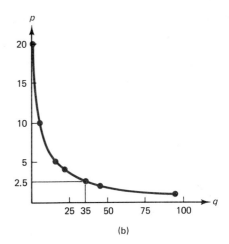

FIGURE 1.42

function. Furthermore, it gives us no information as to the price at which, say, 35 units would be demanded. However, if we connect the points in Fig. 1.42(a) by a smooth curve [see Fig. 1.42(b)], we get a so-called demand curve. From it we could guess that at about \$2.50 per unit, 35 units would be demanded.

Frequently, it is possible and useful to describe a graph, as in Fig. 1.42(b), by means of an equation that defines a continuous function f. Such a function not only gives us a demand equation, $p = f(q)$, which allows us to anticipate corresponding prices and quantities demanded, it also permits a convenient mathematical analysis of the nature and basic properties of demand. Of course, some care must be used in working with equations such as $p = f(q)$. Mathematically, f may be defined when $q = \sqrt{37}$, but from a practical standpoint, a demand of $\sqrt{37}$ units could be meaningless to our particular situation. For example, if a unit is an egg, then a demand of $\sqrt{37}$ eggs makes no sense.

In general, it will be our desire to view practical situations in terms of continuous functions whenever possible so that we may be better able to analyze their nature.

EXERCISE 1.10

In Problems 1–4, determine whether the function is continuous at the given points.

1. $f(x) = \dfrac{x + 4}{x - 2}$; $-2, 0.$

2. $f(x) = \dfrac{x^2 - 4x + 4}{6}$; $2, -2.$

3. $g(x) = \dfrac{x - 3}{x^2 - 9}$; $3, -3.$

4. $h(x) = \dfrac{3}{x^2 + 4}$; $2, -2.$

In Problems 5–8, give a reason why the function is continuous everywhere.

5. $f(x) = 2x^2 - 3.$

6. $f(x) = \dfrac{x + 2}{5}.$

7. $f(x) = \dfrac{x - 1}{x^2 + 4}.$

8. $f(x) = x(1 - x).$

In Problems 9–20, find all points of discontinuity.

9. $f(x) = 3x^2 - 3$.

10. $h(x) = x - 2$.

11. $f(x) = \dfrac{3}{x - 4}$.

12. $f(x) = \dfrac{x^2 + 3x - 4}{x + 4}$.

13. $g(x) = \dfrac{(x^2 - 1)^2}{5}$.

14. $f(x) = 0$.

15. $f(x) = \dfrac{x^2 + 6x + 9}{x^2 + 2x - 15}$.

16. $g(x) = \dfrac{x - 3}{x^2 + x}$.

17. $h(x) = \dfrac{x - 7}{x^3 - x}$.

18. $f(x) = \dfrac{x}{x}$.

19. $p(x) = \dfrac{x}{x^2 + 1}$.

20. $f(x) = \dfrac{x^4}{x^4 - 1}$.

21. Suppose that the long-distance rate for a telephone call from Hazleton, PA, to Los Angeles, CA, is $0.29 for the first minute and $0.20 for each additional minute or fraction thereof. If $y = f(t)$ is a function that indicates the total charge y for a call of t minutes' duration, sketch the graph of f for $0 < t \le 4\frac{1}{2}$. Use your graph to determine the values of t, where $0 < t \le 4\frac{1}{2}$, at which discontinuities occur.

22. The *greatest integer function*, $f(x) = [x]$, is defined to be the greatest integer less than or equal to x, where x can be any real number. For example, $[3] = 3$, $[1.999] = 1$, $[\frac{1}{4}] = 0$, and $[-4.5] = -5$. Sketch the graph of this function for $-3.5 \le x \le 3.5$. Use your sketch to determine the values of x at which discontinuities occur.

1.11 REVIEW

Important Terms and Symbols

Section 1.1	function domain range independent variable dependent variable function value, $f(x)$ demand function supply function
Section 1.2	constant function polynomial function rational function compound function absolute value, $|x|$ r factorial, $r!$
Section 1.3	$f + g$ $f - g$ fg f/g composition function $f \circ g$
Section 1.4	graph of function function-value axis y-intercept x-intercept vertical-line test
Section 1.5	slope point-slope form slope-intercept form general linear equation linearly related
Section 1.6	demand equation demand curve supply equation supply curve linear function
Section 1.7	quadratic function axis of symmetry vertex
Section 1.8	$\lim\limits_{x \to a} f(x)$
Section 1.9	$\lim\limits_{x \to \infty} f(x)$ $\lim\limits_{x \to -\infty} f(x)$
Section 1.10	continuous discontinuous continuous everywhere

Summary

A function f is a rule of correspondence that assigns to each input number x exactly one output number $f(x)$. Usually, a function is specified by an equation that indicates what must be done to an input x to obtain $f(x)$. To obtain a particular function value $f(a)$, we replace each x in the equation by a.

The domain of a function consists of all input numbers, and the range consists of all output numbers. Unless otherwise specified, the domain of f consists of all values of x for which $f(x)$ is a real number.

Some special types of functions are constant functions and polynomial functions. A function that is defined by more than one equation is called a compound function.

Two functions, f and g, can be combined to form a sum, difference, product, quotient, or composition as follows.

$$(f + g)(x) = f(x) + g(x),$$
$$(f - g)(x) = f(x) - g(x),$$
$$(fg)(x) = f(x)g(x),$$
$$\frac{f}{g}(x) = \frac{f(x)}{g(x)},$$
$$(f \circ g)(x) = f(g(x)).$$

We graph a function $y = f(x)$ by plotting a sufficient number of points $(x, f(x))$ and connecting them (where appropriate) so that the basic shape of the graph is apparent. One important point to include is the y-intercept, which is found by letting x be 0 and solving for y. The graph should make it clear what the domain and range are.

The fact that a graph represents a function can be determined by using the vertical-line test. A vertical line cannot cut the graph of a function at more than one point.

The orientation of a nonvertical straight line is characterized by the slope m of the line:

$$m = \frac{y_2 - y_1}{x_2 - x_1},$$

where (x_1, y_1) and (x_2, y_2) are two different points on the line. The slope of a vertical line is not defined, and the slope of a horizontal line is zero. Rising (falling) lines have positive (negative) slopes, and lines with the same slope are parallel. Basic forms for equations of lines are:

Point-slope form:	$y - y_1 = m(x - x_1),$
Slope-intercept form:	$y = mx + b,$
Vertical line:	$x = a,$
Horizontal line:	$y = b,$
General:	$Ax + By + C = 0.$

The linear function $f(x) = ax + b \ (a \neq 0)$ has a straight line for its graph.

In economics, demand functions and supply functions have the form $p = f(q)$ and play an important role. A demand function gives a correspondence between the price p of a product and the number of units q of the product that consumers will purchase at that price during some time period. A supply function gives a correspondence between the price p of a product and the quantity q of the product that manufacturers will supply at that price during some time period. Typically, linear demand curves

have negative slopes and linear supply curves have positive slopes.

A quadratic function has the form

$$f(x) = ax^2 + bx + c \qquad (a \neq 0).$$

Its graph is a parabola that opens upward if $a > 0$ and downward if $a < 0$. The vertex is

$$\left(-\frac{b}{2a}, f\left(-\frac{b}{2a} \right) \right)$$

and the y-intercept is c. The axis of symmetry as well as the x- and y-intercepts are useful in sketching the graph.

The notion of a limit lies at the foundation of calculus. To say that

$$\lim_{x \to a} f(x) = L$$

means that the values of $f(x)$ can be made as close to the number L as we choose by taking x sufficiently close to a. If $\lim\limits_{x \to a} f(x)$ and $\lim\limits_{x \to a} g(x)$ exist and c is a constant, then

1. $\lim\limits_{x \to a} c = c.$

2. $\lim\limits_{x \to a} x^n = a^n.$

3. $\lim\limits_{x \to a} [f(x) \pm g(x)] = \lim\limits_{x \to a} f(x) \pm \lim\limits_{x \to a} g(x).$

4. $\lim\limits_{x \to a} [f(x) \cdot g(x)] = \lim\limits_{x \to a} f(x) \cdot \lim\limits_{x \to a} g(x).$

5. $\lim\limits_{x \to a} [cf(x)] = c \cdot \lim\limits_{x \to a} f(x).$

6. $\lim\limits_{x \to a} \dfrac{f(x)}{g(x)} = \dfrac{\lim\limits_{x \to a} f(x)}{\lim\limits_{x \to a} g(x)}$ if $\lim\limits_{x \to a} g(x) \neq 0.$

7. $\lim\limits_{x \to a} \sqrt[n]{f(x)} = \sqrt[n]{\lim\limits_{x \to a} f(x)}.$

8. If f is a polynomial function, then $\lim\limits_{x \to a} f(x) = f(a).$

The last property means that the limit of a polynomial function can be found by simply substituting a for x. However, with other functions substitution may lead to the meaningless form 0/0. In such cases, algebraic manipulation such as factoring may give a form from which the limit can be determined.

The infinity symbol ∞, which does not represent a number, is used in describing limits. The statement

$$\lim_{x \to \infty} f(x) = L$$

means that as x increases without bound, the values of $f(x)$ approach the number L. A similar statement applies when $x \to -\infty$, which means that x is decreasing without bound. In general, if n is positive, then

$$\lim_{x \to \infty} \frac{1}{x^n} = 0 \quad \text{and} \quad \lim_{x \to -\infty} \frac{1}{x^n} = 0.$$

To say that the limit of a function is ∞ (or $-\infty$) does not mean that the limit exists. In fact, it is a way of saying that the limit does not exist, but it tells *why* there is no limit. For example, if $\lim_{x \to \infty} f(x) = \infty$, then the values of $f(x)$ are increasing without bound as x gets larger and larger, so no limit exists.

There is a rule for evaluating the limit of a rational function (quotient of polynomials) as $x \to \infty$ or $x \to -\infty$. If $f(x)$ is a rational function and $a_n x^n$ and $b_m x^m$ are the terms in the numerator and denominator, respectively, with

the greatest powers of x, then

$$\lim_{x \to \infty} f(x) = \lim_{x \to \infty} \frac{a_n x^n}{b_m x^m}$$

and $\quad \lim_{x \to -\infty} f(x) = \lim_{x \to -\infty} \frac{a_n x^n}{b_m x^m}.$

Moreover, as $x \to \infty$ (or $x \to -\infty$) the limit of a polynomial function is either ∞ or $-\infty$. It is the same as the limit of the term of the polynomial that involves the greatest power of x.

A function that has no break in its graph when x is a is said to be continuous at a. Otherwise, it is discontinuous at a. In particular, a function that is not defined when $x = a$ must be discontinuous there. Polynomial functions are continuous everywhere, and rational functions are discontinuous only at points where the denominator is zero.

Review Problems

In Problems **1–6**, *give the domain of each function.*

1. $f(x) = \dfrac{x}{x^2 - 3x + 2}.$ **2.** $g(x) = x^2 + 3x.$

3. $F(t) = 7t + 4t^2.$ **4.** $G(x) = 18.$

5. $h(x) = \dfrac{\sqrt{x}}{x - 1}.$ **6.** $H(s) = \dfrac{\sqrt{s - 5}}{4}.$

In Problems **7–14**, *find the functional values for the given function.*

7. $f(x) = 3x^2 - 4x + 7;\quad f(0),\quad f(-3),\quad f(5),\quad f(t).$

8. $g(x) = 4;\quad g(4),\quad g(\frac{1}{100}),\quad g(-156),\quad g(x + 4).$

9. $G(x) = \sqrt{x - 1};\quad G(1),\quad G(10),\quad G(t + 1),\quad G(x^2).$

10. $F(x) = \dfrac{x - 3}{x + 4};\quad F(-1),\quad F(0),\quad F(5),\quad F(x + 3).$

11. $h(u) = \dfrac{\sqrt{u + 4}}{u};\quad h(5),\quad h(-4),\quad h(x),\quad h(u - 4).$

12. $H(s) = \dfrac{(s - 4)^2}{3};\quad H(-2),\quad H(7),\quad H(\frac{1}{2}),\quad H(x^2).$

13. $f(x) = \begin{cases} 4, & \text{if } x < 2 \\ 8 - x^2, & \text{if } x > 2 \end{cases};\quad f(4),\quad f(-2),\quad f(0),\quad f(10).$

14. $h(q) = \begin{cases} q, & \text{if } -1 \le q < 0 \\ 3 - q, & \text{if } 0 \le q < 3 \\ 2q^2, & \text{if } 3 \le q \le 5 \end{cases}$;

$h(0),\quad h(4),\quad h(-\frac{1}{2}),\quad h(\frac{1}{2}).$

In Problems **15** *and* **16**, *determine* (a) $f(x + h)$, *and* (b) $\dfrac{f(x + h) - f(x)}{h}$, *and simplify your answers.*

15. $f(x) = 3 - 7x.$ **16.** $f(x) = x^2 + 4.$

17. If $f(x) = 3x - 1$ and $g(x) = 2x + 3$, find the following.
a. $(f + g)(x).$ b. $(f + g)(4).$
c. $(f - g)(x).$ d. $(fg)(x).$
e. $(fg)(1).$ f. $\dfrac{f}{g}(x).$
g. $(f \circ g)(x).$ h. $(f \circ g)(5).$
i. $(g \circ f)(x).$

18. If $f(x) = x^2$ and $g(x) = 2x + 1$, find the following.
a. $(f + g)(x).$ b. $(f - g)(x).$
c. $(f - g)(-3).$ d. $(fg)(x).$
e. $\dfrac{f}{g}(x).$ f. $\dfrac{f}{g}(2).$
g. $(f \circ g)(x).$ h. $(g \circ f)(x).$
i. $(g \circ f)(-4).$

In Problems **19–22**, *find* $(f \circ g)(x)$ *and* $(g \circ f)(x).$

19. $f(x) = \dfrac{1}{x},\quad g(x) = x - 1.$

20. $f(x) = \dfrac{x + 1}{4}, \quad g(x) = \sqrt{x}.$

21. $f(x) = x + 2, \quad g(x) = x^3.$

22. $f(x) = 2, \quad g(x) = 3.$

In Problems 23–36, graph each function and give its domain and range. For those that are linear, also give the slope and the vertical-axis intercept. For those that are quadratic, give all intercepts and the vertex.

23. $y = f(x) = 4 - 2x.$

24. $f(x) = 5.$

25. $y = f(x) = 9 - x^2.$

26. $G(u) = \sqrt{u + 4}.$

27. $y = h(t) = t^2 - 4t - 5.$

28. $f(x) = |x| + 1.$

29. $y = f(x) = \begin{cases} 1 - x, & \text{if } x \le 0, \\ 1, & \text{if } x > 0. \end{cases}$

30. $f(t) = t^2 + 2t.$

31. $y = g(t) = \dfrac{2}{t - 4}.$

32. $s = g(t) = 8 - 2t - t^2.$

33. $y = F(x) = -x^2 - 2x - 3.$

34. $g(t) = \sqrt{4t}.$

35. $p = g(t) = 3t.$

36. $y = f(x) = \dfrac{x}{3} - 2.$

37. The predicted annual sales S (in dollars) of a new product is given by the equation $S = 150{,}000 + 3000t$, where t is the time in years from 1987. Such an equation is called a *trend equation*. Find the predicted annual sales for 1992. Is S a function of t?

38. The slope of the line through $(2, 3)$ and $(k, 3)$ is 0. Find k.

In Problems 39–44, find the slope-intercept form and a general linear form of an equation of the straight line that has the indicated properties.

39. Passes through $(3, -2)$ and has y-intercept 1.

40. Passes through $(-1, -1)$ and is parallel to the line $y = 3x - 4$.

41. Passes through $(10, 4)$ and has slope $\frac{1}{2}$.

42. Passes through $(3, 5)$ and is vertical.

43. Passes through $(-2, 4)$ and is horizontal.

44. Determine whether the point $(0, -7)$ lies on the line through $(1, -3)$ and $(4, 9)$.

45. Suppose that f is a linear function such that $f(1) = 5$ and $f(x)$ decreases by 4 units for every 3-unit increase in x. Find $f(x)$.

46. If f is a linear function such that $f(-1) = 8$ and $f(2) = 5$, find $f(x)$.

47. Suppose that consumers will demand 20 (thousand) pairs of shoes when the price is 35 (dollars per pair) and 25 pairs when the price is 30. Find the demand equation, assuming that it is linear.

48. Celsius temperature C is a linear function of Fahrenheit temperature F.

    a. Use the facts that 32°F is the same as 0°C and 212°F is the same as 100°C to find this function.

    b. Find C when F = 50.

49. In psychology the term *semantic memory* refers to our knowledge of the meaning and relationships of words as well as the means by which we store and retrieve such information [5]. In a network model of semantic memory, there is a hierarchy of levels at which information is stored. In an experiment by Collins and Quillian based on a network model, data were obtained on the reaction time to respond to simple questions about nouns. The graph of the results shows that, on the average, reaction time R (in milliseconds) is a linear function of the level L at which a characterizing property of the noun is stored. At level 0 the reaction time is 1310; at level 2 the reaction time is 1460.

    a. Find the linear function.

    b. Find the reaction time at level 1.

    c. Find the slope and determine its significance.

In Problems 50–67, find the limits.

50. $\displaystyle \lim_{x \to 0} \frac{2x^2 - 3x + 1}{2x^2 - 2}.$

51. $\displaystyle \lim_{x \to -1} (2x^2 + 6x - 1).$

52. $\displaystyle \lim_{x \to -2} \frac{x + 1}{x^2 - 2}.$

53. $\displaystyle \lim_{x \to 3} \frac{x^2 - 9}{x^2 - 3x}.$

54. $\displaystyle \lim_{x \to 2} \frac{x^2 - 4}{x^2 - 3x + 2}.$

55. $\displaystyle \lim_{h \to 0} (x + h).$

56. $\displaystyle \lim_{x \to 1} \frac{x^2 + x - 2}{x - 1}.$

57. $\displaystyle \lim_{x \to 4} \sqrt{4}.$

58. $\lim\limits_{x\to 2}\dfrac{2-x}{x-2}$.

59. $\lim\limits_{x\to -2}\dfrac{x^2-4x-12}{x^2+6x+8}$.

60. $\lim\limits_{x\to\infty}\dfrac{x^2+1}{x^2}$.

61. $\lim\limits_{x\to\infty}\dfrac{2}{x+1}$.

62. $\lim\limits_{x\to -\infty}(x^4+x^5)$.

63. $\lim\limits_{x\to\infty}\dfrac{3x-2}{5x+3}$.

64. $\lim\limits_{x\to -\infty}\dfrac{x^6}{x^5}$.

65. $\lim\limits_{t\to -\infty}\dfrac{2t^3-3}{t-3}$.

66. $\lim\limits_{x\to\infty}\dfrac{x^2-1}{(3x+2)^2}$.

67. $\lim\limits_{x\to\infty}\sqrt{3x}$.

68. For a particular host-parasite relationship, it was determined that when the host density (number of hosts per unit of area) is x, then the number of hosts parasitized over a certain period of time is y, where

$$y=11\left(1-\frac{1}{1+2x}\right).$$

If the host density were to increase without bound, what value would y approach?

69. For a particular predator-prey relationship, it was determined that the number y of prey consumed by an individual predator over a period of time was a function of prey density x (the number of prey per unit of area). Suppose

$$y=f(x)=\frac{10x}{1+0.1x}.$$

If the prey density were to increase without bound, what value would y approach?

70. Is $f(x)=x^2-2$ continuous everywhere? Give a reason for your answer.

71. Is $f(x)=x/4$ continuous everywhere? Give a reason for your answer.

In Problems 72–77, find the points of discontinuity (if any) for each function.

72. $f(x)=\dfrac{0}{x^3}$.

73. $f(x)=\dfrac{x^2}{x+3}$.

74. $f(x)=(3-2x)^2$.

75. $f(x)=\dfrac{x-1}{2x^2+3}$.

76. $f(x)=\dfrac{2x+6}{x^3+x}$.

77. $f(x)=\dfrac{4-x^2}{x^2+3x-4}$.

2

Differentiation

Now we begin our study of calculus. The ideas involved in calculus are completely different from those of algebra and geometry. The power and importance of these ideas and their applications will be evident to you later in the text. The objective of this chapter is not only to convey an understanding of what the so-called "derivative" of a function is, but also to teach techniques of finding derivatives by properly applying rules.

2.1 THE DERIVATIVE

One of the main problems with which calculus deals is finding the slope of the *tangent line* at a point on a curve. In geometry you probably thought of a tangent line, or *tangent,* to a circle as a line that meets the circle at exactly one point (Fig. 2.1). Unfortunately, this idea of a tangent is not very useful for other kinds of curves.

For example, in Fig. 2.2(a) the lines L_1 and L_2 intersect the curve at exactly one point. Although we would not think of L_2 as the tangent at this point, it seems natural that L_1 is. In Fig. 2.2(b) we would consider L_3 to be

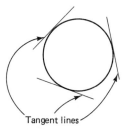

Tangent lines

FIGURE 2.1

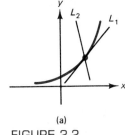

(a)

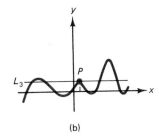

(b)

FIGURE 2.2

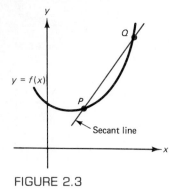

FIGURE 2.3

the tangent at point P even though L_3 intersects the curve at other points. From these examples, you can see that we must drop the idea that a tangent is simply a line that intersects a curve at only one point. To develop a suitable definition of tangent line, we use the limit concept.

The graph of a function $y = f(x)$ appears in Fig. 2.3. Here P and Q are two different points on the curve. A line passing through two different points on a curve is called a **secant line**. Figure 2.3 shows the secant line through P and Q. If Q moves along the curve and approaches P from the right, typical secant lines are PQ', PQ'', and so on as shown in Fig. 2.4. As Q approaches P from the left, they are PQ_1, PQ_2, and so on. *In both cases, the secant lines approach the **same** limiting position.* This common limiting position of the secant lines is defined to be the **tangent line** to the curve at P. This definition seems reasonable and avoids the difficulty mentioned at the beginning of this section.

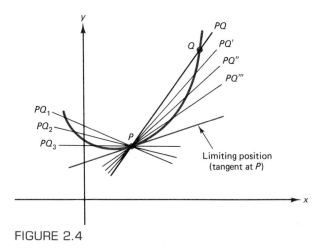

FIGURE 2.4

A curve does not necessarily have a tangent at each of its points. For example, the curve $y = |x|$ does not have a tangent at $(0, 0)$ for the following reason. In Fig. 2.5, a secant line joining $(0, 0)$ to a nearby point to its right must always be the line $y = x$, and one to a nearby point to its left is the line $y = -x$. Thus the limiting position of secant lines through $(0, 0)$ and points on the curve to the right of $(0, 0)$ is the line $y = x$, but the limiting position of secant lines through $(0, 0)$ and points to its left is the line $y = -x$. Since there is no common limiting position, there is no tangent.

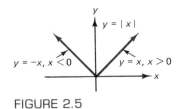

FIGURE 2.5

Now that we have a suitable definition of a tangent to a curve at a point, we can define the *slope of a curve* at a point.

Definition
*The **slope of a curve** at a point P is the slope of the tangent line at P.*

Since the tangent at P is a limiting position of secant lines PQ, the slope of the tangent is the limiting value of the slopes of the secant lines as Q approaches P. We shall find an expression for the slope of the curve $y = f(x)$ at point $P = (x_1, f(x_1))$ shown in Fig. 2.6. If $Q = (x_2, f(x_2))$, the slope of the secant line PQ is

$$m_{PQ} = \frac{f(x_2) - f(x_1)}{x_2 - x_1}.$$

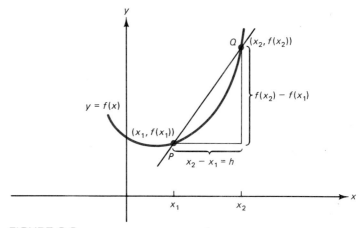

FIGURE 2.6

If the difference $x_2 - x_1$ is called h, we can write x_2 as $x_1 + h$. Here $h \neq 0$, for if $h = 0$, then $x_2 = x_1$ and no secant line exists. Thus

$$m_{PQ} = \frac{f(x_1 + h) - f(x_1)}{(x_1 + h) - x_1} = \frac{f(x_1 + h) - f(x_1)}{h}.$$

As Q moves along the curve toward P, then x_2 approaches x_1. This means that h approaches zero. The limiting value of the slopes of the secant lines—which is the slope of the tangent line at $(x_1, f(x_1))$—is the following limit:

$$m_{\text{tan}} = \lim_{h \to 0} \frac{f(x_1 + h) - f(x_1)}{h}. \tag{1}$$

EXAMPLE 1   *Find the slope of the curve $y = f(x) = x^2$ at the point* (1, 1).

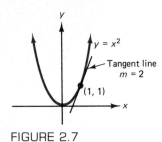

FIGURE 2.7

The slope is the limit in Eq. (1) with $f(x) = x^2$ and $x_1 = 1$.

$$\lim_{h \to 0} \frac{f(1 + h) - f(1)}{h} = \lim_{h \to 0} \frac{(1 + h)^2 - (1)^2}{h}$$

$$= \lim_{h \to 0} \frac{1 + 2h + h^2 - 1}{h} = \lim_{h \to 0} \frac{2h + h^2}{h}$$

$$= \lim_{h \to 0} \frac{h(2 + h)}{h} = \lim_{h \to 0} (2 + h) = 2.$$

Thus the tangent line to $y = x^2$ at $(1, 1)$ has slope 2 (Fig. 2.7).

We can generalize Eq. (1) so it applies to any point $(x, f(x))$ on a curve. Replacing x_1 by x gives a function, called the *derivative* of f, whose input is x and whose output is the slope of the tangent line to the curve at $(x, f(x))$. We thus have the following definition, which forms the basis of differential calculus.

Definition

*The **derivative** of a function f is the function denoted f' (read "f prime") and defined by*

$$f'(x) = \lim_{h \to 0} \frac{f(x + h) - f(x)}{h}$$

*(provided this limit exists). If $f'(x)$ can be found, f is said to be **differentiable** (at x) and $f'(x)$ is called the derivative of f at x or the derivative of f with respect to x. The process of finding the derivative is called **differentiation**.*

EXAMPLE 2   *If $f(x) = x^2$, find the derivative of f.*

Applying the definition above gives

$$f'(x) = \lim_{h \to 0} \frac{f(x + h) - f(x)}{h}$$

$$= \lim_{h \to 0} \frac{(x + h)^2 - x^2}{h} = \lim_{h \to 0} \frac{x^2 + 2xh + h^2 - x^2}{h}$$

$$= \lim_{h \to 0} \frac{2xh + h^2}{h} = \lim_{h \to 0} \frac{h(2x + h)}{h} = \lim_{h \to 0} (2x + h) = 2x.$$

Observe that in taking the limit we treated x as a constant because it was h, not x, that was changing. Also note that $f'(x) = 2x$ defines a function of x, which we can interpret as giving the slope of the tangent line to the graph of f at $(x, f(x))$. For example, if $x = 1$, then the slope is $f'(1) = 2(1) = 2$, which confirms the result in Example 1.

Besides $f'(x)$, other notations for the derivative of $y = f(x)$ at x are

$$\frac{dy}{dx} \qquad \text{(pronounced "dee } y\text{, dee } x\text{"),}$$

Pitfall

$\frac{dy}{dx}$ is not a fraction, but is a single symbol for a derivative. We have not yet attached any meaning to individual symbols such as dy and dx.

$$\frac{d}{dx}[f(x)] \qquad \text{(dee } f(x)\text{, dee } x\text{),}$$

$$y' \qquad\qquad (y \text{ prime}),$$

$$D_x y \qquad\qquad \text{(dee } x \text{ of } y\text{),}$$

$$D_x[f(x)] \qquad \text{(dee } x \text{ of } f(x)\text{)).}$$

If the derivative of $y = f(x)$ can be evaluated at $x = x_1$, the resulting number $f'(x_1)$ is called the derivative of f at x_1 and f is said to be *differentiable* at x_1. Other notations for $f'(x_1)$ are

$$\left.\frac{dy}{dx}\right|_{x=x_1} \qquad \text{and} \quad y'(x_1).$$

It can be shown that *a function that is differentiable at x_1 must be continuous there*. But a function continuous at x_1 is *not* necessarily differentiable there. For example, Fig. 2.5 clearly shows that $y = f(x) = |x|$ is continuous at 0. But we showed that there is no tangent there, so $f'(0)$ does not exist. Thus f is not differentiable at 0.

EXAMPLE 3   *If $f(x) = 2x^2 + 2x + 3$, find $f'(1)$. Then find an equation of the tangent line to the graph of f at $(1, 7)$.*

We shall find $f'(x)$ and evaluate it at $x = 1$.

$$f'(x) = \lim_{h \to 0} \frac{f(x + h) - f(x)}{h}$$

$$= \lim_{h \to 0} \frac{[2(x + h)^2 + 2(x + h) + 3] - (2x^2 + 2x + 3)}{h}$$

$$= \lim_{h \to 0} \frac{2x^2 + 4xh + 2h^2 + 2x + 2h + 3 - 2x^2 - 2x - 3}{h}$$

$$= \lim_{h \to 0} \frac{4xh + 2h^2 + 2h}{h} = \lim_{h \to 0} (4x + 2h + 2)$$

$$= 4x + 2.$$

$$f'(1) = 4(1) + 2 = 6.$$

Thus the tangent to the graph at $(1, 7)$ has slope 6. A point-slope form of the tangent line is $y - 7 = 6(x - 1)$. Simplifying gives $y = 6x + 1$.

Pitfall

In Example 3 it is **not** correct to say that since the derivative is $4x + 2$, the tangent line at $(1, 7)$ is $y - 7 = (4x + 2)(x - 1)$. The derivative must be **evaluated** at the point of tangency to determine the slope of the tangent line at that point.

EXAMPLE 4   *Find the slope of the curve* $y = 2x + 3$ *at the point where* $x = 6.$

Letting $y = f(x) = 2x + 3$, we have

$$\frac{dy}{dx} = \lim_{h \to 0} \frac{f(x + h) - f(x)}{h} = \lim_{h \to 0} \frac{[2(x + h) + 3] - (2x + 3)}{h}$$

$$= \lim_{h \to 0} \frac{2h}{h} = \lim_{h \to 0} 2 = 2.$$

Since $D_x(2x + 3) = 2$, the slope when $x = 6$, or in fact at any point, is 2. Note that the curve is a straight line and thus has the same slope at each point.

If a variable, say p, is a function of some variable, say q, then we speak of the derivative of p with respect to q, written dp/dq.

EXAMPLE 5   *If* $p = f(q) = \dfrac{1}{2q}$, *find* $\dfrac{dp}{dq}.$

$$\frac{dp}{dq} = \frac{d}{dq}\left(\frac{1}{2q}\right) = \lim_{h \to 0} \frac{f(q + h) - f(q)}{h}$$

$$= \lim_{h \to 0} \frac{\dfrac{1}{2(q + h)} - \dfrac{1}{2q}}{h} = \lim_{h \to 0} \frac{\dfrac{q - (q + h)}{2q(q + h)}}{h}$$

$$= \lim_{h \to 0} \frac{q - (q + h)}{h[2q(q + h)]} = \lim_{h \to 0} \frac{-h}{h[2q(q + h)]}$$

$$= \lim_{h \to 0} \frac{-1}{2q(q + h)} = -\frac{1}{2q^2}.$$

Note that when $q = 0$, neither the function nor its derivative exists.

As a final note we point out that the derivative of $y = f(x)$ at x is nothing more than the following limit:

$$\lim_{h \to 0} \frac{f(x + h) - f(x)}{h}.$$

Although we can interpret the derivative as a function that gives the slope of the tangent line to the curve $y = f(x)$ at the point $(x, f(x))$, this interpretation is simply a geometric convenience that assists our understanding. The limit above may exist aside from any geometric consideration at all. As you will see later, there are other useful interpretations.

EXERCISE 2.1

In Problems 1–14, use the definition of the derivative to find each of the following.

1. $f'(x)$ if $f(x) = x$.

2. $f'(x)$ if $f(x) = 4x - 1$.

3. $\dfrac{dy}{dx}$ if $y = 2x + 4$.

4. $\dfrac{dy}{dx}$ if $y = -3x$.

5. $\dfrac{d}{dx}(3 - 2x)$.

6. $\dfrac{d}{dx}\left(4 - \dfrac{x}{2}\right)$.

7. $f'(x)$ if $f(x) = 3$.

8. $f'(x)$ if $f(x) = 7.01$.

9. $D_x(x^2 + 4x - 8)$.

10. $D_x y$ if $y = x^2 + 5$.

11. $\dfrac{dp}{dq}$ if $p = 2q^2 + 5q - 1$.

12. $D_x(x^2 - x - 3)$.

13. $D_x y$ if $y = \dfrac{1}{x}$.

14. $\dfrac{dC}{dq}$ if $C = \dfrac{2}{q - 3}$.

15. Find the slope of the curve $y = x^2 + 4$ at the point $(-2, 8)$.

16. Find the slope of the curve $y = 2 - 3x^2$ at the point $(1, -1)$.

17. Find the slope of the curve $y = 4x^2 - 5$ when $x = 0$.

18. Find the slope of the curve $y = 1/x^2$ when $x = 1$.

In Problems 19–24, find an equation of the tangent line to the curve at the given point.

19. $y = x + 4$; $(3, 7)$.

20. $y = 2x^2 - 5$; $(-2, 3)$.

21. $y = 3x^2 + 3x - 4$; $(-1, -4)$.

22. $y = (x - 1)^2$; $(0, 1)$.

23. $y = \dfrac{3}{x + 1}$; $(2, 1)$.

24. $y = \dfrac{5}{1 - 3x}$; $(2, -1)$.

25. Equations may involve derivatives of functions. In an article on interest rate deregulation, Christofi and Agapos [11] solve the equation

$$r = \left(\frac{\eta}{1 + \eta}\right)\left(r_L - \frac{dC}{dD}\right)$$

for η (the Greek letter "eta"). Here r is the deposit rate paid by commercial banks, r_L is the rate earned by commercial banks, C is the administrative cost of transforming deposits into return-earning assets, D is the savings deposits level, and η is the deposit elasticity with respect to the deposit rate. Find η.

2.2 RULES FOR DIFFERENTIATION

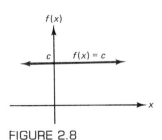

FIGURE 2.8

You would probably agree that differentiating a function by direct use of the definition of a derivative can be tedious. Fortunately, there are rules that give us completely mechanical and efficient procedures for differentiation. They also avoid direct use of limits. We shall look at some rules in this section.

To begin with, recall that the graph of the constant function $f(x) = c$ is a horizontal line (Fig. 2.8), which has a slope of zero everywhere. This means that $f'(x) = 0$, which is our first rule.

RULE 1

If c is a constant, then

$$\frac{d}{dx}(c) = 0.$$

That is, the derivative of a constant function is zero.

EXAMPLE 1

a. $D_x(3) = 0$ because 3 is a constant function.

b. If $g(x) = \sqrt{5}$, then $g'(x) = 0$ because g is a constant function. For example, the derivative of g when $x = 4$ is $g'(4) = 0$.

c. If $s(t) = (1{,}938{,}623)^{807.4}$, then $ds/dt = 0$.

To prove the next rule, we must expand a binomial. Recall that

$$(x + h)^2 = x^2 + 2xh + h^2$$
$$\text{and} \qquad (x + h)^3 = x^3 + 3x^2h + 3xh^2 + h^3.$$

In both expansions the exponents of x decrease from left to right while those of h increase. This is true for the general case $(x + h)^n$, where n is a positive integer. It can be shown that

$$(x + h)^n = x^n + nx^{n-1}h + (\quad)x^{n-2}h^2 + \cdots + (\quad)xh^{n-1} + h^n,$$

where the missing numbers inside the parentheses are certain constants. This formula is used to prove the next rule, which involves the derivative of x raised to a constant power.

RULE 2

If n is any real number, then

$$\frac{d}{dx}(x^n) = nx^{n-1},$$

provided that x^{n-1} is defined. That is, the derivative of a constant power of x is the exponent times x raised to a power one less than the given power.

Proof. We shall give a proof for the case when n is a positive integer. If $f(x) = x^n$, applying the definition of the derivative gives

$$f'(x) = \lim_{h \to 0} \frac{f(x + h) - f(x)}{h} = \lim_{h \to 0} \frac{(x + h)^n - x^n}{h}.$$

By our previous discussion on expanding $(x + h)^n$,

$$f'(x) = \lim_{h \to 0} \frac{x^n + nx^{n-1}h + (\ )x^{n-2}h^2 + \cdots + h^n - x^n}{h}.$$

In the numerator the sum of the first and last terms is 0. Dividing each of the remaining terms by h gives

$$f'(x) = \lim_{h \to 0} [nx^{n-1} + (\ )x^{n-2}h + \cdots + h^{n-1}].$$

Each term after the first term has h as a factor and must approach 0 as $h \to 0$. Hence $f'(x) = nx^{n-1}$.

EXAMPLE 2

a. By Rule 2, $D_x(x^2) = 2x^{2-1} = 2x$.

b. If $F(x) = x = x^1$, then $F'(x) = 1 \cdot x^{1-1} = 1 \cdot x^0 = 1$. Thus the derivative of x with respect to x is 1.

c. To differentiate $y = \sqrt{x}$, we write $\sqrt{x}$ as $x^{1/2}$ so that it has the form x^n. Thus

$$\frac{dy}{dx} = \frac{1}{2}x^{(1/2)-1} = \frac{1}{2}x^{-1/2} = \frac{1}{2\sqrt{x}}.$$

Note that $y = \sqrt{x}$ is defined for $x \geq 0$, but its derivative, $1/(2\sqrt{x})$, is defined only when $x > 0$. From the graph of $y = \sqrt{x}$ in Fig. 2.9, it is clear that when $x = 0$, the tangent is a vertical line for which the slope is not defined.

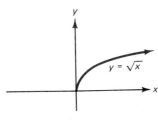

FIGURE 2.9

Pitfall

Do not write $\dfrac{1}{x\sqrt{x}}$ as $\dfrac{1}{x^{3/2}}$ and then just differentiate the denominator.

$$D_x\left(\frac{1}{x^{3/2}}\right) \neq \frac{1}{\frac{3}{2}x^{1/2}}.$$

d. Let $h(x) = \dfrac{1}{x\sqrt{x}}$. To apply Rule 2 we *must* write $h(x)$ as $h(x) = x^{-3/2}$ so that it has the form x^n.

$$D_x\left(\frac{1}{x\sqrt{x}}\right) = D_x(x^{-3/2}) = -\frac{3}{2}x^{(-3/2)-1} = -\frac{3}{2}x^{-5/2}.$$

The next rule involves the derivative of a constant times a function.

RULE 3

If f is a differentiable function and c is a constant, then

$$\frac{d}{dx}[cf(x)] = cf'(x).$$

That is, the derivative of a constant times a function is the constant times the derivative of the function.

Proof. If $g(x) = cf(x)$, applying the definition of the derivative of g gives

$$g'(x) = \lim_{h \to 0} \frac{g(x + h) - g(x)}{h} = \lim_{h \to 0} \frac{cf(x + h) - cf(x)}{h}$$

$$= \lim_{h \to 0} \left[c \cdot \frac{f(x + h) - f(x)}{h} \right] = c \cdot \lim_{h \to 0} \frac{f(x + h) - f(x)}{h}.$$

But $\lim_{h \to 0} \dfrac{f(x + h) - f(x)}{h}$ is $f'(x)$, so $g'(x) = cf'(x)$.

EXAMPLE 3   *Differentiate the following functions.*

a. $g(x) = 5x^3$.

Here g is a constant (5) times a function (x^3).

$$\frac{d}{dx}(5x^3) = 5D_x(x^3) \qquad \text{(Rule 3)}$$

$$= 5(3x^{3-1}) = 15x^2 \qquad \text{(Rule 2)}.$$

b. $f(q) = \dfrac{13q}{5}$.

Because $\dfrac{13q}{5} = \dfrac{13}{5}q$, f is a constant $\left(\dfrac{13}{5}\right)$ times a function (q).

$$f'(q) = \frac{13}{5}D_q(q) \qquad \text{(Rule 3)}$$

$$= \frac{13}{5} \cdot 1 = \frac{13}{5} \qquad \text{(Rule 2)}.$$

c. $y = \dfrac{0.702}{\sqrt[5]{x^2}} = 0.702x^{-2/5}$.

Note that y can be considered a constant times a function.

$$y' = 0.702D_x(x^{-2/5}) \qquad \text{(Rule 3)}$$

$$= 0.702\left(-\frac{2}{5}x^{-7/5}\right) = -0.2808x^{-7/5} \qquad \text{(Rule 2)}.$$

Pitfall

If $f(x) = (4x)^3$, you may be tempted to write $f'(x) = 3(4x)^2$. This is **incorrect**! The reason is that Rule 2 applies to a power of the variable x, **not** a power of an expression involving x such as $4x$. To apply our rules we must get a suitable form for $f(x)$. We can write $(4x)^3$ as 4^3x^3 or $64x^3$. Thus

$$f'(x) = 64D_x(x^3) = 64(3x^2) = 192x^2.$$

The next rule involves derivatives of sums and differences of functions. A proof is indicated in Appendix B.

RULE 4

If f and g are differentiable functions, then

$$\frac{d}{dx}[f(x) + g(x)] = f'(x) + g'(x)$$

and $\qquad \dfrac{d}{dx}[f(x) - g(x)] = f'(x) - g'(x).$

That is, the derivative of the sum (or difference) of two functions is the sum (or difference) of their derivatives.

Rule 4 can be extended to the derivative of any number of sums and differences of functions. For example,

$$\frac{d}{dx}[f(x) - g(x) + h(x) + k(x)] = f'(x) - g'(x) + h'(x) + k'(x).$$

EXAMPLE 4   *Differentiate the following functions.*

a. $F(x) = 3x^5 + \sqrt{x}$.

Here F is the sum of two functions, $3x^5$ and $\sqrt{x}$. Thus

$$F'(x) = D_x(3x^5) + D_x(x^{1/2}) \qquad \text{(Rule 4)}$$

$$= 3D_x(x^5) + D_x(x^{1/2}) \qquad \text{(Rule 3)}$$

$$= 3(5x^4) + \frac{1}{2}x^{-1/2} = 15x^4 + \frac{1}{2\sqrt{x}} \qquad \text{(Rule 2)}.$$

b. $f(z) = \dfrac{z^4}{4} - \dfrac{5}{z^{1/3}}$.

Note that we can write $f(z) = \frac{1}{4}z^4 - 5z^{-1/3}$. Since f is the difference of two functions,

$$
\begin{aligned}
f'(z) &= D_z(\tfrac{1}{4}z^4) - D_z(5z^{-1/3}) &&\text{(Rule 4)} \\
&= \tfrac{1}{4}D_z(z^4) - 5D_z(z^{-1/3}) &&\text{(Rule 3)} \\
&= \tfrac{1}{4}(4z^3) - 5(-\tfrac{1}{3}z^{-4/3}) &&\text{(Rule 2)} \\
&= z^3 + \tfrac{5}{3}z^{-4/3}.
\end{aligned}
$$

c. $y = 6x^3 - 2x^2 + 7x - 8$.

$$
\begin{aligned}
\frac{dy}{dx} &= D_x(6x^3) - D_x(2x^2) + D_x(7x) - D_x(8) \\
&= 6D_x(x^3) - 2D_x(x^2) + 7D_x(x) - D_x(8) \\
&= 6(3x^2) - 2(2x) + 7(1) - 0 \\
&= 18x^2 - 4x + 7.
\end{aligned}
$$

EXAMPLE 5   *Find the derivative of $f(x) = 2x(x^2 - 5x + 2)$ when $x = 2$.*

We multiply and then differentiate each term.

$$
\begin{aligned}
f(x) &= 2x^3 - 10x^2 + 4x. \\
f'(x) &= 2(3x^2) - 10(2x) + 4(1) \\
&= 6x^2 - 20x + 4. \\
f'(2) &= 6(2)^2 - 20(2) + 4 = -12.
\end{aligned}
$$

EXAMPLE 6   *Find an equation of the tangent line to the curve $y = \dfrac{3x^2 - 2}{x}$ when $x = 1$.*

By writing y as a difference of two functions, we have $y = \dfrac{3x^2}{x} - \dfrac{2}{x} = 3x - 2x^{-1}$. Thus

$$
\frac{dy}{dx} = 3(1) - 2[(-1)x^{-2}] = 3 + \frac{2}{x^2}.
$$

The slope of the tangent line to the curve when $x = 1$ is

$$
\frac{dy}{dx}\bigg|_{x=1} = 3 + \frac{2}{1^2} = 5.
$$

To find the y-coordinate of the point on the curve where $x = 1$, we substitute this value of x into the equation of the *curve*. When $x = 1$, then $y =$

$[3(1)^2 - 2]/1 = 1$. Hence the point $(1, 1)$ lies on both the curve and the tangent line. Therefore, an equation of the tangent line is

$$y - 1 = 5(x - 1),$$
$$y = 5x - 4.$$

EXERCISE 2.2

In Problems 1–50, differentiate the functions.

1. $f(x) = 7$.

2. $f(x) = (\frac{933}{465})^{2/3}$.

3. $f(x) = x^8$.

4. $f(x) = 0.7x$.

5. $f(x) = 9x^5$.

6. $f(x) = \sqrt{2}x^{83/4}$.

7. $g(w) = w^{-7}$.

8. $f(t) = 3t^{-2}$.

9. $f(x) = 3x - 2$.

10. $f(w) = 5w - 7(\frac{4}{5})$.

11. $f(p) = \dfrac{13p}{5} + \dfrac{7}{3}$.

12. $q(x) = \dfrac{5x + 2}{8}$.

13. $g(x) = 3x^2 - 5x - 2$.

14. $f(q) = 7q^2 - 5q + 3$.

15. $f(q) = -3q^3 + \frac{9}{2}q^2 + 9q + 9$.

16. $f(x) = 100x^{-3} - 50x^{-1/2} + 10x - 1$.

17. $f(x) = 2x^{501} + 0.2x^{3.4} - 9^{2/3}$.

18. $f(x) = 17 + 8x^{1/7} - 10x^{12} - 3x^{-15}$.

19. $f(x) = 2(13 - x^4)$.

20. $f(s) = 5(s^4 - 3)$.

21. $g(x) = \dfrac{13 - x^4}{3}$.

22. $f(x) = \dfrac{5(x^4 - 3)}{2}$.

23. $f(x) = x^{-4} - 9x^{1/3} + 5x^{-2/5}$.

24. $f(z) = 3z^{1/4} - 12^2 - 8z^{-3/4}$.

25. $h(x) = -2(27x - 14x^5)$.

26. $f(x) = \dfrac{-(1 + x - x^2 + x^3 + x^4 - x^5)}{2}$.

27. $f(x) = -2x^2 + \dfrac{3}{2}x + \dfrac{x^4}{4} + 2$.

28. $p(x) = \dfrac{x^7}{7} + \dfrac{x}{2}$.

29. $f(x) = \dfrac{1}{x}$.

30. $f(x) = \dfrac{7}{x^3}$.

31. $f(s) = \dfrac{1}{4s^5}$.

32. $g(w) = \dfrac{2}{3w^3}$.

33. $f(t) = 4\sqrt{t}$.

34. $f(x) = \dfrac{4}{\sqrt{x}}$.

35. $q(x) = \dfrac{1}{\sqrt[5]{x^3}}$.

36. $f(x) = \dfrac{3}{\sqrt[4]{x^3}} - \sqrt[4]{x^3}$.

37. $f(x) = x(3x^2 - 7x + 7)$.

38. $f(x) = x^3(3x^6 - 5x^2 + 4)$.

39. $g(t) = \dfrac{t^2}{2} - \dfrac{2}{t^2}$.

40. $f(x) = x\sqrt{x}$.

41. $f(x) = x^3(3x)^2$.

42. $f(x) = \sqrt{x}(5 - 6x + 3\sqrt[4]{x})$.

43. $v(x) = x^{-2/3}(x + 5)$.

44. $f(x) = x^{3/5}(x^2 + 7x + 1)$.

45. $f(q) = \dfrac{4q^3 + 7q - 4}{q}$.

46. $f(w) = \dfrac{w - 5}{w^5}$.

47. $f(x) = (x + 1)(x + 3)$.

48. $f(x) = x^2(x - 2)(x + 4)$.

49. $w(x) = \dfrac{x^2 + x^3}{x^2}$.

50. $f(x) = \dfrac{7x^3 + x}{2\sqrt{x}}$.

For each curve in Problems 51–54, find the slope at the given points.

51. $y = 3x^2 + 4x - 8$; $(0, -8)$, $(2, 12)$, $(-3, 7)$.

52. $y = 5 - 6x - 2x^3$;   (0, 5), $(\frac{3}{2}, -\frac{43}{4})$, $(-3, 77)$.

53. $y = 4$;   when $x = -4$, $x = 7$, $x = 22$.

54. $y = 2x - 3\sqrt{x}$;   when $x = 1$, $x = 16$, $x = 25$.

In Problems 55 and 56, find an equation of the tangent line to the curve at the given point.

55. $y = 4x^2 + 5x + 2$;   (1, 11).

56. $y = \dfrac{1 - x^2}{5}$;   (4, −3).

57. Find an equation of the tangent line to the curve $y = 3 + x - 5x^2 + x^4$ when $x = 0$.

58. Repeat Problem 57 for the curve $y = \dfrac{\sqrt{x}(2 - x^2)}{x}$ when $x = 4$.

59. Find all points on the curve $y = \frac{1}{3}x^3 - x^2$ where the tangent line is horizontal.

60. Find all points on the curve $y = x^2 - 5x + 3$ where the slope is 1.

61. Eswaran and Kotwal [12] consider agrarian economies in which there are two types of workers, permanent and casual. Permanent workers are employed on long-term contracts and may receive benefits such as holiday gifts and emergency aid. Casual workers are hired on a daily basis and perform routine and menial tasks such as weeding, harvesting, and threshing. The difference z in the present-value cost of hiring a permanent worker over that of hiring a casual worker is given by

$$z = (1 + b)w_p - bw_c,$$

where w_p and w_c are wage rates for permanent labor and casual labor, respectively, b is a constant, and w_p is a function of w_c. Eswaran and Kotwal claim that

$$\frac{dz}{dw_c} = (1 + b)\left[\frac{dw_p}{dw_c} - \frac{b}{1 + b}\right].$$

Verify this.

2.3   THE DERIVATIVE AS A RATE OF CHANGE

Historically, an important application of the derivative involves motion in a straight line. It gives us a convenient way to interpret the derivative as a *rate of change*. To denote the change in a variable such as x, the symbol Δx (read "delta x") is commonly used. For example, if x changes from 1 to 3, then the change in x is $\Delta x = 3 - 1 = 2$. The new value of x (=3) is the old value plus the change, or $1 + \Delta x$. Similarly, if t increases by Δt, the new value is $t + \Delta t$. We shall use Δ-notation in the following discussion.

   Suppose a particle moves along the number line in Fig. 2.10 according to the equation

$$s = f(t) = t^2,$$

FIGURE 2.10

where s is the position of the particle at time t. This equation is called an *equation of motion*. Assume that t is in seconds and s is in meters. At $t = 1$, the position is $s = f(1) = 1^2 = 1$, and at $t = 3$ the position is $s = f(3) = 3^2 = 9$. Over this two-second time interval, the particle has a change in position, or a *displacement*, of $9 - 1 = 8$ m, and the *average velocity* (v_{ave}) of the particle is defined as

$$v_{ave} = \frac{\text{displacement}}{\text{length of time interval}} \qquad (1)$$

$$= \frac{8}{2} = 4 \text{ m/s}.$$

More generally, over the time interval from t to $t + \Delta t$, the particle moves from position $f(t)$ to position $f(t + \Delta t)$. Thus its displacement is $f(t + \Delta t) - f(t)$, which is a change in s-values, denoted by Δs:

$$\Delta s = f(t + \Delta t) - f(t).$$

Since the given time interval has length Δt, from Eq. (1) the particle's average velocity is

$$v_{\text{ave}} = \frac{\Delta s}{\Delta t} = \frac{f(t + \Delta t) - f(t)}{\Delta t}.$$

The expression $\Delta s / \Delta t$ is also called the *average rate of change* of s with respect to t over the interval from t to $t + \Delta t$.

If Δt were to become smaller and smaller, the average velocity over the interval from t to $t + \Delta t$ would be close to what we might call the *instantaneous velocity* at time t—that is, the velocity at a *point in time* (t) as opposed to velocity over an *interval* of time. We define the limit of the average velocity as $\Delta t \to 0$ to be the **instantaneous velocity** v at time t, or the **instantaneous rate of change** of s with respect to t.

$$v = \lim_{\Delta t \to 0} v_{\text{ave}} = \lim_{\Delta t \to 0} \frac{\Delta s}{\Delta t} = \lim_{\Delta t \to 0} \frac{f(t + \Delta t) - f(t)}{\Delta t}.$$

The last limit is just the definition of the derivative of f with respect to t, because replacing Δt by h gives

$$\lim_{h \to 0} \frac{f(t + h) - f(t)}{h}.$$

Thus

$$v = \frac{ds}{dt}.$$

For the original equation of motion, $s = t^2$, the instantaneous velocity at time t is given by

$$v = \frac{ds}{dt} = 2t.$$

For example, the instantaneous velocity when $t = 1$ is

$$\left. \frac{ds}{dt} \right|_{t=1} = 2(1) = 2 \text{ m/s}.$$

Usually, instantaneous velocity is simply referred to as *velocity*.

EXAMPLE 1   *Suppose the equation of motion of a particle moving along a number line is given by $s = \dfrac{3t^2 + 5}{4}$. Find the velocity when $t = 10$.*

The velocity at any time t is given by

$$v = \frac{ds}{dt} = D_t\left(\frac{3t^2 + 5}{4}\right)$$

$$= \frac{1}{4}D_t(3t^2 + 5) = \frac{1}{4}[D_t(3t^2) + D_t(5)]$$

$$= \frac{1}{4}[6t + 0] = \frac{3}{2}t.$$

When $t = 10$,

$$v = \frac{3}{2} \cdot 10 = 15.$$

Our discussion of rate of change of s with respect to t applies equally well to *any* function $y = f(x)$. This means as follows.

If $y = f(x)$, then

$$\frac{\Delta y}{\Delta x} = \frac{f(x + \Delta x) - f(x)}{\Delta x} = \begin{cases} \textbf{average rate of change} \\ \textbf{of } y \textbf{ with respect to } x \\ \textbf{over the interval from} \\ x \textbf{ to } x + \Delta x \end{cases}$$

and $\quad \dfrac{dy}{dx} = \lim\limits_{\Delta x \to 0} \dfrac{\Delta y}{\Delta x} = \begin{cases} \textbf{instantaneous rate of} \\ \textbf{change of } y \textbf{ with respect to } x. \end{cases}$ $\qquad$ (2)

Because the instantaneous rate of change of $y = f(x)$ at a point is a derivative, it is also the *slope of the tangent line* to the graph of $y = f(x)$ at that point. For convenience we usually refer to instantaneous rate of change simply as *rate of change*.

From Eq. (2), if Δx (a change in x) is close to 0, then $\Delta y/\Delta x$ is close to dy/dx. That is,

$$\frac{\Delta y}{\Delta x} \approx \frac{dy}{dx},$$

where the symbol $\approx$ means "approximately equals." Hence

$$\Delta y \approx \frac{dy}{dx}\Delta x.$$

That is, if x changes by Δx, then the change in y, Δy, is approximately dy/dx times the change in x. In particular, if x changes by 1, an estimate of the change in y is dy/dx.

EXAMPLE 2 $\quad$ *If $y = f(x)$, $f(3) = 5$, and $\dfrac{dy}{dx} = 8$, estimate the change in y if x changes from 3 to 3.5.*

We have $\frac{dy}{dx} = 8$ and $\Delta x = 3.5 - 3 = 0.5$. The change in y is given by Δy and

$$\Delta y \approx \frac{dy}{dx} \Delta x = 8(0.5) = 4.$$

We remark that $f(3.5) = f(3) + \Delta y$ and can be estimated by $5 + 4 = 9$.

EXAMPLE 3   *Find the rate of change of $y = x^4$ with respect to x and evaluate it when $x = 2$ and when $x = -1$.*

The rate of change is dy/dx:

$$\frac{dy}{dx} = 4x^3.$$

When $x = 2$, then $dy/dx = 4(2)^3 = 32$. This means that if x increases by a small amount, then y increases approximately 32 times as much. More simply, we say that y is increasing 32 times as fast as x does. When $x = -1$, then $dy/dx = 4(-1)^3 = -4$. The significance of the minus sign on -4 is that y is *decreasing* 4 times as fast as x increases.

The interpretation of a derivative as a rate of change has applications to business and economics.

EXAMPLE 4   *Let $p = 100 - q^2$ be the demand function for a manufacturer's product. Find the rate of change of price p per unit with respect to quantity q. How fast is the price changing with respect to q when $q = 5$? Assume that p is in dollars.*

The rate of change of p with respect to q is dp/dq.

$$\frac{dp}{dq} = \frac{d}{dq}(100 - q^2) = -2q.$$

$$\left.\frac{dp}{dq}\right|^{q=5} = -2(5) = -10.$$

This means that when 5 units are demanded, an *increase* of one extra unit demanded corresponds to a decrease of approximately \$10 in the price per unit that consumers are willing to pay.

EXAMPLE 5   *A sociologist is studying various suggested programs that can aid in the education of preschool-age children in a certain city. The sociologist believes that x years after the beginning of a particular program,* *f(x) thousand preschoolers will be enrolled, where*

$$f(x) = \frac{10}{9}(12x - x^2), \qquad 0 \le x \le 12.$$

At what rate would enrollment change (a) *after three years from the start of this program and* (b) *after nine years?*

The rate of change of $f(x)$ is $f'(x)$:

$$f'(x) = \frac{10}{9}(12 - 2x).$$

a. After three years the rate of change is

$$f'(3) = \frac{10}{9}[12 - 2(3)] = \frac{10}{9} \cdot 6 = \frac{20}{3} = 6\frac{2}{3}.$$

Thus enrollment would be increasing at the rate of $6\frac{2}{3}$ thousand preschoolers per year.

b. After nine years the rate is

$$f'(9) = \frac{10}{9}[12 - 2(9)] = \frac{10}{9}[-6] = -\frac{20}{3} = -6\frac{2}{3}.$$

Thus enrollment would be *decreasing* at the rate of $6\frac{2}{3}$ thousand preschoolers per year.

A manufacturer's **total cost function**, $c = f(q)$, gives the total cost c of producing and marketing q units of a product. The rate of change of c with respect to q is called **marginal cost**. Thus

$$\textbf{marginal cost} = \frac{dc}{dq}.$$

For example, suppose that $c = f(q) = 0.1q^2 + 3$ is a cost function, where c is in dollars and q is in pounds. Then

$$\frac{dc}{dq} = 0.2q.$$

The marginal cost when 4 lb are produced is dc/dq evaluated when $q = 4$:

$$\left.\frac{dc}{dq}\right|_{q=4} = 0.2(4) = 0.80.$$

This means that if production is increased by one pound, from 4 lb to 5 lb, then the change in cost is approximately $0.80. That is, the additional pound costs about $0.80. In general, *we interpret marginal cost as the approximate cost of one additional unit of output.* [The actual cost of producing one more pound beyond 4 lb is $f(5) - f(4) = \$0.90.$]

If c is the total cost of producing q units of a product, then the **average cost** per unit, $\overline{c}$, is

$$\overline{c} = \frac{c}{q}. \tag{3}$$

For example, if the total cost of 20 units is $100, then the average cost per unit is $\overline{c} = 100/20 = \$5$. By multiplying both sides of Eq. (3) by q, we have

$$c = q\overline{c}.$$

That is, total cost is the product of the number of units produced and the average cost per unit.

EXAMPLE 6   *If a manufacturer's average cost equation is*

$$\overline{c} = 0.0001q^2 - 0.02q + 5 + \frac{5000}{q},$$

find the marginal cost function. What is the marginal cost when 50 units are produced?

We first find the total cost c. Since $c = q\overline{c}$, then

$$c = q\overline{c}$$
$$= q\left[0.0001q^2 - 0.02q + 5 + \frac{5000}{q} \right].$$
$$= 0.0001q^3 - 0.02q^2 + 5q + 5000.$$

Differentiating c, we have the marginal cost function:

$$\frac{dc}{dq} = 0.0001(3q^2) - 0.02(2q) + 5(1) + 0$$
$$= 0.0003q^2 - 0.04q + 5.$$

The marginal cost when 50 units are produced is

$$\frac{dc}{dq}\bigg|_{q=50} = 0.0003(50)^2 - 0.04(50) + 5 = 3.75.$$

If c is in dollars and production is increased by one unit from $q = 50$ to $q = 51$, then the cost of the additional unit is approximately $3.75. If production is increased by $\frac{1}{3}$ unit from $q = 50$, then the cost of the additional output is approximately $(\frac{1}{3})(3.75) = \$1.25$.

Suppose that $r = f(q)$ is the **total revenue function** for a manufacturer. The equation $r = f(q)$ states that the total dollar value received for selling q units of a product is r. The **marginal revenue** is defined as the rate of change of the total dollar value received with respect to the total number of units sold. Hence marginal revenue is merely the derivative of r with respect to q.

$$\textbf{marginal revenue} = \frac{dr}{dq}.$$

Marginal revenue indicates the rate at which revenue changes with respect to units sold. We interpret it as the approximate revenue received from selling one additional unit of output.

EXAMPLE 7

Suppose that a manufacturer sells a product at \$2 per unit. If q units are sold, the total revenue is given by

$$r = 2q.$$

The marginal revenue function is

$$\frac{dr}{dq} = \frac{d}{dq}(2q) = 2,$$

which is a constant function. Thus the marginal revenue is 2 regardless of the number of units sold. This is what we would expect because the manufacturer receives \$2 for each unit sold.

For the total revenue function in Example 7, $r = f(q) = 2q$, we have

$$\frac{dr}{dq} = 2.$$

This means that revenue is changing at the rate of \$2 per unit, regardless of the number of units sold. Although this is valuable information, it may be more significant when compared to r itself. For example, if $q = 50$, then $r = 2(50) = \$100$. Thus the rate of change of revenue is $2/100 = 0.02$ **of r.** On the other hand, if $q = 5000$, then $r = 2(5000) = \$10,000$, so the rate of change of r is $2/10,000 = 0.0002$ **of r.** Although r changes at the same rate at each level, when compared to r itself this rate is relatively smaller when $r = 10,000$ than when $r = 100$. By considering the ratio

$$\frac{dr/dq}{r},$$

we have a means of comparing the rate of change of r with r itself. This ratio is called the *relative rate of change* of r. We have shown above that the relative rate of change when $q = 50$ is

$$\frac{dr/dq}{r} = \frac{2}{100} = 0.02,$$

and when $q = 5000$, it is

$$\frac{dr/dq}{r} = \frac{2}{10,000} = 0.0002.$$

By multiplying relative rates by 100, we obtain *percentage rates of change*. The percentage rate of change when $q = 50$ is $(0.02)(100) = 2$ percent; when $q = 5000$ it is $(0.0002)(100) = 0.02$ percent. Thus, for example, if an additional unit beyond 50 is sold, the revenue increases by approximately 2 percent.

In general, for any function f we have the following definition.

Definition
*The **relative rate of change** of $f(x)$ is*

$$\frac{f'(x)}{f(x)}.$$

*The **percentage rate of change** of $f(x)$ is*

$$\frac{f'(x)}{f(x)} \cdot 100.$$

EXAMPLE 8 *Find the relative and percentage rates of change of $y = f(x) = 3x^2 - 5x + 25$ when $x = 5$.*

Here

$$f'(x) = 6x - 5.$$

Since $f'(5) = 6(5) - 5 = 25$ and $f(5) = 3(5)^2 - 5(5) + 25 = 75$, the relative rate of change of y when $x = 5$ is

$$\frac{f'(5)}{f(5)} = \frac{25}{75} \approx 0.333.$$

Multiplying 0.333 by 100 gives the percentage rate of change: $(0.333)(100) = 33.3$ percent.

EXERCISE 2.3

In each of Problems 1–6 an equation of motion is given. For the given value of t, find (a) the position and (b) the velocity. Assume that t is in seconds and s is in meters.

1. $s = t^2 - 3t$; $t = 4$.

2. $s = \frac{1}{2}t + 1$; $t = 2$.

3. $s = 2t^3 + 6$; $t = 1$.

4. $s = -3t^2 + 2t + 1$; $t = 1$.

5. $s = t^4 - 2t^3 + t$; $t = 2$.

6. $s = t^4 - t^{5/2}$; $t = 0$.

7. Sociologists studied the relation between income and number of years of education for members of a particular urban group. They found that a person with x years of education before seeking regular employment can expect to receive an average yearly income of y dollars per year, where

$$y = 4x^{5/2} + 4900, \qquad 4 \le x \le 16.$$

Find the rate of change of income with respect to number of years of education. Evaluate when $x = 9$.

8. Find the rate of change of the area A of a circle with

respect to its radius r if $A = \pi r^2$. Evaluate when $r = 3$ in.

9. The approximate temperature T of the skin in terms of the temperature T_e of the environment is given in [13] by

$$T = 32.8 + 0.27(T_e - 20),$$

where T and T_e are in degrees Celsius. Find the rate of change of T with respect to T_e.

10. The volume V of a spherical cell is given by $V = \frac{4}{3}\pi r^3$, where r is the radius. Find the rate of change of volume with respect to the radius when $r = 6.5 \times 10^{-4}$ cm.

In Problems 11–16, cost functions are given where c is the cost of producing q units of a product. In each case find the marginal cost function. What is the marginal cost at the given value(s) of q?

11. $c = 500 + 10q$; $q = 100$.

12. $c = 5000 + 6q$; $q = 36$.

13. $c = 0.3q^2 + 2q + 850$; $q = 3$.

14. $c = 0.1q^2 + 3q + 2$; $q = 3$.

15. $c = q^2 + 50q + 1000$; $q = 15, q = 16, q = 17$.

16. $c = 0.03q^3 - 0.6q^2 + 4.5q + 7700$; $q = 10, q = 20, q = 100$.

In Problems 17–20, $\bar{c}$ represents average cost per unit, which is a function of the number q of units produced. Find the marginal cost function and the marginal cost for the indicated values of q.

17. $\bar{c} = 0.01q + 5 + \dfrac{500}{q}$; $q = 50, q = 100$.

18. $\bar{c} = 2 + \dfrac{1000}{q}$; $q = 25, q = 235$.

19. $\bar{c} = 0.00002q^2 - 0.01q + 6 + \dfrac{20,000}{q}$; $q = 100$, $q = 500$.

20. $\bar{c} = 0.001q^2 - 0.3q + 40 + \dfrac{7000}{q}$; $q = 10, q = 20$.

In Problems 21–24, r represents total revenue and is a function of the number q of units sold. Find the marginal revenue function and the marginal revenue for the indicated values of q.

21. $r = 0.7q$; $q = 8, q = 100, q = 200$.

22. $r = q(15 - \frac{1}{30}q)$; $q = 5, q = 15, q = 150$.

23. $r = 250q + 45q^2 - q^3$; $q = 5, q = 10, q = 25$.

24. $r = 2q(30 - 0.1q)$; $q = 10, q = 20$.

25. The total cost function for a hosiery mill is estimated in [14] by

$$c = -10,484.69 + 6.750q - 0.000328q^2,$$

where q is output in dozens of pairs and c is total cost in dollars. Find the marginal cost function and evaluate it when $q = 5000$.

26. The total cost function for an electric light and power plant is estimated in [15] by

$$c = 32.07 - 0.79q + 0.02142q^2 - 0.0001q^3,$$

where q, $20 \le q \le 90$, is eight-hour total output (as percentage of capacity) and c is total fuel cost in dollars. Find the marginal cost function and evaluate it when $q = 70$.

27. Suppose that the 100 largest cities in the United States in 1920 are ranked according to magnitude (areas of cities). From [16] the following relation approximately holds:

$$PR^{0.93} = 5,000,000,$$

where P is the population of the city having respective rank R. This relation is called the *law of urban concentration* for 1920. Solve for P in terms of R and then find how fast population is changing with respect to rank.

28. Under the straight-line method of depreciation, the value v of a certain machine after t years have elapsed is given by $v = 50,000 - 5000t$ where $0 \le t \le 10$. How fast is v changing with respect to t when $t = 2$? $t = 3$? at any time t?

29. A study of the winter moth was made in Nova Scotia (adapted from [17]). The prepupae of the moth fall onto the ground from host trees. It was found that at a distance of x ft from the base of a host tree, the prepupal density (number of prepupae per square foot of soil) was y, where

$$y = 59.3 - 1.5x - 0.5x^2, \qquad 1 \le x \le 9.$$

a. At what rate is the prepupal density changing with respect to distance from the base of the tree when $x = 6$?

b. For what value of x is the prepupal density decreasing at the rate of 6 prepupae per square foot per foot?

30. For the cost function $c = 0.4q^2 + 4q + 5$, find the rate of change of c with respect to q when $q = 2$. Also, what is $\Delta c/\Delta q$ over the interval from $q = 2$ to $q = 3$?

In Problems 31–36, find (a) the rate of change of y with respect to x and (b) the relative rate of change of y. At the given value of x, find (c) the rate of change of y, (d) the relative rate of change of y, and (e) the percentage rate of change of y.

31. $y = f(x) = x + 4$; $x = 5$.

32. $y = f(x) = 4 - 2x$; $x = 3$.

33. $y = 3x^2 + 6$; $x = 2$.

34. $y = 2 - x^2$; $x = 0$.

35. $y = 8 - x^3$; $x = 1$.

36. $y = x^2 + 3x - 4$; $x = -1$.

37. For the cost function $c = 0.2q^2 + 1.2q + 4$, how fast does c change with respect to q when $q = 5$? Determine the percentage rate of change of c with respect to q when $q = 5$.

38. In a discussion of contemporary waters of shallow seas, Odum [18] claims that in such waters the total organic matter y (in milligrams per liter) is a function of species diversity x (in number of species per 1000 individuals). If $y = 100/x$, at what rate is the total organic matter changing with respect to species diversity when $x = 10$? What is the percentage rate of change when $x = 10$?

39. For a certain manufacturer, the revenue r obtained

from the sale of q units of a product is given by $r = 30q - 0.3q^2$.

a. How fast does r change with respect to q?

b. When $q = 10$, find the relative rate of change of r.

c. To the nearest percent, find the percentage rate of change of r when $q = 10$.

40. Repeat Problem 39 if the revenue function is given by $r = 20q - 0.1q^2$ and $q = 100$.

41. The weight W of a limb of a tree is given by $W = 2t^{0.432}$, where t is time. Find the relative rate of change of W with respect to t.

42. A psychological experiment [1] was conducted to analyze human response to electrical shocks (stimuli). The subjects received shocks of various intensities. The response R to a shock of intensity I (in microamperes) was to be a number that indicated the perceived magnitude relative to that of a "standard" shock. The standard shock was assigned a magnitude of 10. Two groups of subjects were tested under slightly different conditions. The responses R_1 and R_2 of the first and second groups to a shock of intensity I were given by

$$R_1 = \frac{I^{1.3}}{1855.24}, \qquad 800 \le I \le 3500,$$

and $\quad R_2 = \dfrac{I^{1.3}}{1101.29}, \qquad 800 \le I \le 3500.$

a. For each group determine the relative rate of change of response with respect to intensity.

b. How do these changes compare?

c. In general, if $f(x) = C_1 x^n$ and $g(x) = C_2 x^n$, where C_1 and C_2 are constants, how do the relative rates of change of f and g compare?

2.4   PRODUCT AND QUOTIENT RULES

The equation $F(x) = (x^2 + 3x)(4x + 5)$ expresses $F(x)$ as a product of two functions: $x^2 + 3x$ and $4x + 5$. To find $F'(x)$ by using only our previous rules, we first multiply the functions, which gives $F(x) = 4x^3 + 17x^2 + 15x$. Then we differentiate term by term:

$$F'(x) = 12x^2 + 34x + 15. \tag{1}$$

However, in many problems that involve differentiating a product of functions, the multiplication is not as simple as it is here. Often it is not even practical to attempt it. Fortunately, there is a rule for differentiating a product, and the rule avoids such multiplications. Since the derivative of a sum of functions is the sum of their derivatives, you might think that the derivative of a product of two functions is the product of their derivatives. This is **not** the case, as the next rule shows. (A proof of this rule is in Appendix B.)

RULE 5

Product Rule. If f and g are differentiable functions, then

$$\frac{d}{dx}[f(x)g(x)] = f(x)g'(x) + g(x)f'(x).$$

That is, the derivative of the product of two functions is the first function times the derivative of the second, plus the second function times the derivative of the first.

EXAMPLE 1   *If $F(x) = (x^2 + 3x)(4x + 5)$, find $F'(x)$.*

Here F can be considered a product of two functions: $f(x) = x^2 + 3x$ and $g(x) = 4x + 5$. By Rule 5, the product rule,

$$
\begin{aligned}
F'(x) &= f(x)g'(x) + g(x)f'(x) \\
&= (x^2 + 3x)\, D_x(4x + 5) + (4x + 5)\, D_x(x^2 + 3x) \\
&= (x^2 + 3x)(4) + (4x + 5)(2x + 3) \\
&= 12x^2 + 34x + 15 \qquad \text{(simplifying)}.
\end{aligned}
$$

This agrees with our previous result [see Eq. (1)].

Pitfall

We repeat: the derivative of the product of two functions **is not** the product of their derivatives. For example, $D_x(x^2 + 3x) = 2x + 3$ and $D_x(4x + 5) = 4$, but from Example 1,

$$D_x[(x^2 + 3x)(4x + 5)] = 12x^2 + 34x + 15 \neq (2x + 3)4.$$

EXAMPLE 2   *Find the slope of the graph of*

$$f(x) = (7x^3 - 5x + 2)(2x^4 + 7)$$

when $x = 1$.

Here $f(x)$ is the product of $7x^3 - 5x + 2$ and $2x^4 + 7$. By the product rule,

$$
\begin{aligned}
f'(x) &= (7x^3 - 5x + 2)\, D_x(2x^4 + 7) + (2x^4 + 7)\, D_x(7x^3 - 5x + 2) \\
&= (7x^3 - 5x + 2)(8x^3) + (2x^4 + 7)(21x^2 - 5).
\end{aligned}
$$

Evaluating $f'(x)$ at $x = 1$ gives the slope of the graph at that point:

$$f'(1) = 4(8) + 9(16) = 176.$$

Note: **We do not have to simplify the derivative before evaluating it.**

Usually, we do not use the product rule when simpler ways are obvious. For example, if $f(x) = 2x(x + 3)$, then it is quicker to write $f(x) = 2x^2 + 6x$, from which $f'(x) = 4x + 6$. Similarly, we do not usually use the product rule to differentiate $y = 4(x^2 - 3)$. Since the 4 is a constant multiplier, by Rule 3 we have $y' = 4(2x) = 8x$.

The next rule is used for differentiating a *quotient* of two functions and a proof is in Appendix B.

RULE 6

Quotient Rule. If f and g are differentiable functions such that $g(x) \neq 0$, then

$$\frac{d}{dx}\left[\frac{f(x)}{g(x)}\right] = \frac{g(x)f'(x) - f(x)g'(x)}{[g(x)]^2}.$$

That is, the derivative of the quotient of two functions is the denominator times the derivative of the numerator, minus the numerator times the derivative of the denominator, all divided by the square of the denominator.

Pitfall

The derivative of the quotient of two functions is **not** the quotient of their derivatives. For example,

$$D_x\left(\frac{x^2}{x+1}\right) \neq \frac{D_x(x^2)}{D_x(x+1)}.$$

EXAMPLE 3

a. *If* $F(x) = \dfrac{4x^2 + 3}{2x - 1}$, *find* $F'(x)$.

Let $f(x) = 4x^2 + 3$ and $g(x) = 2x - 1$. Then $F(x) = f(x)/g(x)$ and by Rule 6, the quotient rule,

$$F'(x) = \frac{g(x)f'(x) - f(x)g'(x)}{[g(x)]^2}$$

$$= \frac{(2x - 1)D_x(4x^2 + 3) - (4x^2 + 3)D_x(2x - 1)}{(2x - 1)^2}$$

$$= \frac{(2x - 1)(8x) - (4x^2 + 3)(2)}{(2x - 1)^2}$$

$$= \frac{8x^2 - 8x - 6}{(2x - 1)^2} = \frac{2(4x^2 - 4x - 3)}{(2x - 1)^2}.$$

b. *Find* $D_x\dfrac{1}{x^2}$.

Although the quotient rule can be used, a simpler and more direct method is to write $1/x^2$ as x^{-2} and then apply the rule for differentiating x^n.

$$D_x \frac{1}{x^2} = D_x x^{-2} = -2x^{-3} = -\frac{2}{x^3}.$$

Alternatively, applying the quotient rule to this problem gives

$$D_x \frac{1}{x^2} = \frac{(x^2) D_x(1) - (1) D_x(x^2)}{(x^2)^2}$$

$$= \frac{x^2(0) - 1(2x)}{x^4} = \frac{-2x}{x^4} = -\frac{2}{x^3}.$$

EXAMPLE 4   *Suppose that the demand equation for a manufacturer's product is $p = 1000/(q + 5)$. Then p is the price per unit (in dollars) at which consumers demand q units of the product. Find the marginal revenue function and evaluate it when $q = 45$.*

The revenue r received for selling q units is given by

$$\textbf{revenue} = \textbf{(price)(quantity)},$$

$$r = pq.$$

Thus the revenue function is

$$r = \left(\frac{1000}{q + 5} \right) q,$$

$$r = \frac{1000q}{q + 5}.$$

The marginal revenue function is dr/dq.

$$\frac{dr}{dq} = \frac{(q + 5) D_q(1000q) - (1000q) D_q(q + 5)}{(q + 5)^2}$$

$$= \frac{(q + 5)(1000) - (1000q)(1)}{(q + 5)^2} = \frac{5000}{(q + 5)^2}.$$

$$\left. \frac{dr}{dq} \right|_{q=45} = \frac{5000}{(45 + 5)^2} = \frac{5000}{2500} = 2.$$

This means that selling one additional unit beyond 45 results in approximately $2 more in revenue.

A function that plays an important role in economic analysis is the **consumption function**. The consumption function $C = f(I)$ expresses a relationship between the total national income I and the total national consumption C. Usually, both I and C are expressed in billions of dollars. The *marginal* *propensity to consume* is defined as the rate of change of consumption with respect to income. It is merely the derivative of C with respect to I.

$$\text{marginal propensity to consume} = \frac{dC}{dI}.$$

If we assume that the difference between income I and consumption C is savings S, then

$$S = I - C.$$

Differentiating both sides with respect to I gives

$$\frac{dS}{dI} = \frac{d}{dI}(I) - \frac{d}{dI}(C) = 1 - \frac{dC}{dI}.$$

We define dS/dI as the **marginal propensity to save**. Thus the marginal propensity to save indicates how fast savings change with respect to income.

EXAMPLE 5   *If the consumption function is given by*

$$C = \frac{5(2\sqrt{I^3} + 3)}{I + 10},$$

determine the marginal propensity to consume and the marginal propensity to save when $I = 100$.

$$\frac{dC}{dI} = \frac{(I + 10)\, D_I[5(2I^{3/2} + 3)] - 5(2\sqrt{I^3} + 3)\, D_I[I + 10]}{(I + 10)^2}$$

$$= \frac{(I + 10)[5(3I^{1/2})] - 5(2\sqrt{I^3} + 3)[1]}{(I + 10)^2}.$$

When $I = 100$ the marginal propensity to consume is

$$\frac{dC}{dI}\bigg|_{I = 100} = \frac{6485}{12,100} \approx 0.536.$$

The marginal propensity to save when $I = 100$ is $1 - 0.536 = 0.464$. This means that if a current income of \$100 billion increases by \$1 billion, the nation will consume approximately 53.6 percent (536/1000) and save 46.4 percent (464/1000) of that increase.

EXERCISE 2.4

In Problems 1–32, differentiate the functions.

1. $f(x) = (4x + 1)(6x + 3)$.

2. $f(x) = (3x - 1)(7x + 2)$.

3. $s(t) = (8 - 7t)(t^2 - 2)$.

4. $Q(x) = (5 - 2x)(x^2 + 1)$.

5. $f(r) = (3r^2 - 4)(r^2 - 5r + 1)$.

6. $C(I) = (2I^2 - 3)(3I^2 - 4I + 1)$.

7. $y = (x^2 + 3x - 2)(2x^2 - x - 3)$.

8. $y = (2 - 3x + 4x^2)(1 + 2x - 3x^2)$.

9. $f(w) = (8w^2 + 2w - 3)(5w^3 + 2)$.

10. $f(x) = (3x - x^2)(3 - x - x^2)$.

11. $y = (x^2 - 1)(3x^3 - 6x + 5) - 4(4x^2 + 2x + 1)$.

12. $h(x) = 4(x^5 - 3)(2x^3 + 4) + 3(8x^2 - 5)(3x + 2)$.

13. $f(p) = \frac{3}{2}(\sqrt{p} - 4)(4p - 5)$.

14. $g(x) = (\sqrt{x} - 3x + 1)(\sqrt[4]{x} - 2\sqrt{x})$.

15. $y = 7 \cdot \frac{2}{3}$.

16. $y = \frac{2x - 3}{4x + 1}$.

17. $y = \frac{x + 2}{x - 1}$.

18. $f(x) = \frac{-2x}{1 - x}$.

19. $h(z) = \frac{5 - 2z}{z^2 - 4}$.

20. $y = \frac{x^2 - 4x + 2}{x^2 + x + 1}$.

21. $y = \frac{8x^2 - 2x + 1}{x^2 - 5x}$.

22. $F(z) = \frac{z^4 + 4}{3z}$.

23. $g(x) = \frac{1}{x^{100} + 1}$.

24. $y = \frac{3}{7x^3}$.

25. $u(v) = \frac{v^5 - 8}{v}$.

26. $y = \frac{x - 5}{2\sqrt{x}}$.

27. $y = \frac{3x^2 - x - 1}{\sqrt[3]{x}}$.

28. $q(x) = 13x^2 + \frac{x - 1}{2x + 3} - \frac{4}{x}$.

29. $y = 7 - \frac{4}{x - 8} + \frac{2x}{3x + 1}$.

30. $y = \frac{4 - 5x}{(2x - 1)(3x + 2)}$.

31. $H(s) = \frac{(s + 2)(s - 4)}{s - 5}$.

32. $f(s) = \frac{17}{s(5s^2 - 10s + 4)}$.

33. Find the slope of the curve

$$y = (4x^2 + 2x - 5)(x^3 + 7x + 4)$$

at $(-1, 12)$.

34. Find the slope of the curve $y = \frac{x^3}{x^4 + 1}$ at $(1, \frac{1}{2})$.

In Problems 35 and 36, find an equation of the tangent line to the curve at the given point.

35. $y = \frac{6}{x - 1}$;　$(3, 3)$.

36. $y = \frac{4x + 5}{x^2}$;　$(-1, 1)$.

In Problems 37 and 38, determine the relative rate of change of y with respect to x for the given value of x.

37. $y = \frac{x}{2x - 6}$;　$x = 1$.　38. $y = \frac{1 - x}{1 + x}$;　$x = 5$.

In Problems 39–42, each equation represents a demand function for a certain product where p denotes price per unit for q units. Find the marginal revenue function in each case. Recall that revenue = pq.

39. $p = 25 - 0.02q$.　40. $p = 500/q$.

41. $p = \frac{108}{q + 2} - 3$.　42. $p = \frac{q + 750}{q + 50}$.

43. For the United States in the period 1922–1942, the consumption function is estimated in [19] by

$$C = 0.672I + 113.1.$$

Find the marginal propensity to consume.

44. Repeat Problem 43 if $C = 0.712I + 95.05$ for the United States for 1929–1941 [19].

In Problems 45 and 46, each equation represents a consumption function. Find the marginal propensity to consume and the marginal propensity to save for the given value of I.

45. $C = 2 + 2\sqrt{I}$;　$I = 9$.

46. $C = 6 + \frac{3I}{4} - \frac{\sqrt{I}}{3}$;　$I = 25$.

47. If the total cost function for a manufacturer is given by

$$c = \frac{5q^2}{q + 3} + 5000,$$

find the marginal cost function.

48. The persistence of sound in a room after the sound source is turned off is called *reverberation*. The *reverberation time RT* of the room is the time it takes for the intensity level of the sound to fall 60 decibels.

In the acoustical design of an auditorium [20] the following formula may be used to compute the RT of the room:

$$RT = \frac{0.05V}{A + xV},$$

where V is the room volume, A is the total room absorption, and x is the air absorption coefficient. Assuming that A and x are positive constants, show that the rate of change of RT with respect to V is always positive. If the total room volume increases by one unit, does reverberation time increase or decrease?

49. In a predator-prey experiment [21], it was statistically determined that the number of prey consumed, y, by an individual predator was a function of prey density x (the number of prey per unit of area), where

$$y = \frac{0.7355x}{1 + 0.02744x}.$$

Determine the rate of change of prey consumed with respect to prey density.

50. For a particular host-parasite relationship, it is determined that when the host density (number of hosts per unit of area) is x, then the number of hosts that are parasitized is y, where

$$y = \frac{900x}{10 + 45x}.$$

At what rate is the number of hosts parasitized changing with respect to host density when $x = 2$?

51. In a discussion of social security benefits, Feldstein [22] differentiates a function of the form

$$f(x) = \frac{a(1 + x) - b(2 + n)x}{a(2 + n)(1 + x) - b(2 + n)x},$$

where a, b, and n are constants. He determines that

$$f'(x) = \frac{-(1 + n)ab}{[a(1 + x) - bx]^2(2 + n)}.$$

Verify this. (*Hint:* For convenience let $2 + n = c$.)

2.5 THE CHAIN RULE AND POWER RULE

Our next rule, the chain rule, is one of the most important rules for finding derivatives. Before stating it, we shall consider the following situation. Suppose that

$$y = u^2 \quad \text{and} \quad u = 2x + 1.$$

Here y is a function of u, and u is a function of x. If we substitute $2x + 1$ for u in the first equation, we can consider y to be a function of x:

$$y = (2x + 1)^2.$$

After expanding we can find dy/dx in the usual way.

$$y = 4x^2 + 4x + 1.$$
$$\frac{dy}{dx} = 8x + 4.$$

From this example you can see that finding dy/dx by first performing a substitution could be quite involved, especially if we had $y = u^{100}$ above instead of $y = u^2$. Fortunately, the chain rule will allow us to handle such situations with ease.

RULE 7

Chain Rule. If y is a differentiable function of u and u is a differentiable function of x, then y is a differentiable function of x and

$$\frac{dy}{dx} = \frac{dy}{du} \cdot \frac{du}{dx}.$$

Let us see why the chain rule is reasonable. Suppose that $y = 8u + 5$ and $u = 2x - 3$. Let x change by one unit. How does u change? Answer: $du/dx = 2$. But for *each* one-unit change in u there is a change in y of $dy/du = 8$. Therefore, what is the change in y if x changes by one unit, that is, what is dy/dx? Answer: $8 \cdot 2$, which is $\dfrac{dy}{du} \cdot \dfrac{du}{dx}$. Thus $\dfrac{dy}{dx} = \dfrac{dy}{du} \cdot \dfrac{du}{dx}$.

EXAMPLE 1

a. *If $y = 2u^2 - 3u - 2$ and $u = x^2 + 4$, find dy/dx.*

By Rule 7, the chain rule,

$$\frac{dy}{dx} = \frac{dy}{du} \cdot \frac{du}{dx} = \frac{d}{du}(2u^2 - 3u - 2) \cdot \frac{d}{dx}(x^2 + 4)$$

$$= (4u - 3)(2x).$$

We can write our answer in terms of x alone by replacing u by $x^2 + 4$.

$$\frac{dy}{dx} = [4(x^2 + 4) - 3](2x) = [4x^2 + 13](2x) = 8x^3 + 26x.$$

b. *If $y = \sqrt{w}$ and $w = 7 - t^3$, find dy/dt.*

Here y is a function of w, and w is a function of t. Hence we can view y as a function of t. By the chain rule,

$$\frac{dy}{dt} = \frac{dy}{dw} \cdot \frac{dw}{dt} = \frac{d}{dw}(\sqrt{w}) \cdot \frac{d}{dt}(7 - t^3)$$

$$= \left(\frac{1}{2}w^{-1/2}\right)(-3t^2) = \frac{1}{2\sqrt{w}}(-3t^2)$$

$$= -\frac{3t^2}{2\sqrt{w}} = -\frac{3t^2}{2\sqrt{7 - t^3}}.$$

The chain rule states that if $y = f(u)$ and $u = g(x)$, then

$$\frac{dy}{dx} = \frac{dy}{du} \cdot \frac{du}{dx}.$$

Actually, the chain rule applies to a composition function because

$$y = f(u) = f(g(x)) = (f \circ g)(x).$$

Thus y, as a function of x, is $f \circ g$. This means that we can use the chain rule to differentiate a function when we recognize the function as a composition. However, we must first break down the function into composite parts.

For example, to differentiate

$$y = (x^3 - x^2 + 6)^{100}$$

we think of the function as a composition. Let

$$y = f(u) = u^{100} \quad \text{and} \quad u = g(x) = x^3 - x^2 + 6.$$

Then $y = (x^3 - x^2 + 6)^{100} = [g(x)]^{100} = f(g(x))$. Now that we have a composition, we differentiate. Since $y = u^{100}$ and $u = x^3 - x^2 + 6$, by the chain rule

$$\frac{dy}{dx} = \frac{dy}{du} \cdot \frac{du}{dx}$$
$$= (100u^{99})(3x^2 - 2x)$$
$$= 100(x^3 - x^2 + 6)^{99}(3x^2 - 2x).$$

We have just used the chain rule to differentiate $y = (x^3 - x^2 + 6)^{100}$, which is a power of a *function* of x, not simply a power of x. The following rule, called the *power rule,* generalizes our result and is a special case of the chain rule.

RULE 8

Power Rule. If u is a differentiable function of x and n is any real number, then

$$\frac{d}{dx}(u^n) = nu^{n-1}\frac{du}{dx}.$$

Proof. Let $y = u^n$. Since y is a differentiable function of u and u is a differentiable function of x, the chain rule gives

$$\frac{dy}{dx} = \frac{dy}{du} \cdot \frac{du}{dx}.$$

But $dy/du = nu^{n-1}$. Thus

$$\frac{dy}{dx} = nu^{n-1}\frac{du}{dx},$$

which is the power rule.

Another way of writing the power rule is

$$\frac{d}{dx}([u(x)]^n) = n[u(x)]^{n-1}u'(x).$$

EXAMPLE 2

a. *If* $y = (x^3 - 1)^7$, *find* y'.

Since y is a power of a *function* of x, the power rule applies. Letting $u(x) = x^3 - 1$ and $n = 7$, we have

$$y' = n[u(x)]^{n-1}u'(x)$$

$$= 7(x^3 - 1)^{7-1}\frac{d}{dx}(x^3 - 1)$$

$$= 7(x^3 - 1)^6(3x^2) = 21x^2(x^3 - 1)^6.$$

b. *If* $y = \sqrt[3]{(4x^2 + 3x - 2)^2}$, *find* dy/dx *when* $x = -2$.

Since

$$y = (4x^2 + 3x - 2)^{2/3},$$

we use the power rule with $u = 4x^2 + 3x - 2$ and $n = \frac{2}{3}$.

$$\frac{dy}{dx} = \frac{2}{3}(4x^2 + 3x - 2)^{(2/3)-1}D_x(4x^2 + 3x - 2)$$

$$= \frac{2}{3}(4x^2 + 3x - 2)^{-1/3}(8x + 3)$$

$$= \frac{2(8x + 3)}{3\sqrt[3]{4x^2 + 3x - 2}}.$$

$$\frac{dy}{dx}\bigg|_{x=-2} = \frac{2(-13)}{3\sqrt[3]{8}} = -\frac{13}{3}.$$

c. *If* $y = \dfrac{1}{x^2 - 2}$, *find* $\dfrac{dy}{dx}$.

Although the quotient rule can be used here, we shall treat the right side as the power $(x^2 - 2)^{-1}$ and use the power rule. Let $u = x^2 - 2$. Then $y = u^{-1}$ and

$$\frac{dy}{dx} = nu^{n-1}\frac{du}{dx}$$

$$= (-1)(x^2 - 2)^{-1-1}D_x(x^2 - 2)$$

$$= (-1)(x^2 - 2)^{-2}(2x)$$

$$= -\frac{2x}{(x^2 - 2)^2}.$$

EXAMPLE 3

a. *If* $z = \left(\dfrac{2s + 5}{s^2 + 1}\right)^4$, *find* $\dfrac{dz}{ds}$.

Since z is a power of a function, we use the power rule.

$$\frac{dz}{ds} = 4\left(\frac{2s + 5}{s^2 + 1}\right)^{4-1} D_s\left(\frac{2s + 5}{s^2 + 1}\right).$$

By the quotient rule,

$$\frac{dz}{ds} = 4\left(\frac{2s + 5}{s^2 + 1}\right)^3 \frac{(s^2 + 1)(2) - (2s + 5)(2s)}{(s^2 + 1)^2}.$$

Simplifying, we have

$$\frac{dz}{ds} = 4 \cdot \frac{(2s + 5)^3}{(s^2 + 1)^3} \cdot \frac{-2s^2 - 10s + 2}{(s^2 + 1)^2}$$

$$= \frac{-8(s^2 + 5s - 1)(2s + 5)^3}{(s^2 + 1)^5}.$$

b. *If* $y = (x^2 - 4)^5(3x + 5)^4$, *find* y'.

Since y is a product, we first apply the product rule.

$$y' = (x^2 - 4)^5 D_x[(3x + 5)^4] + (3x + 5)^4 D_x[(x^2 - 4)^5].$$

Now we can use the power rule.

$$y' = (x^2 - 4)^5[4(3x + 5)^3(3)] + (3x + 5)^4[5(x^2 - 4)^4(2x)].$$

Simplifying gives

$$y' = 12(x^2 - 4)^5(3x + 5)^3 + 10x(3x + 5)^4(x^2 - 4)^4$$
$$= 2(x^2 - 4)^4(3x + 5)^3[6(x^2 - 4) + 5x(3x + 5)] \qquad \text{[factoring]}$$
$$= 2(x^2 - 4)^4(3x + 5)^3(21x^2 + 25x - 24).$$

Table 2.1 lists the differentiation formulas in this chapter. You should know these formulas, know the mechanics involved in applying them, and know both the definition and the interpretations of a derivative. We assume that all functions in the table are differentiable.

Let us now use our knowledge of calculus to develop a concept relevant to economic studies. Suppose a manufacturer hires m employees who produce a total of q units of a product per day. We can think of q as a function of m. If r is the total revenue the manufacturer receives for selling these units, then r can also be considered a function of m. Thus we can look at dr/dm, the rate of change of revenue with respect to the number of employees. Since

$$\text{total revenue} = (\text{price per unit})(\text{number of units sold}),$$

TABLE 2.1
Differentiation Formulas

$\dfrac{d}{dx}(c) = 0$, where c is any constant.

$\dfrac{d}{dx}(x^n) = nx^{n-1}$, where n is any real number.

$\dfrac{d}{dx}[cf(x)] = cf'(x)$.

$\dfrac{d}{dx}[f(x) + g(x)] = f'(x) + g'(x)$.

$\dfrac{d}{dx}[f(x) - g(x)] = f'(x) - g'(x)$.

$\dfrac{d}{dx}[f(x)g(x)] = f(x)g'(x) + g(x)f'(x)$.

$\dfrac{d}{dx}\left[\dfrac{f(x)}{g(x)}\right] = \dfrac{g(x)f'(x) - f(x)g'(x)}{[g(x)]^2}$.

$\dfrac{dy}{dx} = \dfrac{dy}{du} \cdot \dfrac{du}{dx}$, where y is a function of u and u is a function of x.

$\dfrac{d}{dx}(u^n) = nu^{n-1}\dfrac{du}{dx}$.

we have

$$r = pq, \tag{1}$$

where p is the price per unit. Here p is a function of q and is determined by the product's demand equation. With Eq. (1) we can find dr/dm by the product rule:

$$\frac{dr}{dm} = p\frac{d}{dm}(q) + q\frac{d}{dm}(p) = p\frac{dq}{dm} + q\frac{dp}{dm}.$$

But by the chain rule,

$$\frac{dp}{dm} = \frac{dp}{dq} \cdot \frac{dq}{dm}.$$

Therefore,

$$\frac{dr}{dm} = p\frac{dq}{dm} + q\frac{dp}{dq} \cdot \frac{dq}{dm}$$

or

$$\frac{dr}{dm} = \frac{dq}{dm}\left(p + q\frac{dp}{dq}\right). \tag{2}$$

The derivative dr/dm is called the **marginal revenue product**. It is approximately the change in revenue that results when a manufacturer hires an extra employee.

EXAMPLE 4   *A manufacturer determines that m employees will produce a total of q units of a product per day, where $q = 10m^2/\sqrt{m^2 + 19}$. If the demand equation for the product is $p = 900/(q + 9)$, where p is in dollars per unit, determine the marginal revenue product when m = 9.*

First we find dq/dm and dp/dq. Using the quotient and power rules, we obtain

$$\frac{dq}{dm} = \frac{(m^2 + 19)^{1/2} D_m(10m^2) - (10m^2) D_m[(m^2 + 19)^{1/2}]}{[(m^2 + 19)^{1/2}]^2}$$

$$= \frac{(m^2 + 19)^{1/2}(20m) - 10m^2[(1/2)(m^2 + 19)^{-1/2}(2m)]}{m^2 + 19}$$

$$= \frac{10m(m^2 + 19)^{-1/2}[(m^2 + 19)(2) - m(m)]}{m^2 + 19}$$

$$= \frac{10m(m^2 + 38)}{(m^2 + 19)^{3/2}}.$$

Since $p = 900(q + 9)^{-1}$, then by the power rule,

$$\frac{dp}{dq} = 900[(-1)(q + 9)^{-2}(1)] = -\frac{900}{(q + 9)^2}.$$

Substituting into Eq. (2), we have the marginal revenue product:

$$\frac{dr}{dm} = \frac{10m(m^2 + 38)}{(m^2 + 19)^{3/2}}\left[p + q\left(-\frac{900}{[q + 9]^2}\right)\right].$$

When $m = 9$, then $q = 81$, which gives $p = 10$. Thus

$$\left.\frac{dr}{dm}\right|_{m=9} = 10.71.$$

This means that if a tenth employee is hired, revenue will increase by approximately $10.71 per day.

EXERCISE 2.5

In Problems 1–4, use the chain rule.

1. If $y = u^2 - 2u$ and $u = x^2 - x$, find dy/dx.

2. If $y = 2u^3 - 8u$ and $u = 7x - x^3$, find dy/dx.

3. If $y = 1/w^2$ and $w = 2 - x$, find dy/dx.

4. If $y = \sqrt[3]{z}$ and $z = x^6 - x^2 + 1$, find dy/dx.

In Problems 5–34, find y'.

5. $y = (3x + 2)^6$.

6. $y = (5 - x^2)^3$.

7. $y = 3(x^3 - 8x^2 + x)^{100}$.

8. $y = \dfrac{(2x^2 + 1)^4}{2}$.

9. $y = (x^2 - 2)^{-3}$.

10. $y = (7x - x^4)^{-3/2}$.

11. $y = \sqrt{5x^2 - x}$.

12. $y = \sqrt{3x^2 - 7}$.

13. $y = 2\sqrt[5]{(x^3 + 1)^2}$.

14. $y = 3\sqrt[3]{(x^2 + 1)^2}$.

15. $y = \dfrac{6}{2x^2 - x + 1}$.

16. $y = \dfrac{2}{x^4 + 2}$.

17. $y = \dfrac{1}{(x^2 - 3x)^2}$.

18. $y = \dfrac{1}{(1 - x)^3}$.

19. $y = \dfrac{2}{\sqrt{8x - 1}}$.

20. $y = \dfrac{1}{(3x^2 - x)^{2/3}}$.

21. $y = \sqrt[3]{7x} + \sqrt[3]{7}x$.

22. $y = \sqrt{2x} + \dfrac{1}{\sqrt{2x}}$.

23. $y = x^2(x - 4)^5$.

24. $y = x\sqrt{1 - x}$.

25. $y = 2x\sqrt{6x - 1}$.

26. $y = x^2(x^3 - 1)^4$.

27. $y = (8x - 1)^3(2x + 1)^4$.

28. $y = (6x + 1)^7(2x - 3)^3$.

29. $y = \left(\dfrac{x - 7}{x + 4}\right)^{10}$.

30. $y = \left(\dfrac{2x}{x + 2}\right)^4$.

31. $y = \sqrt{\dfrac{x - 2}{x + 3}}$.

32. $y = \sqrt[3]{\dfrac{8x^2 - 3}{x^2 + 2}}$.

33. $y = \dfrac{2x - 5}{(x^2 + 4)^3}$.

34. $y = \dfrac{(2x + 3)^3}{x^2 + 4}$.

In Problems 35 and 36, use the quotient rule and power rule to find y'. Do not simplify your answer.

35. $y = \dfrac{(2x + 1)(3x - 5)^2}{(x^2 - 7)^4}$.

36. $y = \dfrac{\sqrt{x + 2}(4x^2 - 1)^2}{9x - 3}$.

In Problems 37–40, q is the total number of units produced per day by m employees of a manufacturer, and p is the price per unit at which the q units are sold. In each case find the marginal revenue product for the given value of m.

37. $q = 2m$, $p = -0.5q + 20$; $m = 5$.

38. $q = \dfrac{200m - m^2}{20}$, $p = -0.1q + 70$; $m = 40$.

39. $q = \dfrac{10m^2}{\sqrt{m^2 + 9}}$, $p = \dfrac{525}{q + 3}$; $m = 4$.

40. $q = \dfrac{100m}{\sqrt{m^2 + 19}}$, $p = \dfrac{4500}{q + 10}$; $m = 9$.

41. Suppose $p = 100 - \sqrt{q^2 + 20}$ is a demand equation for a manufacturer's product.

   a. Find the rate of change of p with respect to q.

   b. Find the relative rate of change of p with respect to q.

   c. Find the marginal revenue function. [*Hint:* Use the fact that revenue = (price)(quantity).]

42. If $p = k/q$, where k is a constant, is the demand equation for a manufacturer's product, and $q = f(m)$ defines a function that gives the total number of units produced per day by m employees, show that the marginal revenue product is always zero.

43. The cost c of producing q units of a product is given by $c = 4000 + 10q + 0.1q^2$. If the price per unit p is given by the equation $q = 800 - 2.5p$, use the chain rule to find the rate of change of cost with respect to price per unit when $p = 80$.

44. A governmental health agency examined the records of a group of individuals who were hospitalized with a particular illness. It was found that the total proportion who had been discharged at the end of t days of hospitalization is given by $f(t)$, where

$$f(t) = 1 - \left(\dfrac{300}{300 + t}\right)^3.$$

Find $f'(300)$ and interpret your answer.

45. If the total cost function for a manufacturer is given by

$$c = \dfrac{5q^2}{\sqrt{q^2 + 3}} + 5000,$$

find the marginal cost function.

46. For a certain population, if E is the number of years of a person's education and S represents a numerical value of the person's status based on that educational level, then $S = 4\left(\dfrac{E}{4} + 1\right)^2$.

   a. How fast is status changing with respect to education when $E = 16$?

   b. At what level of education does the rate of change of status equal 8?

47. The volume V of a spherical cell is given by $V = \frac{4}{3}\pi r^3$, where r is the radius. At time t seconds, the radius r (in centimeters) is given by $r = 10^{-8}t^2 + 10^{-7}t$. Use the chain rule to find dV/dt when $t = 10$.

48. Under certain conditions, the pressure p developed in body tissue by ultrasonic beams is given in [13] by a function of the intensity I:

$$p = (2\rho VI)^{1/2},$$

where ρ (a Greek letter read "rho") is density, V is velocity of propagation, and I is intensity. Here, ρ and V are constants.

a. Find the rate of change of p with respect to I.

b. Find the relative rate of change.

49. Suppose that for a certain group of 20,000 births, the number l_x of people surviving to age x years is

$$l_x = 2000\sqrt{100 - x}, \qquad 0 \le x \le 100.$$

a. Find the rate of change of l_x with respect to x and evaluate your answer for $x = 36$.

b. Find the relative rate of change of l_x when $x = 36$.

50. A muscle has the ability to shorten when a load, such as a weight, is imposed on it. The equation

$$(P + a)(v + b) = k$$

is called the "fundamental equation of muscle contraction" [13]. Here P is the load imposed on the muscle, v is the velocity of the shortening of the muscle fibres, and a, b, and k are positive constants. Express v as a function of P. Use your result to find dv/dP.

2.6 IMPLICIT DIFFERENTIATION

Suppose that you were given the equation $y - 2x^2 + 4 = 0$ and wanted to find dy/dx. You would probably first solve for y in terms of x to get the form $y = f(x)$.

$$y - 2x^2 + 4 = 0,$$
$$y = 2x^2 - 4 \qquad [\text{form } y = f(x)].$$

Thus

$$\frac{dy}{dx} = 4x.$$

The form $y = f(x)$ expresses y directly in terms of x and is called an *explicit* form of a function. Up to now, all the functions you differentiated were given in that form. But the original equation above, $y - 2x^2 + 4 = 0$, is said to be an *implicit* form of a function. The word *implicit* is used because, although the equation does not directly express y as a function of x, it is assumed or *implied* that the equation determines y as some function of x.

Sometimes you may have an implicit form of a function which is difficult, or perhaps impossible, to solve for y in terms of x. Fortunately, you may still find dy/dx by a method called **implicit differentiation**.

To show you how implicit differentiation works, we shall use it on the equation $x^2 + y^2 - 4 = 0$ to find dy/dx. First, we assume that the equation defines y as some differentiable function of x, say $y = f(x)$. Next, we treat

y as a function of x and differentiate both sides of the equation with respect to x. Finally, we solve the result for dy/dx.

$$\frac{d}{dx}(x^2 + y^2 - 4) = \frac{d}{dx}(0),$$

$$\frac{d}{dx}(x^2) + \frac{d}{dx}(y^2) - \frac{d}{dx}(4) = \frac{d}{dx}(0). \tag{1}$$

We know that $\dfrac{d}{dx}(x^2) = 2x$ and that both $\dfrac{d}{dx}(4)$ and $\dfrac{d}{dx}(0)$ are 0. But $\dfrac{d}{dx}(y^2)$ is **not** $2y$ because we are differentiating with respect to x, not y. Since y is assumed to be a function of x, y^2 has the form u^n, where y plays the role of u. Just as the power rule states that $\dfrac{d}{dx}(u^2) = 2u\dfrac{du}{dx}$, we have $\dfrac{d}{dx}(y^2) = 2y\dfrac{dy}{dx}$ or $2yy'$. Thus Eq. (1) becomes

$$2x + 2yy' = 0.$$

Solving for y' gives

$$2yy' = -2x,$$

$$y' = -\frac{x}{y}.$$

Notice that the expression for y' involves the variable y as well as x. This means that, to find y' at a certain point, both coordinates of the point must be substituted into y'. For example, at $(\sqrt{3}, 1)$ we have $x = \sqrt{3}$ and $y = 1$. So

$$y' \bigg|_{(\sqrt{3},\, 1)} = -\frac{\sqrt{3}}{1} = -\sqrt{3}.$$

EXAMPLE 1   *For each of the following, find y' by implicit differentiation.*

a. $y + y^3 = x$.

We treat y as a (differentiable) function of x and differentiate both sides with respect to x.

$$D_x(y) + D_x(y^3) = D_x(x).$$

Now, $D_x(y)$ can be written y', and $D_x(x) = 1$. By the power rule, $D_x(y^3)$ is $3y^2\dfrac{dy}{dx}$ or $3y^2y'$. Thus

$$y' + 3y^2y' = 1.$$

Solving for y' gives

$$y'(1 + 3y^2) = 1,$$

$$y' = \frac{1}{1 + 3y^2}.$$

b. $\sqrt{x} + \sqrt{y} = 4$.

Since $x^{1/2} + y^{1/2} = 4$, then $D_x(x^{1/2}) + D_x(y^{1/2}) = D_x(4)$:

$$\frac{1}{2x^{1/2}} + \frac{1}{2y^{1/2}}y' = 0.$$

Solving for y' gives

$$\frac{y'}{2y^{1/2}} = -\frac{1}{2x^{1/2}},$$

$$y' = -\frac{y^{1/2}}{x^{1/2}} = -\frac{\sqrt{y}}{\sqrt{x}}.$$

Using the original equation, we can express the answer in terms of x alone.

$$y' = -\frac{4 - \sqrt{x}}{\sqrt{x}} = \frac{\sqrt{x} - 4}{\sqrt{x}}.$$

EXAMPLE 2   *Find the slope of the curve $x^2 + 5xy - 3y^3 - 4 = 0$ at* (2, 0).

We assume that y is a function of x and differentiate both sides with respect to x.

$$D_x(x^2) + 5D_x(xy) - 3D_x(y^3) - D_x(4) = D_x(0).$$

To find $D_x(xy)$ we use the product rule.

$$[2x] + 5[xD_x(y) + yD_x(x)] - 3[3y^2D_x(y)] - 0 = 0,$$
$$[2x] + 5[xy' + y] - 3[3y^2y'] = 0,$$
$$2x + 5xy' + 5y - 9y^2y' = 0.$$

Solving for y' gives

$$5xy' - 9y^2y' = -2x - 5y,$$
$$y'(5x - 9y^2) = -2x - 5y,$$
$$y' = \frac{-2x - 5y}{5x - 9y^2} = \frac{2x + 5y}{9y^2 - 5x}.$$

The slope of the curve at (2, 0) is

$$y'\Big|_{(2, 0)} = \frac{2(2) + 5(0)}{9(0)^2 - 5(2)} = -\frac{2}{5}.$$

EXAMPLE 3   *If $p = 500 - 3q^2$, find dq/dp, the rate of change of q with respect to p.*

Treating q as a function of p and differentiating both sides with respect to p, we have

$$D_p(p) = D_p(500) - D_p(3q^2),$$

$$1 = 0 - 6q \frac{dq}{dp}.$$

Solving for dq/dp gives

$$\frac{dq}{dp} = -\frac{1}{6q}.$$

EXERCISE 2.6

In Problems 1–18, find dy/dx by implicit differentiation.

1. $x^2 + 4y^2 = 4$.

2. $3x^2 + 6y^2 = 1$.

3. $3y^4 - 5x = 0$.

4. $2x^2 - 3y^2 = 4$.

5. $\sqrt{x} + \sqrt{y} = 3$.

6. $x^{1/5} + y^{1/5} = 4$.

7. $x^{3/4} + y^{3/4} = 7$.

8. $y^3 = 4x$.

9. $xy = 4$.

10. $x + xy - 2 = 0$.

11. $xy - y - 4x = 5$.

12. $x^2 + y^2 = 2xy + 3$.

13. $x^3 + y^3 - 12xy = 0$.

14. $2x^3 + 3xy + y^3 = 0$.

15. $x = \sqrt{y} + \sqrt[3]{y}$.

16. $x^3y^3 + x = 9$.

17. $3x^2y^3 - x + y = 25$.

18. $ax^2 - by^2 = c$.

19. If $x + xy + y^2 = 7$, find y' at $(1, 2)$.

20. Find the slope of the curve $4x^2 + 9y^2 = 1$ at the point $(0, \frac{1}{3})$; at the point (x_0, y_0).

21. Find an equation of the tangent line to the curve $x^3 + y^2 = 3$ at the point $(-1, 2)$.

22. Repeat Problem 21 for the curve $y^2 + xy - x^2 = 5$ at the point $(4, 3)$.

For the demand equations in Problems 23–26, find the rate of change of q with respect to p.

23. $p = 100 - q^2$.

24. $p = 400 - \sqrt{q}$.

25. $p = \dfrac{20}{(q + 5)^2}$.

26. $p = \dfrac{20}{q^2 + 5}$.

27. The equation $(P + a)(v + b) = k$ is called the "fundamental equation of muscle contraction" [13]. Here P is the load imposed on the muscle, v is the velocity of the shortening of the muscle fibres, and a, b, and k are positive constants. Use implicit differentiation to show that dv/dP, in terms of P, is given by

$$\frac{dv}{dP} = -\frac{k}{(P + a)^2}.$$

2.7 HIGHER-ORDER DERIVATIVES

As we said earlier, the derivative of a function $y = f(x)$ is itself a function, $f'(x)$. If we differentiate $f'(x)$, the resulting function is called the **second derivative** of f at x. It is denoted $f''(x)$, which is read "f double prime of x." Similarly, the derivative of the second derivative is called the **third derivative**, written $f'''(x)$. Continuing in this way, we get **higher-order derivatives**. Some notations for higher-order derivatives are given in Table 2.2. To avoid clumsy notation, primes are not used beyond the third derivative.

TABLE 2.2

First derivative:	y',	$f'(x)$,	$\dfrac{dy}{dx}$,	$\dfrac{d}{dx}[f(x)]$,	$D_x y$
Second derivative:	y'',	$f''(x)$,	$\dfrac{d^2 y}{dx^2}$,	$\dfrac{d^2}{dx^2}[f(x)]$,	$D_x^2 y$
Third derivative:	y''',	$f'''(x)$,	$\dfrac{d^3 y}{dx^3}$,	$\dfrac{d^3}{dx^3}[f(x)]$,	$D_x^3 y$
Fourth derivative:	$y^{(4)}$,	$f^{(4)}(x)$,	$\dfrac{d^4 y}{dx^4}$,	$\dfrac{d^4}{dx^4}[f(x)]$,	$D_x^4 y$

Pitfall

The symbol $D_x^2 y$ represents the second derivative of y. It is not the same as $[D_x y]^2$, the square of the first derivative of y.

$$D_x^2 y \neq [D_x y]^2.$$

EXAMPLE 1   *If $f(x) = 6x^3 - 12x^2 + 6x - 2$, find all higher-order derivatives.*

Differentiating $f(x)$ gives

$$f'(x) = 18x^2 - 24x + 6.$$

Differentiating $f'(x)$ gives

$$f''(x) = 36x - 24.$$

Similarly,

$$f'''(x) = 36,$$
$$f^{(4)}(x) = 0.$$

All successive derivatives are also 0: $f^{(5)}(x) = 0$, and so on.

EXAMPLE 2   *If $y = (2x^2 + 3)^4$, find y''.*

By the power rule,

$$y' = 4(2x^2 + 3)^3(4x) = 16x(2x^2 + 3)^3.$$

By the product rule,

$$
\begin{aligned}
y'' &= 16[x(3)(2x^2 + 3)^2(4x) + (2x^2 + 3)^3(1)] \\
&= 16(2x^2 + 3)^2[12x^2 + (2x^2 + 3)] \\
&= 16(2x^2 + 3)^2(14x^2 + 3).
\end{aligned}
$$

EXAMPLE 3   *If $y = f(x) = \dfrac{16}{x + 4}$, find $\dfrac{d^2 y}{dx^2}$ and evaluate it when $x = 4$.*

Since $y = 16(x + 4)^{-1}$, the power rule gives

$$\frac{dy}{dx} = -16(x + 4)^{-2},$$

$$\frac{d^2 y}{dx^2} = 32(x + 4)^{-3} = \frac{32}{(x + 4)^3}.$$

Evaluating when $x = 4$,

$$\frac{d^2y}{dx^2}\bigg|_{x=4} = \frac{32}{8^3} = \frac{1}{16}.$$

The second derivative evaluated at $x = 4$ is also denoted $f''(4)$ or $y''(4)$.

EXAMPLE 4   *If $f(t) = 6\sqrt[3]{t}$, find the rate of change of $f'(t)$.*

To find the rate of change of any function, we must find its derivative. Thus we want $D_t[f'(t)]$, which is $f''(t)$. Since $f(t) = 6t^{1/3}$,

$$f'(t) = 6\left(\frac{1}{3}t^{-2/3}\right) = 2t^{-2/3},$$

$$f''(t) = 2\left(-\frac{2}{3}t^{-5/3}\right) = -\frac{4}{3}t^{-5/3}.$$

EXERCISE 2.7

In Problems 1–22, find the indicated derivative.

1. $f(x) = 2x^3, \quad f'''(x)$.

2. $f(x) = \dfrac{x^5}{20}, \quad f'''(x)$.

3. $y = 8, \quad \dfrac{d^2y}{dx^2}$.

4. $y = 2^6, \quad \dfrac{d^2y}{dx^2}$.

5. $y = 3x^3 - 8x^2 + 2x - 1, \quad y'''$.

6. $y = -2x^4 - 3x^2 - 2x + 4, \quad y'''$.

7. $f(t) = t^4 - \dfrac{t^3}{6} + 3t, \quad f''(t)$.

8. $g(p) = 3p^4 + p^2 - 1, \quad g''(p)$.

9. $y = \dfrac{1}{\sqrt{x}}, \quad y^{(4)}$.

10. $f(q) = 8q^{1/4} - 8q^{-1/4}, \quad f'''(q)$.

11. $f(p) = \dfrac{1}{6p^3}, \quad f'''(p)$. $-\dfrac{10}{p^6}$

12. $y = \dfrac{1}{x}, \quad y'''$.

13. $f(r) = \sqrt{1 - r}, \quad f''(r)$.

14. $f(x) = \sqrt{x}, \quad D_x^2[f(x)]$.

15. $y = \dfrac{1}{5x - 6}, \quad \dfrac{d^2y}{dx^2}$.

16. $y = \sqrt{8 + 3^3}, \quad y''$.

17. $y = (x^2 - 4)^{10}, \quad y''$.

18. $y = (2x + 1)^4, \quad y''$.

19. $y = \dfrac{x + 1}{x - 1}, \quad y''$.

20. $y = 2x^{1/2} + (2x)^{1/2}, \quad y''$.

21. $y = x(1 - 2x)^3, \quad y''$.

22. $y = \dfrac{3}{(4x - 3)^4}, \quad y''$.

23. Find the rate of change of $f'(x)$ if $f(x) = (5x - 3)^4$.

24. Find the rate of change of $f''(x)$ if

$$f(x) = 6\sqrt{x} + \frac{1}{6\sqrt{x}}.$$

25. If $c = 0.3q^2 + 2q + 850$ is a cost function, how fast is marginal cost changing when $q = 100$?

26. If $p = 1000 - 45q - q^2$ is a demand equation, how fast is marginal revenue changing when $q = 10$?

27. If $f(x) = x^4 - 6x^2 + 5x - 6$, determine the values of x for which $f''(x) = 0$.

2.8 REVIEW

Important Terms and Symbols

Section 2.1   secant line   tangent line   slope of a curve   derivative, $f'(x)$

Section 2.3 Δx rate of change   total cost function   marginal cost   average cost
total revenue function   marginal revenue   relative rate of change
percentage rate of change

Section 2.4   product rule   quotient rule   consumption function   marginal propensity to consume
marginal propensity to save

Section 2.5   chain rule   power rule   marginal revenue product

Section 2.6   implicit differentiation

Section 2.7   higher-order derivatives, $f''(x)$, $f'''(x)$, and so on

Summary

The tangent line (or tangent) to a curve at point P is the limiting position of secant lines PQ as Q approaches P along the curve. The slope of the tangent at P is called the slope of the curve at P.

If $y = f(x)$, the derivative of f at x is the function defined by the limit

$$f'(x) = \lim_{h \to 0} \frac{f(x + h) - f(x)}{h}.$$

Geometrically, the derivative gives the slope of the curve $y = f(x)$ at the point $(x, f(x))$. An equation of the tangent at a particular point (x_1, y_1) is obtained by evaluating $f'(x_1)$, which is the slope m of the tangent, and substituting into the point-slope form $y - y_1 = m(x - x_1)$.

The basic rules for finding derivatives are as follows, where we assume that all functions are differentiable.

$$\frac{d}{dx}(c) = 0, \text{ where } c \text{ is any constant.}$$

$$\frac{d}{dx}(x^n) = nx^{n-1}, \text{ where } n \text{ is any real number.}$$

$$\frac{d}{dx}[cf(x)] = cf'(x).$$

$$\frac{d}{dx}[f(x) + g(x)] = f'(x) + g'(x).$$

$$\frac{d}{dx}[f(x) - g(x)] = f'(x) - g'(x).$$

$$\frac{d}{dx}[f(x)g(x)] = f(x)g'(x) + g(x)f'(x).$$

$$\frac{d}{dx}\left[\frac{f(x)}{g(x)}\right] = \frac{g(x)f'(x) - f(x)g'(x)}{[g(x)]^2}$$

$$\frac{dy}{dx} = \frac{dy}{du} \cdot \frac{du}{dx}, \text{ where } y \text{ is a function of } u \text{ and } u \text{ is a function of } x.$$

$$\frac{d}{dx}(u^n) = nu^{n-1}\frac{du}{dx}.$$

The derivative dy/dx can also be interpreted as giving the (instantaneous) rate of change of y with respect to x:

$$\frac{dy}{dx} = \lim_{\Delta x \to 0} \frac{\Delta y}{\Delta x} = \lim_{\Delta x \to 0} \frac{\text{change in } y}{\text{change in } x}.$$

In particular, if $s = f(t)$ is an equation of motion, where s is position at time t, then

$$\frac{ds}{dt} = \text{velocity at time } t.$$

In economics, the term *marginal* is used to describe derivatives of specific types of functions. If $c = f(q)$ is a total cost function (c is the total cost of q units of a product), then the rate of change

$$\frac{dc}{dq} \text{ is called marginal cost.}$$

We interpret marginal cost as the approximate cost of one additional unit of output. (Average cost per unit, $\bar{c}$, is related to total cost c by $\bar{c} = c/q$ or $c = \bar{c}q$.)

A total revenue function $r = f(q)$ gives a manufacturer's revenue r for selling q units of product. (Revenue r and price p are related by $r = pq$.) The rate of change

$$\frac{dr}{dq} \text{ is called marginal revenue,}$$

which is interpreted as the approximate revenue obtained from selling one additional unit of output.

If r is the revenue that a manufacturer receives when the total output q of m employees is sold at a price p per unit, then the derivative dr/dm is called the marginal revenue product and is given by

$$\frac{dr}{dm} = \frac{dq}{dm}\left(p + q\frac{dp}{dq}\right).$$

The marginal revenue product gives the approximate change in revenue that results when the manufacturer hires an extra employee.

If $C = f(I)$ is a consumption function, where I is national income and C is national consumption, then

$$\frac{dC}{dI} \text{ is marginal propensity to consume,}$$

$$1 - \frac{dC}{dI} \text{ is marginal propensity to save.}$$

For any function, the relative rate of change of $f(x)$ is

$$\frac{f'(x)}{f(x)},$$

which compares the rate of change of $f(x)$ with $f(x)$ itself. The percentage rate of change is

$$\frac{f'(x)}{f(x)} \cdot 100.$$

If an equation implicitly defines y as a function of x [rather than defining it explicitly in the form $y = f(x)$], then dy/dx can be found by implicit differentiation. With this method, we treat y as a function of x and differentiate both sides of the equation with respect to x. When doing this, keep in mind that $D_x y^n = ny^{n-1}\dfrac{dy}{dx}$. Finally, we solve the resulting equation for dy/dx.

Because the derivative $f'(x)$ of a function $y = f(x)$ is itself a function, it can be successively differentiated to obtain the second derivative $f''(x)$, the third derivative $f'''(x)$, and other higher-order derivatives.

Review Problems

In Problems 1 and 2, use the definition of the derivative to find $f'(x)$.

1. $f(x) = 2 - x^2$.

2. $f(x) = 2x^2 - 3x + 1$.

In Problems 3–40, differentiate.

3. $y = 6^3$.

4. $y = x$.

5. $y = 7x^4 - 6x^3 + 5x^2 + 1$.

6. $y = 4(x^2 + 5) - 7x$.

7. $f(s) = s^2(s^2 + 2)$.

8. $y = \sqrt{x + 3}$.

9. $y = \dfrac{x^2 + 3}{5}$.

10. $y = \dfrac{1}{x^3}$.

11. $y = (x^2 + 6x)(x^3 - 6x^2 + 4)$.

12. $y = (x^2 + 1)^{100}(x - 6)$.

13. $f(x) = (2x^2 + 4x)^{100}$.

14. $f(w) = w\sqrt{w} + w^2$.

15. $y = \dfrac{1}{2x + 1}$.

16. $y = \dfrac{x^4 + 4x^2}{2x}$.

17. $y = (8 + 2x)(x^2 + 1)^4$.

18. $g(z) = (2z)^{3/5} + 5$.

19. $f(z) = \dfrac{z^2 - 1}{z^2 + 1}$.

20. $y = \dfrac{x - 5}{(x + 2)^2}$.

21. $y = \sqrt[3]{4x - 1}$.

22. $h(t) = (1 + 2^2)^0$.

23. $2xy + y^2 = 6$ (find y').

24. $4x^2 - 9y^2 = 4$ (find y').

25. $y = \dfrac{1}{\sqrt{1 - x}}$.

26. $y = \dfrac{x(x + 1)}{2x^2 + 3}$.

27. $h(x) = (x - 6)^4(x + 5)^3$.

28. $y + xy + y^2 = 1$ (find y').

29. $y = \dfrac{5x - 4}{x + 1}$.

30. $y = \dfrac{(x + 3)^5}{x}$.

31. $y = 2x^{-3/8} + (2x)^{-3/8}$. **32.** $f(x) = 5x\sqrt{1 - 2x}$.

33. $y = \dfrac{x^2 + 6}{\sqrt{x^2 + 5}}$. **34.** $y = \sqrt{\dfrac{x}{2}} + \sqrt{\dfrac{2}{x}}$.

35. $y = (x^3 + 6x^2 + 9)^{3/5}$. **36.** $y = \sqrt[3]{(1 - 3x^2)^2}$.

37. $x^2y^2 = 1$ (find y').

38. $y = 0.5x(x + 1)^{-2} + 0.3$.

39. $g(z) = \dfrac{-7z}{(z - 1)^{-1}}$.

40. $g(z) = \dfrac{-3}{4(z^5 + 2z - 5)^4}$.

*In Problems **41–48**, find the indicated derivative at the given point. It is not necessary to simplify the derivative before substituting the coordinates.*

41. $y = x^4 - 2x^3 + 6x$; y''', $(1, 5)$.

42. $y = x^2\sqrt{x}$; y''', $(1, 1)$.

43. $y = \dfrac{x}{\sqrt{x - 1}}$; y'', $(5, \frac{5}{2})$.

44. $y = \dfrac{2}{1 - x}$; y'', $(-2, \frac{2}{3})$.

45. $y = (2x - 3)^4$; y''', $(2, 1)$.

46. $y = \sqrt{x}(2x - 1)$; y'', $(1, 1)$.

47. $y = \dfrac{4x}{x^2 + 4}$; y'', $(2, 1)$.

48. $y = (x + 1)^3(x - 1)$; y'', $(-1, 0)$.

*In Problems **49–54**, find an equation of the tangent line to the curve at the point corresponding to the given value of x.*

49. $y = x^2 - 6x + 4$, $x = 1$.

50. $y = -2x^3 + 6x + 1$, $x = 2$.

51. $y = \sqrt[3]{x}$, $x = 8$.

52. $y = \dfrac{x}{1 - x}$, $x = 3$.

53. $x^2 - y^2 = 9$, $x = 7$, $y > 0$.

54. $xy = 6$, $x = 1$.

55. If $f(x) = 4x^2 + 2x + 8$, find the relative and percentage rates of change of $f(x)$ when $x = 1$.

56. If $f(x) = x/(x + 4)$, find the relative and percentage rates of change of $f(x)$ when $x = 1$.

57. If $r = q(20 - 0.1q)$ is a total revenue function, find the marginal revenue function.

58. If $c = 0.0001q^3 - 0.02q^2 + 3q + 6000$ is a total cost function, find the marginal cost when $q = 100$.

59. If $C = 7 + 0.6I - 0.25\sqrt{I}$ is a consumption function, find the marginal propensity to consume and the marginal propensity to save when $I = 16$.

60. If $p = (q + 14)/(q + 4)$ is a demand equation, find the rate of change of price p with respect to quantity q.

61. If $p = -0.5q + 450$ is a demand equation, find the marginal revenue function.

62. If $\bar{c} = 0.03q + 1.2 + (3/q)$ is an average cost function, find marginal cost when $q = 100$.

63. The total cost function for an electric light and power plant is estimated in **[15]** by

$$c = 16.68 + 0.125q + 0.00439q^2, \qquad 20 \le q \le 90$$

where q is eight-hour total output (as percentage of capacity) and c is total fuel cost in dollars. Find the marginal cost function and evaluate it when $q = 70$.

64. A manufacturer has determined that m employees will produce a total of q units of product per day where $q = m(50 - m)$. If the demand function is given by $p = -0.01q + 9$, find the marginal revenue product when $m = 10$.

65. In a study (adapted from **[17]**) of the winter moth in Nova Scotia, it was determined that the average number of eggs, y, in a female moth was a function of the female's abdominal width x (in millimeters), where

$$y = f(x) = 14x^3 - 17x^2 - 16x + 34,$$

and $1.5 \le x \le 3.5$. At what rate does the number of eggs change with respect to abdominal width when $x = 2$?

66. For a particular host-parasite relationship, it is found that when the host density (number of hosts per unit of area) is x, then the number of hosts that are parasitized is y, where

$$y = 10\left(1 - \frac{1}{1 + 2x}\right), \qquad x \geq 0.$$

For what value of x does dy/dx equal $\frac{1}{5}$?

67. Bacteria are growing in a culture. The time t (in hours) for the number of bacteria to double in number (gen-

eration time) is a function of the temperature T (in degrees Celsius) of the culture and is given by

$$t = f(T) = \begin{cases} \frac{1}{24}T + \frac{11}{4}, & \text{if } 30 \leq T \leq 36, \\ \frac{4}{3}T - \frac{175}{4}, & \text{if } 36 < T \leq 39. \end{cases}$$

Find dt/dT when (a) $T = 38$ and (b) $T = 35$.

Curve Sketching

3.1 SYMMETRY

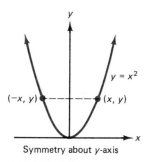

Symmetry about *y*-axis

FIGURE 3.1

Examining the graphical behavior of equations is a basic part of mathematics and has applications to other areas of study. In this section we examine equations to determine whether their graphs have *symmetry*.

Consider the graph of $y = x^2$ in Fig. 3.1. The portion to the left of the *y*-axis is the reflection (or mirror image) through the *y*-axis of that portion to the right of the *y*-axis, and vice versa. This graph is said to be **symmetric about the y-axis**. That means if (x, y) is any point on the graph, then the point $(-x, y)$ must also lie on the graph.

There is a simple test for *y*-axis symmetry. The graph of an equation is symmetric about the *y*-axis if replacing *x* with $-x$ results in an equation that is equivalent to the original equation.

EXAMPLE 1 *Show that the graph of $y = x^2$ is symmetric about the y-axis.*

In $y = x^2$ we replace *x* with $-x$.

$$y = (-x)^2,$$
$$y = x^2.$$

The result is the same as the original equation, so the graph is symmetric about the *y*-axis.

Another type of symmetry is shown by the graph of $x = y^2$ in Fig. 3.2. Here the portion below the *x*-axis is the reflection through the *x*-axis of that portion above the *x*-axis, and vice versa. If the point (x, y) lies on the graph, then $(x, -y)$ also lies on it. This graph is said to be **symmetric about the x-axis**. In general, the graph of an equation is symmetric about the x-axis if replacing y by $-y$ results in an equivalent equation. For example, applying this test to the graph of $x = y^2$ shown in Fig. 3.2 gives

$$x = (-y)^2,$$
$$x = y^2,$$

which is equivalent to the original equation. Thus the graph is symmetric about the x-axis.

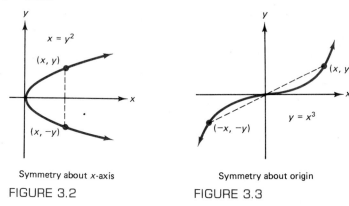

Symmetry about x-axis   Symmetry about origin

FIGURE 3.2   FIGURE 3.3

A third type of symmetry, **symmetry about the origin**, is illustrated by the graph of $y = x^3$ (Fig. 3.3). Whenever the point (x, y) lies on the graph, then $(-x, -y)$ also lies on it. As a result, the line segment joining (x, y) and $(-x, -y)$ is bisected by the origin. In general, the graph of an equation is symmetric about the origin if replacing x by $-x$ and y by $-y$ results in an equivalent equation. For example, applying this test to the graph of $y = x^3$ shown in Fig. 3.3 gives

$$-y = (-x)^3,$$
$$-y = -x^3,$$
$$y = x^3,$$

which is equivalent to the original equation. Thus the graph is symmetric about the origin.

Table 3.1 summarizes the tests for symmetry. When we know that a graph has symmetry, we can sketch it by plotting fewer points than would otherwise be needed.

TABLE 3.1
Tests for Symmetry

Symmetry about x-axis	Replace y by $-y$ in given equation; symmetric if equivalent equation is obtained
Symmetry about y-axis	Replace x by $-x$ in given equation; symmetric if equivalent equation is obtained
Symmetry about origin	Replace x by $-x$ and y by $-y$ in given equation; symmetric if equivalent equation is obtained

EXAMPLE 2   *Test $y = f(x) = 1 - x^4$ for symmetry about the x-axis, y-axis, or the origin and sketch the graph.*

Symmetry   *x-axis:* Replacing y by $-y$ in $y = 1 - x^4$ gives

$$-y = 1 - x^4 \quad \text{or} \quad y = -1 + x^4,$$

which is not equivalent to the given equation. Thus the graph is *not* symmetric about the x-axis.

y-axis: Replacing x by $-x$ in $y = 1 - x^4$ gives

$$y = 1 - (-x)^4 \quad \text{or} \quad y = 1 - x^4,$$

which is the same as the given equation. Thus the graph *is* symmetric about the y-axis.

Origin: Replacing x by $-x$ and y by $-y$ in $y = 1 - x^4$ gives

$$-y = 1 - (-x)^4, \qquad -y = 1 - x^4, \qquad y = -1 + x^4,$$

which is not equivalent to the given equation. Thus the graph is *not* symmetric about the origin.

Discussion   Recall that, when you graph a function, a convenient point to plot is the *y-intercept*. It is the point $(0, y)$ where the graph intersects the y-axis. The y-value is found by setting x equal to 0 and solving for y. Thus for $y = 1 - x^4$ the y-intercept is 1. After plotting some points to the right of the y-axis, we can sketch the *entire* graph by using the property of symmetry about the y-axis (Fig. 3.4).

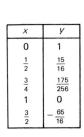

x	y
0	1
$\frac{1}{2}$	$\frac{15}{16}$
$\frac{3}{4}$	$\frac{175}{256}$
1	0
$\frac{3}{2}$	$-\frac{65}{16}$

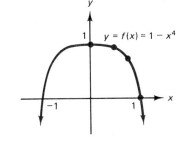

FIGURE 3.4

In Example 2 we showed that the graph of $y = f(x) = 1 - x^4$ does not have x-axis symmetry. With the exception of the constant function $f(x) = 0$, *the graph of any **function** $y = f(x)$ cannot be symmetric about the x-axis,* because such symmetry implies two y-values with the same x-value.

EXAMPLE 3   *Test the graph of $4x^2 + 9y^2 = 36$ for symmetry and sketch.*

Symmetry   Testing for x-axis symmetry, we replace y by $-y$:

$$4x^2 + 9(-y)^2 = 36 \quad \text{or} \quad 4x^2 + 9y^2 = 36.$$

Since we obtain the original equation, there is symmetry about the x-axis.

Testing for y-axis symmetry, we replace x by $-x$:

$$4(-x)^2 + 9y^2 = 36 \quad \text{or} \quad 4x^2 + 9y^2 = 36.$$

Again we have the original equation, so there is also symmetry about the y-axis.

Testing for symmetry about the origin, we replace x by $-x$ and y by $-y$:

$$4(-x)^2 + 9(-y)^2 = 36 \quad \text{or} \quad 4x^2 + 9y^2 = 36.$$

Since this is the original equation, the graph is also symmetric about the origin.

Discussion To find the y-intercepts, we have

$$4(0)^2 + 9y^2 = 36, \qquad 9y^2 = 36, \qquad y^2 = 4, \qquad y = \pm 2.$$

Thus the y-intercepts are 2 and -2. For this graph we shall plot not only the y-intercepts but also the x-intercepts. Recall that an x-intercept is a point $(x, 0)$ where the graph intersects the x-axis. The x-value is found by replacing y with 0 in $4x^2 + 9y^2 = 36$ and solving for x.

$$4x^2 + 9(0)^2 = 36, \qquad 4x^2 = 36, \qquad x^2 = 9, \qquad x = \pm 3.$$

Thus the x-intercepts are 3, and -3. Although x-intercepts are not always as easy to find as they are here, they are often useful in graphing.

In Fig. 3.5 the intercepts and some points in the first quadrant are plotted. The points in that quadrant are then connected by a smooth curve. By symmetry about the x-axis, the points in the fourth quadrant are obtained. Then by symmetry about the y-axis the complete graph is found. There are other ways

x	y
± 3	0
0	± 2
1	$\dfrac{4\sqrt{2}}{3}$
2	$\dfrac{2\sqrt{5}}{3}$
$\dfrac{5}{2}$	$\dfrac{\sqrt{11}}{3}$

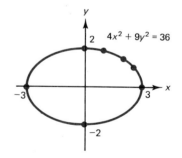

FIGURE 3.5

of graphing the equation by using symmetry. For example, after plotting the intercepts and some points in the first quadrant, then by symmetry about the origin we can obtain the points in the third quadrant. By symmetry about the x-axis (or y-axis) we can then obtain the entire graph.

In Example 3 the graph is symmetric about the *x*-axis, the *y*-axis, and the origin. It can be shown that **for any graph, if any two of the three types of symmetry exist, then the remaining type must also exist.**

EXERCISE 3.1

In Problems **1–12**, *test for symmetry about the x-axis, the y-axis, or the origin. Do not sketch the graph.*

1. $y = 5x$.

2. $y = f(x) = x^2 - 4$.

3. $2x^2 + y^2 x^4 = 8 - y$.

4. $x = y^3$.

5. $4x^2 - 9y^2 = 36$.

6. $y = 7$.

7. $x = -y^{-4}$.

8. $y = \sqrt{x^2 - 4}$.

9. $x - 4y - y^2 + 21 = 0$.

10. $x^3 - xy + y^2 = 0$.

11. $y = f(x) = \dfrac{x^3}{x^2 + 5}$.

12. $x^2 + xy + y^2 = 0$.

In Problems **13–20**, *test for symmetry about the x-axis, the y-axis, or the origin. Then sketch the graph.*

13. $y = x^2 + 1$.

14. $x = y^4$.

15. $2x + y^2 = 4$.

16. $y = x - x^3$.

17. $y = f(x) = 3x - x^3$.

18. $x^2 + y^2 = 16$.

19. $4x^2 + y^2 = 16$.

20. $x^2 - y^2 = 1$.

3.2 ASYMPTOTES

Besides having symmetry, the graph of a function may also have *asymptotes*. To describe this concept, we must first consider certain types of limits.

In Fig. 3.6 you can see that as *x* approaches 0 through values *greater*

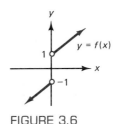

FIGURE 3.6

than 0 (or from the *right-hand side* of 0), *f(x)* approaches 1. We symbolize this **one-sided limit** by

$$\lim_{x \to 0^+} f(x) = 1,$$

where $x \to 0^+$ is read "*x* approaches 0 *from the right*." On the other hand, as *x* approaches 0 *from the left* (that is, through values less than 0), *f(x)* approaches -1. We write this one-sided limit as

$$\lim_{x \to 0^-} f(x) = -1.$$

Thus as $x \to 0$, some function values approach 1, while others approach -1. They do not approach the same number. For the limit to exist as $x \to 0$, $f(x)$

must approach *one* number only, regardless of from which side x approaches 0. Thus

$$\lim_{x \to 0} f(x) \text{ does not exist.}$$

One-sided limits are useful in describing the behavior of curves. In Fig. 3.7(a) notice that as $x \to a^+$, no limit exists because $f(x)$ increases without

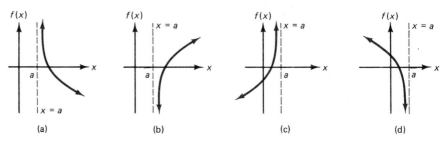

FIGURE 3.7

bound, that is, becomes positively infinite:

$$\lim_{x \to a^+} f(x) = \infty.$$

In Fig. 3.7(b), as $x \to a^+$, $f(x)$ becomes negatively infinite:

$$\lim_{x \to a^+} f(x) = -\infty.$$

In Figs. 3.7(c) and (d), we have

$$\lim_{x \to a^-} f(x) = \infty \quad \text{and} \quad \lim_{x \to a^-} f(x) = -\infty,$$

respectively. In each of these four situations, the indicated one-sided limit does not exist since $f(x)$ does not approach a finite number. The symbols ∞ and $-\infty$ indicate *why* there are no limits.

Furthermore, observe that each graph in Fig. 3.7 has an "explosion" around the vertical line $x = a$ in the sense that a one-sided limit of $f(x)$ at a is either ∞ or $-\infty$. The line $x = a$ is called a *vertical asymptote* for the graph. A vertical asymptote is not part of the graph but is a useful aid in sketching it because part of the graph approaches the asymptote. The explosion around $x = a$ causes the function to be discontinuous at a.

Definition

*The line $x = a$ is a **vertical asymptote** for the graph of the function f if and only if at least one of the following is true:*

$$\lim_{x \to a^+} f(x) = \infty \ (\text{or} -\infty)$$

$$or \quad \lim_{x \to a^-} f(x) = \infty \ (\text{or} -\infty).$$

To determine vertical asymptotes, we must find values of x around which $f(x)$ increases or decreases without bound. For a rational function (a quotient of two polynomials), these x-values are precisely those for which the denominator is zero but the numerator is not zero. For example, consider the rational function

$$f(x) = \frac{x + 1}{x}.$$

When x is 0, the denominator is 0 but the numerator is not. In Table 3.2 observe that the closer x is to 0 from the right, the larger $\frac{x + 1}{x}$ is, so

$$\lim_{x \to 0^+} \frac{x + 1}{x} = \infty.$$

Thus the line $x = 0$ (the y-axis) is a vertical asymptote (Fig. 3.8). Moreover, Table 3.2 shows that the closer x is to 0 from the left, the more negative

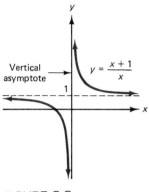

Vertical asymptote $y = \dfrac{x + 1}{x}$

FIGURE 3.8

TABLE 3.2

x	$f(x)$	x	$f(x)$
0.01	101	-0.01	-101
0.001	1,001	-0.001	$-1,001$
0.0001	10,001	-0.0001	$-10,001$

$\dfrac{x + 1}{x}$ is. Thus

$$\lim_{x \to 0^-} \frac{x + 1}{x} = -\infty.$$

We conclude that $f(x)$ increases without bound as $x \to 0^+$ and decreases without bound as $x \to 0^-$.

In summary, we have a rule for vertical asymptotes.

VERTICAL ASYMPTOTE RULE FOR RATIONAL FUNCTIONS

Suppose that $f(x) = P(x)/Q(x)$, where P and Q are polynomial functions. The line $x = a$ is a vertical asymptote for the graph of f if and only if $Q(a) = 0$ and $P(a) \neq 0$.

EXAMPLE 1   *Determine vertical asymptotes for the graph of*

$$f(x) = \frac{x^2 - 4x}{x^2 - 4x + 3}.$$

Since f is a rational function, the vertical asymptote rule applies. We have

$$f(x) = \frac{x(x-4)}{(x-3)(x-1)},$$

so the denominator is 0 when x is 3 or 1. Neither of these values makes the numerator 0. Thus the lines $x = 3$ and $x = 1$ are vertical asymptotes (see Fig. 3.9).

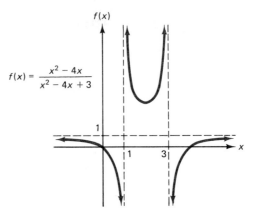

$$f(x) = \frac{x^2 - 4x}{x^2 - 4x + 3}$$

FIGURE 3.9

A curve $y = f(x)$ may have another kind of asymptote. In Fig. 3.10(a), as x increases without bound ($x \to \infty$), the graph approaches the horizontal

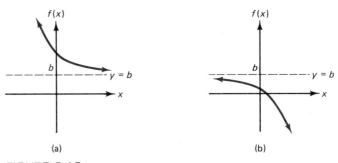

(a)   (b)

FIGURE 3.10

line $y = b$. That is,

$$\lim_{x \to \infty} f(x) = b.$$

In Fig. 3.10(b), as x becomes negatively infinite, the graph approaches the horizontal line $y = b$. That is,

$$\lim_{x \to -\infty} f(x) = b.$$

In each case the line $y = b$ is called a *horizontal asymptote* for the graph. It is a horizontal line around which the graph "settles" either as $x \to \infty$ or as $x \to -\infty$.

Although the graph of a horizontal line settles around itself as $x \to \infty$ or as $x \to -\infty$, a line is not considered to have asymptotes. In summary, we have the following definition.

Definition

Let f be a nonlinear function. The line $y = b$ is a **horizontal asymptote** *for the graph of f if and only if at least one of the following is true:*

$$\lim_{x \to \infty} f(x) = b \quad or \quad \lim_{x \to -\infty} f(x) = b.$$

To determine horizontal asymptotes, we must find the limits of $f(x)$ as $x \to \infty$ and as $x \to -\infty$. To illustrate, we again consider

$$f(x) = \frac{x + 1}{x}.$$

Since this is a rational function, we can use the procedures of Sec. 1.9 to find these limits.

$$\lim_{x \to \infty} \frac{x + 1}{x} = \lim_{x \to \infty} \frac{x}{x} = \lim_{x \to \infty} 1 = 1.$$

Thus the line $y = 1$ is a horizontal asymptote (refer back to Fig. 3.8). Also,

$$\lim_{x \to -\infty} \frac{x + 1}{x} = \lim_{x \to -\infty} \frac{x}{x} = \lim_{x \to -\infty} 1 = 1.$$

Therefore, the graph settles down near the horizontal line $y = 1$ both as $x \to \infty$ and $x \to -\infty$.

EXAMPLE 2   *Find horizontal asymptotes for the graph of*
$f(x) = \dfrac{x^2 - 4x}{x^2 - 4x + 3}.$

We have

$$\lim_{x \to \infty} \frac{x^2 - 4x}{x^2 - 4x + 3} = \lim_{x \to \infty} \frac{x^2}{x^2} = \lim_{x \to \infty} 1 = 1.$$

Thus the line $y = 1$ is a horizontal asymptote. The same result is obtained as $x \to -\infty$ (refer back to Fig. 3.9).

When the numerator of a rational function has degree greater than that of the denominator, no limit exists as $x \to \infty$ or $x \to -\infty$. From this we conclude that *whenever the degree of the numerator of a rational function is greater than the degree of the denominator, the graph of the function cannot have a horizontal asymptote.*

EXAMPLE 3   *Find vertical and horizontal asymptotes for the graph of* $y = f(x) = x^3 + 2x$.

This is a rational function with denominator 1, which is never zero. By the vertical asymptote rule, there are no vertical asymptotes. Because the degree of the numerator (3) is greater than the degree of the denominator (0), there are no horizontal asymptotes. However, let us examine the behavior of the graph of f as $x \to \infty$ and $x \to -\infty$.

$$\lim_{x \to \infty} (x^3 + 2x) = \lim_{x \to \infty} x^3 = \infty$$

and $$\lim_{x \to -\infty} (x^3 + 2x) = \lim_{x \to -\infty} x^3 = -\infty.$$

Thus as $x \to \infty$, the graph must extend indefinitely upward, and as $x \to -\infty$, the graph must extend indefinitely downward (see Fig. 3.11).

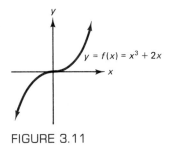

FIGURE 3.11

The results in Example 3 can be generalized to any polynomial function.

> A polynomial function has neither a horizontal nor a vertical asymptote.

EXAMPLE 4   *Sketch the graph of* $y = \dfrac{x^2}{x^2 - 1}$ *with the aid of symmetry and asymptotes.*

Symmetry Since y is a function of x and it is not the zero function, the graph is *not* symmetric about the x-axis.
    Testing for y-axis symmetry, we replace x by $-x$:

$$y = \frac{(-x)^2}{(-x)^2 - 1}, \qquad y = \frac{x^2}{x^2 - 1}.$$

The graph *is* symmetric about the y-axis. *Symmetry about exactly one axis implies that the graph cannot be symmetric about the origin.* (This follows from the last paragraph of Sec. 3.1.)

Asymptotes *Horizontal:* Since y is a rational function,

$$\lim_{x \to \infty} \frac{x^2}{x^2 - 1} = \lim_{x \to \infty} \frac{x^2}{x^2} = \lim_{x \to \infty} 1 = 1.$$

Therefore, as $x \to \infty$ the graph approaches the line $y = 1$, a horizontal asymptote. By symmetry, as $x \to -\infty$ the graph again approaches the line $y = 1$.

Vertical: We use the vertical asymptote rule. Since

$$y = \frac{x^2}{x^2 - 1} = \frac{x^2}{(x + 1)(x - 1)},$$

when x is -1 or 1 the denominator is zero and the numerator is not zero. Thus the lines $x = -1$ and $x = 1$ are vertical asymptotes.

Discussion We plot the y-intercept and some other points on the graph where $x > 0$, especially points near the asymptotes. By using properties of symmetry and asymptotes, it is relatively easy to sketch the graph (see Fig. 3.12). If there is any doubt as to the behavior of the graph around an asymptote, further analysis of limits may be required.

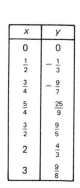

x	y
0	0
$\frac{1}{2}$	$-\frac{1}{3}$
$\frac{3}{4}$	$-\frac{9}{7}$
$\frac{5}{4}$	$\frac{25}{9}$
$\frac{3}{2}$	$\frac{9}{5}$
2	$\frac{4}{3}$
3	$\frac{9}{8}$

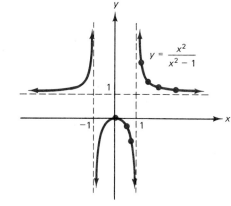

FIGURE 3.12

EXERCISE 3.2

In Problems 1–12, find the horizontal and vertical asymptotes for the graphs of the functions. Do not sketch the graphs.

1. $f(x) = \dfrac{x - 1}{2x + 3}$.

2. $y = \dfrac{2x + 1}{2x - 1}$.

3. $y = \dfrac{4}{x}$.

4. $y = -\dfrac{4}{x^2}$.

5. $y = x^3 - 5x + 8$.

6. $y = \dfrac{x^3}{x^2 - 9}$.

7. $f(x) = \dfrac{2x^2}{x^2 + x - 6}$.

8. $f(x) = \dfrac{x^2}{5}$.

9. $y = \dfrac{4}{x - 6} + 4$.

10. $f(x) = \dfrac{5}{2x^2 - 9x + 4}$.

11. $f(x) = \dfrac{3 - x^4}{x^3 + x^2}$.

12. $f(x) = \dfrac{x^2 + x}{x}$.

In Problems 13–20, determine whether the graphs are symmetric about the x-axis, y-axis, or origin; find horizontal and vertical asymptotes; sketch the graphs.

13. $y = \dfrac{3}{x - 1}$.

14. $y = \dfrac{x}{4 - x}$.

15. $f(x) = \dfrac{8}{x^3}$.

16. $f(x) = \dfrac{1}{x^4}$.

17. $f(x) = \dfrac{1}{x^2 - 1}$.

18. $f(x) = \dfrac{x^2}{x^2 - 4}$.

19. $y = \dfrac{x^2 - 1}{x^2 - 4}$.

20. $y = \dfrac{x^3 - x}{x}$.

21. In discussing the time pattern of purchasing, Mantell and Sing **[23]** use the curve

$$y = \frac{x}{a + bx}$$

as a mathematical model. They claim that $y = 1/b$ is an asymptote. Verify this.

22. The population N of a certain small city t years from now is predicted to be given by

$$N = 20{,}000 + \frac{10{,}000}{(t + 2)^2}.$$

The graph of this function has a horizontal asymptote, which corresponds to the population in the long run. Find this asymptote.

23. Let $\bar{c} = (5000 + 6q)/q$ be an average cost function. By finding $\lim\limits_{q \to \infty} \bar{c}$, show that the average cost approaches a level of stability if the producer continually increases output. What is the limiting value of the average cost? Sketch the graph of the average cost function. Note that $q > 0$.

3.3 RELATIVE EXTREMA

In our next discussion, we shall use the term *interval* to describe certain sets of (real) numbers. For example, if $a < b$, the set of all numbers x between a and b is denoted (a, b) and is called an **open interval**. The numbers a and b are called **endpoints** of the interval, but they are not in the set. Thus the notation (a, b) represents all x such that $a < x < b$. The set of all x where $a \le x \le b$ is called a **closed interval** and is denoted $[a, b]$. The brackets indicate that the endpoints *are* included. Figure 3.13 shows both types of intervals on a number line.

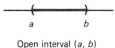

Open interval (a, b) Closed interval $[a, b]$

FIGURE 3.13

Other types of intervals are shown in Fig. 3.14, where the symbols ∞ and $-\infty$ are not numbers but merely a convenience for indicating that an interval extends indefinitely in some direction. The interval $(-\infty, \infty)$ is the set of all real numbers. If a function is continuous at each point in an interval, then the function is said to be *continuous on that interval*.

In curve sketching, just plotting points may not give enough information about a curve's shape. For example, the points $(-1, 0)$, $(0, -1)$, and $(1, 0)$ satisfy the equation $y = (x + 1)^3 (x - 1)$. Based on these points you might

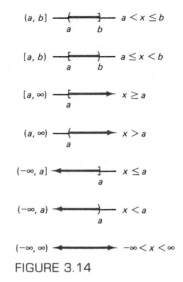

FIGURE 3.14

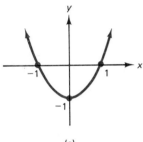

(a)

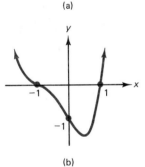

(b)

FIGURE 3.15

hastily conclude that the graph should appear as in Fig. 3.15(a), but in fact the actual shape is given in Fig. 3.15(b). Even using information about symmetry and asymptotes would be of little help here. In this section and the next one, we shall explore the powerful role that differentiation plays in analyzing a function so that we may determine the true shape and behavior of its graph.

We begin by analyzing the graph of the function $y = f(x)$ in Fig. 3.16. Notice that as x increases (goes from left to right) on the interval I_1 between a and b, the values of $f(x)$ increase and the curve is rising. Symbolically, this observation means that if x_1 and x_2 are any two points in I_1 where $x_1 < x_2$, then $f(x_1) < f(x_2)$. Here f is said to be an *increasing function* on I_1. On the other hand, as x increases on the interval I_2 between c and d, the curve is falling. Here $x_3 < x_4$ implies $f(x_3) > f(x_4)$, and f is said to be a *decreasing function* on I_2. In general we have the following definition.

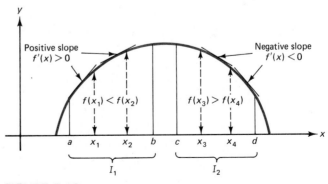

FIGURE 3.16

Definition

*A function f is an **increasing function** on the interval I if and only if for any two points x_1, x_2 in I such that $x_1 < x_2$, then $f(x_1) < f(x_2)$. A function f is a **decreasing function** on the interval I if and only if for any two points x_1, x_2 in I such that $x_1 < x_2$, then $f(x_1) > f(x_2)$.*

A function is said to be increasing (decreasing) at a *point* x_0 if there is an open interval containing x_0 on which the function is increasing (decreasing). Thus in Fig. 3.16, f is increasing at x_1 and is decreasing at x_3.

Turning again to Fig. 3.16, we note that over the interval I_1, tangent lines to the curve have positive slopes, so $f'(x)$ must be positive for all x in I_1. Over the interval I_2, tangent lines have negative slopes, so $f'(x) < 0$ for all x in I_2. These facts lead to the following rule, which allows us to use the derivative to determine when a function is increasing or decreasing.

RULE 1

If $f'(x) > 0$ for all x in an interval I, then f is an increasing function on I. If $f'(x) < 0$ for all x in I, then f is a decreasing function on I.

To determine when a function is increasing or decreasing by applying Rule 1, we must determine intervals on which the derivative is either always positive or always negative. To do this, we make use of an important fact: *If a function g is continuous on an interval and g is never zero on that interval, then g(x) has the same sign throughout the interval.* That is, either $g(x) > 0$ or $g(x) < 0$ for all x in the interval. As a result, we can determine the sign of $g(x)$ over the entire interval simply by determining the sign of $g(x)$ at any point in the interval. For example, in Fig. 3.15(a), on the interval $(-1, 1)$ g is continuous and $g(x)$ is not zero. Because 0 is in $(-1, 1)$ and $g(0)$ is negative (-1), it follows that $g(x)$ is negative *throughout* $(-1, 1)$.

To further illustrate these ideas, we shall use Rule 1 to find the intervals on which $y = 18x - \frac{2}{3}x^3$ is increasing or decreasing. Letting $y = f(x)$ we have

$$f'(x) = 18 - 2x^2 = 2(9 - x^2) = 2(3 + x)(3 - x).$$

From the factored form of $f'(x)$ we see that $f'(x)$ is 0 only when x is -3 or 3. Therefore $f'(x)$ is not 0 on the intervals where $x < -3$, $-3 < x < 3$, or $x > 3$ (see Fig. 3.17). On any particular one of these intervals, the function f' is continuous and is not 0, so $f'(x)$ must have the same sign throughout the interval. For example, -5 is in $(-\infty, -3)$ and

$$f'(-5) = 2(3 - 5)(3 + 5) = 2(-)(+) = -.$$

[Note how we conveniently found the sign of $f'(x)$ by using the signs of its factors.] Since $f'(-5)$ is negative, $f'(x)$ is negative *throughout* $(-\infty, -3)$.

FIGURE 3.17

Thus by Rule 1, f is decreasing on $(-\infty, -3)$. Similarly, 0 is in the interval $(-3, 3)$ and

$$f'(0) = 2(3 + 0)(3 - 0) = 2(+)(+) = +.$$

Since $f'(0)$ is positive, $f'(x)$ is positive *throughout* $(-3, 3)$, so f is increasing on $(-3, 3)$. Finally, 5 is in $(3, \infty)$ and

$$f'(5) = 2(3 + 5)(3 - 5) = 2(+)(-) = -.$$

Thus f is decreasing *throughout* $(3, \infty)$ [see Fig. 3.18(a)].

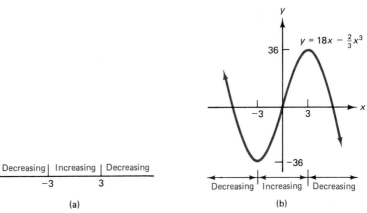

(a)

(b)

FIGURE 3.18

In summary, f is decreasing on $(-\infty, -3)$ and $(3, \infty)$, and is increasing on $(-3, 3)$. This corresponds to the rising and falling nature of the graph in Fig. 3.18(b). These results could be sharpened. Actually, by definition, f is decreasing on $(-\infty, -3]$ and $[3, \infty)$, and increasing on $[-3, 3]$. However, for our purposes open intervals are sufficient. *It will be our practice to determine* ***open*** *intervals on which a function is increasing or decreasing.*

Look now at the graph of $y = f(x)$ in Fig. 3.19. Some observations can be made. First, there is something special about the points P_1, P_2, and P_3.

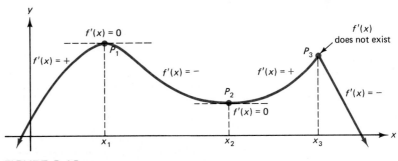

FIGURE 3.19

Notice that P_1 is *higher* than any other "nearby" point on the curve—likewise for P_3. The point P_2 is *lower* than any other "nearby" point on the curve. Since P_1, P_2, and P_3 may not necessarily be the highest or lowest points on the *entire* curve, we simply say that the graph of f has a *relative maximum* (*point*) when $x = x_1$ and when $x = x_3$, and has a *relative minimum* (*point*) when $x = x_2$. On the function level, f has a *relative maximum* (*value*) when $x = x_1$ and $x = x_3$, and has a *relative minimum* (*value*) when $x = x_2$. When we refer to a relative maximum or minimum, it is understood to refer to a point or value, depending on the context. Turning back to the graph, we see that there is an *absolute maximum* (highest point on the entire curve) when $x = x_1$, but there is no *absolute minimum* (lowest point on the entire curve), since the curve is assumed to extend downward indefinitely. We define these new terms as follows:

Definition

*A function f has a **relative maximum** when x = x_0 if there is an open interval containing x_0 on which $f(x_0) \geq f(x)$ for all x in the interval. The relative maximum is $f(x_0)$. A function f has a **relative minimum** when x = x_0 if there is an open interval containing x_0 on which $f(x_0) \leq f(x)$ for all x in the interval. The relative minimum is $f(x_0)$.*

Definition

*A function f has an **absolute maximum** when x = x_0 if $f(x_0) \geq f(x)$ for all x in the domain of f. The absolute maximum is $f(x_0)$. A function f has an **absolute minimum** when x = x_0 if $f(x_0) \leq f(x)$ for all x in the domain of f. The absolute minimum is $f(x_0)$.*

We refer to either a relative maximum or a relative minimum as a **relative extremum** (plural: *relative extrema*). Similarly, we speak of **absolute extrema.**

When dealing with relative extrema, we compare the function value at a point to those of nearby points; however, when dealing with absolute extrema, we compare the function value at a point to all others determined by the domain. Thus relative extrema are "local" in nature, while absolute extrema are "global" in nature.

In Fig. 3.19 we observe that at a relative extremum the derivative may not be defined (as when $x = x_3$). But whenever it is defined at a relative extremum, it is 0 (as when $x = x_1$ and $x = x_2$), and hence the tangent line is horizontal. We may state:

RULE 2

If f has a relative extremum when $x = x_0$, then $f'(x_0) = 0$ or $f'(x_0)$ is not defined.

From Rule 2 it should be clear that relative extrema *may* occur at points on the graph of f where $f'(x) = 0$ or where f' is not defined. These points are called *critical points* and their x-coordinates are called *critical values*.

Definition

*If x_0 is in the domain of f and either $f'(x_0) = 0$ or $f'(x_0)$ is not defined, then x_0 is called a **critical value** of f. If x_0 is a critical value, then $(x_0, f(x_0))$ is called a **critical point**.*

Thus at a critical point, there may be a relative maximum, a relative minimum, or neither. Moreover, from Fig. 3.19, we observe that each relative extremum occurs at a point around which the sign of $f'(x)$ is changing. For the relative maximum when $x = x_1$, $f'(x)$ goes from $+$ for $x < x_1$ to $-$ for $x > x_1$, *as long as x is near x_1*. At the relative minimum when $x = x_2$, $f'(x)$ goes from $-$ to $+$, and at the relative maximum when $x = x_3$, it again goes from $+$ to $-$. Thus *around relative maxima, f is increasing and then decreasing, and the reverse holds for relative minima*.

RULE 3

If x_0 is a critical value of f and $f'(x)$ changes from positive to negative as x increases through x_0, then f has a relative maximum when $x = x_0$. If $f'(x)$ changes from negative to positive as x increases through x_0, then f has a relative minimum when $x = x_0$.

Pitfall

Not every critical value corresponds to a relative extremum. For example, if $y = f(x) = x^3$, then $f'(x) = 3x^2$. Since $f'(0) = 0$ and $f(0)$ is defined, 0 is a critical value. If $x < 0$, then $3x^2 > 0$. If $x > 0$, then $3x^2 > 0$. Since $f'(x)$ does not change sign, no relative maximum or minimum exists. Indeed, since $f'(x) \geq 0$ for all x, the graph of f never falls and f is said to be *nondecreasing* (see Fig. 3.20).

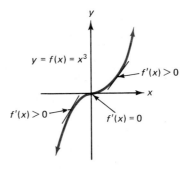

FIGURE 3.20

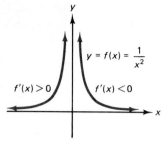

$y = f(x) = \dfrac{1}{x^2}$

$f'(x) > 0$ $f'(x) < 0$

FIGURE 3.21

It is important to understand that not every value of x where $f'(x)$ does not exist is a critical value. For example, if $y = f(x) = 1/x^2$, then $f'(x) = -2/x^3$. Although $f'(x)$ is not defined when $x = 0$, 0 is not a critical value because 0 is not in the domain of f. That is, no y-value corresponds to $x = 0$. Thus a relative extremum cannot occur when $x = 0$. Nevertheless, the derivative may change sign around any x-value where $f'(x)$ is not defined, so such values are important in determining intervals over which f is increasing or decreasing. If $x < 0$, then $f'(x) = -2/x^3 > 0$. If $x > 0$, $f'(x) = -2/x^3 < 0$. Thus f is increasing on $(-\infty, 0)$ and decreasing on $(0, \infty)$ (see Fig. 3.21).

From our discussions and the "Pitfall" above, you should realize that a critical value is only a "candidate" for a relative extremum. It may correspond to a relative maximum, a relative minimum, or neither.

Summarizing the results of this section, we have the *first-derivative test* for the relative extrema of $y = f(x)$:

FIRST-DERIVATIVE TEST FOR RELATIVE EXTREMA

1. Find $f'(x)$.

2. Determine all values of x where $f'(x) = 0$ or $f'(x)$ is not defined.

3. On the intervals suggested by the values in step 2, determine whether f is increasing ($f'(x) > 0$) or decreasing ($f'(x) < 0$).

4. For each critical value x_0, determine whether $f'(x)$ changes sign as x increases through x_0. There is a relative maximum when $x = x_0$ if $f'(x)$ changes from $+$ to $-$ going from left to right, and a relative minimum if $f'(x)$ changes from $-$ to $+$ going from left to right. If $f'(x)$ does not change sign, there is no relative extremum when $x = x_0$.

EXAMPLE 1   *If $y = f(x) = x + \dfrac{4}{x + 1}$, use the first-derivative test to find when relative extrema occur.*

1. $f(x) = x + 4(x + 1)^{-1}$, so

$$f'(x) = 1 + 4(-1)(x + 1)^{-2} = 1 - \frac{4}{(x + 1)^2}$$

$$= \frac{(x + 1)^2 - 4}{(x + 1)^2} = \frac{x^2 + 2x - 3}{(x + 1)^2} = \frac{(x + 3)(x - 1)}{(x + 1)^2}.$$

Note that we expressed $f'(x)$ as a quotient with numerator and denominator factored. This enables us in step 2 to easily determine when $f'(x)$ is 0 or not defined.

2. Setting $f'(x) = 0$ gives $x = -3, 1$. The denominator of $f'(x)$ is 0 when x is -1, so $f'(-1)$ does not exist. The values -3 and 1 are critical values, but -1 is not because $f(-1)$ is not defined.

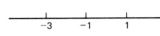

FIGURE 3.22

3. The three values in step 2 lead us to consider four intervals (Fig. 3.22). [On each of these intervals, $f'(x)$ is continuous and is not zero.]

$$\text{If } x < -3, \text{ then } f'(x) = \frac{(-)(-)}{(+)} = +, \text{ so } f \text{ is increasing;}$$

$$\text{if } -3 < x < -1, \text{ then } f'(x) = \frac{(+)(-)}{(+)} = -, \text{ so } f \text{ is decreasing;}$$

$$\text{if } -1 < x < 1, \text{ then } f'(x) = \frac{(+)(-)}{(+)} = -, \text{ so } f \text{ is decreasing;}$$

$$\text{if } x > 1, \text{ then } f'(x) = \frac{(+)(+)}{(+)} = +, \text{ so } f \text{ is increasing (Fig. 3.23).}$$

Thus f is increasing on the intervals $(-\infty, -3)$ and $(1, \infty)$, and is decreasing on $(-3, -1)$ and $(-1, 1)$.

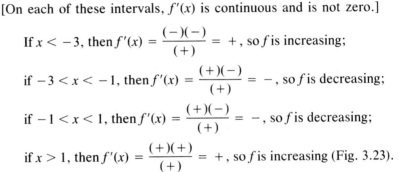

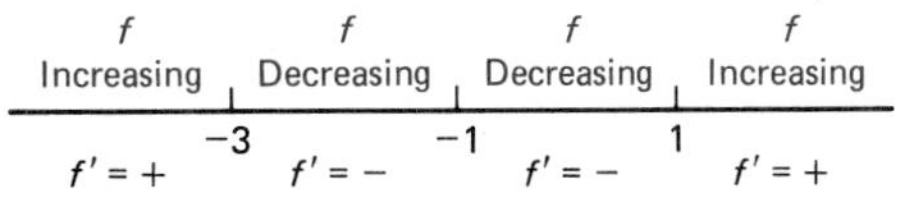

f	f	f	f
Increasing	Decreasing	Decreasing	Increasing

-3 -1 1

$f' = +$ $f' = -$ $f' = -$ $f' = +$

FIGURE 3.23

4. When $x = -3$, there is a relative maximum since $f'(x)$ changes from $+$ to $-$. [This relative maximum value is $f(-3) = -3 + (4/-2) = -5$.] When $x = 1$, there is a relative minimum since $f'(x)$ changes from $-$ to $+$. This relative value is $f(1) = 3$. We ignore $x = -1$ since -1 is not a critical value. The graph is shown in Fig. 3.24.

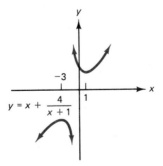

$$y = x + \frac{4}{x + 1}$$

FIGURE 3.24

EXAMPLE 2   *Test $y = f(x) = x^{2/3}$ for relative extrema.*

We have $f'(x) = \frac{2}{3}x^{-1/3} = 2/(3\sqrt[3]{x})$. When $x = 0$, then $f'(x)$ is not defined but $f(x)$ is defined. Thus 0 is a critical value and there are no other critical values. If $x < 0$, then $f'(x) < 0$. If $x > 0$, then $f'(x) > 0$. Therefore, there is a relative (as well as an absolute) minimum when $x = 0$ (see Fig. 3.25).

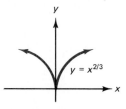

FIGURE 3.25

EXAMPLE 3   *Sketch the graph of $y = f(x) = 2x^2 - x^4$.*

Symmetry   Testing for y-axis symmetry, we have

$$y = 2(-x)^2 - (-x)^4, \quad \text{or} \quad y = 2x^2 - x^4.$$

Since this is the original equation, there is y-axis symmetry. Since y is a function (and not the zero function), there is no x-axis symmetry, and hence no symmetry about the origin.

Asymptotes   No horizontal or vertical asymptotes exist, since f is a polynomial function.

First-Derivative Test

1. $y' = 4x - 4x^3 = 4x(1 - x^2) = 4x(1 + x)(1 - x)$.

2. Setting $y' = 0$ gives the critical values $x = 0, \pm 1$. The critical points are $(-1, 1)$, $(0, 0)$, and $(1, 1)$. The y-coordinates of these points were found by substituting $x = 0, \pm 1$ into the *original* equation, $y = 2x^2 - x^4$.

3. There are four intervals to consider in Fig. 3.26:

$$\begin{array}{ccc} \underline{} & \underline{} & \underline{} \\ -1 & 0 & 1 \end{array}$$

FIGURE 3.26

If $x < -1$, then $y' = 4(-)(-)(+) = +$ and f is increasing;
if $-1 < x < 0$, then $y' = 4(-)(+)(+) = -$ and f is decreasing;
if $0 < x < 1$, then $y' = 4(+)(+)(+) = +$ and f is increasing;
if $x > 1$, then $y' = 4(+)(+)(-) = -$ and f is decreasing (Fig. 3.27).

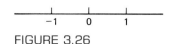

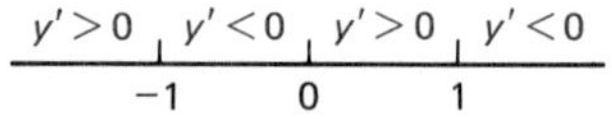

FIGURE 3.27

4. Relative maxima occur at $(-1, 1)$ and $(1, 1)$; a relative minimum occurs at $(0, 0)$.

Discussion In Fig. 3.28(a) we have plotted the horizontal tangents at the relative maximum and minimum points. We know the curve rises from the left, has a relative maximum, then falls, has a relative minimum, then rises to a relative maximum, and falls thereafter. A sketch is shown in Fig. 3.28(b).

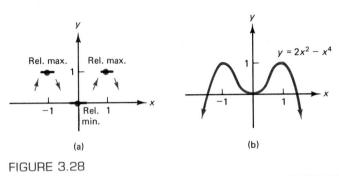

(a)   (b)

FIGURE 3.28

In Example 3 absolute maxima occur at $x = \pm 1$ [see Fig. 3.28(b)]. There is no absolute minimum.

If the domain of a function is an interval that contains an endpoint, to determine *absolute* extrema we must not only examine the function for relative extrema, but we must also take into consideration the values of $f(x)$ at the endpoints. Although endpoints are not considered when we look for relative maxima or minima, they may yield *absolute* maxima or minima. Example 4 will illustrate.

EXAMPLE 4   *Find when extrema (relative and absolute) occur for* $y = f(x) = x^2 - 4x + 5$ *on the closed interval* $[1, 4]$.

1. $f'(x) = 2x - 4 = 2(x - 2)$.

2. Setting $f'(x) = 0$ gives the critical value $x = 2$.

3. The intervals to consider are when $x < 2$ and when $x > 2$. If $x < 2$, then $f'(x) < 0$ and f is decreasing; if $x > 2$, then $f'(x) > 0$ and f is increasing.

4. Thus there is a relative minimum when $x = 2$. It occurs on the graph at the point $(2, 1)$ (see Fig. 3.29).

5. Since f is decreasing for $x < 2$, an absolute maximum may *possibly* occur at the left-hand endpoint of the domain of f, that is, when $x = 1$. Similarly, since f is increasing for $x > 2$, an absolute maximum may *possibly* occur at the right-hand endpoint, that is, when $x = 4$. Testing the endpoints, we have $f(1) = 2$ and $f(4) = 5$. Noting that $f(4) > f(1)$, we conclude that an absolute maximum occurs when $x = 4$. When $x = 2$ there is an absolute, as well as a relative, minimum.

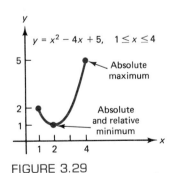

FIGURE 3.29

You have seen that if f is a function such that $f'(x) > 0$ on an interval I, then f is increasing there. In fact, f' gives the rate of increase of f. However, this rate itself might be increasing or decreasing. That is, $[f'(x)]'$, or more simply $f''(x)$, may be positive or negative. If $f''(x) > 0$ on I, then we say that f is increasing *at an increasing rate* there. If $f''(x) < 0$, then f is increasing *at a decreasing rate*. On the other hand, if $f'(x) < 0$ on I, then f is decreasing there. We say that f is decreasing at an increasing or decreasing rate depending, respectively, on whether we have $f''(x) > 0$ or $f''(x) < 0$ on I. In the next section you will see how these concepts relate to the graph of a function.

EXAMPLE 5   *Show that* $y = f(x) = \sqrt{x}$ *is increasing for* $x > 0$ *and determine whether it is increasing at an increasing or decreasing rate there.*

Since $f'(x) = \frac{1}{2}x^{-1/2} = 1/(2\sqrt{x})$, then $f'(x) > 0$ for $x > 0$. Thus f is increasing on $(0, \infty)$. To determine if f is increasing at an increasing or decreasing rate, we must consider the sign of $f''(x)$ on $(0, \infty)$.

$$f''(x) = -\frac{1}{4}x^{-3/2} = -\frac{1}{4\sqrt{x^3}}.$$

Since $f''(x) < 0$ for $x > 0$, then f is increasing at a *decreasing* rate on $(0, \infty)$.

EXERCISE 3.3

*In Problems **1–22**, determine when the function is increasing or decreasing and determine when relative maxima and minima occur. Do not sketch the graph.*

1. $y = x^2 + 2$.

2. $y = x^2 + 4x + 3$.

3. $y = x - x^2 + 2$.

4. $y = 4x - x^2$.

5. $y = -\dfrac{x^3}{3} - 2x^2 + 5x - 2$.

6. $y = 4x^3 - 3x^4$.

7. $y = x^4 - 2x^2$.

8. $y = -2 + 12x - x^3$.

9. $y = x^3 - 6x^2 + 9x$.

10. $y = x^3 - 6x^2 + 12x - 6$.

11. $y = 3x^5 - 5x^3$.

12. $y = 5x - x^5$.

13. $y = -x^5 - 5x^4 + 200$.

14. $y = 3x^4 - 4x^3 + 1$.

15. $y = \dfrac{1}{x - 1}$.

16. $y = \dfrac{3}{x}$.

17. $y = \dfrac{10}{\sqrt{x}}$.

18. $y = \dfrac{x}{x + 1}$.

19. $y = \dfrac{x^2}{1 - x}$.

20. $y = x + \dfrac{4}{x}$.

21. $y = (x + 2)^3(x - 5)^2$.

22. $y = x^2(x + 3)^4$.

*In Problems **23–34**, determine: intervals on which the functions are increasing or decreasing; relative maxima and minima; symmetry; horizontal and vertical asymptotes. Then sketch the graphs.*

23. $y = x^2 - 6x - 7$.

24. $y = 2x^2 - 5x - 12$.

25. $y = 3x - x^3$.

26. $y = x^4 - 16$.

27. $y = 2x^3 - 9x^2 + 12x$.

28. $y = x^3 - 9x^2 + 24x - 19$.

29. $y = x^4 + 4x^3 + 4x^2$.

30. $y = x^5 - \frac{5}{4}x^4$.

31. $y = \dfrac{x + 1}{x - 1}$.

32. $y = \dfrac{x^2}{x^2 + 1}$.

33. $y = \dfrac{x^2}{x + 3}$.

34. $y = x + \dfrac{1}{x}$.

*In Problems **35–40**, find when absolute maxima and minima occur for the given function on the given interval.*

35. $f(x) = x^2 - 2x + 3$, $[-1, 2]$.

36. $f(x) = -2x^2 - 6x + 5$, $[-2, 3]$.

37. $f(x) = \frac{1}{3}x^3 - x^2 - 3x + 1$, $[0, 2]$.

38. $f(x) = \frac{1}{4}x^4 - \frac{3}{2}x^2$, $[0, 1]$.

39. $f(x) = 4x^3 + 3x^2 - 18x + 3$, $[\frac{1}{2}, 3]$.

40. $f(x) = x^{4/3}$, $[-8, 8]$.

41. A manufacturer's *fixed costs* are those costs that under normal conditions do not depend on the level of production. Some examples are rent, officers' salaries, and normal maintenance. If fixed costs are $c_f = 25,000$, show that the average fixed cost function $\overline{c}_f = c_f/q$ is a decreasing function for $q > 0$. Thus as output q increases, each unit's portion of fixed cost declines.

42. If $c = 4q - q^2 + 2q^3$ is a cost function, when is marginal cost increasing?

43. Given the demand function $p = 400 - 2q$, find when marginal revenue is increasing.

44. For the cost function $c = \sqrt{q}$, show that marginal and average costs are always decreasing for $q > 0$.

45. A psychological experiment (adapted from [1]) was conducted to analyze human response to electrical shocks (stimuli). The subjects received shocks of various intensities. The response R to a shock of intensity I (in microamperes) was to be a number that indicated the perceived magnitude relative to that of a "standard" shock. The standard shock was assigned a magnitude of 10. If

$$R = \frac{I^{4/3}}{2500}, \qquad 500 \le I \le 3500,$$

show that R increases with respect to I and that it is increasing at an increasing rate.

46. In a discussion of thermal pollution in [10], the efficiency E of a power plant is given by

$$E = 0.71\left(1 - \frac{T_c}{T_h}\right),$$

where T_c and T_h are the respective absolute temperatures of the hotter and colder reservoirs. Assume that T_c is a positive constant and that T_h is positive. Using calculus, show that as T_h increases, the efficiency increases. Also show that E increases with respect to T_h at a decreasing rate.

47. To study a predator-prey relationship, Holling [21] conducted an experiment in which a blindfolded subject, the "predator," stood in front of a 3-foot square table on which uniform sandpaper disks, the "prey," were placed. For 1 minute the "predator" searched for the disks by tapping with a finger. As a disk was found, it was removed and searching resumed. The experiment was repeated at various disk densities (number of disks per 9 square feet). It was estimated that y disks were picked up in 1 minute when x disks were on the table, where

$$y = \frac{0.70x}{1 + 0.03x}, \qquad x \ge 0.$$

Holling states that y increases at a decreasing rate as the density of the disks (x) increases.

a. Verify this claim.

b. Sketch the graph of the equation.

48. For the predator-prey relationship of Problem 47, y/x gives the proportion of prey "destroyed" at a given prey density ($x > 0$). Show that y/x is a decreasing function of x, and sketch the graph of this function.

49. For a certain population, if E is the number of years of a person's education and S represents a numerical value of status based on that educational level, then $S = 0.25 E^2$.

a. Determine the rate of change and the relative rate of change of status with respect to education when $E = 10$.

b. Show that as E increases, then the rate of change of status with respect to education increases, but the relative rate of change decreases.

50. In a discussion of the pricing of local telephone service, Renshaw[24] determines that total revenue r is given by

$$r = 2F + \left(1 - \frac{a}{b}\right)p - p^2 + \frac{a^2}{b},$$

where p is an indexed price per call, and a, b, and F are constants. Determine the value of p that maximizes revenue.

51. Eswaran and Kotwal[12] consider agrarian economies in which there are two types of workers, permanent and casual. Permanent workers are employed on long-term contracts and may receive benefits such as holiday gifts and emergency aid. Casual workers are hired on a daily basis and perform routine and menial tasks such as weeding, harvesting, and threshing. The difference z in the present-value cost of hiring a permanent worker over that of hiring a casual worker is given by

$$z = (1 + b)w_p - bw_c,$$

where w_p and w_c are wage rates for permanent labor and casual labor, respectively; b is a positive constant; and w_p is a function of w_c.

a. Show that

$$\frac{dz}{dw_c} = (1 + b)\left[\frac{dw_p}{dw_c} - \frac{b}{1 + b}\right].$$

b. If $dw_p/dw_c < b/(1 + b)$, show that z is a decreasing function of w_c.

3.4 CONCAVITY

You have seen that the first derivative provides much information for sketching curves. It is used to determine when a function is increasing or decreasing and to locate relative maxima and minima. However, to be sure that we know the true shape of a curve, we may need more information. For example, consider the curve $y = f(x) = x^2$. Since $f'(x) = 2x$, $x = 0$ is a critical value. If $x < 0$, then $f'(x) < 0$ and f is decreasing; if $x > 0$, then $f'(x) > 0$ and f is increasing. Thus there is a relative minimum when $x = 0$. Figure 3.30(a) and (b) both meet the preceding conditions. But which one truly describes the curve? This question will easily be settled by using the second derivative and the notion of *concavity*.

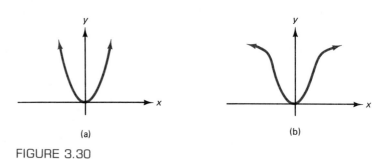

(a)   (b)

FIGURE 3.30

In Fig. 3.31 note that each curve $y = f(x)$ "bends" (or opens) upward. This means that if tangent lines are drawn to each curve, the curves lie above

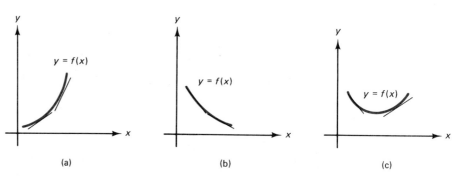

(a)   (b)   (c)

FIGURE 3.31

them. Moreover, the slopes of the tangent lines *increase* in value as x increases. In part (a) the slopes go from small positive values to larger values; in part (b) they are negative and approaching zero (thus increasing); in part (c) they pass from negative values to positive values. Since $f'(x)$ gives the slope at a point, an increasing slope means that f' itself must be an increasing function. To describe this, each curve (or function f) is said to be *concave up*.

In Fig. 3.32 it can be seen that each curve lies below the tangent lines and the curves are bending downward. As x increases, the slopes of the tangent lines are *decreasing*. Thus f' must be a decreasing function here, and we say that f is *concave down*.

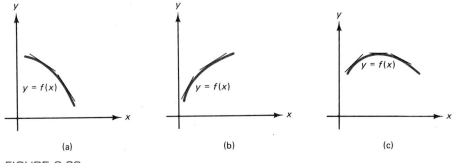

(a)   (b)   (c)

FIGURE 3.32

Definition

*A function f is said to be **concave up** [**concave down**] on an interval I if f' is increasing [decreasing] on I.*

Remember: *If f is concave up on an interval I, then geometrically its graph is bending upward there. If f is concave down, then its graph is bending downward.*

Since f' is increasing when its derivative $f''(x)$ is positive and f' is decreasing when $f''(x)$ is negative, we can state the following rule:

RULE 4

If $f''(x) > 0$ for all x in an interval I, then f is concave up on I. If $f''(x) < 0$ for all x in I, then f is concave down on I.

A function f is also said to be concave up at a *point* x_0 if there exists an open interval around x_0 on which f is concave up. In fact, for the functions that we shall consider, if $f''(x_0) > 0$, then f is concave up at x_0.* Similarly, f is concave down at x_0 if $f''(x_0) < 0$.

---

* This is guaranteed for functions f such that f'' is continuous.

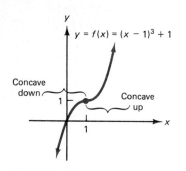

FIGURE 3.33

EXAMPLE 1

a. *Test* $y = f(x) = (x - 1)^3 + 1$ *for concavity.*

To apply Rule 4, we must examine the signs of y''. Now, $y' = 3(x - 1)^2$, so $y'' = 6(x - 1)$. Thus f is concave up when $6(x - 1) > 0$; that is, when $x > 1$. And f is concave down when $6(x - 1) < 0$; that is, when $x < 1$ (see Fig. 3.33).

b. *Test* $y = x^2$ *for concavity.*

We have $y' = 2x$ and $y'' = 2$. Because y'' is always positive, the graph of $y = x^2$ must always be concave up, as in Fig. 3.30(a). The graph cannot appear as in Fig. 3.30(b), for that curve is sometimes concave down.

A point on a graph, such as (1, 1) in Fig. 3.33, where concavity changes from downward to upward, or vice versa, is called an *inflection point*. Around such a point the sign of $f''(x)$ must go from $-$ to $+$ or from $+$ to $-$.

Definition

*A function f has an **inflection point** when $x = x_0$ if and only if x_0 is in the domain of f and f changes concavity at x_0.*

In summary, we have the following procedure for determining concavity and inflection points.

> To test a function for concavity and inflection points, first find the values of x where $f''(x)$ is 0 or undefined. These values of x determine intervals. On each interval determine whether $f''(x) > 0$ (f is concave up) or $f''(x) < 0$ (f is concave down). If $f''(x)$ is 0 or undefined at $x = x_0$, and x_0 also is in the domain of f, then f has an inflection point at $x = x_0$ provided that concavity changes around x_0.

EXAMPLE 2   *Test $y = 6x^4 - 8x^3 + 1$ for concavity and inflection points.*

We have $y' = 24x^3 - 24x^2$, so

$$y'' = 72x^2 - 48x = 24x(3x - 2).$$

To find when $y'' = 0$, we set each factor in y'' equal to 0. This gives $x = 0$, $\frac{2}{3}$. We also note that y'' is never undefined. Thus there are three intervals to consider (Fig. 3.34):

FIGURE 3.34

If $x < 0$, then $y'' = 24(-)(-) = +$, so the curve is concave up;

if $0 < x < \frac{2}{3}$, then $y'' = 24(+)(-) = -$, so the curve is concave down;

if $x > \frac{2}{3}$, then $y'' = 24(+)(+) = +$, so the curve is concave up (see Fig. 3.35).

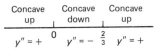

Concave up		Concave down		Concave up
$y'' = +$	0	$y'' = -$	$\frac{2}{3}$	$y'' = +$

FIGURE 3.35

Since concavity changes at $x = 0$ and $x = \frac{2}{3}$, which are in the domain of y, inflection points occur there (see Fig. 3.36). In summary, the curve is concave up on $(-\infty, 0)$ and $(\frac{2}{3}, \infty)$ and is concave down on $(0, \frac{2}{3})$. Inflection points occur when $x = 0$ or $x = \frac{2}{3}$. These points are $(0, 1)$ and $(\frac{2}{3}, -\frac{5}{27})$.

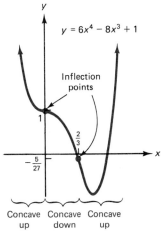

FIGURE 3.36

Pitfall

If $f''(x_0) = 0$ or f'' is not defined at x_0, this does not prove that the graph of f has an inflection point when $x = x_0$. For example, if $f(x) = x^4$, then $f''(x) = 12x^2$ and $f''(0) = 0$. But $x < 0$ implies $f''(x) > 0$, and $x > 0$ implies $f''(x) > 0$. Thus concavity does not change and there is no inflection point at $x = 0$ (see Fig. 3.37).

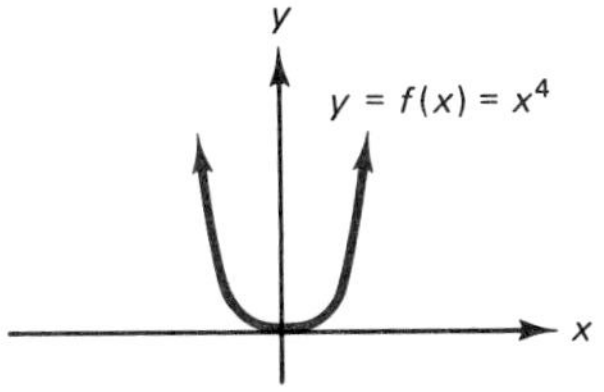

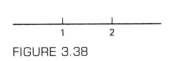

FIGURE 3.37

EXAMPLE 3 *Sketch the graph of* $y = 2x^3 - 9x^2 + 12x$.

Symmetry   None.

Asymptotes   Because y is a polynomial function, there are no asymptotes.

Relative Maxima and Minima   Letting $y = f(x)$, we have

$$f'(x) = 6x^2 - 18x + 12 = 6(x^2 - 3x + 2) = 6(x - 1)(x - 2).$$

The critical values are $x = 1, 2$ (see Fig. 3.38).

If $x < 1$, then $f'(x) = 6(-)(-) = +$, so f is increasing;
if $1 < x < 2$, then $f'(x) = 6(+)(-) = -$, so f is decreasing;
if $x > 2$, then $f'(x) = 6(+)(+) = +$, so f is increasing (see Fig. 3.39).

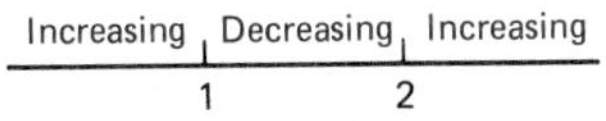

There is a relative maximum when $x = 1$ and a relative minimum when $x = 2$.

Concavity

$$f''(x) = 12x - 18 = 6(2x - 3).$$

FIGURE 3.38

FIGURE 3.39

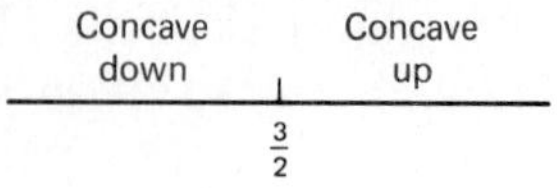

Concave down	Concave up

$$\frac{3}{2}$$

FIGURE 3.40

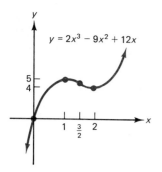

$y = 2x^3 - 9x^2 + 12x$

$\frac{5}{4}$

1   $\frac{3}{2}$ 2

FIGURE 3.41

Setting $f''(x) = 0$ gives a possible inflection point at $x = \frac{3}{2}$. When $x < \frac{3}{2}$, then $f''(x) < 0$ and f is concave down. When $x > \frac{3}{2}$, then $f''(x) > 0$ and f is concave up (see Fig. 3.40). Since concavity changes, there is an inflection point when $x = \frac{3}{2}$.

Discussion   We now find the coordinates of the important points on the graph (and any other useful points, such as the y-intercept.)

x	0	1	$\frac{3}{2}$	2
y	0	5	$\frac{9}{2}$	4

As x increases, the function is first concave down and increases to a relative maximum at $(1, 5)$; it then decreases to $(\frac{3}{2}, \frac{9}{2})$; it then becomes concave up but continues to decrease until it reaches a relative minimum at $(2, 4)$; thereafter it increases and is still concave up (see Fig. 3.41).

EXAMPLE 4   *Sketch the graph of $y = \dfrac{1}{4 - x^2}$.*

Symmetry   There is symmetry only about the y-axis: replacing x by $-x$ gives

$$y = \frac{1}{4 - (-x)^2} \quad \text{or} \quad y = \frac{1}{4 - x^2},$$

which is the same as the original equation.

Asymptotes   Testing for horizontal asymptotes, we have

$$\lim_{x \to \infty} \frac{1}{4 - x^2} = \lim_{x \to \infty} \frac{1}{-x^2} = -\lim_{x \to \infty} \frac{1}{x^2} = 0.$$

Similarly,

$$\lim_{x \to -\infty} \frac{1}{4 - x^2} = 0.$$

Thus $y = 0$ (the x-axis) is a horizontal asymptote. Since the denominator of $1/(4 - x^2)$ is 0 when $x = \pm 2$, and the numerator is not 0 for these values of x, the lines $x = 2$ and $x = -2$ are vertical asymptotes.

Relative Maxima and Minima   Since $y = (4 - x^2)^{-1}$,

$$y' = -1(4 - x^2)^{-2}(-2x) = \frac{2x}{(4 - x^2)^2}.$$

We see that y' is 0 when $x = 0$ and y' is undefined when $x = \pm 2$. But only 0 is a critical value. If $x < -2$, then $y' < 0$; if $-2 < x < 0$, then $y' < 0$; if $0 < x < 2$, then $y' > 0$; if $x > 2$, then $y' > 0$. The function is decreasing on

$(-\infty, -2)$ and $(-2, 0)$ and increasing on $(0, 2)$ and $(2, \infty)$ (see Fig. 3.42). There is a relative minimum when $x = 0$.

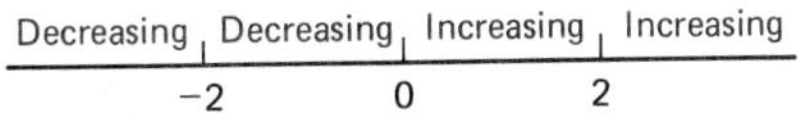

| Decreasing | Decreasing | Increasing | Increasing |

$$-2 \qquad 0 \qquad 2$$

FIGURE 3.42

Concavity

$$y'' = \frac{(4 - x^2)^2(2) - (2x)2(4 - x^2)(-2x)}{(4 - x^2)^4}$$
$$= \frac{2(4 - x^2)[(4 - x^2) - (2x)(-2x)]}{(4 - x^2)^4} = \frac{2(4 + 3x^2)}{(4 - x^2)^3}.$$

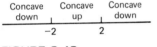

| Concave down | Concave up | Concave down |

$$-2 \qquad 2$$

FIGURE 3.43

Setting $y'' = 0$, we get no real roots. But y'' is undefined when $x = \pm 2$. Thus concavity may change around these values. If $x < -2$, then $y'' < 0$; if $-2 < x < 2$, then $y'' > 0$; if $x > 2$, then $y'' < 0$. The graph is concave up on $(-2, 2)$ and concave down on $(-\infty, -2)$ and $(2, \infty)$ (see Fig. 3.43). Although concavity changes around $x = \pm 2$, these values of x are not in the domain of the original function and hence do not give inflection points.

Discussion   Plotting the points in the table in Fig. 3.44, some arbitrarily chosen, and using the above information, we get the indicated graph. Due to symmetry, our table has only $x \geq 0$.

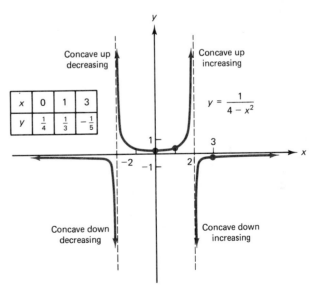

FIGURE 3.44

EXERCISE 3.4

*In Problems **1–10**, determine concavity and the x-values where inflection points occur. Do not sketch the graphs.*

1. $y = -2x^2 + 4x$.

2. $y = 3x^2 - 6x + 5$.

3. $y = 4x^3 + 12x^2 - 12x$.

4. $y = x^3 - 6x^2 + 9x + 1$.

5. $y = x^4 - 6x^2 + 5x - 6$.

6. $y = -\dfrac{x^4}{4} + \dfrac{9x^2}{2} + 2x$.

7. $y = \dfrac{x+1}{x-1}$.

8. $y = x + \dfrac{1}{x}$.

9. $y = \dfrac{x^2}{x+3}$.

10. $y = \dfrac{x^2}{x^2+1}$.

*In Problems **11–32**, sketch each curve. Determine: intervals on which the function is increasing, decreasing, concave up, concave down; relative maxima and minima; inflection points; symmetry; horizontal and vertical asymptotes.*

11. $y = x^2 + 4x + 3$.

12. $y = x^2 + 2$.

13. $y = 4x - x^2$.

14. $y = x - x^2 + 2$.

15. $y = x^3 - 9x^2 + 24x - 19$.

16. $y = 3x - x^3$.

17. $y = \dfrac{x^3}{3} - 4x$.

18. $y = x^3 - 6x^2 + 9x$.

19. $y = x^3 - 3x^2 + 3x - 3$.

20. $y = 2x^3 - 9x^2 + 12x$.

21. $y = 4x^3 - 3x^4$.

22. $y = -\dfrac{x^3}{3} - 2x^2 + 5x - 2$.

23. $y = 5x - x^5$.

24. $y = (3 + 2x)^3$.

25. $y = 3x^4 - 4x^3 + 1$.

26. $y = \dfrac{x^5}{100} - \dfrac{x^4}{20}$.

27. $y = \dfrac{3}{x}$.

28. $y = \dfrac{1}{x-1}$.

29. $y = \dfrac{x}{x+1}$.

30. $y = \dfrac{10}{\sqrt{x}}$.

31. $y = x^2 + \dfrac{1}{x^2}$.

32. $y = \dfrac{x^2}{1-x}$.

33. Show that the graph of the demand equation $p = 100/(q + 2)$ is decreasing and concave up for $q > 0$.

34. For the cost function $c = 3q^2 + 5q + 6$, show that the graph of the average cost function $\bar{c}$ is always concave up for $q > 0$.

35. The number of species of plants on a plot may depend on the size of the plot. For example, in Fig. 3.45 we see that on 1-square-meter plots there are three species (A, B, and C on the left plot; A, B, and D on the right

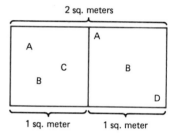

FIGURE 3.45

plot), and on a 2-square-meter plot there are four species (A, B, C, and D).

In a study (adapted from [**25**]) of rooted plants in a certain geographic region, it was determined that the average number of species, S, occurring on plots of size A (in square meters) is given by

$$S = f(A) = 12\sqrt[4]{A}, \qquad 0 \le A \le 625.$$

Sketch the graph of f. (*Note:* Your graph should be rising and concave down. Thus the number of species is increasing with respect to area, but at a decreasing rate.)

36. In a model of the effect of contraception on birth rate [**4**], the equation

$$R = f(x) = \dfrac{x}{4.4 - 3.4x}, \qquad 0 \le x \le 1$$

gives the proportional reduction R in the birth rate as a function of the efficiency x of a contraception method.

An efficiency of 0.2 (or 20%) means that the probability of becoming pregnant is 80% of the probability of becoming pregnant without the contraceptive. Find the reduction (as a percentage) when efficiency is (a) 0, (b) 0.5, and (c) 1. Find dR/dx and d^2R/dx^2 and sketch the graph of the equation.

37. If you were to recite members of a category, such as four-legged animals, the words that you utter would probably occur in "chunks" with distinct pauses between such chunks. For example, you might say the following for the category of four-legged animals:

dog, cat, mouse, rat,

(pause)

horse, donkey, mule,

(pause)

cow, pig, goat, lamb

etc.

The pauses may occur because one may have to mentally search for subcategories (animals around the house, beasts of burden, farm animals, etc.)

The elapsed time between onsets of successive words is called *interresponse time*. A function [26] has been used to analyze the length of time for pauses and chunk size (number of words in a chunk). This function f is such that

$$f(t) = \begin{cases} \text{the average number of words} \\ \text{that occur in succession with} \\ \text{interresponse times less than } t. \end{cases}$$

The graph of f has a shape similar to that in Fig. 3.46

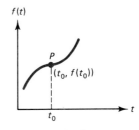

FIGURE 3.46

and is best fit by a third-degree polynomial:

$$f(t) = At^3 + Bt^2 + Ct + D.$$

The point P has special meaning. It is such that the value t_0 separates interresponse time *within* chunks from those *between* chunks. Mathematically, P is a critical point that is also a point of inflection. Assume these two conditions and show that (a) $t_0 = -B/(3A)$ and (b) $B^2 = 3AC$.

3.5 THE SECOND-DERIVATIVE TEST

The second derivative may be used to test certain critical values for relative extrema. Observe in Fig. 3.47 that when $x = x_0$ there is a horizontal tangent; that is, $f'(x_0) = 0$. This implies the possibility of a relative maximum or minimum. Moreover, we see that the curve is bending upward there [that is, $f''(x_0) > 0$]. This leads us to conclude that there is a relative minimum at x_0. On the other hand, $f'(x_1) = 0$ but the curve is bending downward at x_1 (that is, $f''(x_1) < 0$). From this we conclude that a relative maximum exists there.

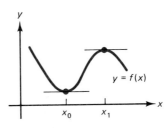

FIGURE 3.47

This technique of examining the second derivative at points where the first derivative is 0 is called the *second-derivative test* for relative extrema.

SECOND-DERIVATIVE TEST FOR RELATIVE EXTREMA

Suppose that $f'(x_0) = 0$.

If $f''(x_0) < 0$, then f has a relative maximum at x_0;
if $f''(x_0) > 0$, then f has a relative minimum at x_0.

The second-derivative test does *not* apply when $f''(x_0) = 0$. Under this condition, at x_0 there may be a relative maximum, a relative minimum, or neither. In such cases, the first-derivative test should be used to analyze what is happening at x_0.

EXAMPLE 1   *Test the following for relative maxima and minima. Use the second-derivative test if possible.*

a. $y = 18x - \frac{2}{3}x^3$.

$$y' = 18 - 2x^2 = 2(9 - x^2) = 2(3 + x)(3 - x).$$
$$y'' = -4x.$$

Solving $y' = 0$ gives critical values $x = \pm 3$. If $x = 3$, then $y'' = -4(3) = -12 < 0$. So there is a relative maximum when $x = 3$. If $x = -3$, then $y'' = -4(-3) = 12 > 0$, so there is a relative minimum when $x = -3$ (see Fig. 3.18).

b. $y = 6x^4 - 8x^3 + 1$.

$$y' = 24x^3 - 24x^2 = 24x^2(x - 1).$$
$$y'' = 72x^2 - 48x.$$

Solving $y' = 0$ gives the critical values $x = 0, 1$. If $x = 1$, then $y'' > 0$, so there is a relative minimum when $x = 1$. If $x = 0$, then $y'' = 0$ and the second-derivative test does not apply. We now turn to the first-derivative test to analyze what is happening at 0. If $x < 0$, then $y' < 0$; if $0 < x < 1$, then $y' < 0$. Thus no relative maximum or minimum exists when $x = 0$ (see Fig. 3.36).

c. $y = x^4$.

$$y' = 4x^3.$$
$$y'' = 12x^2.$$

Solving $y' = 0$ gives the critical value $x = 0$. If $x = 0$, then $y'' = 0$ and the second-derivative test does not apply. Since $y' < 0$ for $x < 0$, and $y' > 0$ for $x > 0$, by the first-derivative test there is a relative minimum when $x = 0$ (see Fig. 3.37).

If a continuous function has *exactly one* relative extremum on an interval, it can be shown that the relative extremum must also be an *absolute* extremum on the interval. To illustrate, in Example 1(c), $y = x^4$ has a relative minimum when $x = 0$ and there are no other relative extrema. Since $y = x^4$ is continuous, this relative minimum is also an absolute minimum for the function.

EXAMPLE 2   *If* $y = f(x) = x^3 - 3x^2 - 9x + 5$, *determine when absolute extrema occur on the interval* $(0, \infty)$.

We have

$$f'(x) = 3x^2 - 6x - 9 = 3(x^2 - 2x - 3)$$
$$= 3(x + 1)(x - 3).$$

The only critical value on the interval $(0, \infty)$ is 3. Applying the second-derivative test at this point gives

$$f''(x) = 6x - 6.$$
$$f''(3) = 6(3) - 6 = 12 > 0.$$

Thus there is a relative minimum when $x = 3$. Since this is the only relative extremum on $(0, \infty)$ and f is continuous on this interval, we conclude by our previous discussion that there is an *absolute* minimum when $x = 3$

EXERCISE 3.5

*In Problems **1–10**, test for relative maxima and minima. Use the second-derivative test if possible. In Problems **1–4**, state whether the relative extrema are also absolute extrema.*

1. $y = x^2 - 5x + 6$.

2. $y = -2x^2 + 6x + 12$.

3. $y = -4x^2 + 2x - 8$.

4. $y = 3x^2 - 5x + 6$.

5. $y = x^3 - 27x + 1$.

6. $y = x^3 - 12x + 1$.

7. $y = -x^3 + 3x^2 + 1$.

8. $y = x^4 - 2x^2 + 4$.

9. $y = 2x^4 + 2$.

10. $y = -x^7$.

3.6 REVIEW

Important Terms and Symbols

Section 3.1 x-axis symmetry y-axis symmetry   symmetry about origin x-intercept y-intercept

Section 3.2 $\lim\limits_{x \to a^+} f(x)$ $\lim\limits_{x \to a^-} f(x)$ vertical asymptote
vertical asymptote rule for rational functions   horizontal asymptote

Section 3.3   intervals   increasing function   decreasing function   relative maximum
relative minimum   relative extrema   absolute extrema   critical value
critical point   first-derivative test

Section 3.4   concave up   concave down   inflection point

Section 3.5   second-derivative test

Summary

When the graph of an equation has symmetry, the mirror-image effect allows us to sketch the graph by plotting fewer points than would otherwise be needed. The tests for symmetry are:

Symmetry about x-axis	Replace y by $-y$ in given equation. Symmetric if equivalent equation is obtained.
Symmetry about y-axis	Replace x by $-x$ in given equation. Symmetric if equivalent equation is obtained.
Symmetry about origin	Replace x by $-x$ and y by $-y$ in given equation. Symmetric if equivalent equation is obtained.

Intercepts and asymptotes are also aids in curve sketching. Graphs "explode" near vertical asymptotes, and they "settle down" near horizontal asymptotes. The line $x = a$ is a vertical asymptote for the graph of a function f if $\lim f(x) = \infty$ or $-\infty$ as x approaches a from the right $(x \to a^+)$ or from the left $(x \to a^-)$. For the case of a rational function, $f(x) = P(x)/Q(x)$, we can find vertical asymptotes without evaluating limits. If $Q(a) = 0$ but $P(a) \neq 0$, then the line $x = a$ is a vertical asymptote.

The line $y = b$ is a horizontal asymptote for the graph of a nonlinear function f if at least one of the following is true:

$$\lim_{x \to \infty} f(x) = b \quad \text{or} \quad \lim_{x \to -\infty} f(x) = b.$$

In particular, a polynomial function has neither a horizontal nor a vertical asymptote. Moreover, a rational function whose numerator has degree greater than that of the denominator does not have a horizontal asymptote.

The first derivative is used to determine when a function is increasing or decreasing and to locate relative maxima and minima. If $f'(x)$ is positive throughout an interval, then over that interval f is increasing and its graph rises (from left to right). If $f'(x)$ is negative throughout an interval, then over that inverval f is decreasing and its graph is falling.

A point (x_0, y_0) on the graph at which $f'(x)$ is 0 or is not defined is a candidate for a relative extremum, and

x_0 is called a critical value. For a relative extremum to occur at x_0, the first derivative must change sign around x_0. The following procedure is the first-derivative test for the relative extrema of $y = f(x)$:

First-Derivative Test for Relative Extrema

1. Find $f'(x)$.

2. Determine all values of x where $f'(x) = 0$ or $f'(x)$ is not defined.

3. On the intervals suggested by the values in Step 2, determine whether f is increasing ($f'(x) > 0$) or decreasing ($f'(x) < 0$).

4. For each critical value x_0, determine whether $f'(x)$ changes sign as x increases through x_0. There is a relative maximum when $x = x_0$ if $f'(x)$ changes from $+$ to $-$, and a relative minimum if $f'(x)$ changes from $-$ to $+$. If $f'(x)$ does not change sign, there is no relative extremum when $x = x_0$.

If the domain of f is a closed interval, then to locate absolute extrema we not only consider where relative extrema occur, but also examine $f(x)$ at the endpoints of the interval.

The second derivative is used to determine concavity and points of inflection. If $f''(x) > 0$ throughout an interval, then f is concave up over that interval and its graph bends upward. If $f''(x) < 0$ over an interval, then throughout that interval f is concave down and its graph bends downward. A point on the graph where concavity changes is an inflection point. The point (x_0, y_0) on the graph is a possible point of inflection if $f''(x_0)$ is 0 or is not defined.

The second derivative also provides a means for testing certain critical values for relative extrema.

Second-Derivative Test for Relative Extrema

Suppose that $f'(x_0) = 0$.

If $f''(x_0) < 0$, then f has a relative maximum at x_0; if $f''(x_0) > 0$, then f has a relative minimum at x_0.

Review Problems

In Problems **1** *and* **2**, *test for symmetry.*

1. $y = 2x - 3x^3$.

2. $\dfrac{xy^2}{x^2 + 1} = 4$.

In Problems **3** *and* **4**, *find horizontal and vertical asymptotes.*

3. $y = \dfrac{3x^2}{x^2 - 16}$.

4. $y = \dfrac{x + 1}{4x - 2x^2}$.

In Problems **5** *and* **6**, *find critical values.*

5. $f(x) = \dfrac{x^2}{2 - x}$.

6. $f(x) = (x - 1)^2(x + 6)^4$.

In Problems **7** *and* **8**, *find intervals on which the function is increasing or decreasing.*

7. $f(x) = -x^3 + 6x^2 - 9x$.   **8.** $f(x) = \dfrac{x^2}{(x + 1)^2}$.

In Problems **9** *and* **10**, *find intervals on which the function is concave up or concave down.*

9. $f(x) = x^4 - x^3 - 14$.   **10.** $f(x) = \dfrac{x + 1}{x - 1}$.

In Problems **11** *and* **12**, *test for relative extrema.*

11. $f(x) = \dfrac{x^6}{6} + \dfrac{x^3}{3}$.   **12.** $f(x) = \dfrac{x^2}{x^2 - 4}$.

In Problems **13** *and* **14**, *find where inflection points occur.*

13. $y = x^5 - 5x^4 + 3x$.   **14.** $y = \dfrac{x^2 + 1}{x}$.

In Problems **15** *and* **16**, *test for absolute extrema on the given interval.*

15. $y = 3x^4 - 4x^3$, $[0, 2]$.

16. $y = 2x^3 - 15x^2 + 36x$, $[0, 3]$.

In Problems **17–24**, *sketch the graph of the function. Indicate intervals on which the function is increasing, decreasing, concave up, concave down; indicate relative maximum points, relative minimum points, points of inflection, horizontal asymptotes, vertical asymptotes and symmetry.*

17. $y = x^2 - 2x - 24$.   **18.** $y = x^3 - 27x$.

19. $y = x^3 - 12x + 20$.

20. $y = x^4 - 4x^3 - 20x^2 + 150$.

21. $y = x^3 + x$.   **22.** $y = \dfrac{x + 2}{x - 3}$.

23. $f(x) = \dfrac{100(x + 5)}{x^2}$.   **24.** $y = \dfrac{x^2 - 4}{x^2 - 1}$.

25. Over what interval is $y = f(x) = x^3 - x^2 - x + 10$ decreasing at an increasing rate?

26. Over what interval is $y = f(x) = (x^2 - 1)/(x^2 + 1)$ increasing at an increasing rate?

Applications of Differentiation

4.1 APPLIED MAXIMA AND MINIMA

By using techniques from Chapter 3, we can solve problems that involve maximizing or minimizing a quantity. For example, we might want to maximize profit or minimize cost. The crucial part is expressing the quantity to be maximized or minimized as a function of some variable involved in the problem. Then we differentiate and test the resulting critical values. For this, the first-derivative test or the second-derivative test may be used, although it is often obvious from the nature of the problem whether or not a critical value represents an appropriate answer. Because our interest is in *absolute* maxima and minima, sometimes we must examine endpoints of the domain of the function.

EXAMPLE 1   *The demand equation for a manufacturer's product is* $p = (80 - q)/4$, *where q is the number of units and p is price per unit. At what value of q will there be maximum revenue? What is the maximum revenue?*

Let r be total revenue, the quantity to be maximized. Then revenue = (price)(quantity). Thus

$$r = pq = \frac{80 - q}{4} \cdot q = \frac{80q - q^2}{4},$$

where $q \geq 0$. Setting $dr/dq = 0$:

$$\frac{dr}{dq} = \frac{80 - 2q}{4} = 0,$$

$$q = 40.$$

Thus 40 is a critical value. Examining the first derivative gives $dr/dq > 0$ for $0 \le q < 40$, so r is increasing. If $q > 40$, then $dr/dq < 0$, so r is decreasing. Because to the left of 40 we have r increasing and to the right r is decreasing, we conclude that $q = 40$ gives the *absolute* maximum revenue. This revenue is $[80(40) - (40)^2]/4 = 400$.

EXAMPLE 2   *A manufacturer's total cost function is given by*

$$c = \frac{q^2}{4} + 3q + 400,$$

where q is the number of units produced. At what level of output will average cost per unit be a minimum? What is this minimum?

The quantity to be minimized is average cost $\overline{c}$. The average cost function is

$$\overline{c} = \frac{c}{q} = \frac{\dfrac{q^2}{4} + 3q + 400}{q} = \frac{q}{4} + 3 + \frac{400}{q}. \tag{1}$$

Here q must be positive. To minimize $\overline{c}$ we differentiate.

$$D_q \overline{c} = \frac{1}{4} - \frac{400}{q^2} = \frac{q^2 - 1600}{4q^2}.$$

To get the critical values, we solve $D_x \overline{c} = 0$.

$$q^2 - 1600 = 0,$$
$$(q - 40)(q + 40) = 0,$$
$$q = 40 \qquad \text{(since } q > 0\text{)}.$$

To determine if this level of output gives a relative minimum, we shall use the second-derivative test.

$$D_q^2 \overline{c} = \frac{800}{q^3},$$

which is positive for $q = 40$. Thus $\overline{c}$ has a relative minimum when $q = 40$. We note that $\overline{c}$ is continuous for $q > 0$. Since $q = 40$ is the only relative extremum, we conclude that this relative minimum is indeed an absolute minimum. Substituting $q = 40$ in Eq. (1) gives the minimum average cost $\overline{c} = 23$.

EXAMPLE 3   *An enzyme is a protein that acts as a catalyst for increasing the rate of a chemical reaction that occurs in cells. In a certain reaction, an enzyme is converted to another enzyme called the product. The product acts as a catalyst for its own formation. The rate R at which the product is formed (with respect to time) is given by*

$$R = kp(l - p),$$

where l is the total initial amount of both enzymes, p is the amount of the product enzyme, and k is a positive constant. For what value of p will R be a maximum?

We can write $R = k(pl - p^2)$. Setting $dR/dp = 0$ and solving for p gives

$$\frac{dR}{dp} = k(l - 2p) = 0.$$

$$p = \frac{l}{2}.$$

Now, $d^2R/dp^2 = -2k$. Since $k > 0$, the second derivative is always negative. Hence, $p = l/2$ gives a relative maximum. Moreover, since R is a continuous function of p, we conclude that we indeed have an absolute maximum when $p = l/2$.

Calculus can be applied to inventory decisions, as the following example shows.

EXAMPLE 4   *A manufacturer annually produces and sells* 10,000 *units of a product. Sales are uniformly distributed throughout the year. He wishes to determine the number of units to be manufactured in each production run in order to minimize total annual setup costs and carrying costs. The same number of units is produced in each run. This number is referred to as the* **economic lot size** *or* **economic order quantity.** *The production cost of each unit is* $20 *and carrying costs (insurance, interest, storage, etc.) are estimated to be* 10 *percent of the value of the average inventory. Setup costs per production run are* $40. *Find the economic lot size.*

Let q be the number of units in a production run. Since sales are distributed at a uniform rate, we shall assume that inventory varies uniformly from q to 0 between production runs. Thus we take the average inventory to be $q/2$ units. The production costs are $20 per unit, so the value of the average inventory is $20(q/2)$. Carrying costs are 10 percent of this value:

$$0.10(20)\left(\frac{q}{2}\right).$$

The number of production runs per year is 10,000/q. Thus the total setup costs are

$$40\left(\frac{10,000}{q}\right).$$

Hence the total annual carrying costs and setup costs C are given by

$$C = 0.10(20)\left(\frac{q}{2}\right) + 40\left(\frac{10{,}000}{q}\right)$$

$$= q + \frac{400{,}000}{q} \qquad (q > 0).$$

$$\frac{dC}{dq} = 1 - \frac{400{,}000}{q^2} = \frac{q^2 - 400{,}000}{q^2}.$$

Setting $dC/dq = 0$, we get

$$q^2 = 400{,}000.$$

Since $q > 0$, we choose

$$q = \sqrt{400{,}000} = 200\sqrt{10} \approx 632.5.$$

To determine if this value of q minimizes C, we shall examine the first derivative. If $0 < q < \sqrt{400{,}000}$, then $dC/dq < 0$. If $q > \sqrt{400{,}000}$, then $dC/dq > 0$. We conclude that there is an *absolute* minimum at $q = 632.5$. The number of production runs is $10{,}000/632.5 \approx 15.8$. For practical purposes, there would be 16 lots, each having an economic lot size of 625 units.

EXAMPLE 5   *The Vista TV Cable Co. currently has 3500 subscribers who are each paying a monthly rate of $8. A survey reveals that there will be 50 more subscribers for each $0.10 decrease in the rate. At what rate will maximum revenue be obtained, and how many subscribers will there be at this rate?*

Let x be the new rate (in dollars). Then the total decrease in the old rate is $8 - x$, and the number of $0.10 decreases is $\dfrac{8 - x}{0.10}$. For *each* of these decreases there will be 50 more subscribers. Thus the total of *new* subscribers is $50\left(\dfrac{8 - x}{0.10}\right)$ and the total of all subscribers is

$$3500 + 50\left(\frac{8 - x}{0.10}\right), \tag{2}$$

where $0 \le x \le 8$. The quantity that we want to maximize is revenue r, which is given by $r = $ (rate)(number of subscribers):

$$r = x\left[3500 + 50\left(\frac{8 - x}{0.10}\right)\right]$$

$$= x[3500 + 4000 - 500x]$$

$$= 7500x - 500x^2.$$

Setting $r' = 0$, we have

$$r' = 7500 - 1000x = 0.$$
$$x = 7.50.$$

If $0 \le x < 7.5$, then $r' > 0$; if $7.5 < x \le 8$, then $r' < 0$. We conclude that when the rate is \$7.50 there is an absolute maximum. Substituting $x = 7.50$ into (2) gives 3750 subscribers.

EXAMPLE 6   *In an article in a sociology journal, it was stated that if a particular health-care program for the elderly were initiated, then t years after its start, n thousand elderly people would receive direct benefits, where*

$$n = \frac{t^3}{3} - 6t^2 + 32t, \qquad 0 \le t \le 12.$$

For what value of t does the maximum number receive benefits?

Setting $dn/dt = 0$, we have

$$\frac{dn}{dt} = t^2 - 12t + 32 = 0,$$
$$(t - 4)(t - 8) = 0,$$
$$t = 4 \quad \text{or} \quad t = 8.$$

Now, $d^2n/dt^2 = 2t - 12$, which is negative for $t = 4$ and positive for $t = 8$. Thus there is a relative maximum when $t = 4$. This gives $n = 53\frac{1}{3}$. To determine whether this is an absolute maximum, we must find n at the endpoints of the domain. If $t = 0$, then $n = 0$. If $t = 12$, then $n = 96$. Thus an absolute maximum occurs when $t = 12$. A graph of the function is given in Fig. 4.1.

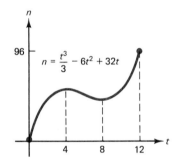

$$n = \frac{t^3}{3} - 6t^2 + 32t$$

FIGURE 4.1

Pitfall

Example 6 illustrates that you should not ignore endpoints when finding absolute extrema on a closed interval.

EXAMPLE 7   *For insurance purposes a manufacturer plans to fence in a 10,800-sq-ft rectangular storage area adjacent to a building by using the building as one side of the enclosed area (see Fig. 4.2). The fencing parallel to the building faces a highway and will cost \$3 per ft. installed, while the fencing for the other two sides costs \$2 per ft. installed. Find the amount of each type of fence so that the total cost of the fence will be a minimum. What is the minimum cost?*

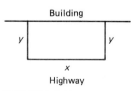

FIGURE 4.2

In Fig. 4.2 we have labeled the length of the side parallel to the building as x and the lengths of the other two sides as y, where x and y are in feet. The cost (in dollars) of the fencing along the highway is $3x$, and along each of the other sides it is $2y$. Thus the total cost C of the fencing is

$$C = 3x + 2y + 2y = 3x + 4y.$$

We wish to minimize C. In order to differentiate, we first express C in terms of one variable only. To do this we find a relationship between x and y. Since the storage area xy must be 10,800,

$$xy = 10,800$$

$$\text{or} \quad y = \frac{10,800}{x}.$$

By substitution we have

$$C = 3x + 4\left(\frac{10,800}{x}\right) = 3x + \frac{43,200}{x},$$

where $x > 0$. To minimize C we set $dC/dx = 0$ and solve for x:

$$\frac{dC}{dx} = 3 - \frac{43,200}{x^2} = 0,$$

$$3 = \frac{43,200}{x^2},$$

from which

$$x^2 = \frac{43,200}{3} = 14,400.$$

$$x = 120 \quad (\text{since } x > 0).$$

Now, $d^2C/dx^2 = 86,400/x^3 > 0$ for $x = 120$, so we conclude that $x = 120$ indeed gives the minimum value of C. When $x = 120$, then $y = 10,800/120 = 90$. Thus 120 ft. of the \$3 fencing and 180 ft. of the \$2 fencing are needed. This gives a cost of \$720.

In the next example we use the word *monopolist*. Under a situation of monopoly, there is only one seller of a product for which there are no similar substitutes, and the seller, that is, the monopolist, controls the market. By considering the demand equation for the product, the monopolist may set the price (or volume of output) so that maximum profit will be obtained.

EXAMPLE 8   *The demand equation for a monopolist's product is* $p = 400 - 2q$ *and the average cost function is* $\overline{c} = 0.2q + 4 + (400/q)$, *where* q *is the number of units, and both* p *and* $\overline{c}$ *are expressed in dollars per unit.*

a. *Determine the level of output at which profit is maximized.*

b. *Determine the price at which maximum profit occurs.*

c. *Determine the maximum profit.*

d. *If, as a regulatory device, the government imposes a tax of $22 per unit on the monopolist, what is the new price for profit maximization?*

$$\text{Profit} = \text{total revenue} - \text{total cost.}$$

Since total revenue $r = pq = 400q - 2q^2$ and total cost $c = q\bar{c} = 0.2q^2 + 4q + 400$, profit P is

$$P = r - c = 400q - 2q^2 - (0.2q^2 + 4q + 400).$$
$$P = 396q - 2.2q^2 - 400, \tag{3}$$

where $q > 0$.

a. To maximize profit we set $dP/dq = 0$.

$$\frac{dP}{dq} = 396 - 4.4q = 0.$$

$$q = 90.$$

Since $d^2P/dq^2 = -4.4 < 0$, we conclude that $q = 90$ gives maximum profit.

b. Setting $q = 90$ in the demand equation gives $p = 400 - 2(90) = 220$.

c. Replacing q by 90 in Eq. (3) gives $P = 17,420$.

d. The tax of $22 per unit means that for q units the total cost increases by $22q$. The new cost function is $c_1 = 0.2q^2 + 4q + 400 + 22q$, and the profit P_1 is given by

$$P_1 = 400q - 2q^2 - (0.2q^2 + 4q + 400 + 22q).$$
$$P_1 = 374q - 2.2q^2 - 400.$$

Setting $dP_1/dq = 0$ gives

$$\frac{dP_1}{dq} = 374 - 4.4q = 0.$$

$$q = 85.$$

Since $d^2P_1/dq^2 = -4.4 < 0$, we conclude that to maximize profit, the monopolist restricts output to 85 units at a higher price of $p_1 = 400 - 2(85) = 230$. Since this price is only $10 more than before, only part of the tax has been shifted to the consumer, and the monopolist must bear the cost of the balance. The profit now is $15,495, which is less than the former profit.

We conclude this section by using calculus to develop an important principle in economics. Suppose that $p = f(q)$ is the demand function for a firm's product, where p is price per unit and q is the number of units produced and sold. Then the total revenue $r = qp = qf(q)$ is a function of q. Let the total cost c of producing q units be given by the cost function $c = g(q)$. Thus the total profit P, which is total revenue − total cost, is also a function of q:

$$P = r - c = qf(q) - g(q).$$

Let us consider the most profitable output for the firm. Ignoring special cases, we know that profit is maximized when $dP/dq = 0$.

$$\frac{dP}{dq} = \frac{d}{dq}(r - c) = \frac{dr}{dq} - \frac{dc}{dq} = 0.$$

Thus

$$\frac{dr}{dq} = \frac{dc}{dq}.$$

But dr/dq is marginal revenue MR, and dc/dq is marginal cost MC. Thus under typical conditions, *to maximize profit it is necessary that*

$$MR = MC.$$

EXERCISE 4.1

In this set of problems, p is price per unit (in dollars) and q is output per unit of time. Fixed costs refer to costs that remain constant at all levels of production in a given time period. (An example is rent.)

1. A manufacturer finds that the total cost c of producing a product is given by the cost function $c = 0.05q^2 + 5q + 500$. At what level of output will average cost per unit be at a minimum?

2. The cost per hour C (in dollars) of operating an automobile is given by

$$C = 0.12s - 0.0012s^2 + 0.08, \qquad 0 \le s \le 60,$$

where s is the speed in miles per hour. At what speed is the cost per hour a minimum?

3. The demand equation for a monopolist's product is $p = -5q + 30$. At what price will revenue be maximized?

4. For a monopolist's product, the revenue function is given by $r = 240q + 57q^2 - q^3$. Determine the output for maximum revenue.

5. In a laboratory an experimental antibacterial agent is applied to a population of 100 bacteria. Data indicate that the number N of bacteria, t hours after the agent is introduced, is given by

$$N = \frac{14,400 + 120t + 100t^2}{144 + t^2}.$$

For what value of t does the maximum number of bacteria in the population occur? What is this maximum number?

6. In his model for storage and shipping costs of materials for a manufacturing process, Lancaster[27] derives the following cost function:

$$C(k) = 100\left[100 + 9k + \frac{144}{k}\right], \qquad 1 \le k \le 100,$$

where $C(k)$ is the total cost (in dollars) of storage and transportation for 100 days of operation if a load of k tons of material is moved every k days.

a. Find $C(1)$.

b. For what value of k does $C(k)$ have a minimum?

c. What is the minimum value?

7. A group of biologists studied the nutritional effects on rats that were fed a diet containing 10 percent protein (adapted from [9]). The protein consisted of yeast and cottonseed flour. By varying the percent p of yeast in

the protein mix, the group found that the (average) weight gain (in grams) of a rat over a period of time was given by

$$f(p) = 160 - p - \frac{900}{p + 10}, \qquad 0 \le p \le 100.$$

a. Find the maximum weight gain.

b. Find the minimum weight gain.

8. The severity R of the reaction of the human body to an initial dose D of a drug is given in [28] by

$$R = f(D) = D^2\left(\frac{C}{2} - \frac{D}{3}\right),$$

where the constant C denotes the maximum amount of the drug that may be given. Show that R has a maximum *rate of change* when $D = C/2$.

9. For a monopolist's product, the demand equation is $p = 72 - 0.04q$ and the cost function is $c = 500 + 30q$. At what level of output will profit be maximized? At what price does this occur and what is the profit?

10. For a monopolist, the cost per unit of producing a product is $3 and the demand equation is $p = 10/\sqrt{q}$. What price will give the greatest profit?

11. For a monopolist's product, the demand equation is $p = 42 - 4q$ and the average cost function is given by $\bar{c} = 2 + (80/q)$. Find the profit-maximizing price.

12. For a monopolist's product, the demand function is $p = 50/\sqrt{q}$ and the average cost function is $\bar{c} = 0.50 + (1000/q)$. Find the profit-maximizing price and output.

13. For XYZ Manufacturing Co., total fixed costs are $1200, material and labor costs combined are $2 per unit, and the demand equation is $p = 100/\sqrt{q}$. What level of output will maximize profit? What is the price at profit maximization?

14. A real estate firm owns 100 garden-type apartments. At $400 per month, each apartment can be rented. However, for each $10 per month increase, there will be two vacancies with no possibility of filling them. What rent per apartment will maximize monthly revenue?

15. A TV cable company has 1000 subscribers who are paying $5 per month. It can get 100 more subscribers for each $0.10 decrease in the monthly rate. What rate will yield maximum revenue, and what will this revenue be?

16. A manufacturer finds that for the first 500 units of a product that are produced and sold, the profit is $50 per unit. The profit on each of the units beyond 500 is decreased by $0.10 times the number of additional units produced. For example, the total profit when 502 units are produced and sold is 500(50) + 2(49.80). What level of output will maximize profit?

17. Find two numbers whose sum is 40 and whose product is a maximum.

18. Find two nonnegative numbers whose sum is 20 and such that the product of twice one number and the square of the other number will be a maximum.

19. A company has set aside $3000 to fence in a rectangular portion of land adjacent to a stream by using the stream for one side of the enclosed area. The cost of the fencing parallel to the stream is $5 per foot installed, and the fencing for the remaining two sides is $3 per foot installed. Find the dimensions of the maximum enclosed area.

20. The owner of the Laurel Nursery Garden Center wants to fence in 1000 sq ft of land in a rectangular plot to be used for different types of shrubs. The plot is to be divided into four equal plots with three fences parallel to the same pair of sides as shown in Fig. 4.3. What is the least number of feet of fence needed?

FIGURE 4.3

21. A container manufacturer is designing a rectangular box, open at the top and with a square base, that is to have a volume of 32 cu ft. If the box is to require the least amount of material, what must be the dimensions of the box? See Fig. 4.4.

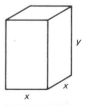

FIGURE 4.4

22. An open-top box with a square base is to be constructed from 192 sq ft of material. What should be the dimensions of the box if the volume is to be a maximum? What is the maximum volume? See Fig. 4.4.

23. An open box is to be made by cutting equal squares from each corner of a 12-in. square piece of cardboard and then folding up the sides. Find the length of the side of the square that must be cut out if the volume of the box is to be maximized. What is the maximum volume? See Fig. 4.5.

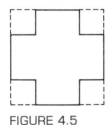

FIGURE 4.5

24. A rectangular cardboard poster is to have 150 sq in. for printed matter. It is to have a 3-in. margin at the top and bottom and a 2-in. margin on each side. Find the dimensions of the poster so that the amount of cardboard used is minimized (see Fig. 4.6). (*Hint:* First find the values of x and y in Fig. 4.6 that minimize the amount of cardboard.)

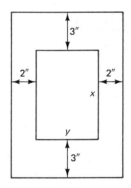

FIGURE 4.6

25. A cylindrical can, open at the top, is to have a fixed volume of K. Show that if the least amount of material is to be used, then both the radius and height are equal to $\sqrt[3]{K/\pi}$ (see Fig. 4.7).

Volume = $\pi r^2 h$
Surface area = $2\pi r h + \pi r^2$

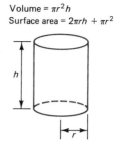

Open at top

FIGURE 4.7

26. A cylindrical can, open at the top, is to be made from a fixed amount of material, K. If the volume is to be a maximum, show that both the radius and height are equal to $\sqrt{K/(3\pi)}$ (see Fig. 4.7).

27. Suppose that the demand equation for a monopolist's product is $p = 600 - 2q$ and the total cost function is $c = 0.2q^2 + 28q + 200$. Find the profit-maximizing output and price, and determine the corresponding profit. If the government were to impose a tax of $22 per unit on the manufacturer, what would be the new profit-maximizing output and price? What is the profit now?

28. Use the *original* data in Problem 27 and assume that the government imposes a license fee of $100 on the manufacturer. This is a lump-sum amount without regard to output. Show that the profit-maximizing price and output remain the same. Show, however, that there will be less profit.

29. A manufacturer has to produce annually 1000 units of a product that is sold at a uniform rate during the year. The production cost of each unit is $10 and carrying costs (insurance, interest, storage, etc.) are estimated to be 12.8 percent of the value of average inventory. Set-up costs per production run are $40. Find the economic lot size.

30. For a monopolist's product, the cost function is $c = 0.004q^3 + 20q + 5000$ and the demand function is $p = 450 - 4q$. Find the profit-maximizing output.

31. Imperial Educational Services (I.E.S.) is considering offering a workshop in resource allocation to key personnel at Acme Corp. To make the offering economically feasible, I.E.S. feels that at least 30 persons must attend at a cost of $50 each. Moreover, I.E.S. will

agree to reduce the charge for *everybody* by $1.25 for each person over the 30 who attends. How many people should be in the group for I.E.S. to maximize revenue? Assume that the maximum allowable number in the group is 40.

32. The Kiddie Toy Company plans to lease an electric motor to be used 90,000 horsepower-hours per year in manufacturing. One horsepower-hour is the work done in one hour by a one-horsepower motor. The annual cost to lease a suitable motor is $150 plus $0.60 per horsepower. The cost per horsepower-hour of operating the motor is $0.006/N$, where N is the horsepower. What size motor, in horsepower, should be leased in order to minimize cost?

33. The cost of operating a truck on a throughway (excluding the salary of the driver) is $0.11 + (s/300)$ dollars per mile, where s is the (steady) speed of the truck in miles per hour. The truck driver's salary is $12 per hour. At what speed should the truck driver operate the truck to make a 700-mile trip most economical?

34. For a manufacturer, the cost of making a part is $3 per unit for labor and $1 per unit for materials; overhead is fixed at $2000 per week. If more than 5000 units are made each week, labor is $4.50 per unit for those units in excess of 5000. At what level of production will average cost per unit be at a minimum?

35. Suppose that each day a company produces x tons of chemical A ($x \le 4$) and y tons of chemical B where $y = (24 - 6x)/(5 - x)$. The profit on chemical A is $2000 per ton, and on B it is $1000 per ton. How much of chemical A should be produced per day to maximize profit? Answer the same question if the profit on A is P per ton and that on B is $P/2$ per ton.

36. To erect an office building, fixed costs are $250,000 and include land, architect's fee, basement, foundation, etc. If x floors are to be constructed, the cost (excluding fixed costs) is $c = (x/2)[100,000 + 5000(x - 1)]$. The revenue per month is $5000 per floor. Find the number of floors that will yield a maximum rate of return on

investment (rate of return = total revenue/total cost).

37. In a model by Smith[29] for power output P of an animal at a given speed as a function of its movement or *gait j*, the following relation is derived:

$$P(j) = Aj\frac{L^4}{V} + B\frac{V^3L^2}{1 + j}.$$

Here A and B are constants, j is a measure of the "jumpiness" of the gait, L is a constant representing linear dimension, and V is a constant forward speed. Assume that P is a minimum when $dP/dj = 0$. Show that when this occurs, then

$$(1 + j)^2 = \frac{BV^4}{AL^2}.$$

As a passing comment, Smith indicates, ". . . at top speed, j is zero for an elephant, 0.3 for a horse, and 1 for a greyhound, approximately."

38. In a model of traffic flow on a lane of a freeway, the number N of cars the lane can carry per unit time is given in [10] by

$$N = \frac{-2a}{-2at_r + v - \dfrac{2al}{v}},$$

where a is the acceleration of a car when stopping ($a < 0$), t_r is reaction time to begin braking, v is average speed of the cars, and l is length of a car. Assume that a, t_r, and l are constant. To find at most how many cars a lane can carry, we want to find the speed v that maximizes N. To maximize N it suffices to minimize the denominator $-2at_r + v - (2al/v)$.

a. Find the value of v that minimizes the denominator.

b. Evaluate your answer in part (a) when $a = -19.6$ (ft/sec^2), $l = 20$ (ft) and $t_r = 0.5$ (sec). Your answer will be in feet per second.

c. Find the corresponding value of N to one decimal place. Your answer will be in cars per second. Convert your answer to cars per hour.

4.2 NEWTON'S METHOD

It is quite easy to solve equations of the form $f(x) = 0$ when f is a linear or quadratic function. For example, we can solve $x^2 + 3x - 2 = 0$ by the quadratic formula. However, if $f(x)$ has degree greater than two (or is not a polynomial), it may be difficult, or even impossible, to find solutions (or roots) of $f(x) = 0$ by the methods to which you are accustomed. For this reason we may settle for approximate solutions, which can be obtained in a variety of efficient ways. In this section you will learn how the derivative may be used to approximate the real roots of the equation $f(x) = 0$ (provided f is differentiable). The procedure we shall develop, called *Newton's Method*, is well-suited for a calculator or computer.

One way of locating a root of $f(x) = 0$ is by first sketching the graph of $y = f(x)$ and estimating the root from the graph. A point on the graph where $y = 0$ is an x-intercept, and the x-value of this point is a root of $f(x) = 0$. Another way of locating a root is based on the following fact:

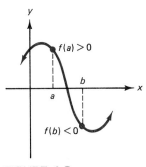

FIGURE 4.8

> If f is continuous on the interval $[a, b]$ and $f(a)$ and $f(b)$ have opposite signs, then the equation $f(x) = 0$ has at least one real root between a and b.

Figure 4.8 depicts this situation. The x-intercept between a and b corresponds to a root of $f(x) = 0$, and we can use either a or b to approximate this root.

Assuming that we have an approximation to a root, we turn to a way of improving such an approximation. In Fig. 4.9 you can see that $f(r) = 0$, so r is a root of the equation $f(x) = 0$. Suppose x_1 is an initial approximation

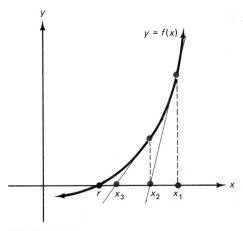

FIGURE 4.9

to r (and one that is close to r). Observe that the tangent line to the curve at $(x_1, f(x_1))$ intersects the x-axis at the point $(x_2, 0)$, and x_2 is a better approximation to r than is x_1.

We can find x_2 from the equation of the tangent line. The slope of the tangent line is $f'(x_1)$, so a point-slope form for this line is

$$y - f(x_1) = f'(x_1)(x - x_1). \tag{1}$$

Since $(x_2, 0)$ is on the tangent line, its coordinates must satisfy Eq. (1). This gives

$$0 - f(x_1) = f'(x_1)(x_2 - x_1),$$

$$-\frac{f(x_1)}{f'(x_1)} = x_2 - x_1 \qquad [\text{if } f'(x_1) \neq 0].$$

Thus

$$x_2 = x_1 - \frac{f(x_1)}{f'(x_1)}. \tag{2}$$

To get a better approximation to r, we again perform the procedure described above, but this time we use x_2 as our starting point. This gives the approximation x_3, where

$$x_3 = x_2 - \frac{f(x_2)}{f'(x_2)}. \tag{3}$$

Continuing in this way we hope to obtain better approximations in the sense that the sequence of values

$$x_1, x_2, x_3, \ldots$$

will approach r. In actual practice we terminate the process when we have reached a desired degree of accuracy.

If you analyze Eqs. (2) and (3), you can see how x_2 is obtained from x_1, and how x_3 is obtained from x_2. In general, x_{n+1} is obtained from x_n by means of the following general formula, called **Newton's Method**:

NEWTON'S METHOD

$$x_{n+1} = x_n - \frac{f(x_n)}{f'(x_n)}, \qquad n = 1, 2, 3, \ldots. \tag{4}$$

A formula, like Eq. (4), that indicates how one number in a sequence is obtained from the preceding one is called a **recursion formula**.

EXAMPLE 1   *Approximate the root of $x^4 - 4x + 1 = 0$ that lies between 0 and 1. Continue the approximation procedure until two successive approximations differ by less than 0.0001.*

Letting $f(x) = x^4 - 4x + 1$, we have

$$f(0) = 0 - 0 + 1 = 1$$
$$\text{and} \qquad f(1) = 1 - 4 + 1 = -2.$$

(Note the sign change.) Since $f(0)$ is closer to 0 than is $f(1)$, we choose 0 to be our first approximation, x_1. Now,

$$f'(x) = 4x^3 - 4,$$

so

$$f(x_n) = x_n^4 - 4x_n + 1 \quad \text{and} \quad f'(x_n) = 4x_n^3 - 4.$$

Substituting into Eq. (4) gives the recursion formula

$$x_{n+1} = x_n - \frac{f(x_n)}{f'(x_n)} = x_n - \frac{x_n^4 - 4x_n + 1}{4x_n^3 - 4}. \tag{5}$$

Since $x_1 = 0$, letting $n = 1$ in Eq. (5) gives

$$x_2 = x_1 - \frac{x_1^4 - 4x_1 + 1}{4x_1^3 - 4}$$

$$= 0 - \frac{0^4 - 4(0) + 1}{4(0)^3 - 4} = 0.25.$$

Letting $n = 2$ in Eq. (5) gives

$$x_3 = x_2 - \frac{x_2^4 - 4x_2 + 1}{4x_2^3 - 4}$$

$$= 0.25 - \frac{(0.25)^4 - 4(0.25) + 1}{4(0.25)^3 - 4} = 0.25099.$$

Letting $n = 3$ in Eq. (5) gives

$$x_4 = x_3 - \frac{x_3^4 - 4x_3 + 1}{4x_3^3 - 4}$$

$$= 0.25099 - \frac{(0.25099)^4 - 4(0.25099) + 1}{4(0.25099)^3 - 4}$$

$$= 0.25099.$$

The data obtained thus far are displayed in Table 4.1.

TABLE 4.1

n	x_n	x_{n+1}
1	0.00000	0.25000
2	0.25000	0.25099
3	0.25099	0.25099

Since the values of x_3 and x_4 differ by less than 0.0001, we take the root to be 0.25099 (that is, x_4).

EXAMPLE 2   *Approximate the root of $x^3 = 3x - 1$ that lies between -1 and -2. Continue the approximation procedure until two successive approximations differ by less than 0.0001.*

Letting $f(x) = x^3 - 3x + 1$ [we need the form $f(x) = 0$], we find that
$$f(-1) = (-1)^3 - 3(-1) + 1 = 3$$
and $$f(-2) = (-2)^3 - 3(-2) + 1 = -1.$$

(Note the sign changes.) Since $f(-2)$ is closer to 0 than is $f(-1)$, we choose -2 to be our first approximation, x_1. Now
$$f'(x) = 3x^2 - 3,$$
so
$$f(x_n) = x_n^3 - 3x_n + 1 \quad\text{and}\quad f'(x_n) = 3x_n^2 - 3.$$

Substituting into Eq. (4) gives the recursion formula
$$x_{n+1} = x_n - \frac{f(x_n)}{f'(x_n)} = x_n - \frac{x_n^3 - 3x_n + 1}{3x_n^2 - 3}. \tag{6}$$

Since $x_1 = -2$, letting $n = 1$ in Eq. (6) gives
$$x_2 = x_1 - \frac{x_1^3 - 3x_1 + 1}{3x_1^2 - 3}$$
$$= (-2) - \frac{(-2)^3 - 3(-2) + 1}{3(-2)^2 - 3} = -1.88889.$$

Continuing in this way we obtain Table 4.2.

TABLE 4.2

n	x_n	x_{n+1}
1	-2.00000	-1.88889
2	-1.88889	-1.87945
3	-1.87945	-1.87939

Since the values of x_3 and x_4 differ by 0.00006, which is less than 0.0001, we take the root to be -1.87939 (that is, x_4).

In case your choice for the initial approximation, x_1, gives the derivative a value of zero, choose a better approximation to the desired root. A graph of f could be helpful in this situation. Finally, we remark that there are times when the sequence of approximations does not approach the root. A discussion of such situations is beyond the scope of this book.

EXERCISE 4.2

Use Newton's method to approximate the indicated root of the given equation. Continue the approximation procedure until the difference of two successive approximations is less than 0.0001.

1. $x^3 - 4x + 1 = 0$;   root between 0 and 1.

2. $x^3 + 2x^2 - 1 = 0$;   root between 0 and 1.

3. $x^3 - x - 1 = 0$;   root between 1 and 2.

4. $x^3 - 9x + 6 = 0$;   root between 2 and 3.

5. $x^3 + x + 16 = 0$;   root between -3 and -2.

6. $x^3 = 2x + 5$;   root between 2 and 3.

7. $x^4 = 3x - 1$;   root between 0 and 1.

8. $x^4 + 4x - 1 = 0$;   root between -2 and -1.

9. $x^4 - 2x^3 + x^2 - 3 = 0$;   root between 1 and 2.

10. $x^4 - x^3 + x - 2 = 0$;   root between 1 and 2.

4.3 DIFFERENTIALS

We shall soon give you a reason for using the symbol dy/dx to denote the derivative of y with respect to x. To do this, we introduce the notion of the *differential* of a function.

Definition
*Let $y = f(x)$ be a differentiable function of x and let Δx denote a change in x, where Δx can be any real number. Then the **differential of y**, denoted dy or $d[f(x)]$, is given by*

$$dy = f'(x)\, \Delta x.$$

Note that dy is a function of two variables, namely x and Δx.

EXAMPLE 1   *Find the differential of $y = x^3 - 2x^2 + 3x - 4$ and evaluate it when $x = 1$ and $\Delta x = 0.04$.*

$$dy = D_x(x^3 - 2x^2 + 3x - 4)\, \Delta x$$
$$= (3x^2 - 4x + 3)\, \Delta x.$$

When $x = 1$ and $\Delta x = 0.04$, then

$$dy = [3(1)^2 - 4(1) + 3](0.04) = 0.08.$$

If $y = x$, then $dy = d(x) = 1\,\Delta x = \Delta x$. Hence the differential of $f(x) = x$ is Δx. We abbreviate $d(x)$ by dx. Thus $dx = \Delta x$. From now on, it will be our practice to write dx for Δx when finding a differential. For example,

$$d(x^2 + 5) = D_x(x^2 + 5)\, dx = 2x\, dx.$$

Summarizing, if $y = f(x)$ defines a differentiable function of x, then

$$\boxed{dy = f'(x)\, dx,}$$

where dx is any real number. Provided that $dx \neq 0$, we can divide both sides by dx:

$$\frac{dy}{dx} = f'(x).$$

That is, dy/dx can be viewed either as the quotient of two differentials, namely dy divided by dx, or as one symbol for the derivative of f at x. It is for this reason that we introduced the symbol dy/dx to denote the derivative.

EXAMPLE 2

a. If $f(x) = \sqrt{x}$, then

$$d(\sqrt{x}) = D_x(\sqrt{x})\, dx = \frac{1}{2}x^{-1/2}\, dx = \frac{1}{2\sqrt{x}}\, dx.$$

b. If $u = (x^2 + 3)^5$, then

$$du = 5(x^2 + 3)^4(2x)\, dx = 10x(x^2 + 3)^4\, dx.$$

The differential can be interpreted geometrically. In Fig. 4.10 the point P, where $P = (x, f(x))$, is on the curve $y = f(x)$. Suppose x changes by dx, a real number, to the value $x + dx$. Then the function value is $f(x + dx)$, and the corresponding point on the curve is $Q = (x + dx, f(x + dx))$. Passing through P and Q are horizontal and vertical lines, respectively, that intersect at S. A line L tangent to the curve at P intersects segment QS at R, forming the right triangle PRS. Observe that the graph of f near P is approximated

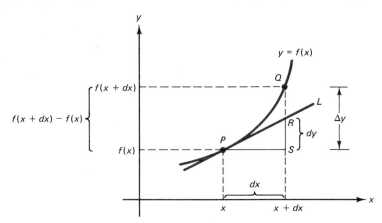

FIGURE 4.10

by the tangent line at P. The slope of L is $f'(x)$ or, equivalently, it is $\overline{SR}/\overline{PS}$:

$$f'(x) = \frac{\overline{SR}}{\overline{PS}}.$$

Since $dy = f'(x)\, dx$, and $dx = \overline{PS}$,

$$dy = f'(x)\, dx = \frac{\overline{SR}}{\overline{PS}} \cdot \overline{PS} = \overline{SR}.$$

Thus if dx is a change in x at P, then dy is the corresponding vertical change along the **tangent line** at P. Note that for the same dx, the vertical change along the **curve** is $\Delta y = \overline{SQ} = f(x + dx) - f(x)$. Do not confuse Δy with dy. However, it is apparent from Fig. 4.10 that

when dx is close to 0, dy is an approximation to Δy.

Thus $\Delta y \approx dy$. This fact is useful in estimating Δy, a change in y, as Example 3 shows.

EXAMPLE 3   *A governmental health agency examined the records of a group of individuals who were hospitalized with a particular illness. It was found that the total proportion P who were discharged at the end of t days of hospitalization is given by*

$$P = P(t) = 1 - \left(\frac{300}{300 + t}\right)^3.$$

Use differentials to approximate the change in the proportion discharged if t changes from 300 to 305.

The change in t from 300 to 305 is $\Delta t = dt = 305 - 300 = 5$. The change in P is $\Delta P = P(305) - P(300)$. We approximate ΔP by dP.

$$\Delta P \approx dP = P'(t)\, dt = -3\left(\frac{300}{300 + t}\right)^2 \left[-\frac{300}{(300 + t)^2}\right] dt.$$

When $t = 300$ and $dt = 5$,

$$dP = -3\left(\frac{300}{600}\right)^2 \left[-\frac{300}{(600)^2}\right] 5$$

$$= -3\left(\frac{1}{2}\right)^2 \left[-\frac{1}{2(600)}\right] 5 = \frac{1}{320} \approx 0.0031.$$

For a comparison, the actual value of ΔP is $P(305) - P(300) = 0.87807 - 0.87500 = 0.00307$ (to 5 decimal places).

We said that if $y = f(x)$, then $\Delta y \approx dy$ if dx is small. Thus

$$\Delta y = f(x + dx) - f(x) \approx dy$$

or

$$\boxed{f(x + dx) \approx f(x) + dy.} \qquad (1)$$

This formula gives us a way of estimating a function value, $f(x + dx)$, as the next example shows.

EXAMPLE 4   *Approximate $\sqrt[3]{126}$ by using the differential.*

Letting $y = f(x) = \sqrt[3]{x}$, we need to approximate the function value $f(126)$. Since $d(\sqrt[3]{x}) = (1/3)x^{-2/3}\,dx$, from Eq. (1) we have

$$f(x + dx) \approx f(x) + dy,$$

$$\sqrt[3]{x + dx} \approx \sqrt[3]{x} + \frac{1}{3x^{2/3}}\,dx.$$

We know the exact value of $\sqrt[3]{125}$, so we shall let $x = 125$ and $dx = 1$. Then $x + dx = 126$ and dx is small.

$$\sqrt[3]{125 + 1} \approx \sqrt[3]{125} + \frac{1}{3(125)^{2/3}}(1),$$

$$\sqrt[3]{126} \approx 5 + \frac{1}{3(25)} = 5\frac{1}{75}.$$

Thus $\sqrt[3]{126}$ is estimated by $5\frac{1}{75}$, which is approximately 5.01333. The actual value of $\sqrt[3]{126}$ to five decimal places is 5.01330.

EXAMPLE 5   *The demand function for a product is given by*

$$p = f(q) = 20 - \sqrt{q},$$

where p is the price per unit in dollars for q units. By using differentials, approximate the price when 99 units are demanded.

We want to approximate $f(99)$. By Eq. (1),

$$f(q + dq) \approx f(q) + dp,$$

$$\text{where} \qquad dp = -\frac{1}{2\sqrt{q}}\,dq.$$

We choose $q = 100$ and $dq = -1$ because $q + dq = 99$, dq is small, and it is easy to compute $f(100) = 20 - \sqrt{100} = 10$.

$$f(99) = f[100 + (-1)] \approx f(100) - \frac{1}{2\sqrt{100}}(-1)$$

$$\approx 10 + 0.05 = 10.05.$$

Thus the price per unit when 99 units are demanded is approximately \$10.05.

EXERCISE 4.3

In Problems 1–8, find the differential of the function in terms of x and dx.

1. $y = 3x - 4$.

2. $y = 2$.

3. $f(x) = \sqrt{x^4 + 2}$.

4. $f(x) = (4x^2 - 5x + 2)^3$.

5. $u = \dfrac{1}{x^2}$.

6. $u = \dfrac{1}{\sqrt{x}}$.

7. $p = x^2(4x + 3)$.

8. $p = x\sqrt{x}$.

In Problems 9–12, find Δy and dy for the given values of x and dx.

9. $y = 4 - 7x$; $x = 3$, $dx = 0.02$.

10. $y = 4x^2 - 3x + 10$; $x = -1$, $dx = 0.25$.

11. $y = \sqrt{25 - x^2}$; $x = 3$, $dx = -0.1$. (Use the fact that $\sqrt{16.59} \approx 4.073$.)

12. $y = (3x + 2)^2$; $x = -1$, $dx = -0.03$.

In Problems 13–16, approximate each expression by using differentials.

13. $\sqrt{101}$.

14. $\sqrt{120}$.

15. $\sqrt[3]{63}$.

16. $\sqrt[4]{17}$.

17. Suppose that the profit P (in dollars) of producing q units of a product is

$$P = 396q - 2.2q^2 - 400.$$

Using differentials, find the approximate change in profit if the level of production changes from $q = 80$ to $q = 81$. Find the actual change.

18. Given the revenue function

$$r = 250q + 45q^2 - q^3,$$

use differentials to find the approximate change in revenue if the number of units increases from $q = 40$ to $q = 41$. Find the actual change.

19. The demand equation for a product is $p = 10/\sqrt{q}$. Using differentials, approximate the price when 24 units are demanded.

20. Answer the same question in Problem 19 if 101 units are demanded.

21. If $y = f(x)$, then the *proportional change* in y is defined to be $\Delta y/y$, which can be approximated with differentials by dy/y. Use this last form to approximate the proportional change in the cost function

$$c = f(q) = \dfrac{q^4}{2} + 3q + 400$$

when $q = 10$ and $dq = 2$. Give your answer to one decimal place.

22. Suppose that S is a numerical value of status based on a person's annual income I (in thousands of dollars). For a certain population, suppose $S = 20\sqrt{I}$. Use differentials to appproximate S for a person with an annual income of $15,000, that is, $I = 15$.

23. The volume V of a spherical cell is given by $V = \frac{4}{3}\pi r^3$, where r is the radius. Estimate the change in volume when the radius changes from 6.5×10^{-4} cm to 6.6×10^{-4} cm.

24. The equation $(P + a)(v + b) = k$ is called the "fundamental equation of muscle contraction" [13]. Here P is the load imposed on the muscle, v is the velocity of the shortening of the muscle fibers, and a, b, and k are positive constants. Find v in terms of P and then use the differential to approximate the change in v due to a small change in P.

25. In a study of rooted plants in a certain region (adapted from [25]), it was determined that the average number of species S occurring on plots of size A (in square meters) was given by

$$S = 12\sqrt[4]{A}, \qquad 0 \le A \le 900.$$

Use differentials to approximate the average number of species in an 80-square-meter plot.

4.4 ELASTICITY OF DEMAND

Elasticity of demand is a means by which economists measure how a change in the price of a product will affect the quantity demanded. That is, it refers to consumer response to price changes. Loosely speaking, elasticity of demand is the ratio of the resulting percentage change in quantity demanded to a given percentage change in price:

$$\dfrac{\text{percentage change in quantity}}{\text{percentage change in price}}.$$

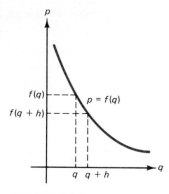

FIGURE 4.11

For example, if for a price increase of 5%, quantity demanded were to decrease by 2%, we would loosely say that elasticity of demand is $-2/5$.

To be more general, suppose that $p = f(q)$ is the demand function for a product. Consumers will demand q units at a price of $f(q)$ per unit, and will demand $q + h$ units at a price of $f(q + h)$ per unit (Fig. 4.11). The *percentage change in quantity* demanded from q to $q + h$ is

$$\frac{(q + h) - q}{q} \cdot 100 = \frac{h}{q} \cdot 100.$$

The corresponding percentage change in price per unit is

$$\frac{f(q + h) - f(q)}{f(q)} \cdot 100.$$

The ratio of these percentage changes is

$$\frac{\dfrac{h}{q} \cdot 100}{\dfrac{f(q + h) - f(q)}{f(q)} \cdot 100} = \frac{h}{q} \cdot \frac{f(q)}{f(q + h) - f(q)}$$

$$= \frac{f(q)}{q} \cdot \frac{h}{f(q + h) - f(q)}$$

$$= \frac{\dfrac{f(q)}{q}}{\dfrac{f(q + h) - f(q)}{h}}. \qquad (1)$$

If f is differentiable, then as $h \to 0$ the limit of $[f(q + h) - f(q)]/h$ is $f'(q) = dp/dq$. Thus the limit of (1) is

$$\frac{\dfrac{f(q)}{q}}{\dfrac{dp}{dq}} \qquad \text{or} \qquad \frac{\dfrac{p}{q}}{\dfrac{dp}{dq}},$$

which is called *point elasticity*.

Definition
*If $p = f(q)$ is a differentiable demand function, the **point elasticity of demand**, denoted by the Greek letter η (eta), at (q, p) is given by*

$$\eta = \frac{\dfrac{p}{q}}{\dfrac{dp}{dq}}.$$

To illustrate, let us find the point elasticity of demand for the demand function $p = 1200 - q^2$.

$$\eta = \frac{\dfrac{p}{q}}{\dfrac{dp}{dq}} = \frac{\dfrac{1200 - q^2}{q}}{-2q} = -\frac{1200 - q^2}{2q^2} = -\left[\frac{600}{q^2} - \frac{1}{2}\right]. \qquad (2)$$

For example, if $q = 10$, then $\eta = -[(600/10^2) - \frac{1}{2}] = -5\frac{1}{2}$. This means that if price were increased by 1% when $q = 10$, the quantity demanded would decrease by approximately $5\frac{1}{2}$%. Similarly, increasing price by $\frac{1}{2}$% results in a decrease in demand of approximately 2.75%.

Note that when elasticity is evaluated, no units are attached to it—it is nothing more than a real number. For normal behavior of demand, a price increase (decrease) corresponds to a quantity decrease (increase). Thus dp/dq will always be negative or 0, and η (where defined) will always be negative or 0. Some economists disregard the minus sign; in the above situation they would consider the elasticity to be $5\frac{1}{2}$. We shall not adopt this practice.

There are three categories of elasticity:

1. When $|\eta| > 1$, demand is **elastic**.

2. When $|\eta| = 1$, demand has **unit elasticity**.

3. When $|\eta| < 1$, demand is **inelastic**.

In Eq. (2) we found that $|\eta| = 5\frac{1}{2}$ when $q = 10$, so demand is elastic. If $q = 20$, then $|\eta| = |-[(600/20^2) - \frac{1}{2}]| = 1$, so demand has unit elasticity. If $q = 25$, then $|\eta| = |-\frac{23}{50}|$ and demand is inelastic.

Loosely speaking, for a given percentage change in price, there is a greater percentage change in quantity demanded if demand is elastic, a smaller percentage change if demand is inelastic, and an equal percentage change if demand has unit elasticity.

EXAMPLE 1 *Find the point elasticity of demand for the demand equation $p = k/q$ where $k > 0$.*

$$\eta = \frac{\dfrac{p}{q}}{\dfrac{dp}{dq}} = \frac{\dfrac{k}{q^2}}{\dfrac{-k}{q^2}} = -1.$$

Thus the demand has unit elasticity. The graph of $p = k/q$ is called an *equilateral hyperbola* and is often found in economics texts in discussions of elasticity. See Fig. 1.8 for the graph of such a curve.

Point elasticity for a *linear* demand equation is quite interesting. Suppose that the equation has the form

$$p = mq + b \quad \text{where} \quad m < 0 \text{ and } b > 0$$

(see Fig. 4.12). We assume that $q > 0$; thus $p < b$. The point elasticity of demand is

$$\eta = \frac{\dfrac{p}{q}}{\dfrac{dp}{dq}} = \frac{\dfrac{p}{q}}{m} = \frac{p}{mq} = \frac{p}{p - b}.$$

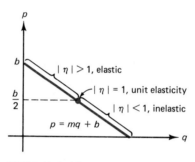

FIGURE 4.12

By considering $d\eta/dp$, we shall show that η is a decreasing function of p. By the quotient rule,

$$\frac{d\eta}{dp} = \frac{(p - b) - p}{(p - b)^2} = -\frac{b}{(p - b)^2}.$$

Since $b > 0$ and $(p - b)^2 > 0$, then $d\eta/dp < 0$, so η is a decreasing function of p—as p increases, η must decrease. However, p ranges between 0 and b, and at the midpoint of this range, $b/2$,

$$\eta = \frac{\dfrac{b}{2}}{\dfrac{b}{2} - b} = \frac{\dfrac{b}{2}}{-\dfrac{b}{2}} = -1.$$

Therefore, if $p < b/2$, then $\eta > -1$; if $p > b/2$, then $\eta < -1$. Because we must have $\eta \le 0$, we can state these facts another way. When $p < b/2$, $|\eta| < 1$ and demand is inelastic; when $p = b/2$, $|\eta| = 1$ and demand has unit elasticity; when $p > b/2$, $|\eta| > 1$ and demand is elastic. This shows that the slope of a demand curve is not a measure of elasticity. The slope of the foregoing line is m everywhere, but elasticity varies with the point on the line.

Turning to a different situation, we can relate how elasticity of demand affects changes in revenue (marginal revenue). If $p = f(q)$ is a manufacturer's demand function, total revenue r is given by

$$r = pq.$$

To find marginal revenue, dr/dq, we differentiate r by using the product rule.

$$\frac{dr}{dq} = p + q \frac{dp}{dq}. \qquad (3)$$

Factoring the right side of Eq. (3), we have

$$\frac{dr}{dq} = p\left(1 + \frac{q}{p}\frac{dp}{dq}\right).$$

But

$$\frac{q}{p}\frac{dp}{dq} = \frac{\dfrac{dp}{dq}}{\dfrac{p}{q}} = \frac{1}{\eta}.$$

Thus

$$\frac{dr}{dq} = p\left(1 + \frac{1}{\eta}\right). \qquad (4)$$

If demand is elastic, then $\eta < -1$ and $1 + \dfrac{1}{\eta} > 0$. If demand is inelastic, then $\eta > -1$ and $1 + \dfrac{1}{\eta} < 0$. Let us assume that $p > 0$. From Eq. (4) we can conclude that $dr/dq > 0$ on intervals for which demand is elastic; hence, total revenue r is increasing there. On the other hand, marginal revenue is negative on intervals for which demand is inelastic; hence, total revenue is decreasing there.

Thus we conclude from the above argument that as more units are sold, a manufacturer's total revenue increases if demand is elastic but decreases if demand is inelastic. That is, if demand is elastic, a lower price will increase revenue. This means that a lower price will cause a large enough increase in demand to actually increase revenue. If inelastic, a lower price will decrease revenue. For unit elasticity, a lower price leaves total revenue unchanged.

EXERCISE 4.4

In Problems 1–8, find the point elasticity of the demand equation for the indicated value of q and determine whether demand is elastic, inelastic, or has unit elasticity.

1. $p = 40 - 2q$; $q = 5$.

2. $p = 12 - 0.03q$; $q = 300$.

3. $p = 6 - 0.01q$; $q = 300$.

4. $p = 100 - q$; $q = 50$.

5. $p = \dfrac{1000}{q}$; $q = 288$.   6. $p = \dfrac{1000}{q^2}$; $q = 156$.

7. $p = \dfrac{500}{q + 2}$; $q = 100$.   8. $p = \dfrac{800}{2q + 1}$; $q = 25$.

9. For the linear demand equation $p = 13 - 0.05q$, verify that demand is elastic when $p = 10$, is inelastic when $p = 3$, and has unit elasticity when $p = 6.50$.

10. For what value (or values) of q do the following demand equations have unit elasticity?

   a. $p = 26 - 0.10q$.

   b. $p = 1200 - q^2$.

11. Verify that $\dfrac{dr}{dq} = p\left(1 + \dfrac{1}{\eta}\right)$ if $p = 40 - 2q$.

12. Repeat Problem 11 for $p = 1000/q^2$.

13. Let $p = mq + b$ be a linear demand equation where $m < 0$ and $b > 0$. Show that $\eta = 0$ when $p = 0$.

4.5 REVIEW

Important Terms and Symbols

Section 4.2   Newton's method

Section 4.3   differential dy dx

Section 4.4   point elasticity of demand   elastic   inelastic   unit elasticity

Summary

In a practical sense, the power of calculus is that it allows us to maximize or minimize quantities. For example, in the area of economics we can maximize profit or minimize cost. Some important relationships that are used in economics problems are:

$$\bar{c} = \frac{c}{q}, \qquad \text{average cost per unit} = \frac{\text{total cost}}{\text{quantity}},$$

$$r = pq, \qquad \text{revenue} = (\text{price})\,(\text{quantity}),$$

$$P = r - c, \qquad \text{profit} = \text{total revenue} - \text{total cost}.$$

Newton's Method is the name given to the following formula which is used to approximate the roots of the equation $f(x) = 0$, provided f is differentiable:

$$x_{n+1} = x_n - \frac{f(x_n)}{f'(x_n)}, \qquad n = 1, 2, 3, \ldots.$$

If $y = f(x)$ is a differentiable function of x, we define the differential dy by

$$dy = f'(x)\, dx,$$

where dx (or Δx) is a change in x and can be any real number. If dx is close to zero, then dy is an approximation to Δy, a change in y:

$$\Delta y \approx dy.$$

Moreover, dy can be used to estimate a function value. We use the relationship

$$f(x + dx) \approx f(x) + dy.$$

Here $f(x + dx)$ is the value to be estimated; x and dx are chosen so that $f(x)$ is easy to compute and dx is small.

Point elasticity of demand is a number that measures how consumer demand is affected by price change. It is given by

$$\eta = \frac{p/q}{dp/dq},$$

where $p = f(q)$ is a demand function. The three categories of elasticity are:

1. $|\eta| > 1$: demand is elastic.

2. $|\eta| = 1$: unit elasticity.

3. $|\eta| < 1$: demand is inelastic.

To put it simply, for a given percentage change in price, there is a greater percentage change in quantity demanded if demand is elastic, a smaller percentage change if demand is inelastic, and an equal percentage change if demand has unit elasticity.

Review Problems

1. A manufacturer determines that m employees on a certain production line will produce q units per month, where $q = 80m^2 - 0.1m^4$. To obtain maximum monthly production, how many employees should be assigned to the production line?

2. The demand function for a monopolist's product is $p = \sqrt{600 - q}$. If the monopolist wants to produce at least 100 units but not more than 300 units, how many units should be produced to maximize total revenue?

3. The demand function for a monopolist's product is $p = 400 - 2q$ and the average cost per unit for producing q units is $\bar{c} = q + 160 + (2000/q)$, where p and $\bar{c}$ are in dollars per unit. Find the maximum profit that the monopolist can achieve.

4. If $c = 0.01q^2 + 5q + 100$ is a cost equation, find the average cost equation. At what level of production q is there a minimum average cost?

5. A rectangular box is to be made by cutting out equal squares from each corner of a piece of cardboard 10 in. by 16 in. and then folding up the sides. What must be the length of the side of the square cut out if the volume of the box is to be a maximum?

6. A rectangular field is to be enclosed by a fence and divided equally into three parts by two fences parallel to one pair of the sides. If a total of 800 ft of fencing is to be used, find the dimensions of the field if its area is to be maximized.

7. A rectangular poster having an area of 500 square inches is to have a 4-in. margin at each side and at the bottom and a 6-in. margin at the top. The remainder of the poster is for printed matter. Find the dimensions of the poster so that the area for the printed matter is maximized.

In Problems **8** and **9**, determine the differential of the function in terms of x and dx.

8. $u = \dfrac{x}{x - 7}$.

9. $f(x) = (1 - 3x)^4$.

10. If $p = q^2 + 8q$, use differentials to estimate Δp if q changes from 4 to 4.02.

In Problems **11** and **12**, approximate the expression by using differentials.

11. $\sqrt{25.5}$.

12. $\sqrt{143}$.

For the demand equation in Problems **13–15**, determine whether demand is elastic, inelastic, or has unit elasticity for the indicated value of q.

13. $p = 18 - 0.02q$; $q = 600$.

14. $p = \dfrac{500}{q}$; $q = 200$.

15. $p = 900 - q^2$; $q = 10$.

16. Fahrenheit temperature F and Celsius temperature C are related by $F = \frac{9}{5}C + 32$. Using differentials, find how much F would change due to a change in C of $\frac{1}{2}°$.

17. The equation $x^3 - 2x - 2 = 0$ has a root between 1 and 2. Use Newton's method to estimate the root. Continue the approximation procedure until the difference of two successive approximations is less than 0.0001. Round your answer to 4 decimal places.

Integration

Chapters 2–4 dealt with differential calculus. We differentiated a function and obtained another function, its derivative. *Integral calculus* is concerned with the reverse process. We are given the derivative of a function and must find the original function. The need for doing this arises in a natural way. For example, we may have a marginal revenue function and want to find the revenue function from it. Integral calculus also involves a limit concept that allows us to take the limit of a special kind of sum as the number of terms in the sum becomes infinite. This is the real power of integral calculus! With this notion we may find the area of a region that cannot be found by any other convenient method.

5.1 THE INDEFINITE INTEGRAL

Given a function f, if F is a function such that

$$F'(x) = f(x), \tag{1}$$

then F is called an *antiderivative* of f. Thus an antiderivative of f is simply a function whose derivative is f. Multiplying both sides of Eq. (1) by the differential dx gives $F'(x)\, dx = f(x)\, dx$. However, because $F'(x)\, dx$ is the differential of F, we have $dF = f(x)\, dx$. Thus we can think of an antiderivative of f as a function whose differential is $f(x)\, dx$.

Definition
*An **antiderivative** of a function f is a function F such that*

$$F'(x) = f(x)$$

or equivalently, in differential notation,

$$dF = f(x)\, dx.$$

For example, since $D_x(x^2) = 2x$, x^2 is an antiderivative of $2x$. However, it is not the only antiderivative of $2x$. Since

$$D_x(x^2 + 1) = 2x \qquad \text{and} \qquad D_x(x^2 - 5) = 2x,$$

both $x^2 + 1$ and $x^2 - 5$ are also antiderivatives of $2x$. It can be shown that *any* antiderivative of $2x$ must have the form $x^2 + C$, where C is a constant. Thus *any two antiderivatives of $2x$ differ only by a constant.*

The most general antiderivative of $2x$ is denoted $\displaystyle\int 2x\, dx$, which is read "the *indefinite integral* of $2x$ with respect to x." Since all antiderivatives of $2x$ have the form $x^2 + C$, we write

$$\int 2x\, dx = x^2 + C.$$

The symbol $\displaystyle\int$ is called the **integral sign**, $2x$ is the **integrand**, and C is the **constant of integration**. The dx is part of the integral notation and indicates the variable involved. Here x is the **variable of integration**.

More generally, the **indefinite integral** of any function f with respect to x is written $\displaystyle\int f(x)\, dx$ and denotes the most general antiderivative of f. It can be shown that all antiderivatives of f differ only by a constant. Thus if F is any antiderivative of f, then

$$\int f(x)\, dx = F(x) + C, \qquad \text{where } C \text{ is a constant.}$$

To *integrate f* means to find $\displaystyle\int f(x)\, dx$. In summary,

$$\boxed{\int f(x)\, dx = F(x) + C \text{ if and only if } F'(x) = f(x).}$$

EXAMPLE 1   *Find $\displaystyle\int 5\, dx$.*

First we must find (perhaps better words are "guess at") a function whose derivative is 5. Since we know that $D_x(5x) = 5$, $5x$ is an antiderivative of 5. Thus

$$\int 5\, dx = 5x + C.$$

Pitfall

It is **incorrect** to write $\int 5\ dx =$ 5x. **Do not forget the constant of integration.**

Using differentiation formulas from Chapter 2, we have compiled a list of basic integration formulas in Table 5.1. These formulas are easily verified.

TABLE 5.1
Basic Integration Formulas

1. $\displaystyle\int k\ dx = kx + C,\quad k$ a constant.

2. $\displaystyle\int x^n\ dx = \frac{x^{n+1}}{n+1} + C,\quad n \neq -1.$

3. $\displaystyle\int k\,f(x)\ dx = k\int f(x)\ dx,\quad k$ a constant.

4. $\displaystyle\int [f(x) \pm g(x)]\ dx = \int f(x)\ dx \pm \int g(x)\ dx.$

For example, Formula 2 is true because the derivative of $x^{n+1}/(n+1)$ is x^n for $n \neq -1$. (We must have $n \neq -1$ because the denominator is zero when $n = -1$.) Formula 2 states that the indefinite integral of a power of x (except x^{-1}) is obtained by increasing the exponent for x by one, dividing by the new exponent, and adding on the constant of integration. The case for x^{-1} will be discussed in Chapter 6.

To verify Formula 3 we must show that the derivative of $k\int f(x)\ dx$ is $kf(x)$. Since the derivative of $k\int f(x)\ dx$ is k times the derivative of $\int f(x)\ dx$, which is $f(x)$, Formula 3 is verified. You should verify the other formulas. Formula 4 can be extended to any number of sums or differences.

EXAMPLE 2   *Find the following indefinite integrals.*

a. $\int 1\ dx.$

By Formula 1 with $k = 1$,

$$\int 1\ dx = 1x + C = x + C.$$

Usually, we write $\int 1\ dx$ as $\int dx$. Thus $\int dx = x + C.$

b. $\int x^5\ dx.$

By Formula 2 with $n = 5$,

$$\int x^5\ dx = \frac{x^{5+1}}{5+1} + C = \frac{x^6}{6} + C.$$

c. $\int 7x\ dx.$

By Formula 3 with $k = 7$ and $f(x) = x$,

$$\int 7x \, dx = 7 \int x \, dx.$$

Since x is x^1, by Formula 2 we have

$$\int x^1 \, dx = \frac{x^{1+1}}{1+1} + C_1 = \frac{x^2}{2} + C_1,$$

where C_1 is the constant of integration. Therefore,

$$\int 7x \, dx = 7 \int x \, dx = 7 \left[\frac{x^2}{2} + C_1 \right] = \frac{7}{2}x^2 + 7C_1.$$

Since $7C_1$ is just an arbitrary constant, for simplicity we shall replace it by C. Thus

$$\int 7x \, dx = \frac{7}{2}x^2 + C.$$

It is not necessary to write all intermediate steps when integrating. More simply we write

$$\int 7x \, dx = (7)\frac{x^2}{2} + C = \frac{7}{2}x^2 + C.$$

Pitfall

Only a constant factor of the integrand can "jump" in front of an integral sign. Because x is not a constant,

$$\int 7x \, dx \neq 7x \int dx = (7x)(x + C) = 7x^2 + 7Cx.$$

EXAMPLE 3   *Find the following indefinite integrals.*

a. $\int \dfrac{1}{\sqrt{t}} \, dt.$

Here t is the variable of integration. We rewrite the integrand so that a basic formula can be used. Since $1/\sqrt{t} = t^{-1/2}$, applying Formula 2 gives

$$\int \frac{1}{\sqrt{t}} \, dt = \int t^{-1/2} \, dt = \frac{t^{(-1/2)+1}}{-\frac{1}{2}+1} + C = \frac{t^{1/2}}{\frac{1}{2}} + C = 2\sqrt{t} + C.$$

b. $\int \dfrac{1}{6x^3} \, dx.$

$$\int \frac{1}{6x^3} \, dx = \frac{1}{6} \int x^{-3} \, dx = \left(\frac{1}{6}\right)\frac{x^{-3+1}}{-3+1} + C$$

$$= -\frac{x^{-2}}{12} + C = -\frac{1}{12x^2} + C.$$

EXAMPLE 4   *Find the following indefinite integrals.*

a. $\int (x^2 + 2x)\ dx.$

By Formula 4,

$$\int (x^2 + 2x)\ dx = \int x^2\ dx + \int 2x\ dx.$$

Now,

$$\int x^2\ dx = \frac{x^{2+1}}{2+1} + C_1 = \frac{x^3}{3} + C_1,$$

and $$\int 2x\ dx = 2\int x\ dx = (2)\frac{x^{1+1}}{1+1} + C_2 = x^2 + C_2.$$

Thus

$$\int (x^2 + 2x)\ dx = \frac{x^3}{3} + x^2 + C_1 + C_2.$$

Replacing $C_1 + C_2$ by C, we have

$$\int (x^2 + 2x)\ dx = \frac{x^3}{3} + x^2 + C.$$

Omitting intermediate steps, we simply write

$$\int (x^2 + 2x)\ dx = \frac{x^3}{3} + (2)\frac{x^2}{2} + C = \frac{x^3}{3} + x^2 + C.$$

b. $\int (2\sqrt[5]{x^4} - 7x^3 + 1)\ dx.$

$$\int (2\sqrt[5]{x^4} - 7x^3 + 1)\ dx = 2\int x^{4/5}\ dx - 7\int x^3\ dx + \int dx$$

$$= (2)\frac{x^{9/5}}{\frac{9}{5}} - (7)\frac{x^4}{4} + x + C$$

$$= \frac{10}{9}x^{9/5} - \frac{7}{4}x^4 + x + C.$$

Although there are many antiderivatives of a function, we may want to find a *particular* one that satisfies a certain condition. The following examples illustrate.

EXAMPLE 5   *Suppose that $y' = x^2 - 6$. If $y(3) = 1$ (this notation means that $y = 1$ when $x = 3$), find the function y. Determine $y(-3)$.*

Since $y' = x^2 - 6$, y is an antiderivative of $x^2 - 6$. Thus

$$y = \int (x^2 - 6)\ dx = \frac{x^3}{3} - 6x + C.$$

Therefore, *any* function of the general form

$$y = \frac{x^3}{3} - 6x + C \tag{2}$$

satisfies $y' = x^2 - 6$. To determine the *particular* function for which $y(3) = 1$, we must find the value that C must assume. Substituting $x = 3$ and $y = 1$ into Eq. (2) gives

$$1 = \frac{3^3}{3} - 6(3) + C,$$

$$= 9 - 18 + C.$$

$$1 = -9 + C,$$

from which $C = 10$. Thus the function we are looking for is

$$y = \frac{x^3}{3} - 6x + 10. \tag{3}$$

To find $y(-3)$, we let $x = -3$ in Eq. (3):

$$y(-3) = \frac{(-3)^3}{3} - 6(-3) + 10 = 19.$$

EXAMPLE 6 *For a particular urban group, sociologists studied the average yearly income y (in dollars) that a person can expect to receive with x years of education before seeking regular employment. They estimated that the rate at which income changes with respect to education is given by*

$$\frac{dy}{dx} = 10x^{3/2}, \quad 4 \le x \le 16,$$

where $y = 5872$ when $x = 9$. Find y.

Here y is an antiderivative of $10x^{3/2}$. Thus

$$y = \int 10x^{3/2}\, dx = 10 \int x^{3/2}\, dx$$

$$= (10)\frac{x^{5/2}}{\frac{5}{2}} + C.$$

$$y = 4x^{5/2} + C. \tag{4}$$

Since $y = 5872$ when $x = 9$, by substituting these values into Eq. (4) we can determine the value of C:

$$5872 = 4(9)^{5/2} + C$$

$$= 4(243) + C$$

$$= 972 + C.$$

Therefore, $C = 4900$ and

$$y = 4x^{5/2} + 4900.$$

EXAMPLE 7   *If the marginal revenue function for a manufacturer's product is*

$$\frac{dr}{dq} = 2000 - 20q - 3q^2,$$

find the demand function.

Since dr/dq is the derivative of total revenue r,

$$r = \int (2000 - 20q - 3q^2)\, dq$$

$$= 2000q - (20)\frac{q^2}{2} - (3)\frac{q^3}{3} + C.$$

$$r = 2000q - 10q^2 - q^3 + C. \tag{5}$$

We assume that *when no units are sold, total revenue is* 0; that is, $r = 0$ when $q = 0$. Substituting these values into Eq. (5) gives

$$0 = 2000(0) - 10(0)^2 - 0^3 + C.$$

Hence $C = 0$ and

$$r = 2000q - 10q^2 - q^3.$$

To find the demand function, we use this result together with the general relationship that $r = pq$, where p is the price per unit. Solving $r = pq$ for p and substituting for r gives the demand function:

$$p = \frac{r}{q} = \frac{2000q - 10q^2 - q^3}{q}.$$

$$p = 2000 - 10q - q^2.$$

EXAMPLE 8   *In the manufacture of a product, fixed costs per week are* $4000. *Fixed costs are costs, such as rent and insurance, that remain constant at all levels of production in a given time period. If the marginal cost function* dc/dq *is*

$$\frac{dc}{dq} = 0.000001(0.002q^2 - 25q) + 0.2,$$

where c is the total cost (in dollars) of producing q pounds of product per week, find the cost of producing 10,000 pounds in one week.

Since dc/dq is the derivative of total cost c,

$$c = \int [0.000001(0.002q^2 - 25q) + 0.2]\, dq$$

$$= 0.000001 \int (0.002q^2 - 25q)\, dq + \int 0.2\, dq.$$

$$c = 0.000001\left(\frac{0.002q^3}{3} - \frac{25q^2}{2}\right) + 0.2q + C.$$

Fixed costs are constant regardless of output. Therefore, when no product is produced, total cost equals fixed cost. That is, when $q = 0$, then $c = 4000$. By substitution we find that $C = 4000$, so

$$c = 0.000001\left(\frac{0.002q^3}{3} - \frac{25q^2}{2}\right) + 0.2q + 4000. \qquad (6)$$

From Eq. (6), when $q = 10,000$ then $c = 5416\frac{2}{3}$. The total cost for producing 10,000 pounds of product in one week is \$5416.67.

EXERCISE 5.1

*In Problems **1–30**, find the indefinite integrals.*

1. $\int 3\ dx.$

2. $\int \frac{1}{2}\ dx.$

3. $\int x^8\ dx.$

4. $\int 2x^{25}\ dx.$

5. $\int 5x^{-7}\ dx.$

6. $\int \frac{z^{-3}}{3}\ dz.$

7. $\int \frac{1}{x^{10}}\ dx.$

8. $\int \frac{7}{x^4}\ dx.$

9. $\int (8 + u)\ du.$

10. $\int (r^3 + 2r)\ dr.$

11. $\int (y^5 - 5y)\ dy.$

12. $\int (7 - 3w - 2w^2)\ dw.$

13. $\int (3t^2 - 4t + 5)\ dt.$

14. $\int (1 + u + u^2 + u^3)\ du.$

15. $\int \left(\frac{x}{7} - \frac{3}{4}x^4\right)\ dx.$

16. $\int \left(\frac{2x^2}{7} - \frac{8}{3}x^4\right)\ dx.$

17. $\int (x^{8.3} - 9x^6 + 3x^{-4} + x^{-3})\ dx.$

18. $\int (0.3y^4 - 8y^{-3} + 2)\ dy.$

19. $\int \frac{-2\sqrt{x}}{3}\ dx.$

20. $\int dw.$

21. $\int \frac{1}{4\sqrt[8]{x^7}}\ dx.$

22. $\int \frac{-3}{4\sqrt[8]{x}}\ dx.$

23. $\int \left(\frac{x^3}{3} - \frac{3}{x^3}\right)\ dx.$

24. $\int \left(\frac{1}{2x^3} - \frac{1}{x^4}\right)\ dx.$

25. $\int \left(\frac{3w^2}{2} - \frac{2}{3w^2}\right)\ dw.$

26. $\int \frac{2}{7^{-1}}\ ds.$

27. $\int (2\sqrt{x} - 3\sqrt[4]{x})\ dx.$

28. $\int 0\ dx.$

29. $\int \left(-\frac{\sqrt[3]{x^2}}{5} - \frac{7}{2\sqrt{x}} + 6x\right)\ dx.$

30. $\int \left(\sqrt[3]{x} - \frac{1}{\sqrt[3]{x}}\right)\ dx.$

*In Problems **31** and **32**, find y subject to the given conditions.*

31. $dy/dx = 3x - 4,\quad y(-1) = \frac{13}{2}.$

32. $dy/dx = x^2 - x,\quad y(3) = 4.$

*In Problems **33** and **34**, if y satisfies the given conditions, find $y(x)$ for the given value of x.*

33. $y' = 4/\sqrt{x},\quad y(4) = 10;\quad x = 9.$

34. $y' = -x^2 + 2x,\quad y(2) = 1;\quad x = 1.$

*In Problems **35–38**, dc/dq is a marginal cost function and fixed costs are indicated in braces. For Problems 35 and 36, find the total cost function. For Problems 37 and 38, find the total cost for the given value of q.*

35. $dc/dq = 1.35,\quad \{200\}.$

36. $dc/dq = 2q + 50,\quad \{1000\}.$

37. $dc/dq = 0.09q^2 - 1.2q + 4.5,\quad \{7700\};\quad q = 10.$

38. $dc/dq = 0.000102q^2 - 0.034q + 5,\quad \{10,000\};\quad q = 100.$

In Problems 39–42, dr/dq is a marginal revenue function. Find the demand function.

39. $dr/dq = 0.7$.

40. $dr/dq = 15 - \frac{1}{15}q$.

41. $dr/dq = 275 - q - 0.3q^2$.

42. $dr/dq = 10,000 - 2(2q + q^3)$.

43. A group of biologists studied the nutritional effects on rats that were fed a diet containing 10 percent protein (adapted from [9]). The protein consisted of yeast and corn flour. Over a period of time, the group found that the (approximate) rate of change of the average weight gain G (in grams) of a rat with respect to the percentage P of yeast in the protein mix is given by

$$\frac{dG}{dP} = -\frac{P}{25} + 2, \qquad 0 \le P \le 100.$$

If $G = 38$ when $P = 10$, find G.

44. A manufacturer has determined that the marginal cost function is $dc/dq = 0.003q^2 - 0.4q + 40$, where q is the number of units produced. If marginal cost is $27.50 when $q = 50$, and fixed costs are $5000, what is the *average* cost of producing 100 units?

45. A study of the winter moth was made in Nova Scotia (adapted from [17]). The prepupae of the moth fall onto the ground from host trees. It was found that the (approximate) rate at which prepupal density y (number of prepupae per square foot of soil) changes with respect to distance x (in feet) from the base of a host tree is given by

$$\frac{dy}{dx} = -1.5 - x, \qquad 1 \le x \le 9.$$

If $y = 57.3$ when $x = 1$, find y.

46. In the study of flow of fluid in a tube of constant radius R, such as blood flow in portions of the body, one can think of the tube as consisting of concentric tubes of radius r, where $0 \le r \le R$. The velocity v of the fluid is a function of r and is given in [13] by

$$v = \int -\frac{(P_1 - P_2)r}{2l\eta}\, dr,$$

where P_1 and P_2 are pressures at the ends of the tube, η (a Greek letter read "eta") is fluid viscosity, and l is the length of the tube. If $v = 0$ when $r = R$, show that

$$v = \frac{(P_1 - P_2)(R^2 - r^2)}{4l\eta}.$$

5.2 TECHNIQUES OF INTEGRATION

In order to apply the basic integration formulas, it may be necessary to rewrite the integrand. Example 1 shows some techniques.

EXAMPLE 1   *Find the following indefinite integrals.*

a. $\displaystyle\int \frac{(2x - 1)(x + 3)}{6}\, dx.$

By factoring out $\frac{1}{6}$ and multiplying the binomials, we get

$$\int \frac{(2x - 1)(x + 3)}{6}\, dx = \frac{1}{6}\int (2x^2 + 5x - 3)\, dx$$

$$= \frac{1}{6}\left[(2)\frac{x^3}{3} + (5)\frac{x^2}{2} - 3x \right] + C$$

$$= \frac{x^3}{9} + \frac{5x^2}{12} - \frac{x}{2} + C.$$

b. $\displaystyle\int \frac{x^4 - 1}{x^2}\, dx.$

We rewrite the integrand by dividing each term in the numerator by the denominator.

$$\int \frac{x^4 - 1}{x^2}\, dx = \int \left[x^2 - \frac{1}{x^2} \right] dx = \frac{x^3}{3} - \int x^{-2}\, dx$$

$$= \frac{x^3}{3} - \frac{x^{-1}}{-1} + C = \frac{x^3}{3} + \frac{1}{x} + C.$$

The formula

$$\int x^n\, dx = \frac{x^{n+1}}{n+1} + C, \qquad \text{if } n \neq -1,$$

which applies to a power of x, can be generalized to handle a power of a *function* of x. Let u be a differentiable function of x. By the power rule for differentiation, if $n \neq -1$, then

$$\frac{d}{dx} \left(\frac{[u(x)]^{n+1}}{n+1} \right) = \frac{(n+1)[u(x)]^n \cdot u'(x)}{n+1} = [u(x)]^n \cdot u'(x).$$

Thus

$$\int [u(x)]^n \cdot u'(x)\, dx = \frac{[u(x)]^{n+1}}{n+1} + C, \qquad n \neq -1.$$

We call this formula the *power rule for integration*. Since $u'(x)\, dx$ is the differential of u, namely du, for mathematical shorthand we can replace $u(x)$ by u and $u'(x)\, dx$ by du:

POWER RULE FOR INTEGRATION

If u is differentiable, then

$$\int u^n\, du = \frac{u^{n+1}}{n+1} + C, \qquad \text{if } n \neq -1.$$

It is essential that you realize the difference between the power rule for integration and the formula for $\int x^n\, dx$. In the power rule, u represents a function, whereas in $\int x^n\, dx$, x is a variable.

EXAMPLE 2   *Use the power rule for integration to find the following.*

a. $\int (x + 1)^{20} \, dx$.

Since the integrand is a power of the function $x + 1$, we shall set $u = x + 1$. Then $du = dx$ and $\int (x + 1)^{20} \, dx$ has the form $\int u^{20} \, du$. By the power rule for integration,

$$\int (x + 1)^{20} \, dx = \int u^{20} \, du = \frac{u^{21}}{21} + C = \frac{(x + 1)^{21}}{21} + C.$$

Note that we give our answer not in terms of u but explicitly in terms of x.

b. $\int 3x^2(x^3 + 7)^3 \, dx$.

Let $u = x^3 + 7$. Then $du = 3x^2 \, dx$. Fortunately, $3x^2$ appears as a factor in the integrand and can be used as part of du.

$$\int 3x^2(x^3 + 7)^3 \, dx = \int (x^3 + 7)^3 [3x^2 \, dx] = \int u^3 \, du$$

$$= \frac{u^4}{4} + C = \frac{(x^3 + 7)^4}{4} + C.$$

EXAMPLE 3   *Find* $\int x\sqrt{x^2 + 5} \, dx$.

We can write this as $\int x(x^2 + 5)^{1/2} \, dx$. If $u = x^2 + 5$, then $du = 2x \, dx$. Since the *constant* factor 2 in du does *not* appear in the integrand, this integral does not have the form $\int u^n \, du$. However, we can put the given integral in this form by first multiplying and dividing the integrand by 2. This does not change its value. Thus

$$\int x(x^2 + 5)^{1/2} \, dx = \int \frac{2}{2} x(x^2 + 5)^{1/2} \, dx = \int \frac{1}{2}(x^2 + 5)^{1/2}[2x \, dx].$$

Moving the *constant* factor $\frac{1}{2}$ in front of the integral sign, we have

$$\int x(x^2 + 5)^{1/2} \, dx = \frac{1}{2}\int (x^2 + 5)^{1/2}[2x \, dx] \tag{1}$$

$$= \frac{1}{2}\int u^{1/2} \, du = \frac{1}{2}\left[\frac{u^{3/2}}{\frac{3}{2}}\right] + C.$$

Going back to x gives

$$\int x\sqrt{x^2 + 5} \, dx = \frac{(x^2 + 5)^{3/2}}{3} + C.$$

In Example 3 we needed the factor 2 in the integrand. In Eq. (1) it was inserted and the integral was simultaneously multiplied by $\frac{1}{2}$. More generally, if c is a nonzero constant, then

$$\int f(x)\, dx = \int \frac{c}{c} f(x)\, dx = \frac{1}{c}\int cf(x)\, dx.$$

In effect, we can multiply the integrand by a nonzero constant c as long as we compensate for this by multiplying the entire integral by $1/c$. Such a manipulation **cannot** be done with *variable* factors.

Pitfall

When using the form $\int u^n\, du$, do not neglect du. For example,

$$\int (4x + 1)^2\, dx \neq \frac{(4x + 1)^3}{3} + C.$$

The proper way of doing the problem is as follows. Letting $u = 4x + 1$, we have $du = 4\, dx$. Thus

$$\int (4x + 1)^2\, dx = \frac{1}{4}\int (4x + 1)^2 [4\, dx] = \frac{1}{4}\int u^2\, du$$

$$= \frac{1}{4} \cdot \frac{u^3}{3} + C = \frac{(4x + 1)^3}{12} + C.$$

EXAMPLE 4   Find the following indefinite integrals.

a. $\int \dfrac{1}{(1 - w)^2}\, dw.$

The integrand can be rewritten as $(1 - w)^{-2}$. If $u = 1 - w$, then $du = -dw$. Thus we insert a factor of -1 and adjust for it with a factor of -1 in front of the integral.

$$\int (1 - w)^{-2}\, dw = -\int (1 - w)^{-2}[-dw].$$

The integral has the form $\int u^{-2}\, du$, so

$$\int (1 - w)^{-2}\, dw = -\frac{(1 - w)^{-1}}{-1} + C = \frac{1}{1 - w} + C.$$

b. $\int \dfrac{2x^3 + 3x}{(x^4 + 3x^2 + 7)^4}\, dx.$

We can write this as

$$\int (x^4 + 3x^2 + 7)^{-4}(2x^3 + 3x)\, dx.$$

If $u = x^4 + 3x^2 + 7$, then $du = (4x^3 + 6x)\, dx$, which is two times the quantity $(2x^3 + 3x)\, dx$ in the integral. Thus we insert a factor of 2 and adjust for it with a factor of $\frac{1}{2}$ in front of the integral.

$$\int (x^4 + 3x^2 + 7)^{-4}(2x^3 + 3x)\, dx = \frac{1}{2}\int (x^4 + 3x^2 + 7)^{-4}[2(2x^3 + 3x)\, dx]$$

$$= \frac{1}{2}\int (x^4 + 3x^2 + 7)^{-4}[(4x^3 + 6x)\, dx]$$

$$= \frac{1}{2}\int u^{-4}\, du = \frac{1}{2}\cdot\frac{u^{-3}}{-3} + C = -\frac{1}{6u^3} + C$$

$$= -\frac{1}{6(x^4 + 3x^2 + 7)^3} + C.$$

c. $\displaystyle\int \frac{1}{\sqrt{x}\,(\sqrt{x} - 2)^3}\, dx.$

We can write this integral as $\displaystyle\int \frac{(\sqrt{x} - 2)^{-3}}{\sqrt{x}}\, dx.$ Let $u = \sqrt{x} - 2$. Then

$du = \dfrac{1}{2\sqrt{x}}\, dx$ and

$$\int \frac{(\sqrt{x} - 2)^{-3}}{\sqrt{x}}\, dx = 2\int (\sqrt{x} - 2)^{-3}\left[\frac{1}{2\sqrt{x}}\, dx\right]$$

$$= 2\int u^{-3}\, du = 2\left(\frac{u^{-2}}{-2}\right) + C$$

$$= -u^{-2} + C = -(\sqrt{x} - 2)^{-2} + C$$

$$= -\frac{1}{(\sqrt{x} - 2)^2} + C.$$

When using the power rule for integration, take care when making your choice for u. In Example 4(b) you would *not* be able to proceed very far if, for instance, you let $u = 2x^3 + 3x$. At times you may find it necessary to try many choices. So don't just sit and look at an integral. Try something even if it is wrong, because it may give you a hint as to what might work. **Skill at integration comes only after many hours of practice and conscientious study.**

EXAMPLE 5   *Find* $\displaystyle\int 4x^2(x^4 + 1)^2\, dx.$

If we set $u = x^4 + 1$, then $du = 4x^3\, dx$. To get du in the integral, we need an additional factor of the *variable* x. However, we can only adjust for **constant** factors. Thus we cannot use the power rule. To find the integral, we shall first expand $(x^4 + 1)^2$.

$$\int 4x^2(x^4 + 1)^2 \, dx = 4 \int x^2(x^8 + 2x^4 + 1) \, dx$$

$$= 4 \int (x^{10} + 2x^6 + x^2) \, dx$$

$$= 4\left(\frac{x^{11}}{11} + \frac{2x^7}{7} + \frac{x^3}{3}\right) + C.$$

EXAMPLE 6   *For a certain country, the marginal propensity to consume is given by*

$$\frac{dC}{dI} = \frac{3}{4} - \frac{1}{2\sqrt{3I}},$$

where consumption C is a function of national income I. Here I is expressed in billions of slugs (50 slugs = \$0.01). Determine the consumption function for the country if it is known that consumption is 10 billion slugs (C = 10) when I = 12.

Since the marginal propensity to consume is the derivative of C, we have

$$C = \int \left(\frac{3}{4} - \frac{1}{2\sqrt{3I}}\right) dI = \int \frac{3}{4} \, dI - \frac{1}{2}\int (3I)^{-1/2} \, dI$$

$$= \frac{3}{4}I - \frac{1}{2}\int (3I)^{-1/2} \, dI.$$

If we let $u = 3I$, then $du = 3 \, dI$ and

$$C = \frac{3}{4}I - \left(\frac{1}{2}\right)\frac{1}{3}\int (3I)^{-1/2}[3 \, dI]$$

$$= \frac{3}{4}I - \frac{1}{6} \cdot \frac{(3I)^{1/2}}{\frac{1}{2}} + C_1.$$

$$C = \frac{3}{4}I - \frac{\sqrt{3I}}{3} + C_1.$$

When $I = 12$, then $C = 10$, so

$$10 = \frac{3}{4}(12) - \frac{\sqrt{3(12)}}{3} + C_1$$

$$= 9 - 2 + C_1.$$

$$10 = 7 + C_1.$$

Thus $C_1 = 3$, so the consumption function is

$$C = \frac{3}{4}I - \frac{\sqrt{3I}}{3} + 3.$$

EXERCISE 5.2

*In Problems **1–48**, find the indefinite integrals.*

1. $\int (x^2 + 5)(x - 3)\, dx.$

2. $\int x^4(x^3 + 3x^2 + 7)\, dx.$

3. $\int \sqrt{x}\,(x + 3)\, dx.$

4. $\int (\sqrt{z} + 2)^2\, dz.$

5. $\int (u^2 + 1)^2\, du.$

6. $\int \left(\frac{1}{\sqrt[3]{x}} + 1 \right)^2 dx.$

7. $\int v^{-2}(2v^4 + 3v^2 - 2v^{-3})\, dv.$

8. $\int \frac{\sqrt{x}\,(x^5 - \sqrt[3]{x} + 2)}{3}\, dx.$

9. $\int \frac{3x^3 + x^2 - 1}{x^2}\, dx.$

10. $\int \frac{3x^2 - 7x}{4x}\, dx.$

11. $\int (x + 4)^8\, dx.$

12. $\int 2(x + 3)^3\, dx.$

13. $\int 2x(x^2 + 16)^3\, dx.$

14. $\int (3x^2 + 14x)(x^3 + 7x^2 + 1)\, dx.$

15. $\int (3y^2 + 6y)(y^3 + 3y^2 + 1)^{2/3}\, dy.$

16. $\int (-4z^3 - 6)(-z^4 - 6z + 3)^{18}\, dz.$

17. $\int \frac{3}{(3x - 1)^3}\, dx.$

18. $\int \frac{4x}{(2x^2 - 7)^{10}}\, dx.$

19. $\int \sqrt{x + 10}\, dx.$

20. $\int \frac{1}{\sqrt{x - 2}}\, dx.$

21. $\int (7x - 6)^4\, dx.$

22. $\int x^2(3x^3 + 7)^3\, dx.$

23. $\int x(x^2 + 3)^{12}\, dx.$

24. $\int x\sqrt{1 + 2x^2}\, dx.$

25. $\int x^4(27 + x^5)^{1/3}\, dx.$

26. $\int (3 - 2x)^{10}\, dx.$

27. $\int \frac{6z}{(z^2 - 6)^5}\, dz.$

28. $\int \frac{1}{(8y - 3)^3}\, dy.$

29. $\int \sqrt{5x}\, dx.$

30. $\int \frac{1}{(4x)^7}\, dx.$

31. $\int \frac{x}{\sqrt{x^2 - 4}}\, dx.$

32. $\int (t^2 + 4t)(t^3 + 6t^2)^6\, dt.$

33. $\int (x + 1)(3 - 3x^2 - 6x)^3\, dx.$

34. $\int \sqrt{4x - 3}\, dx.$

35. $\int 4\sqrt[3]{y + 1}\, dy.$

36. $\int \frac{x^2}{\sqrt[3]{2x^3 + 9}}\, dx.$

37. $\int x\sqrt{(7 - 5x^2)^3}\, dx.$

38. $\int -(x^2 - 2x^5)(x^3 - x^6)^{-10}\, dx.$

39. $\int \frac{dx}{\sqrt{2x}}.$

40. $\int (2x^3 + x)(x^4 + x^2)\, dx.$

41. $\int r(r^3 + 5)^2\, dr.$

42. $\int \frac{18 + 12x}{(4 - 9x - 3x^2)^5}\, dx.$

43. $\int \frac{8x^3 - 6x - x^4}{3x^3}\, dx.$

44. $\int \frac{2x^4 - 8x^3 - 6x + 4}{x^3}\, dx.$

45. $\int \frac{(\sqrt{x} + 2)^2}{3\sqrt{x}}\, dx.$

46. $\int \frac{4}{\sqrt{2 - 3x}}\, dx.$

47. $\int \sqrt{t}\,(5 - t\sqrt{t})^{0.4}\, dt.$

48. $\int \sqrt{x}\,\sqrt{(8x)^{3/2} + 3}\, dx.$

*In Problems **49** and **50**, find y subject to the given conditions.*

49. $D_x y = (3 - 2x)^2;\quad y(0) = 1.$

50. $D_x y = \frac{x}{(x^2 + 4)^2};\quad y(1) = 0.$

*In Problems **51** and **52**, dr/dq is a marginal revenue function. Find the demand function.*

51. $\dfrac{dr}{dq} = \dfrac{200}{(q + 2)^2}.$

52. $\dfrac{dr}{dq} = \dfrac{900}{(2q + 3)^3}.$

In Problems 53–55, dC/dI represents marginal propensity to consume. Find the consumption function subject to the given condition.

54. $\dfrac{dC}{dI} = \dfrac{3}{4} - \dfrac{1}{2\sqrt{3I}}$; $C(3) = \dfrac{11}{4}$.

53. $\dfrac{dC}{dI} = \dfrac{1}{\sqrt{I}}$; $C(9) = 8$.

55. $\dfrac{dC}{dI} = \dfrac{3}{4} - \dfrac{1}{6\sqrt{I}}$; $C(25) = 23$.

5.3 AREA

In this section we are interested in computing the areas of regions. We will show that area is related to a limiting process as well as to antidifferentiation. In later sections we shall generalize our method for finding area and obtain a powerful mathematical tool, called the *definite integral*, which can be applied to many types of non-area problems that occur in economics and the life and social sciences.

In discussing how to find the area of a region, we shall be working with certain sums. To conveniently indicate a sum we introduce *sigma notation*, so named because the Greek letter Σ (sigma) is used. For example, the symbol

$$\sum_{i=1}^{3} (2i + 5)$$

denotes the sum of those numbers obtained from the expression $2i + 5$ by first replacing i by 1, then by 2, and finally by 3. Thus

$$\sum_{i=1}^{3} (2i + 5) = [2(1) + 5] + [2(2) + 5] + [2(3) + 5]$$

$$= 7 + 9 + 11 = 27.$$

The letter i is called the *index of summation;* the numbers 1 and 3 are the *limits of summation* (1 is the *lower limit* and 3 is the *upper limit*). The values of the index begin at the lower limit and progress through integer values to the upper limit. The symbol used for the index is a "dummy" symbol in the sense that it does not affect the sum of the terms. Any other letter can be used. For example,

$$\sum_{k=1}^{3} (2k + 5) = 7 + 9 + 11 = \sum_{i=1}^{3} (2i + 5).$$

We shall now use sigma notation in considering the area of a region in the plane. Figure 5.1 shows the graph of a function f such that over the interval $[a, b]$, f is continuous and $f(x) \geq 0$ (that is, the graph does not fall below the x-axis). Let us consider the area A of the shaded region bounded by $y = f(x)$ and the lines $y = 0$ (the x-axis), $x = a$, and $x = b$.

In Fig. 5.2, we have divided the interval $[a, b]$ into four subintervals of equal length Δx by using the equally spaced points $x_0 = a$, x_1, x_2, x_3, and

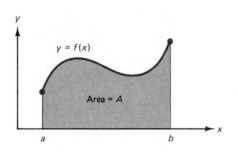

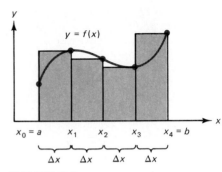

FIGURE 5.1   FIGURE 5.2

$x_4 = b$. Corresponding to each subinterval is a rectangle whose width is Δx and whose height is the value of $f(x)$ at the right-hand endpoint of the subinterval. The sum of the areas of these rectangles is an approximation to A:

$$f(x_1)\,\Delta x + f(x_2)\,\Delta x + f(x_3)\,\Delta x + f(x_4)\,\Delta x \approx A,$$

or, with sigma notation,

$$\sum_{i=1}^{4} f(x_i)\,\Delta x \approx A.$$

In Fig. 5.3, $[a, b]$ is divided into more subintervals, and the sum of the areas of the rectangles is a better approximation to A. This diagram suggests that

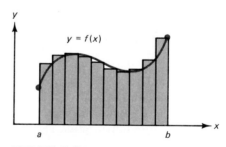

FIGURE 5.3

the more rectangles, the better the approximation is. In fact, it can be shown that if n rectangles are used, the limit of the sum of their areas as $n \to \infty$ is A.

More specifically, we divide $[a, b]$ into n subintervals of equal length Δx. Then we have $\Delta x = (b - a)/n$. The endpoints of the subintervals are $x_0, x_1, \ldots, x_n$ as indicated in Fig. 5.4. Corresponding to an arbitrary subinterval—

FIGURE 5.4

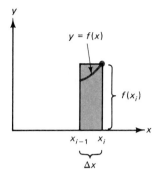

FIGURE 5.5

let's call it the ith subinterval—is a rectangle having width Δx and height $f(x_i)$ (see Fig. 5.5). Thus the area of the rectangle is $f(x_i) \, \Delta x$. The limit of the sum of the areas of all such rectangles is A:*

$$\lim_{n \to \infty} \sum_{i=1}^{n} f(x_i) \, \Delta x = A.$$

Because this method of finding area can be extremely awkward to use, let us take a different approach which, as you will see, involves antidifferentiation. Suppose that there is a function $S = S(x)$, which we shall refer to as an "area" function, that gives the area of the region below the graph of f and above the x-axis from a to x, where $a \le x \le b$. This region is shaded in Fig. 5.6. Two properties of S are as follows:

1. $S(a) = 0$ since the area from a to a is 0.

2. $S(b)$ is the area from a to b; that is, $S(b) = A$.

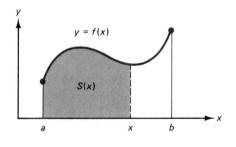

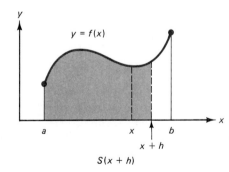

FIGURE 5.6   FIGURE 5.7

If x is increased by h units, then $S(x + h)$ is the area of the shaded region in Fig. 5.7. Hence $S(x + h) - S(x)$ is the difference of the areas in Figs. 5.7 and 5.6: namely, the area of the shaded region in Fig. 5.8. For h sufficiently

* For convenience, our subintervals have the same length. However, in general it is not necessary that this condition be imposed. Moreover, it is not necessary that the height of the ith rectangle be $f(x_i)$; it can be $f(x)$ where x is any point in the ith subinterval.

close to zero, the area of this region is the same as the area of a rectangle (Fig. 5.9) whose width is h and whose height is some value $\bar{y}$ between $f(x)$

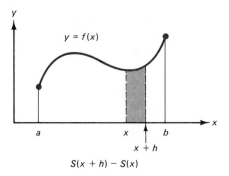

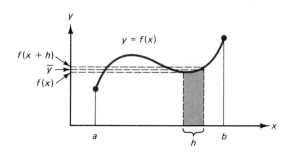

FIGURE 5.8   FIGURE 5.9

and $f(x + h)$. Here $\bar{y}$ is a function of h. Thus the area of the rectangle is, on the one hand, $S(x + h) - S(x)$, and on the other hand, $h\bar{y}$:

$$S(x + h) - S(x) = h\bar{y}.$$

Dividing by h,

$$\frac{S(x + h) - S(x)}{h} = \bar{y}.$$

If $h \to 0$, then $\bar{y}$ approaches the number $f(x)$, so

$$\lim_{h \to 0} \frac{S(x + h) - S(x)}{h} = f(x). \qquad (1)$$

Observe that the left side is the derivative of S. Thus Eq. (1) becomes

$$S'(x) = f(x).$$

We conclude that the area function S has the additional property that its derivative S' is f. That is, S is an antiderivative of f. Now, suppose that F is *any* antiderivative of f. Since both S and F are antiderivatives of the same function, they can differ at most by a constant C:

$$S(x) = F(x) + C. \qquad (2)$$

Recall that $S(a) = 0$. Evaluating both sides of Eq. (2) when $x = a$ gives

$$0 = F(a) + C,$$

so

$$C = -F(a).$$

Thus Eq. (2) becomes

$$S(x) = F(x) - F(a). \qquad (3)$$

Now, recall that $S(b) = A$. Thus from Eq. (3),

$$S(b) = F(b) - F(a),$$
$$A = F(b) - F(a).$$

In summary, *if f is continuous and f(x) ≥ 0 on [a, b], then the area A of the region bounded by the curve y = f(x) and the x-axis from x = a to x = b is given by*

$$A = F(b) - F(a),$$

where F is any antiderivative of f. The difference $F(b) - F(a)$ has a special notation:

$$F(x) \Big|_a^b \quad \text{denotes the difference} \quad F(b) - F(a).$$

Thus

$$A = F(x) \Big|_a^b.$$

We have considered two methods of determining area: one involves a limit of a sum, while the other involves an antiderivative. From a computational point of view, the antiderivative method is easier. However, the limit method has an important mathematical significance, which will be discussed in the next section. The following examples illustrate the antiderivative method to find area.

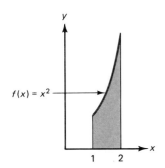

$f(x) = x^2$

1   2

FIGURE 5.10

EXAMPLE 1   *Find the area of the region bounded by y = f(x) = x²* *and the lines y = 0, x = 1, and x = 2 (see Fig. 5.10).*

Notice that $f(x) \geq 0$ and f is continuous on $[1, 2]$. Thus we can apply the antiderivative method. An antiderivative of $f(x) = x^2$ is $F(x) = x^3/3$. Hence

$$\text{area} = F(x) \Big|_a^b = \frac{x^3}{3} \Big|_1^2 = \frac{2^3}{3} - \frac{1^3}{3} = \frac{7}{3} \text{ square units.}$$

Alternatively, if we chose $F(x)$ to be $(x^3/3) + C$, then

$$F(x) \Big|_1^2 = \left(\frac{x^3}{3} + C \right) \Big|_1^2 = \left(\frac{2^3}{3} + C \right) - \left(\frac{1^3}{3} + C \right) = \frac{7}{3},$$

as before. Since the choice of the value of C is immaterial as far as the answer is concerned, for convenience we shall always choose C to be 0, as originally done.

EXAMPLE 2   *Find the area of the region bounded by $y = 8 - 2x^2$ and the x-axis from $x = -1$ to $x = 2$ (see Fig. 5.11).*

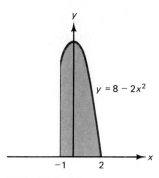

$y = 8 - 2x^2$

−1   2

FIGURE 5.11

An antiderivative of $8 - 2x^2$ is $8x - (2x^3/3)$.

$$\text{Area} = \left(8x - \frac{2x^3}{3} \right) \Bigg|_{-1}^{2}$$

$$= \left[8(2) - \frac{2(2)^3}{3} \right] - \left[8(-1) - \frac{2(-1)^3}{3} \right]$$

$$= 16 - \frac{16}{3} + 8 - \frac{2}{3} = 18 \text{ square units.}$$

EXERCISE 5.3

In Problems 1–12, sketch the region bounded by the given curves and use the antiderivative method to find the area of the region.

1. $y = 2x,\quad y = 0,\quad x = 2,\quad x = 4.$

2. $y = 10 - 3x,\quad y = 0,\quad x = 1,\quad x = 2.$

3. $y = 4 - x,\quad y = 0,\quad x = -3,\quad x = 1.$

4. $y = x^2 - 4,\quad y = 0,\quad x = 2,\quad x = 4.$

5. $y = x^2 + x,\quad y = 0,\quad x = 0,\quad x = 2.$

6. $y = 3 - 2x - x^2,\quad y = 0,\quad x = -2,\quad x = 0.$

7. $y = -4 - 5x - x^2,\quad y = 0,\quad x = -3,\quad x = -2.$

8. $y = (2x + 1)^2,\quad y = 0,\quad x = 3,\quad x = 4.$

9. $y = 8 - x^3,\quad y = 0,\quad x = -1,\quad x = 1.$

10. $y = \dfrac{2(3x^3 + 1)}{5},\quad y = 0,\quad x = 0,\quad x = 2.$

11. $y = \sqrt{x},\quad y = 0,\quad x = 1,\quad x = 4.$

12. $y = \dfrac{1}{x^2},\quad y = 0,\quad x = -5,\quad x = -2.$

In Problems 13 and 14, use the antiderivative method to find the area of the region bounded by the given curves. Then use geometric formulas to confirm your result.

13. $y = 4,\quad y = 0,\quad x = 0,\quad x = 2.$

14. $y = 2x,\quad y = 0,\quad x = 0,\quad x = 4.$

5.4   THE DEFINITE INTEGRAL

In the last section, we saw that if f is continuous and $f(x) \geq 0$ on $[a, b]$, then the area of the region bounded by $y = f(x)$ and the x-axis from $x = a$ to $x = b$ is given by

$$\lim_{n \to \infty} \sum_{i=1}^{n} f(x_i)\, \Delta x,$$

or, alternatively, by

$$F(x) \Big|_a^b = F(b) - F(a),$$

where F is any antiderivative of f. Thus

$$\lim_{n \to \infty} \sum_{i=1}^{n} f(x_i)\, \Delta x = F(b) - F(a). \tag{1}$$

If we do not require that $f(x) \geq 0$, it can be shown that Eq. (1) is still true. In general, if f is any continuous function, then $\lim_{n \to \infty} \sum_{i=1}^{n} f(x_i)\, \Delta x$ is called the **definite integral** of f over the interval $[a, b]$ and is denoted $\int_a^b f(x)\, dx$:

$$\int_a^b f(x)\, dx = \lim_{n \to \infty} \sum_{i=1}^{n} f(x_i)\, \Delta x.$$

The numbers a and b appearing with the integral sign are called **limits of integration**; a is the **lower limit** and b is the **upper limit**. Note that the definite integral is a limit of a sum of the form $\Sigma\, f(x)\, \Delta x$. In fact, one can think of the integral sign as an elongated "S", the first letter of "Summation." In summary, we have the following result known as the *Fundamental Theorem of Integral Calculus*.

FUNDAMENTAL THEOREM OF INTEGRAL CALCULUS

If f is continuous on the interval $[a, b]$ and F is any antiderivative of f, then

$$\int_a^b f(x)\, dx = F(b) - F(a).$$

The Fundamental Theorem gives a connection between antidifferentiation and (definite) integration. A *definite integral*, which is a limit of a special sum, can be conveniently determined by a difference of function values of an antiderivative. Thus we can evaluate definite integrals without using limits. It is important that you understand the difference between an indefinite integral and a definite integral. An indefinite integral is a *function*, but a definite integral is a *number*.

Obviously, if $f(x) \geq 0$ on $[a, b]$, then $\int_a^b f(x)\, dx$ represents the area below the graph of $y = f(x)$ and above the x-axis from $x = a$ to $x = b$. However, if $f(x) < 0$ for some x in $[a, b]$, then $\int_a^b f(x)\, dx$ does not represent area. The reason is that if $f(x_i) < 0$, then $f(x_i)\, \Delta x$ is not the area of a rectangle; it is the negative of the area of a rectangle.

EXAMPLE 1   *Evaluate* $\int_0^2 (4 - x^2)\, dx$.

We apply the Fundamental Theorem with $f(x) = 4 - x^2$, $a = 0$, and $b = 2$. An antiderivative of $4 - x^2$ is $F(x) = 4x - (x^3/3)$.

$$\int_0^2 (4 - x^2)\, dx = F(x)\Big|_a^b = \left(4x - \frac{x^3}{3}\right)\Big|_0^2$$

$$= \left(8 - \frac{8}{3}\right) - (0) = \frac{16}{3}.$$

As in the preceding section, we chose $F(x)$ with constant of integration zero.

For $\int_a^b f(x)\, dx$ we have assumed that $a < b$. We now define the cases in which $a > b$ or $a = b$. If $a > b$,

$$\int_a^b f(x)\, dx = -\int_b^a f(x)\, dx.$$

That is, interchanging the limits of integration changes an integral's sign. Thus, using the result of Example 1, we have

$$\int_2^0 (4 - x^2)\, dx = -\int_0^2 (4 - x^2)\, dx = -\frac{16}{3}.$$

If the limits of integration are equal, we have

$$\int_a^a f(x)\, dx = 0.$$

Some properties of the definite integral deserve mention. The first property restates more formally our comment concerning area.

1. If f is continuous and $f(x) \geq 0$ on $[a, b]$, then $\int_a^b f(x) \, dx$ gives the area of the region bounded by the curve $y = f(x)$, the x-axis, and the lines $x = a$ and $x = b$.

2. $\int_a^b kf(x) \, dx = k \int_a^b f(x) \, dx$, where k is a constant.

3. $\int_a^b [f(x) \pm g(x)] \, dx = \int_a^b f(x) \, dx \pm \int_a^b g(x) \, dx$.

Properties 2 and 3 are similar to rules for indefinite integrals because a definite integral may be evaluated by the Fundamental Theorem in terms of an antiderivative. Two more properties of definite integrals are as follows.

4. $\int_a^b f(x) \, dx = \int_a^b f(t) \, dt$. The variable of integration is a "dummy variable" in the sense that any other variable produces the same result, that is, the same number. You may verify, for example, that $\int_0^2 x^2 \, dx = \int_0^2 t^2 \, dt$.

5. If f is continuous on an interval I and a, b, and c are in I, then

$$\int_a^c f(x) \, dx = \int_a^b f(x) \, dx + \int_b^c f(x) \, dx.$$

This means that the definite integral over an interval can be expressed in terms of definite integrals over subintervals. Thus

$$\int_0^2 (4 - x^2) \, dx = \int_0^1 (4 - x^2) \, dx + \int_1^2 (4 - x^2) \, dx.$$

We shall give more examples of definite integration now and compute some areas in the next section.

EXAMPLE 2   *Evaluate the following definite integrals.*

a. $\int_{-1}^3 (3x^2 - x + 6) \, dx$.

$$\int_{-1}^3 (3x^2 - x + 6) \, dx = \left(x^3 - \frac{x^2}{2} + 6x \right) \Bigg|_{-1}^3$$

$$= \left[3^3 - \frac{3^2}{2} + 6(3) \right] - \left[(-1)^3 - \frac{(-1)^2}{2} + 6(-1) \right]$$

$$= \left(\frac{81}{2} \right) - \left(-\frac{15}{2} \right) = 48.$$

b. $\displaystyle\int_0^1 \frac{x^3}{\sqrt{1+x^4}}\,dx.$

To find an antiderivative of the integrand, we shall apply the power rule for integration.

$$\int_0^1 \frac{x^3}{\sqrt{1+x^4}}\,dx = \int_0^1 x^3(1+x^4)^{-1/2}\,dx$$

$$= \frac{1}{4}\int_0^1 (1+x^4)^{-1/2}[4x^3\,dx] = \left(\frac{1}{4}\right)\frac{(1+x^4)^{1/2}}{\frac{1}{2}}\Bigg|_0^1$$

$$= \frac{1}{2}(1+x^4)^{1/2}\Bigg|_0^1 = \frac{1}{2}(2)^{1/2} - \frac{1}{2}(1)^{1/2}$$

$$= \frac{1}{2}(\sqrt{2} - 1).$$

c. $\displaystyle\int_1^2 [4t + t(t^2+1)^3]\,dt.$

$$\int_1^2 [4t + t(t^2+1)^3]\,dt = 4\int_1^2 t\,dt + \frac{1}{2}\int_1^2 (t^2+1)^3[2t\,dt]$$

$$= (4)\frac{t^2}{2}\Bigg|_1^2 + \left(\frac{1}{2}\right)\frac{(t^2+1)^4}{4}\Bigg|_1^2$$

$$= 2(4 - 1) + \frac{1}{8}(5^4 - 2^4)$$

$$= 6 + \frac{609}{8} = \frac{657}{8}.$$

EXAMPLE 3   *Evaluate* $\displaystyle\int_{-2}^1 x^3\,dx.$

$$\int_{-2}^1 x^3\,dx = \frac{x^4}{4}\Bigg|_{-2}^1 = \frac{1^4}{4} - \frac{(-2)^4}{4} = \frac{1}{4} - \frac{16}{4} = -\frac{15}{4}.$$

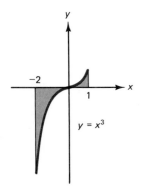

FIGURE 5.12

The reason the result is negative is clear from the graph of $y = x^3$ on the interval $[-2, 1]$ (see Fig. 5.12). For $-2 \le x < 0$, $f(x)$ is negative. Since a definite integral is a limit of a sum of the form $\Sigma\, f(x)\,\Delta x$, then $\displaystyle\int_{-2}^0 x^3\,dx$ is not only a negative number, but it is also the negative of the area of the shaded region in the third quadrant. On the other hand, $\displaystyle\int_0^1 x^3\,dx$ is the area of the shaded region in the first quadrant since $f(x) \ge 0$ on $[0, 1]$. The definite integral over the entire interval $[-2, 1]$ is the *algebraic* sum of these numbers since

$$\int_{-2}^{1} x^3 \, dx = \int_{-2}^{0} x^3 \, dx + \int_{0}^{1} x^3 \, dx.$$

Thus $\int_{-2}^{1} x^3 \, dx$ does not represent the area between the curve and the x-axis. However, the area can be given in the form

$$\left| \int_{-2}^{0} x^3 \, dx \right| + \int_{0}^{1} x^3 \, dx.$$

Pitfall

Remember that $\int_{a}^{b} f(x) \, dx$ is a limit of a sum. In some cases this limit represents area. In others it does not. When $f(x) \geq 0$ on $[a, b]$, then it represents the area between the graph of f and the x-axis from $x = a$ to $x = b$.

Since a function f is an antiderivative of f', by the Fundamental Theorem we have

$$\int_{a}^{b} f'(x) \, dx = f(b) - f(a). \tag{2}$$

But $f'(x)$ is the rate of change of f with respect to x. Thus if we know the rate of change of f and want to find the difference in function values $f(b) - f(a)$, it suffices to evaluate $\int_{a}^{b} f'(x) \, dx$.

EXAMPLE 4   *A manufacturer's marginal cost function is*

$$\frac{dc}{dq} = 0.6q + 2.$$

If production is presently set at $q = 80$ units per week, how much more would it cost to increase production to 100 units per week?

The total cost function is $c = c(q)$, and we want to find the difference $c(100) - c(80)$. The rate of change of c is dc/dq, so by Eq. (2),

$$c(100) - c(80) = \int_{80}^{100} \frac{dc}{dq} \, dq = \int_{80}^{100} (0.6q + 2) \, dq$$

$$= \left(\frac{0.6q^2}{2} + 2q \right) \Bigg|_{80}^{100} = [0.3q^2 + 2q] \Bigg|_{80}^{100}$$

$$= [0.3(100)^2 + 2(100)] - [0.3(80)^2 + 2(80)]$$

$$= 3200 - 2080 = 1120.$$

If c is in dollars, then the cost of increasing production from 80 units to 100 units is \$1120.

EXERCISE 5.4

In Problems 1–28, evaluate the definite integral.

1. $\int_0^3 4\ dx.$

2. $\int_1^3 (2 + 0.2)\ dx.$

3. $\int_1^2 3x\ dx.$

4. $\int_0^2 -5x\ dx.$

5. $\int_{-2}^1 (4x - 6)\ dx.$

6. $\int_1^3 (2w^2 + 1)\ dw.$

7. $\int_{-2}^{-1} (3w^2 - w - 1)\ dw.$

8. $\int_8^9 dt.$

9. $\int_{-1}^1 \sqrt[3]{x^5}\ dx.$

10. $\int_1^2 \frac{x^{-2}}{2}\ dx.$

11. $\int_{1/2}^3 \frac{1}{x^2}\ dx.$

12. $\int_4^9 \left(\frac{1}{\sqrt{x}} - 2\right)\ dx.$

13. $\int_{-1}^1 (z + 1)^5\ dz.$

14. $\int_1^8 (x^{1/3} - x^{-1/3})\ dx.$

15. $\int_0^1 2x^2(x^3 - 1)^3\ dx.$

16. $\int_1^3 (x + 3)^3\ dx.$

17. $\int_1^2 (x + 1)(x^2 + 2x - 5)^3\ dx.$

18. $\int_0^1 (3x^2 + 4x)(x^3 + 2x^2)^4\ dx.$

19. $\int_4^5 \frac{2}{(q - 3)^3}\ dq.$

20. $\int_0^6 \sqrt{2x + 4}\ dx.$

21. $\int_{1/3}^2 \sqrt{10 - 3p}\ dp.$

22. $\int_{-1}^1 q\sqrt{q^2 + 3}\ dq.$

23. $\int_0^1 x^2\sqrt[3]{7x^3 + 1}\ dx.$

24. $\int_0^{\sqrt{7}} \left[2x - \frac{x}{(x^2 + 1)^{5/3}}\right]\ dx.$

25. $\int_1^2 (x + x^{-2} - x^{-3})\ dx.$

26. $\int_a^b (m + ny)\ dy.$

27. $\int_0^1 \frac{6x^3 + 3x}{(x^2 + x^4 + 1)^2}\ dx.$

28. $\int_1^2 \left(6\sqrt{x} - \frac{1}{\sqrt{2x}}\right)\ dx.$

29. In a discussion of gene mutation [30] the following integral occurs:

$$\int_0^{10^{-4}} x^{-1/2}\ dx.$$

Evaluate.

30. Imagine a "one-dimensional" country of length $2R$ (see Fig. 5.13) [31]. Suppose that the production of

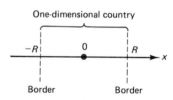

One-dimensional country

FIGURE 5.13

goods for this country is continuously distributed from border to border. If the amount produced each year per unit of distance is $f(x)$, then the country's total yearly production G is given by

$$G = \int_{-R}^R f(x)\ dx.$$

Evaluate G if $f(x) = i$ where i is constant.

31. In discussing traffic safety, Shonle[10] considers how much acceleration a person can tolerate in a crash so that there is no major injury. The *severity index* is defined as follows:

$$\text{severity index} = \int_0^T \alpha^{5/2}\ dt,$$

where α (a Greek letter read "alpha") is considered a constant involved with a weighted average acceleration, and T is the duration of the crash. Find the severity index.

32. In statistics, the mean μ (a Greek letter read "mu") of the continuous probability density function f defined on the interval $[a, b]$ is given by

$$\mu = \int_a^b [x \cdot f(x)]\ dx,$$

and the variance σ^2 (σ is a Greek letter read "sigma") is given by

$$\sigma^2 = \int_a^b (x - \mu)^2 f(x)\, dx.$$

Compute μ and then σ^2 if $a = 0$, $b = 1$, and $f(x) = 1$.

33. For a certain population, suppose that l is a function such that $l(x)$ is the number of persons who reach the age of x in any year of time. This function is called a *life-table function*. Under appropriate conditions, the integral

$$\int_x^{x+n} l(t)\, dt$$

gives the expected number of people in the population between the exact ages of x and $x + n$, inclusive. If $l(x) = 10,000\sqrt{100 - x}$, determine the number of people between the exact ages of 36 and 64 inclusive. Give your answer to the nearest integer, since fractional answers make no sense.

34. The economist Pareto (see [32]) stated an empirical law of distribution of higher incomes that gives the number N of persons receiving x or more dollars. If $dN/dx = -Ax^{-B}$, where A and B are constants, set up a definite integral that gives the total number of persons with incomes between a and b, where $a < b$.

35. A manufacturer's marginal cost function is $dc/dq = 0.2q + 3$. If c is in dollars, determine the cost involved to increase production from 60 to 70 units.

36. Repeat Problem 35 if $dc/dq = 0.003q^2 - 0.6q + 40$ and production increases from 100 to 200 units.

37. A manufacturer's marginal revenue function is given by $dr/dq = 1000/\sqrt{100q}$. If r is in dollars, find the change in the manufacturer's total revenue if production is increased from 400 to 900 units.

38. Repeat Problem 37 if $dr/dq = 250 + 90q - 3q^2$ and production is increased from 10 to 20 units.

39. A sociologist is studying the crime rate in a certain city. She estimates that t months after the beginning of next year, the total number of crimes committed will increase at the rate of $8t + 10$ crimes per month. Determine the total number of crimes that can be expected to be committed next year. How many crimes can be expected to be committed during the last 6 months of that year?

40. For a group of hospitalized individuals, suppose the discharge rate is given by

$$f(t) = \frac{81 \times 10^6}{(300 + t)^4},$$

where $f(t)$ is the proportion of the group discharged per day at the end of t days. What proportion has been discharged by the end of 700 days?

41. In studying the flow of a fluid in a tube of constant radius R, such as blood flow in portions of the body, one can think of the tube as consisting of concentric tubes of radius r, where $0 \le r \le R$. The velocity v of the fluid is a function of r and is given in [13] by

$$v = \frac{(P_1 - P_2)(R^2 - r^2)}{4\eta l},$$

where P_1 and P_2 are pressures at the ends of the tube, η (a Greek letter read "eta") is the fluid viscosity, and l is the length of the tube. The volume rate of flow, Q, through the tube is given by

$$Q = \int_0^R 2\pi r v\, dr.$$

Show that $Q = \dfrac{\pi R^4 (P_1 - P_2)}{8\eta l}$. Note that R occurs as a factor to the fourth power. Thus doubling the radius of the tube has the effect of increasing the flow by a factor of 16. The formula that you derived for the volume rate of flow is called Poiseuille's law, after the French physiologist Jean Poiseuille.

42. In a discussion of a delivered price of a good from a mill to a customer, DeCanio[33] claims that the average delivered price A paid by consumers is given by

$$A = \frac{\int_0^R (m + x)[1 - (m + x)]\, dx}{\int_0^R [1 - (m + x)]\, dx},$$

where m is mill price, x is distance, and R is the maximum distance to the point of sale. DeCanio determines that

$$A = \frac{m + \dfrac{R}{2} - m^2 - mR - \dfrac{R^2}{3}}{1 - m - \dfrac{R}{2}}.$$

Verify this.

5.5 AREA AND THE DEFINITE INTEGRAL

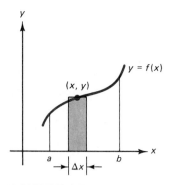

FIGURE 5.14

In Sec. 5.3 we saw that the area of a region could be found by evaluating the limit of a sum of the form $\Sigma f(x)\,\Delta x$, where $f(x)\,\Delta x$ represents the area of a rectangle. From the preceding section, this limit is a special case of a definite integral, so it can easily be found by using the Fundamental Theorem.

When using the definite integral to determine area, you should make a rough sketch of the region involved. Let us consider the area of the region bounded by $y = f(x)$ and the x-axis from $x = a$ to $x = b$, as shown in Fig. 5.14. To set up the integral, a sample rectangle should be included in the sketch because the area of the region is a limit of sums of areas of rectangles. This will not only help you understand the integration process, it will also help you find areas of more complicated regions. Such a rectangle (see Fig. 5.14) is called a **vertical element of area** (or a **vertical strip**). In the diagram, the width of the vertical element is Δx. The height is the y-value of the curve. Hence the rectangle has area $y\,\Delta x$ or $f(x)\,\Delta x$. The area of the entire region is found by summing the areas of all such elements between $x = a$ and $x = b$ and finding the limit of this sum, which is the definite integral. Symbolically, we have

$$\Sigma f(x)\,\Delta x \to \int_a^b f(x)\,dx = \text{area.}$$

Example 1 will illustrate.

EXAMPLE 1   *Find the area of the region bounded by the curve*

$$y = 6 - x - x^2$$

and the x-axis.

First we must sketch the curve so that we can visualize the region. Since

$$y = -(x^2 + x - 6) = -(x - 2)(x + 3),$$

the x-intercepts are 2 and -3. Using techniques of graphing that were previously discussed, we obtain the graph shown in Fig. 5.15. *With this region it is crucial*

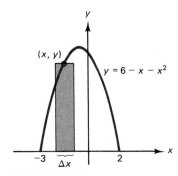

FIGURE 5.15

that the x-intercepts of the curve be found because they determine the interval over which the areas of the elements must be summed. That is, *these x-values are the limits of integration.* The vertical element shown has width Δx and height y. Hence the area of the element is $y \, \Delta x$. Summing the areas of all such elements from $x = -3$ to $x = 2$ and taking the limit via the definite integral gives the area.

$$\sum y \, \Delta x \to \int_{-3}^{2} y \, dx = \text{area}.$$

To evaluate the integral we must express the integrand in terms of the variable of integration, x. Since $y = 6 - x - x^2$,

$$\text{area} = \int_{-3}^{2} (6 - x - x^2) \, dx = \left(6x - \frac{x^2}{2} - \frac{x^3}{3} \right) \Bigg|_{-3}^{2}$$

$$= \left(12 - \frac{4}{2} - \frac{8}{3} \right) - \left(-18 - \frac{9}{2} - \frac{-27}{3} \right) = \frac{125}{6} \text{ square units.}$$

EXAMPLE 2   *Find the area of the region bounded by*

$$y = x^2 + 2x + 2,$$

the x-axis, and the lines $x = -2$ *and* $x = 1$.

A sketch of the region is given in Fig. 5.16.

$$\text{Area} = \int_{-2}^{1} y \, dx = \int_{-2}^{1} (x^2 + 2x + 2) \, dx$$

$$= \left(\frac{x^3}{3} + x^2 + 2x \right) \Bigg|_{-2}^{1} = \left(\frac{1}{3} + 1 + 2 \right) - \left(-\frac{8}{3} + 4 - 4 \right)$$

$$= 6 \text{ square units.}$$

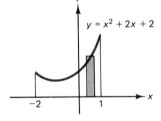

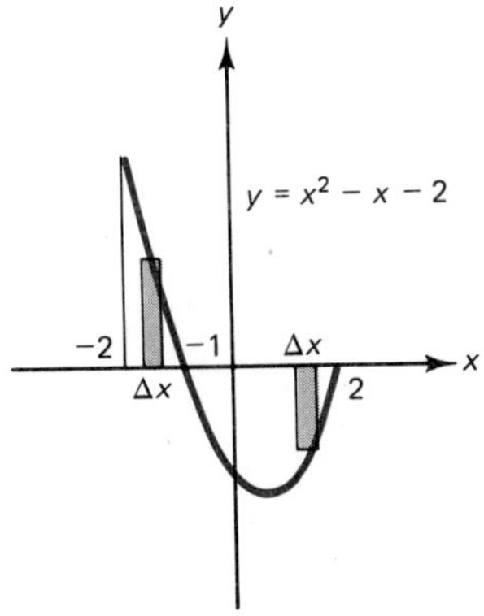

$y = x^2 + 2x + 2$

FIGURE 5.16

EXAMPLE 3   *Find the area of the region bounded by the curves*

$$y = x^2 - x - 2$$

and $y = 0$ *(the x-axis) from* $x = -2$ *to* $x = 2$.

A sketch of the region is given in Fig. 5.17. Notice that the x-intercepts are -1 and 2. It is wrong to write hastily that the area is $\int_{-2}^{2} y \, dx$ for the following reason. The height of the rectangle on the left is y. However, for the rectangle on the right, y is negative, so its height is the positive number $-y$. This points out the importance of sketching the region. On the interval $[-2, -1]$ the area of the element is

$$y \, \Delta x = (x^2 - x - 2) \, \Delta x.$$

$y = x^2 - x - 2$

FIGURE 5.17

On $[-1, 2]$ it is

$$-y \, \Delta x = -(x^2 - x - 2) \, \Delta x.$$

Thus

$$\text{area} = \int_{-2}^{-1} (x^2 - x - 2) \, dx + \int_{-1}^{2} -(x^2 - x - 2) \, dx$$

$$= \left(\frac{x^3}{3} - \frac{x^2}{2} - 2x \right) \Bigg|_{-2}^{-1} - \left(\frac{x^3}{3} - \frac{x^2}{2} - 2x \right) \Bigg|_{-1}^{2}$$

$$= \left[\left(-\frac{1}{3} - \frac{1}{2} + 2 \right) - \left(-\frac{8}{3} - \frac{4}{2} + 4 \right) \right] -$$

$$\left[\left(\frac{8}{3} - \frac{4}{2} - 4 \right) - \left(-\frac{1}{3} - \frac{1}{2} + 2 \right) \right]$$

$$= \frac{19}{3} \text{ square units.}$$

The next example shows the use of area as a probability in statistics.

EXAMPLE 4   *In statistics, a (probability)* **density function** *f of a variable x, where x assumes all values in the interval* $[a, b]$, *has the following properties:*

1. $f(x) \geq 0.$

2. $\displaystyle\int_{a}^{b} f(x) \, dx = 1.$

3. *The probability that x assumes a value between c and d, which is written* $P \, (c \leq x \leq d)$, *where* $a \leq c \leq d \leq b$, *is represented by the area of the region bounded by the graph of f and the x-axis between* $x = c$ *and* $x = d$. *Hence (see Fig. 5.18)*

$$P(c \leq x \leq d) = \int_{c}^{d} f(x) \, dx.$$

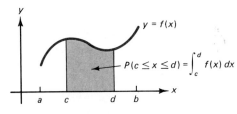

FIGURE 5.18

For the density function given by $f(x) = 6(x - x^2)$, *where* $0 \leq x \leq 1$, *find* (a) $P \, (0 \leq x \leq \frac{1}{4})$ *and* (b) $P(x \geq \frac{1}{2}).$

a. Here $[a, b]$ is $[0, 1]$, c is 0, and d is $\frac{1}{4}$. By Property 3 we have

$$P(0 \le x \le \tfrac{1}{4}) = \int_0^{1/4} 6(x - x^2)\, dx = 6 \int_0^{1/4} (x - x^2)\, dx$$

$$= 6\left(\frac{x^2}{2} - \frac{x^3}{3}\right)\Bigg|_0^{1/4} = (3x^2 - 2x^3)\Bigg|_0^{1/4}$$

$$= \left[3\left(\frac{1}{4}\right)^2 - 2\left(\frac{1}{4}\right)^3\right] - 0 = \frac{5}{32}.$$

b. Since the domain of f is $0 \le x \le 1$, to say that $x \ge \frac{1}{2}$ means $\frac{1}{2} \le x \le 1$. Thus

$$P(x \ge \tfrac{1}{2}) = \int_{1/2}^1 6(x - x^2)\, dx = 6 \int_{1/2}^1 (x - x^2)\, dx$$

$$= 6\left(\frac{x^2}{2} - \frac{x^3}{3}\right)\Bigg|_{1/2}^1 = (3x^2 - 2x^3)\Bigg|_{1/2}^1 = \frac{1}{2}.$$

EXERCISE 5.5

In Problems 1–26, use a definite integral to find the area of the region bounded by the given curve, the x-axis, and the given lines. In each case, first sketch the region. Watch out for areas of regions that are below the x-axis.

1. $y = 3x + 2$, $x = 2$, $x = 3$.

2. $y = 3x + 1$, $x = 0$, $x = 4$.

3. $y = x - 1$, $x = 5$.

4. $y = x + 5$, $x = 2$, $x = 4$.

5. $y = x^2$, $x = 2$, $x = 3$.

6. $y = 2x^2$, $x = 1$, $x = 2$.

7. $y = x^2 + 2$, $x = -1$, $x = 2$.

8. $y = 2x^2 - x$, $x = -2$, $x = -1$.

9. $y = x^2 - 2x$, $x = -3$, $x = -1$.

10. $y = 2x + x^3$, $x = 1$.

11. $y = 9 - x^2$.

12. $y = 3x^2 - 4x$, $x = -2$, $x = -1$.

13. $y = 1 - x - x^3$, $x = -2$, $x = 0$.

14. $y = 3$, $x = -1$, $x = 1$.

15. $y = 3 + 2x - x^2$.

16. $y = 8 - 2x^2$.

17. $y = \sqrt{x + 9}$, $x = -9$, $x = 0$.

18. $y = \dfrac{1}{x^2}$, $x = 2$, $x = 3$.

19. $y = \sqrt[3]{x}$, $x = 2$.

20. $y = x^2 - 2x$, $x = 1$, $x = 3$.

21. $y = x^2 - 2x - 8$.

22. $y = x^3 + 3x^2$, $x = -2$, $x = 2$.

23. $y = x^3$, $x = -2$, $x = 4$.

24. $y = x^2 - 4$, $x = -2$, $x = 2$.

25. $y = 2x - x^2$, $x = 1$, $x = 3$.

26. $y = \sqrt{2x - 1}$, $x = 1$, $x = 5$.

27. Suppose that $f(x) = x/8$, where $0 \le x \le 4$, and that f is a density function (see Example 4).
  a. Find $P(0 \le x \le 1)$.
  b. Find $P(2 \le x \le 4)$.
  c. Find $P(x \ge 3)$.

28. Suppose that $f(x) = 3(1 - x)^2$, where $0 \le x \le 1$, and that f is a density function (see Example 4).
  a. Find $P(\frac{1}{2} \le x \le 1)$.

b. Find $P(\frac{1}{3} \le x \le \frac{1}{2})$.

c. Find $P(x \le \frac{1}{3})$.

d. Use your result from part (c) to find $P(x \ge \frac{1}{3})$.

29. Suppose that $f(x) = 1/(8\sqrt{x})$, where $1 \le x \le 25$, and that f is a density function (see Example 4).

   a. Find $P(4 \le x \le 9)$.

   b. Find $P(x \le 4)$.

c. Find $P(x \ge 9)$.

d. Verify that $P(1 \le x \le 25) = 1$.

30. Under conditions of a continuous uniform distribution, a topic in statistics, the proportion of persons with incomes between a and t, where $a \le t \le b$, is the area of the region between the curve $y = 1/(b - a)$ and the x-axis from $x = a$ to $x = t$. Sketch the graph of the curve and determine the area of the given region.

5.6 AREA BETWEEN CURVES

We shall now find the area of a region enclosed by several curves. As before, our procedure will be to draw a sample element of area and use the definite integral to "add together" the areas of all such elements.

For example, consider the area of the region in Fig. 5.19 that is bounded on the top and bottom by the curves $y = f(x)$ and $y = g(x)$ and on the sides

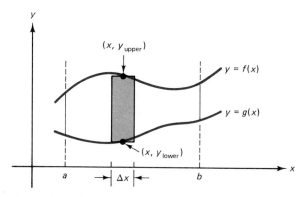

FIGURE 5.19

by the lines $x = a$ and $x = b$. The width of the indicated element is Δx, and the height is the y-value of the upper curve minus the y-value of the lower curve, which we shall write as $y_{upper} - y_{lower}$. Thus the area of the element is

$$[y_{upper} - y_{lower}] \, \Delta x$$

or

$$[f(x) - g(x)] \, \Delta x.$$

Summing the areas of all such elements from $x = a$ to $x = b$ by the definite integral gives the area of the region:

$$\sum [f(x) - g(x)] \, \Delta x \rightarrow \int_a^b [f(x) - g(x)] \, dx = \text{area}.$$

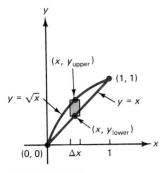

FIGURE 5.20

EXAMPLE 1   *Find the area of the region bounded by the curves $y = \sqrt{x}$ and $y = x$.*

A sketch of the region appears in Fig. 5.20. To determine where the curves intersect, we solve the system formed by the equations $y = \sqrt{x}$ and $y = x$. Eliminating y by substitution, we obtain

$$\sqrt{x} = x,$$
$$x = x^2 \quad \text{(squaring both sides)},$$
$$0 = x^2 - x = x(x - 1).$$
$$x = 0 \quad \text{or} \quad x = 1.$$

If $x = 0$, then $y = 0$; if $x = 1$, then $y = 1$. Thus the curves intersect at $(0, 0)$ and $(1, 1)$. The width of the indicated element of area is Δx. The height is the y-value on the upper curve minus the y-value on the lower curve:

$$y_{\text{upper}} - y_{\text{lower}} = \sqrt{x} - x.$$

Thus the area of the element is $(\sqrt{x} - x)\,\Delta x$. Summing the areas of all such elements from $x = 0$ to $x = 1$ by the definite integral, we get the area of the entire region.

$$\text{Area} = \int_0^1 (\sqrt{x} - x)\,dx$$
$$= \int_0^1 (x^{1/2} - x)\,dx = \left(\frac{x^{3/2}}{\frac{3}{2}} - \frac{x^2}{2} \right) \Bigg|_0^1$$
$$= \left(\frac{2}{3} - \frac{1}{2} \right) - (0 - 0) = \frac{1}{6} \text{ square unit.}$$

It should be obvious to you that knowing the points of intersection is important in determining the limits of integration.

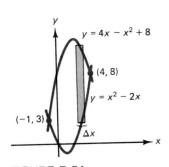

FIGURE 5.21

EXAMPLE 2   *Find the area of the region bounded by the curves $y = 4x - x^2 + 8$ and $y = x^2 - 2x$.*

A sketch of the region appears in Fig. 5.21. To find where the curves intersect, we solve the system of equations $y = 4x - x^2 + 8$ and $y = x^2 - 2x$.

$$4x - x^2 + 8 = x^2 - 2x,$$
$$-2x^2 + 6x + 8 = 0,$$
$$x^2 - 3x - 4 = 0,$$
$$(x + 1)(x - 4) = 0.$$
$$x = -1 \quad \text{or} \quad x = 4.$$

When $x = -1$, then $y = 3$; when $x = 4$, then $y = 8$. Thus the curves intersect at $(-1, 3)$ and $(4, 8)$. The width of the indicated element is Δx. The height is the y-value on the upper curve minus the y-value on the lower curve:

$$y_{upper} - y_{lower} = (4x - x^2 + 8) - (x^2 - 2x).$$

Thus the area of the element is

$$[(4x - x^2 + 8) - (x^2 - 2x)] \, \Delta x = (-2x^2 + 6x + 8) \, \Delta x.$$

Summing all such areas from $x = -1$ to $x = 4$, we have

$$\text{area} = \int_{-1}^{4} (-2x^2 + 6x + 8) \, dx = 41\tfrac{2}{3} \text{ square units.}$$

Sometimes area can more easily be determined by summing areas of horizontal elements rather than vertical elements. In the following example an area will be found by both methods. In each case the element of area determines the form of the integral.

EXAMPLE 3   *Find the area of the region bounded by the curve* $y^2 = 4x$ *and the lines* $y = 3$ *and* $x = 0$ *(the y-axis).*

The region is sketched in Fig. 5.22. When the curves $y = 3$ and $y^2 = 4x$ intersect, then $9 = 4x$, so $x = \tfrac{9}{4}$. Thus the point of intersection is $(\tfrac{9}{4}, 3)$. Since the width of the vertical strip is Δx, we integrate with respect to the variable x. Thus y_{upper} and y_{lower} must be expressed as functions of x. For the curve $y^2 = 4x$, we have $y = \pm 2\sqrt{x}$. But for the portion of this curve that bounds the region we must have $y \geq 0$, so we use $y = 2\sqrt{x}$. Thus the height of the strip is $y_{upper} - y_{lower} = 3 - 2\sqrt{x}$. Therefore, the strip has an area of $(3 - 2\sqrt{x}) \, \Delta x$, and we wish to sum up all such areas from $x = 0$ to $x = \tfrac{9}{4}$.

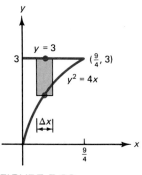

FIGURE 5.22

$$\text{Area} = \int_{0}^{9/4} (3 - 2\sqrt{x}) \, dx = \left(3x - \frac{4x^{3/2}}{3} \right) \Bigg|_{0}^{9/4}$$

$$= \left[3\left(\frac{9}{4}\right) - \frac{4}{3}\left(\frac{9}{4}\right)^{3/2} \right] - (0)$$

$$= \frac{27}{4} - \frac{4}{3}\left[\left(\frac{9}{4}\right)^{1/2} \right]^3 = \frac{27}{4} - \frac{4}{3}\left(\frac{3}{2}\right)^3 = \frac{9}{4} \text{ square units.}$$

Let us now approach this problem from the point of view of a **horizontal element of area** (or **horizontal strip**) as shown in Fig. 5.23. The width of the element is Δy. The length of the element is *the x-value on the right curve minus the x-value on the left curve.* Thus the area of the element is

$$(x_{right} - x_{left}) \, \Delta y.$$

We wish to sum all such areas from $y = 0$ to $y = 3$.

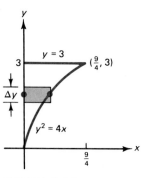

FIGURE 5.23

$$\sum (x_{right} - x_{left}) \, \Delta y \rightarrow \int_{0}^{3} (x_{right} - x_{left}) \, dy.$$

Since the variable of integration is y, we must express x_{right} and x_{left} as functions of y. The right curve is $y^2 = 4x$ or, equivalently, $x = y^2/4$. The left curve is $x = 0$. Thus

$$\text{area} = \int_0^3 (x_{\text{right}} - x_{\text{left}})\, dy$$

$$= \int_0^3 \left(\frac{y^2}{4} - 0\right) dy = \frac{y^3}{12}\Big|_0^3 = \frac{9}{4} \text{ square units.}$$

Note that for this region, it is easier to compute the area by using horizontal strips than by using vertical strips. In any case, remember that **the limits of integration are those limits for the variable of integration**.

EXAMPLE 4 *Find the area of the region bounded by the graphs of $y^2 = x$ and $x - y = 2$.*

A sketch of the region appears in Fig. 5.24. The curves intersect when $y^2 - y = 2$. Thus $y^2 - y - 2 = 0$, or equivalently $(y + 1)(y - 2) = 0$, from which $y = -1$ or $y = 2$. The points of intersection are $(1, -1)$ and $(4, 2)$. Let us consider vertical elements of area [see Fig. 5.24(a)]. Solving $y^2 = x$ for y gives $y = \pm\sqrt{x}$. As seen in Fig. 5.24(a), to the *left* of $x = 1$ the upper end of the element lies on $y = \sqrt{x}$ and the lower end lies on $y = -\sqrt{x}$. To the *right* of $x = 1$, the upper curve is $y = \sqrt{x}$ and the lower curve is the line $x - y = 2$ (or $y = x - 2$). Thus with vertical strips *two* integrals are needed to evaluate the area.

$$\text{Area} = \int_0^1 [\sqrt{x} - (-\sqrt{x})]\, dx + \int_1^4 [\sqrt{x} - (x - 2)]\, dx.$$

Let us consider horizontal strips to see if we can simplify our work. In Fig. 5.24(b), the width of the strip is Δy. The rightmost curve is *always* $x - y = 2$ (or $x = y + 2$) and the leftmost curve is *always* $y^2 = x$ (or $x = y^2$). Thus the area of the horizontal strip is $[(y + 2) - y^2]\,\Delta y$ and the total area is given by

$$\text{area} = \int_{-1}^2 (y + 2 - y^2)\, dy = \frac{9}{2} \text{ square units.}$$

Clearly, the use of horizontal strips is the more desirable approach for the problem.

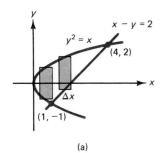

(a)

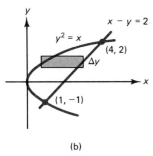

(b)

FIGURE 5.24

EXERCISE 5.6

In Problems 1–22, find the area of the region bounded by the graphs of the given equations. Be sure to find any needed points of intersection. Consider whether the use of horizontal strips makes the integral simpler than that for vertical strips.

1. $y = x^2$, $y = 2x$.

2. $y = x$, $y = -x + 3$, $y = 0$.

3. $y = x^2$, $x = 0$, $y = 4$ $(x \geq 0)$.

4. $y = x^2$, $y = x$.

5. $y = x^2 + 3$, $y = 9$.

6. $y^2 = x$, $x = 2$.

7. $x = 8 + 2y$, $x = 0$, $y = -1$, $y = 3$.

8. $y = x - 4$, $y^2 = 2x$.

9. $y = 4 - x^2$, $y = -3x$.

10. $x = y^2 + 2$, $x = 6$.

11. $y^2 = x$, $y = x - 2$.

12. $y = x^2$, $y = x + 2$.

13. $2y = 4x - x^2$, $2y = x - 4$.

14. $y = \sqrt{x}$, $y = x^2$.

15. $y^2 = x$, $3x - 2y = 1$.

16. $y = 2 - x^2$, $y = x$.

17. $y = 8 - x^2$, $y = x^2$, $x = -1$, $x = 1$.

18. $y^2 = 4 - x$, $y = x + 2$.

19. $y = x^2$, $y = 2$, $y = 5$.

20. $y = x^3 - x$, x-axis.

21. $y = x^3$, $y = x$.

22. $y = x^3$, $y = \sqrt{x}$.

23. A *Lorentz curve* is used in studying income distributions. If x is the cumulative percentage of income recipients, ranked from poorest to richest, and y is the cumulative percentage of income, then equality of income distribution is given by the line $y = x$ in Fig. 5.25, where x and y are expressed as decimals. For example, 10 percent of the people receive 10 percent of total income, 20 percent of the people receive 20 percent of the

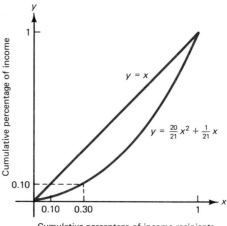

FIGURE 5.25

income, and so on. Suppose the actual distribution is defined by the Lorentz curve $y = \frac{20}{21}x^2 + \frac{1}{21}x$. Note, for example, that 30 percent of the people receive only 10 percent of total income. The degree of deviation from equality is measured by the *coefficient of inequality* [34] for a Lorentz curve. This coefficient is defined to be the area between the curve and the diagonal, divided by the area under the diagonal:

$$\frac{\text{area between curve and diagonal}}{\text{area under diagonal}}.$$

For example, when all incomes are equal, the coefficient of inequality is zero. Find the coefficient of inequality for the Lorentz curve defined above.

24. Find the coefficient of inequality as in Problem 23 for the Lorentz curve defined by $y = \frac{11}{12}x^2 + \frac{1}{12}x$.

5.7 CONSUMERS' AND PRODUCERS' SURPLUS

Determining the area of a region has applications in economics. Figure 5.26 shows a *supply curve* for a product. It indicates the price p at which the manufacturer will sell (or supply) q units. Also shown in Fig. 5.26 is the demand curve for the product. It indicates the price p at which consumers will purchase (or demand) q units. The point (q_0, p_0) where these curves intersect is called the *point of equilibrium*. Here p_0 is the price per unit at which consumers will purchase the same quantity q_0 of a product that producers wish to sell at that price. In short, p_0 is the price at which stability in the producer-consumer relationship occurs.

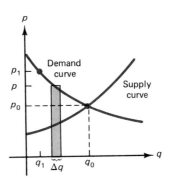

FIGURE 5.26

Let us assume that the market is at equilibrium and the price per unit of the product is p_0. According to the demand curve, there are consumers who would be willing to pay *more* than p_0. For example, at the price per unit of p_1, consumers would buy q_1 units. These consumers are benefiting from the lower equilibrium price p_0.

The vertical strip in Fig. 5.26 has area $p \, \Delta q$. This expression can also be thought of as the total amount of money that consumers would spend by buying Δq units of the product if the price per unit were p. Since the price is actually p_0, these consumers spend only $p_0 \, \Delta q$ for these Δq units and thus benefit by the amount $p \, \Delta q - p_0 \, \Delta q$. This can be written $(p - p_0) \, \Delta q$, which is the area of a rectangle of width Δq and length $p - p_0$ (see Fig. 5.27).

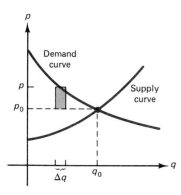

FIGURE 5.27

Summing the areas of all such rectangles from $q = 0$ to $q = q_0$ by integration, we have $\displaystyle\int_0^{q_0} (p - p_0) \, dq$. This integral, under certain conditions, represents the total gain to consumers who are willing to pay more than the equilibrium price. This total gain is called **consumers' surplus**, abbreviated CS. If the demand function is given by $p = f(q)$, then

$$CS = \int_0^{q_0} [f(q) - p_0] \, dq.$$

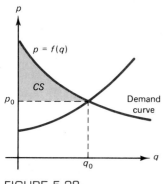

FIGURE 5.28

Geometrically (see Fig. 5.28), consumers' surplus is represented by the area between the line $p = p_0$ and the demand curve $p = f(q)$ from $q = 0$ to $q = q_0$.

Some of the producers also benefit from the equilibrium price, since they are willing to supply the product at prices *less* than p_0. Under certain conditions the total gain to the producers is represented geometrically in Fig. 5.29 by the area between the line $p = p_0$ and the supply curve $p = g(q)$ from $q =$

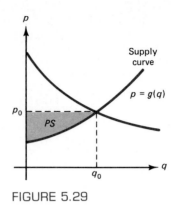

FIGURE 5.29

0 to $q = q_0$. This gain, called **producers' surplus** and abbreviated PS, is given by

$$PS = \int_0^{q_0} [p_0 - g(q)]\, dq.$$

EXAMPLE 1   *The demand function for a product is*

$$p = f(q) = 100 - 0.05q,$$

where p is the price per unit (in dollars) for q units. The supply function is $p = g(q) = 10 + 0.1q$. Determine consumers' surplus and producers' surplus under market equilibrium.

First we must find the equilibrium point by solving the system formed by $p = 100 - 0.05q$ and $p = 10 + 0.1q$.

$$10 + 0.1q = 100 - 0.05q,$$
$$0.15q = 90,$$
$$q = 600.$$

When $q = 600$, then $p = 10 + 0.1(600) = 70$. Thus $q_0 = 600$ and $p_0 = 70$. Consumers' surplus is

$$CS = \int_0^{q_0} [f(q) - p_0]\, dq = \int_0^{600} (100 - 0.05q - 70)\, dq$$

$$= \left(30q - 0.05\frac{q^2}{2}\right)\Bigg|_0^{600} = 18{,}000 - 9000 = 9000.$$

Producers' surplus is

$$PS = \int_0^{q_0} [p_0 - g(q)]\, dq = \int_0^{600} [70 - (10 + 0.1q)]\, dq$$

$$= \left(60q - 0.1\frac{q^2}{2}\right)\Bigg|_0^{600} = 36{,}000 - 18{,}000 = 18{,}000.$$

Thus consumers' surplus is $9000 and producers' surplus is $18,000.

EXERCISE 5.7

In Problems 1–6, the first equation is a demand equation and the second is a supply equation for a product. In each case find the consumers' surplus and producers' surplus under market equilibrium.

1. $p = 20 - 0.8q,$
$p = 4 + 1.2q.$

2. $p = 900 - q^2,$
$p = 100 + q^2.$

3. $p = 68 - 5q,$
$p = q^2 + 2.$

4. $p = 400 - q^2,$
$p = 20q + 100.$

5. $p = 10 - \dfrac{q}{100},$
$p = \dfrac{q}{80} + 1.$

6. $p = 100 - q^2,$
$p = 2q + 20.$

5.8 REVIEW

Important Terms and Symbols

Section 5.1   antiderivative   indefinite integral, $\int f(x)\, dx$ integral sign   integrand
constant of integration   variable of integration   fixed costs

Section 5.2   power rule for integration

Section 5.3 $\sum$ $F(x)\Big|_a^b$

Section 5.4   definite integral, $\int_a^b f(x)\, dx$ lower limit of integration   upper limit of integration
Fundamental Theorem of Integral Calculus

Section 5.5   vertical element of area

Section 5.6   horizontal element of area

Section 5.7   consumers' surplus   producers' surplus

Summary

An antiderivative of a function f is a function F such that $F'(x) = f(x)$. Any two antiderivatives of f differ only by a constant. The most general antiderivative of f is called the indefinite integral of f and is denoted $\int f(x)\, dx$. Thus

$$\int f(x)\, dx = F(x) + C,$$

where C is called the constant of integration.

Some basic integration formulas are as follows:

$$\int k\, dx = kx + C, \qquad k \text{ a constant,}$$

$$\int x^n\, dx = \frac{x^{n+1}}{n+1} + C, \qquad n \neq -1,$$

$$\int k f(x)\, dx = k \int f(x)\, dx, \qquad k \text{ a constant,}$$

and

$$\int [f(x) \pm g(x)]\, dx = \int f(x)\, dx \pm \int g(x)\, dx.$$

Another formula is the power rule for integration:

$$\int u^n\, du = \frac{u^{n+1}}{n+1} + C, \qquad \text{if } n \neq -1.$$

Here u represents a differentiable function of x and du is its differential. When applying the power rule to a given integral, it is important that the integral is written in a form that precisely matches the power rule.

If the rate of change of a function f is known, that is, f' is known, then f is an antiderivative of f'. If the value of $f(x)$ for a given value of x is also known, we can find the particular antiderivative that satisfies this condition. For example, if a marginal cost function dc/dq is given to us, then by integration we can find c. The form of c that we obtain involves a constant of integration. However, if we are also given fixed costs (that is, the costs involved when $q = 0$), we can determine the value of the constant of integration and thus find the particular cost function c. Similarly, if we are given a marginal revenue function dr/dq, then by integration and by using the fact that $r = 0$ when $q = 0$, we can determine the particular revenue function r. Once r is known, the corresponding demand equation can be found by using the equation $p = r/q$.

If f is continuous on $[a, b]$ and $[a, b]$ is divided into n subintervals of equal length Δx, then the definite integral of f over $[a, b]$ is denoted $\int_a^b f(x)\, dx$ and is given by

$$\int_a^b f(x)\, dx = \lim_{n \to \infty} \sum_{i=1}^n f(x_i)\, \Delta x,$$

where x_i is the right-hand endpoint of the ith subinterval. If $f(x) \geq 0$ on $[a, b]$, then geometrically $f(x_i)\, \Delta x$ represents the area of a rectangle of length $f(x_i)$ and width Δx. Thus the area of the region bounded by the graph of $y = f(x)$ and the x-axis, from $x = a$ to $x = b$, is $\int_a^b f(x)\, dx$. A definite integral may be conveniently evaluated by the Fundamental Theorem of Integral Calculus:

$$\int_a^b f(x)\, dx = F(b) - F(a),$$

where F is any antiderivative of f.

Some properties of the definite integral are

$$\int_a^b k f(x)\, dx = k \int_a^b f(x)\, dx, \quad \text{where } k \text{ is a constant,}$$

$$\int_a^b [f(x) \pm g(x)]\, dx = \int_a^b f(x)\, dx \pm \int_a^b g(x)\, dx,$$

and $$\int_a^c f(x)\, dx = \int_a^b f(x)\, dx + \int_b^c f(x)\, dx.$$

If the rate of change of a function f is known, a change in function values of f can be easily found by the formula

$$\int_a^b f'(x)\, dx = f(b) - f(a).$$

As already mentioned, if $f(x) \geq 0$ and continuous on $[a, b]$, the definite integral may be used to find the area of the region bounded by $y = f(x)$ and the x-axis from $x = a$ to $x = b$. The definite integral may also be used to find areas of more complicated regions. In these situations, an element of area should be drawn in the region. This will allow you to set up the proper definite integral. In some situations, vertical elements should be considered, while in others, horizontal elements are more advantageous.

One application of finding area involves consumers' surplus and producers' surplus. Suppose that the market for a product is at equilibrium and (q_0, p_0) is the equilibrium point (the point of intersection of the supply and demand curves for the product). Consumers' surplus, CS, corresponds to the area, from $q = 0$ to $q = q_0$, bounded above by the demand curve, and below by the line $p = p_0$. Thus

$$CS = \int_0^{q_0} [f(q) - p_0]\, dq,$$

where f is the demand function. Producers' surplus, PS, corresponds to the area, from $q = 0$ to $q = q_0$, bounded above by the line $p = p_0$, and below by the supply curve. Thus

$$PS = \int_0^{q_0} [p_0 - g(q)]\, dq,$$

where g is the supply function.

Review Problems

Determine the integrals in Problems 1–26.

1. $\int (x^3 + 2x - 7)\, dx.$

2. $\int dx.$

3. $\int_0^9 (\sqrt{x} + x)\, dx.$

4. $\int \dfrac{5 - 3x}{2}\, dx.$

5. $\int \dfrac{2}{(x + 5)^3}\, dx.$

6. $\int_4^{12} (y - 8)^{501}\, dy.$

7. $\int \dfrac{7x}{(3x^2 - 8)^3}\, dx.$

8. $\int_0^2 \dfrac{2}{\sqrt{4x + 1}}\, dx.$

9. $\int_0^1 \sqrt[3]{3t + 8}\, dt.$

10. $\int \dfrac{4x^2 - x}{x}\, dx.$

11. $\int x^2 \sqrt{3x^3 + 2}\, dx.$

12. $\int (2x^3 + x)(x^4 + x^2)^{3/4}\, dx.$

13. $\int y(y + 1)^2 \, dy.$

14. $\int \dfrac{8x}{3\sqrt[3]{7 - 2x^2}} \, dx.$

15. $\int \left(\dfrac{1}{x^3} + \dfrac{2}{x^2} \right) dx.$

16. $\int_0^1 10^{-8} \, dx.$

17. $\int_{-2}^1 (y^4 - y + 1) \, dy.$

18. $\int_7^{70} dx.$

19. $\int_{\sqrt{3}}^2 7x\sqrt{4 - x^2} \, dx.$

20. $\int_0^1 (2x + 1)(x^2 + x)^4 \, dx.$

21. $\int_0^1 \left[2x - \dfrac{1}{(x + 1)^{2/3}} \right] dx.$

22. $\int_2^8 (\sqrt{2x} - x + 4) \, dx.$

23. $\int \dfrac{\sqrt{t} - 3}{t^2} \, dt.$

24. $\int \dfrac{\sqrt[4]{z} - \sqrt[3]{z}}{\sqrt{z}} \, dz.$

25. $\int \dfrac{(0.5x - 0.1)^4}{0.4} \, dx.$

26. $\int \dfrac{(x^2 + 4)^2}{x^2} \, dx.$

In Problems 27–34, find the area of the region bounded by the given curve, the x-axis, and the given lines.

27. $y = x^2 - 1, \quad x = 2 \quad (y \geq 0).$

28. $y = 6x - 8 - x^2.$

29. $y = \sqrt{x + 4}, \quad x = 0.$

30. $y = x^2 - x - 2, \quad x = -2, \quad x = 2.$

31. $y = 5x - x^2.$

32. $y = \sqrt[4]{x}, \quad x = 1, \quad x = 16.$

33. $y = \dfrac{1}{x^2} + 3, \quad x = 1, \quad x = 3.$

34. $y = x^3 - 1, \quad x = -1.$

In Problems 35–40, find the area of the region bounded by the given curves.

35. $y^2 = 4x, \quad x = 0, \quad y = 2.$

36. $y = 2x^2, \quad x = 0, \quad y = 2 \quad (x \geq 0).$

37. $y = x^2 + 4x - 5, \quad y = 0.$

38. $y = 2x^2, \quad y = x^2 + 9.$

39. $y = x^2 - 2x, \quad y = 12 - x^2.$

40. $y = \sqrt{x}, \quad x = 0, \quad y = 3.$

41. If marginal revenue is given by $dr/dq = 100 - (3/2)\sqrt{2q}$, find the corresponding demand equation.

42. If marginal cost is given by $dc/dq = q^2 + 7q + 6$, and fixed costs are 2500, find the total cost for producing 6 units. Assume costs are in dollars.

43. A manufacturer's marginal revenue function is given by $dr/dq = 275 - q - 0.3q^2$. If r is in dollars, find the increase in the manufacturer's total revenue if production is increased from 10 to 20 units.

44. A manufacturer's marginal cost function is $dc/dq = 500/\sqrt{2q + 25}$. If c is in dollars, find the cost involved to increase production from 100 to 300 units.

45. For a product, suppose that the demand equation is $p = 0.01q^2 - 1.1q + 30$ and its supply equation is $p = 0.01q^2 + 8$. Determine consumers' surplus and producers' surplus under market equilibrium.

46. Dizygotic twins are twins resulting from two fertilized ova. Under appropriate assumptions and over a certain time interval $[0, T_p]$, the mean (average) probability T that both twins have the same sex is given in **[35]** by

$$T = \frac{1}{T_p} \int_0^{T_p} \{[p(m)]^2 + [p(f)]^2\} \, dt,$$

where

$$p(m) = a - \frac{a - b}{T_p} t,$$

$$p(f) = 1 - p(m),$$

and both a and b are constants.

a. Show that

$$T = a^2 + (1 - a)^2 + (a - b)(1 - 2a) + \tfrac{2}{3}(a - b)^2.$$

b. Evaluate T if $a = 0.7$ and $b = 0.3$.

Exponential and Logarithmic Functions

6.1 EXPONENTIAL FUNCTIONS

In this chapter we shall look at two special types of functions that have wide applications in many areas of study. The first function involves a constant raised to a variable power, such as $f(x) = 2^x$. We call such functions *exponential functions*.

Definition
The function f defined by

$$f(x) = b^x,$$

where $b > 0$, $b \neq 1$, and the exponent x is any real number, is called an **exponential function** *with base b.**

Do not confuse the exponential function $y = 2^x$ with the *power function* $y = x^2$, which has a variable base and a constant exponent.

Since the exponent in b^x can be any real number, you may wonder how we assign a value to something like $2^{\sqrt{2}}$. Stated simply, we use approximations. First, $2^{\sqrt{2}}$ is approximately $2^{1.4} = 2^{7/5} = \sqrt[5]{2^7}$, which *is* defined. Better approximations are $2^{1.41} = 2^{141/100} = \sqrt[100]{2^{141}}$, and so on. In this way the meaning of $2^{\sqrt{2}}$ becomes clear.

When you work with exponential functions, it may be necessary to apply rules for exponents. These rules are as follows, where m and n are real numbers and a and b are positive.

---

* If $b = 1$, then $f(x) = 1^x = 1$. This function is so uninteresting that we do not call it an exponential function.

1. $a^m a^n = a^{m+n}$.	**2.** $\dfrac{a^m}{a^n} = a^{m-n}$.
3. $(a^m)^n = a^{mn}$.	**4.** $(ab)^n = a^n b^n$.
5. $\left(\dfrac{a}{b}\right)^n = \dfrac{a^n}{b^n}$.	**6.** $a^1 = a$.
7. $a^0 = 1$.	**8.** $a^{-n} = \dfrac{1}{a^n}$.

Some functions that do not appear to have the exponential form b^x can be put in that form by applying the above rules. For example, $2^{-x} = 1/2^x = (\frac{1}{2})^x$ and $3^{2x} = (3^2)^x = 9^x$.

Graphs of some exponential functions are shown in Fig. 6.1. Notice that each function is continuous. We also make the following observations.

1. The domain of an exponential function is all real numbers.

2. The range is all positive real numbers.

3. Since $b^0 = 1$ for every base b, each graph has y-intercept $(0, 1)$. There is no x-intercept.

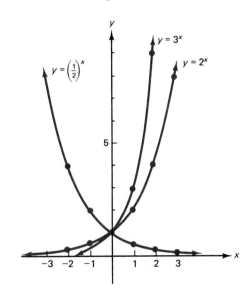

x	2^x	3^x	$\left(\frac{1}{2}\right)^x$
-2	$\frac{1}{4}$	$\frac{1}{9}$	4
-1	$\frac{1}{2}$	$\frac{1}{3}$	2
0	1	1	1
1	2	3	$\frac{1}{2}$
2	4	9	$\frac{1}{4}$
3	8	27	$\frac{1}{8}$

FIGURE 6.1

4. If $b > 1$, then $y = b^x$ is an increasing function; its graph *rises* from left to right. Also,

$$\lim_{x \to \infty} b^x = \infty \quad \text{and} \quad \lim_{x \to -\infty} b^x = 0.$$

Thus as $x \to -\infty$, the graph has the x-axis as a horizontal asymptote. In Quadrant I, the greater the value of b, the more quickly the graph rises. [Compare the graphs of $y = 2^x$ and $y = 3^x$.]

5. If $0 < b < 1$, then $y = b^x$ is a decreasing function; its graph *falls* from left to right. [See the graph of $y = (\frac{1}{2})^x$.] Also,

$$\lim_{x \to \infty} b^x = 0 \quad \text{and} \quad \lim_{x \to -\infty} b^x = \infty.$$

Thus as $x \to \infty$, the graph has the x-axis as a horizontal asymptote.

EXAMPLE 1   *Find each of the following limits.*

a. $\lim\limits_{x \to \infty} \dfrac{5}{2^x}$.

As $x \to \infty$, then $2^x \to \infty$ since $b = 2 > 1$. Dividing 5 by arbitrarily large numbers results in numbers arbitrarily close to 0. Thus

$$\lim_{x \to \infty} \frac{5}{2^x} = 0.$$

b. $\lim\limits_{t \to \infty} [7 + 5(0.6)^{2t+1}]$.

As $t \to \infty$, then $2t + 1 \to \infty$. Thus $(0.6)^{2t+1} \to 0$ since $b = 0.6$ and $0 < 0.6 < 1$. Hence

$$\lim_{t \to \infty} [7 + 5(0.6)^{2t+1}] = \lim_{t \to \infty} 7 + 5 \lim_{t \to \infty} (0.6)^{2t+1}$$

$$= 7 + 5(0) = 7.$$

c. $\lim\limits_{x \to -\infty} \left[\dfrac{1}{x} - 2(10)^{-x} \right]$.

As $x \to -\infty$, then $-x \to \infty$. Hence $10^{-x} \to \infty$ since $b = 10 > 1$. Also, as $x \to -\infty$, then $1/x \to 0$. Thus

$$\lim_{x \to -\infty} \left[\frac{1}{x} - 2(10)^{-x} \right] = -\infty.$$

One of the numbers most useful for a base in an exponential function is a certain irrational number denoted by the letter e in honor of the Swiss mathematician Leonhard Euler (1707–1783):

e is approximately 2.71828.

To be precise, e can be defined as a limit. The graph of $f(x) = (1 + x)^{1/x}$ in Fig. 6.2 makes it clear that as $x \to 0$ the limit of $(1 + x)^{1/x}$ exists. This

x	$(1 + x)^{1/x}$	x	$(1 + x)^{1/x}$
0.5	2.2500	−0.5	4.0000
0.1	2.5937	−0.1	2.8680
0.01	2.7048	−0.01	2.7320
0.001	2.7169	−0.001	2.7196

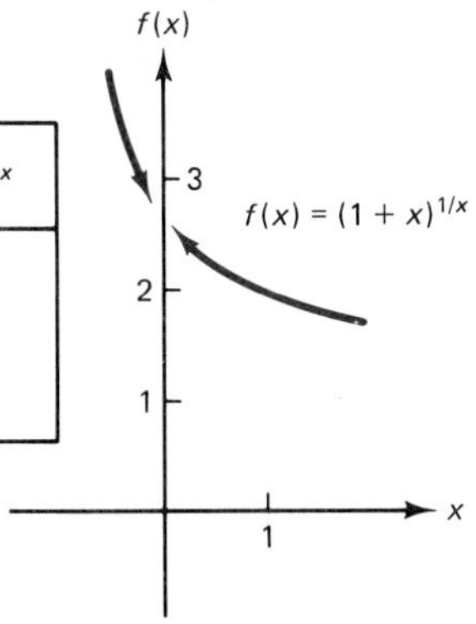

$f(x) = (1 + x)^{1/x}$

FIGURE 6.2

limit is defined to be the number e.

$$\lim_{x \to 0} (1 + x)^{1/x} = e.$$

Although e may seem to be a strange base for an exponential function, it arises quite naturally in calculus (as you will see later). It also occurs in economic analysis and problems involving natural growth (or decay), such as compound interest and population studies. A table of (approximate) values of e^x and e^{-x} is in Appendix D. These values can also be obtained with many calculators. The graph of $y = e^x$ is shown in Fig. 6.3.

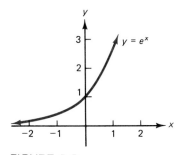

$y = e^x$

FIGURE 6.3

EXAMPLE 2   *The projected population P of a city is given by*

$$P = 100{,}000e^{0.05t},$$

where t is the number of years after 1987. *Predict the population for the year* 2007.

The number of years from 1987 to 2007 is 20, so let $t = 20$. Then

$$P = 100{,}000e^{0.05(20)} = 100{,}000e^1 = 100{,}000e.$$

Since $e \approx 2.71828$,

$$P \approx 100{,}000(2.71828) = 271{,}828.$$

Many forecasts are based on population studies.

In statistics an important function used as a model in describing events occurring in nature is the **Poisson distribution function**:

$$f(x) = \frac{e^{-\mu}\mu^x}{x!}, \qquad x = 0, 1, 2, \ldots.$$

The symbol μ (read "mu") is a Greek letter. In certain situations $f(x)$ gives the probability that exactly x events will occur in an interval of time or space.

The constant μ is the mean, or average number, of occurrences in the interval. The next example illustrates the Poisson distribution.

EXAMPLE 3   *A hemacytometer is a counting chamber divided into squares and is used in studying the number of microscopic structures in a liquid. In a well-known experiment [36], yeast cells were diluted and thoroughly mixed in a liquid, and the mixture was placed in a hemacytometer. With a microscope the number of yeast cells on each square were counted. The probability that there were exactly x yeast cells on a hemacytometer square was found to fit a Poisson distribution with $\mu = 1.8$. Find the probability that there were exactly four cells per square.*

We use the Poisson distribution function with $\mu = 1.8$ and $x = 4$.

$$f(x) = \frac{e^{-\mu}\mu^x}{x!}.$$

$$f(4) = \frac{e^{-1.8}(1.8)^4}{4!}.$$

From the table in Appendix D, $e^{-1.8} \approx 0.16530$, so

$$f(4) \approx \frac{(0.16530)(10.4976)}{24} \approx 0.072.$$

For example, this means that in 400 squares we would *expect* $400(0.072) \approx$ 29 squares to contain exactly 4 cells. (In the experiment, in 400 squares the actual number observed was 30.)

Practically everyone is familiar with **compound interest**, whereby the interest earned by an invested sum of money (or **principal**) is reinvested so that it, too, earns interest. That is, the interest is converted (or *compounded*) into principal and hence there is "interest on interest."

More specifically, suppose that a principal of P dollars is invested at a rate of $100r$ percent compounded annually (for example, at 5 percent, r is 0.05). At the end of one year, the interest is Pr. Thus the accumulated amount, or **compound amount**, is the principal plus interest, that is, $P + Pr$ or, upon factoring,

$$P(1 + r).$$

Under compound interest, this acts as the principal for the second year. Thus the interest at the end of the second year is $[P(1 + r)]r$, and the compound amount is

$$P(1 + r) + [P(1 + r)]r$$
$$= P(1 + r)[1 + r] \qquad \text{(factoring)}$$
$$= P(1 + r)^2.$$

On the other hand, suppose that the principal is invested at an annual rate of r compounded *twice* a year. Then in two years there are four interest periods or **conversion periods**, where the interest rate per period is $r/2$. The compound amount is (following the pattern above)

$$P\left(1 + \frac{r}{2}\right)^4.$$

More generally, if interest is compounded k times a year at an annual rate of r, then the rate per conversion period is r/k. In t years there are kt periods and the compound amount S is

$$S = P\left(1 + \frac{r}{k}\right)^{kt}.$$

If $k \to \infty$, the number of conversion periods increases indefinitely, and the length of each period approaches 0. In this case we say that interest is **compounded continuously**, that is, at every instant of time. The compound amount is

$$\lim_{k \to \infty} P\left(1 + \frac{r}{k}\right)^{kt},$$

which may be written

$$P\left[\lim_{k \to \infty}\left(1 + \frac{r}{k}\right)^{k/r}\right]^{rt}.$$

If we let $x = r/k$, then as $k \to \infty$ we have $x \to 0$. Thus the limit inside the brackets has the form $\lim_{x \to 0} (1 + x)^{1/x}$ which, as we saw before, is e. Therefore,

$$S = Pe^{rt}$$

is the compound amount S of a principal of P dollars after t years at an annual interest rate r compounded continuously.

EXAMPLE 4   *If* $100 *is invested at an annual rate of* 9 *percent compounded continuously, find the compound amount at the end of* (a) 1 *year and* (b) 5 *years.*

a. Here $P = 100$, $r = 0.09$, and $t = 1$.

$$S = Pe^{rt} = 100e^{(0.09)(1)} \approx 100(1.0942) = \$109.42.$$

We can compare this value with the value after one year of a $100 investment at an annual rate of 9 percent compounded semiannually—namely, $100(1.045)^2$ $\approx \$109.20$. The difference is not significant.

b. Here $P = 100$, $r = 0.09$, and $t = 5$.

$$S = 100e^{(0.09)(5)} = 100e^{0.45} \approx 100(1.5683) = \$156.83.$$

Solving $S = Pe^{rt}$ for P gives $P = S/e^{rt}$ or

$$P = Se^{-rt}.$$

Here P is the principal that must be invested now at an annual rate of r compounded continuously so that, at the end of t years, the compound amount will be S. We call P the **present value** of S.

EXAMPLE 5   *A trust fund is being set up by a single payment so that, at the end of 20 years, there will be \$25,000 in the fund. If interest is compounded continuously, at an annual rate of 12 percent, how much money should be initially paid into the fund?*

We want the present value of \$25,000 due in 20 years.

$$P = Se^{-rt} = 25,000e^{-(0.12)(20)}$$
$$= 25,000e^{-2.4} \approx 25,000(0.09072)$$
$$= 2268.$$

Thus \$2268 should be paid into the fund.

EXERCISE 6.1

In Problems 1–6, graph each function.

1. $y = f(x) = 4^x$.

2. $y = f(x) = (\frac{1}{3})^x$.

3. $y = f(x) = 2(\frac{1}{4})^x$.

4. $y = f(x) = \frac{1}{2}(3^{x/2})$.

5. $y = f(x) = 2^x - 1$.

6. $y = f(x) = 2^{x-1}$.

In Problems 7–14, find the limits.

7. $\lim\limits_{x \to \infty} 12^x$.

8. $\lim\limits_{x \to -\infty} (0.4)^x$.

9. $\lim\limits_{x \to \infty} (4 - 3^{-x})$.

10. $\lim\limits_{x \to \infty} \dfrac{9}{7(5)^x}$.

11. $\lim\limits_{p \to -\infty} \dfrac{6}{2 - 2^p}$.

12. $\lim\limits_{t \to \infty} (2^t + t^2)$.

13. $\lim\limits_{x \to \infty} \dfrac{2 + 5e^{1-2x}}{8}$.

14. $\lim\limits_{x \to \infty} \dfrac{e^{-x}}{e^{2x-1}}$.

In Problems 15–18, use the table in Appendix D to find the approximate value of each expression.

15. $e^{1.5}$.

16. $e^{3.4}$.

17. $e^{-0.4}$.

18. $e^{-3/4}$.

19. The projected population P of a city is given by $P = 125,000(1.12)^{t/20}$, where t is the number of years after 1984. Predict the population in 2004.

20. For a certain city the population P grows at the rate of 2 percent per year. The formula $P = 1,000,000(1.02)^t$ gives the population t years after 1980. Find the population in (a) 1981 and (b) 1982.

21. The probability P that a telephone operator will receive exactly x calls during a certain time period is given by

$$P = \dfrac{e^{-3}3^x}{x!}.$$

Find the probability that exactly two calls will be received. Give your answer to four decimal places.

22. Express e^{kt} in the form b^t.

23. At a certain time there are 100 milligrams of a radioactive substance. It decays so that after t years the number of milligrams present, A, is given by $A = 100e^{-0.035t}$. How many milligrams are present after 20 years? Give your answer to the nearest milligram.

24. In a psychological experiment involving learning [2], subjects were asked to give particular responses after being shown certain stimuli. Each stimulus was a pair of letters, and each response was either the digit 1 or 2. After each response the subject was told the correct answer. In this so-called *paired-associate* learning experiment, the theoretical probability P that a subject makes a correct response on the nth trial is given by

$$P = 1 - \tfrac{1}{2}(1 - c)^{n-1}, \qquad n \ge 1, \qquad 0 < c < 1.$$

Find P when $n = 1$, and find $\lim_{n \to \infty} P$.

25. In a study of the effects of food deprivation on hunger [37], an insect was fed until its appetite was completely satisfied. Then it was deprived of food for t hr (deprivation period). At the end of this period the insect was again fed until its appetite was completely satisfied. The approximate weight H (in grams) of the food that was consumed at this time was found to be a function of t, where

$$H = 1.00[1 - e^{-(0.0464t + 0.0670)}].$$

Here H is a measure of hunger. If it is assumed that the deprivation period can increase without bound and that the insect does not die, to what value would H tend?

26. A mail-order company advertises in a national magazine. The company finds that of all small towns, the percentage (given as a decimal) in which exactly x people respond to an ad fits a Poisson distribution with $\mu = 0.5$. From what percentage of small towns can the company expect exactly two people to respond? Give your answer to four decimal places.

27. Suppose that the number x of patients admitted into a hospital emergency room during a certain hour of the day has a Poisson distribution with mean 4. Find the probability that during that hour there will be exactly two emergency patients. Give your answer to four decimal places.

28. In a classic study [38], records of 10 army corps of Prussian cavalrymen were analyzed over a period of 20 years. There were 200 readings in terms of corps-years. It was found that the probability of x cavalrymen per year per corps being kicked to death by a horse fit a Poisson distribution with $\mu = 0.61$.
   a. Find the probability that $x = 2$.
   b. To the nearest unit, how many of the 200 corps-years does the Poisson distribution predict with this number of deaths?

29. If $1000 is invested at an annual rate of 6 percent compounded continuously, find the compound amount at the end of eight years.

30. If $100 is deposited in a savings account that earns interest at an annual rate of $5\frac{1}{2}$ percent compounded continuously, what is the value of the account at the end of two years?

31. The board of directors of a corporation agrees to redeem some of its callable preferred stock in five years. At that time $1,000,000 will be required. If the corporation can invest money at an annual interest rate of 10 percent compounded continuously, how much should it presently invest so that the future value is sufficient to redeem the shares?

32. A trust fund is being set up by a single payment so that, at the end of 30 years, there will be $50,000 in the fund. If interest is compounded continuously at an annual rate of 9 percent, how much money should be paid into the fund initially?

33. The formula $A = Pe^{-rn}$ gives the amount A at the end of n years of a principal P that depreciates at the rate of r (expressed as a decimal) per year compounded continuously. At the end of ten years, what is the value of $60,000 of machinery that depreciates at a rate of 8 percent, compounded continuously? Give your answer to the nearest dollar.

34. An important function used in decision making is the *normal distribution density function*, which in standard form is

$$f(x) = \frac{1}{\sqrt{2\pi}} e^{-x^2/2}.$$

Evaluate $f(0)$, $f(-1)$, and $f(1)$ by using $\frac{1}{\sqrt{2\pi}} = 0.399$.

Give your answers to three decimal places.

35. The demand equation for a new toy is $q = 10,000(0.95123)^p$. It is desired to evaluate q when $p = 10$. To convert the equation into a more desirable computational form, use Appendix D to show that $q = 10,000e^{-0.05p}$. Then evaluate and give your answer to the nearest integer. (*Hint:* Find a number x such that $0.95123 \approx e^{-x}$.)

6.2 LOGARITHMIC FUNCTIONS

In this section the functions of interest to us are *logarithmic functions*, which are related to exponential functions. Figure 6.4(a) shows the graph of the exponential function $s = f(t) = 2^t$. Here f sends an input number t into a *positive* output number s:

$$f : t \rightarrow s \quad \text{where} \quad s = 2^t.$$

For example, f sends 2 into 4.

Looking at the same curve in Fig. 6.4(b), you can see from the small

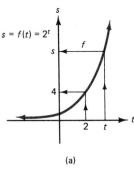

(a)

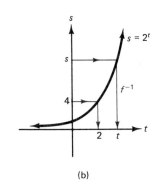

(b)

FIGURE 6.4

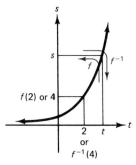

FIGURE 6.5

arrows that, with each positive number s on the vertical axis, we can associate exactly one value of t. With $s = 4$ we associate $t = 2$. By thinking of s as an input and t as an output, we have a function that sends s's into t's. We shall denote this function by f^{-1} (read "f inverse"):*

$$f^{-1} : s \rightarrow t \quad \text{where} \quad s = 2^t.$$

Thus $f^{-1}(s) = t$. The domain of f^{-1} is the range of f (all positive real numbers), and the range of f^{-1} is the domain of f (all real numbers).

The functions f and f^{-1} are related. Figure 6.5 shows that f^{-1} *reverses* the action of f, and vice versa. For example,

$$f \text{ sends 2 into 4}, \quad \text{and} \quad f^{-1} \text{ sends 4 into 2}.$$

More generally, $f(t) = s$ and $f^{-1}(s) = t$. In terms of composition, when either

---

* The -1 is not an exponent, so f^{-1} does *not* mean $\dfrac{1}{f}$.

$f^{-1} \circ f$ or $f \circ f^{-1}$ is applied to an input number, that same number is obtained for output because of the reversing effects of f and f^{-1}. That is,

$$(f^{-1} \circ f)(t) = f^{-1}(f(t)) = f^{-1}(s) = t$$

$$\text{and} \quad (f \circ f^{-1})(s) = f(f^{-1}(s)) = f(t) = s.$$

We give a special name to f^{-1}. It is called the **logarithmic function with base 2** and is written as $\log_2$ [read "logarithm (or log) base 2"]. Therefore, $f^{-1}(4) = \log_2 4 = 2$, and we say that the *logarithm* base 2 of 4 is 2.

In summary,

$$\text{if } s = 2^t, \text{ then } t = \log_2 s. \tag{1}$$

We now generalize our discussion to other bases. In (1), replacing 2 by b, s by x, and t by y gives the following definition.

Definition
*The **logarithmic function** with base b, where $b > 0$ and $b \neq 1$, is denoted $\log_b$ and is defined by*

$$y = \log_b x \quad \text{if and only if} \quad b^y = x.$$

The domain of $\log_b$ is all positive real numbers and its range is all real numbers.

Because a logarithmic function reverses the action of the corresponding exponential function, and vice versa, each logarithmic function is called the *inverse* of its corresponding exponential function, and that exponential function is the inverse of its corresponding logarithmic function.

Remember: When we say that y is the logarithm base b of x, we mean that b raised to the y power is x.

$$\boxed{y = \log_b x \quad \text{means} \quad b^y = x.}$$

In this sense *the logarithm of a number is an exponent:* $\log_b x$ is the power to which we must raise b to get x. For example,

$$\log_2 8 = 3 \quad \text{because} \quad 2^3 = 8.$$

We say that $\log_2 8 = 3$ is the **logarithmic form** of the **exponential form** $2^3 = 8$.

EXAMPLE 1

a. Since $5^2 = 25$, then $\log_5 25 = 2$.

b. Since $10^0 = 1$, then $\log_{10} 1 = 0$.

c. $\log_{64} 8 = \frac{1}{2}$ means that $64^{1/2} = 8$.

d. $\log_2 \frac{1}{16} = -4$ means that $2^{-4} = \frac{1}{16}$.

EXAMPLE 2   *Graph the function $y = \log_2 x$.*

It can be awkward to substitute values of x and then find corresponding values of y. For example, if $x = 3$, then $y = \log_2 3$, which is not easily determined. An easier way to plot points is to use the equivalent exponential form $x = 2^y$. We choose values of y and find the corresponding values of x. If $y = 0$, then $x = 1$. This gives the point $(1, 0)$. Other points are shown in Fig. 6.6.

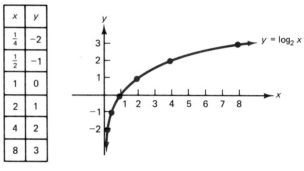

x	y
$\frac{1}{4}$	-2
$\frac{1}{2}$	-1
1	0
2	1
4	2
8	3

FIGURE 6.6

From the graph, you can see that the domain is all positive real numbers. Negative numbers and 0 do not have logarithms. The range is all real numbers. Numbers between 0 and 1 have negative logarithms, and the closer such a number is to 0, the more negative is its logarithm. Numbers greater than 1 have positive logarithms. The log of 1 is 0, which corresponds to the x-intercept $(1, 0)$. There is no y-intercept. The function is continuous and increasing for $x > 0$. The y-axis is a vertical asymptote because as $x \to 0^+$, $\log_2 x \to -\infty$. Moreover, as $x \to \infty$, $\log_2 x \to \infty$. This graph is typical for a logarithmic function with $b > 1$.

It is obvious in Fig. 6.6 that points with different x-values have different y-values (logarithms). More generally, we have the following property.

> If $\log_b m = \log_b n$, then $m = n$.

There is a similar property for the exponential function.

> If $b^m = b^n$, then $m = n$.

Logarithms with base 10 are called **common logarithms**. They were frequently used for computational purposes before the calculator age. The subscript 10 is usually omitted from the notation:

$$\boxed{\log x \quad \text{means} \quad \log_{10} x.}$$

Important in calculus are logarithms with base e, called **natural logarithms**. We use the notation "ln" for such logarithms:

$$\boxed{\ln x \quad \text{means} \quad \log_e x.}$$

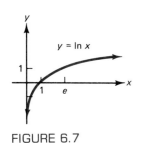

FIGURE 6.7

The symbol ln x may be read "ell-en of x." Appendix E gives a table of approximate values of natural logarithms and instructions on how to use the table. For example, you can see that ln $2 \approx 0.69315$. This means $e^{0.69315} \approx 2$. Figure 6.7 shows the graph of $y = \ln x$. It has the same shape as does Fig. 6.6. Natural and common logarithms can be found with many calculators.

EXAMPLE 3   *Find each of the following.*

a. log 100.

Here the base is 10. Thus log 100 is the power to which we must raise 10 to get 100. Since $10^2 = 100$, log $100 = 2$.

b. ln 1.

Here the base is e. Because $e^0 = 1$, ln $1 = 0$.

c. log 0.1.

Since $0.1 = \frac{1}{10} = 10^{-1}$, log $0.1 = -1$.

d. ln e^{-1}.

Since ln e^{-1} is the power to which e must be raised to get e^{-1}, clearly ln $e^{-1} = -1$.

e. $\log_{36} 6$.

Because $36^{1/2}$ (or $\sqrt{36}$) is 6, $\log_{36} 6 = \frac{1}{2}$.

EXAMPLE 4   *Solve each equation for x.*

a. $\log_2 x = 4$.

In exponential form, $2^4 = x$, so $x = 16$.

b. $\ln(x + 1) = 7$.

The exponential form gives $e^7 = x + 1$. Thus $x = e^7 - 1$.

c. $\log_x 49 = 2$.

In exponential form, $x^2 = 49$, so $x = 7$. We rejected $x = -7$ in solving $x^2 = 49$ because a negative number cannot be a base of a logarithmic function.

d. $e^{5x} = 4$.

In logarithmic form, $\ln 4 = 5x$. Thus $x = \dfrac{\ln 4}{5}$.

EXAMPLE 5   *If interest is compounded continuously at an annual rate r, how long would it take for a principal of P to double?*

The compound amount S after t years at an annual rate r is $S = Pe^{rt}$. When P doubles, then $S = 2P$. Thus $2P = Pe^{rt}$ and so $2 = e^{rt}$. This means that $rt = \ln 2$ or

$$t = \frac{\ln 2}{r} \approx \frac{0.69315}{r}.$$

For example, if $r = 0.05$, then money would double in $0.69315/0.05 \approx 13.9$ years.

The logarithmic function has many important properties. For example, the logarithm of a product of two numbers is the sum of the logarithms of the numbers. Symbolically, $\log_b(mn) = \log_b m + \log_b n$. To prove this, we first let $x = \log_b m$ and $y = \log_b n$. Then $b^x = m$, $b^y = n$, and

$$mn = b^x b^y = b^{x+y}.$$

Thus $mn = b^{x+y}$. In logarithmic form, this means that $\log_b(mn) = x + y$. Therefore $\log_b(mn) = \log_b m + \log_b n$. Other basic properties of logarithms are given in Table 6.1, where m and n are positive and r is any real number. The proofs of Properties 2 and 3 are similar to that of Property 1. Properties 4 and 5 follow from the facts that $b^0 = 1$ and $b^1 = b$.

TABLE 6.1
Properties of Logarithms

1. $\log_b(mn) = \log_b m + \log_b n$.   **5.** $\log_b b = 1$.

2. $\log_b \dfrac{m}{n} = \log_b m - \log_b n$.   **6.** $\log_b b^r = r$.

3. $\log_b m^r = r \log_b m$.   **7.** $b^{\log_b m} = m$.

4. $\log_b 1 = 0$.   **8.** $\log_b m = \dfrac{\log_a m}{\log_a b}$.

To prove Property 6, we have

$$\begin{aligned}
\log_b b^r &= r \log_b b \qquad \text{(Property 3)}\\
&= r \cdot 1 \qquad\quad \text{(Property 5)}\\
&= r.
\end{aligned}$$

Property 7 is true because it merely states in logarithmic form that $\log_b m = \log_b m$. To prove Property 8, let $x = \log_b m$. Then $b^x = m$. Taking logarithms base a of both sides gives

$$\log_a b^x = \log_a m,$$

$$x \log_a b = \log_a m \qquad \text{(Property 3)},$$

$$x = \frac{\log_a m}{\log_a b},$$

$$\text{or} \qquad \log_b m = \frac{\log_a m}{\log_a b}.$$

Property 8 is called the **change of base formula**. It allows conversion from one base (a) to another (b).

EXAMPLE 6

a. *Write* $\log \dfrac{1}{x}$ *in terms of* $\log x$.

By Property 3,

$$\log \frac{1}{x} = \log x^{-1} = -1 \log x = -\log x.$$

In general, $\log_b \dfrac{1}{m} = -\log_b m$.

b. *Write* $\ln \dfrac{x}{zw}$ *in terms of* $\ln x$, $\ln z$, *and* $\ln w$.

$$\ln \frac{x}{zw} = \ln x - \ln (zw) \qquad \text{(Property 2)}$$

$$= \ln x - (\ln z + \ln w) \qquad \text{(Property 1)}$$

$$= \ln x - \ln z - \ln w.$$

c. *Write* $\ln \sqrt[3]{\dfrac{x^5(x-2)^8}{x-3}}$ *in terms of* $\ln x$, $\ln(x-2)$, *and* $\ln(x-3)$.

$$\ln \sqrt[3]{\frac{x^5(x-2)^8}{x-3}} = \ln \left[\frac{x^5(x-2)^8}{x-3}\right]^{1/3} = \frac{1}{3} \ln \frac{x^5(x-2)^8}{x-3}$$

$$= \frac{1}{3} \{\ln[x^5(x-2)^8] - \ln(x-3)\}$$

$$= \frac{1}{3} [\ln x^5 + \ln(x-2)^8 - \ln(x-3)]$$

$$= \frac{1}{3} [5 \ln x + 8 \ln(x-2) - \ln(x-3)].$$

EXAMPLE 7

a. *Find* $\ln e^{3x}$.

By Property 6 with $b = e$, we have $\ln e^{3x} = 3x$. Alternatively, by Properties 3 and 5,

$$\ln e^{3x} = 3x \ln e = 3x(1) = 3x.$$

b. *Find* $\log 1 + \log 1000$.

By Property 4, $\log 1 = 0$. Thus

$$\log 1 + \log 1000 = 0 + \log 10^3$$

$$= 3 \log 10 \qquad \text{(Property 3)}$$

$$= 3(1) = 3 \qquad \text{(Property 5)}.$$

c. *Find* $e^{\ln x^2}$.

By Property 7, $e^{\ln x^2} = x^2$.

EXAMPLE 8   *Express* $\log x$ *in terms of natural logarithms.*

We must transform from base 10 into base e. Thus we use the change of base formula (Property 8) with $b = 10$, $m = x$, and $a = e$.

$$\log x = \frac{\ln x}{\ln 10}.$$

EXAMPLE 9   *An experiment was conducted with a particular type of small animal* [25]. *The logarithm of the amount of oxygen consumed per hour was determined for a number of the animals and was plotted against the logarithms of the weights of the animals. It was found that*

$$\log y = \log 5.934 + 0.885 \log x,$$

where y *was the number of microliters of oxygen consumed per hour, and* x *was the weight of the animal (in grams). Solve for* y.

We first combine the terms on the right side into a single logarithm.

$$\log y = \log 5.934 + 0.885 \log x$$

$$= \log 5.934 + \log x^{0.885} \qquad \text{(Property 3)}.$$

$$\log y = \log(5.934x^{0.885}) \qquad \text{(Property 1)}.$$

By the property that $m = n$ when $\log_b m = \log_b n$, we have

$$y = 5.934x^{0.885}.$$

EXERCISE 6.2

In Problems 1–8, express each logarithmic form exponentially and each exponential form logarithmically.

1. $10^4 = 10,000$.

2. $2 = \log_{12} 144$.

3. $\log_2 64 = 6$.

4. $8^{2/3} = 4$.

5. $e^2 = 7.3891$.

6. $e^{0.33647} = 1.4$.

7. $\ln 3 = 1.09861$.

8. $\log 5 = 0.6990$.

In Problems 9 and 10, graph the functions.

9. $y = f(x) = \log_3 x$.

10. $y = f(x) = \log_{1/2} x$.

In Problems 11–32, evaluate.

11. $\log_6 36$.

12. $\log_2 32$.

13. $\log_3 27$.

14. $\log_{16} 4$.

15. $\log_7 7$.

16. $\log 10,000$.

17. $\log 0.01$.

18. $\log_2 \sqrt{2}$.

19. $\log_5 1$.

20. $\log_5 \frac{1}{25}$.

21. $\log_2 \frac{1}{8}$.

22. $\log_4 \sqrt[5]{4}$.

23. $\log_7 7^{48}$.

24. $\ln \frac{1}{e}$.

25. $\log_2 \frac{2^6}{2^{10}}$.

26. $\log_3 (3^5 \cdot 3^4)^6$.

27. $\log 10 + \ln(e^2)$.

28. $(\log 100)(\ln \sqrt{e})$.

29. $\ln e^{4x}$.

30. $\log 10^x$.

31. $e^{\ln(2x)}$.

32. $10^{\log 3}$.

In Problems 33–46, solve for x.

33. $\log x = -1$.

34. $\log_4 x = 0$.

35. $\ln x = 2$.

36. $\ln x = 1$.

37. $\log_x 8 = 3$.

38. $\log_x 3 = \frac{1}{2}$.

39. $e^{3x} = 2$.

40. $\log_8 64 = x - 1$.

41. $5(3^x - 6) = 10$.

42. $0.1e^{0.1x} = 0.5$.

43. $e^{2x-5} + 1 = 4$.

44. $3e^{2x} - 1 = \frac{1}{2}$.

45. $e^{\ln(2x)} = 5$.

46. $e^{3 \ln x} = 8$.

In Problems 47–50, use Appendix E to find the approximate value of each expression.

47. $\ln 5$.

48. $\ln 3.12$.

49. $\ln 7.39$.

50. $\ln 9.98$.

In Problems 51–62, write the expression in terms of $\ln x$, $\ln(x + 1)$, and/or $\ln(x + 2)$.

51. $\ln[x(x + 1)^2]$.

52. $\ln \frac{\sqrt{x}}{x + 1}$.

53. $\ln \frac{x^2}{(x + 1)^3}$.

54. $\ln[x(x + 1)]^3$.

55. $\ln\left(\frac{x}{x + 1}\right)^3$.

56. $\ln \sqrt{x(x + 1)}$.

57. $\ln \frac{x}{(x + 1)(x + 2)}$.

58. $\ln \frac{x^2(x + 1)}{x + 2}$.

59. $\ln \frac{\sqrt{x}}{(x + 1)^2(x + 2)^3}$.

60. $\ln \frac{1}{x(x + 1)(x + 2)}$.

61. $\ln\left[\frac{1}{x + 2} \sqrt[5]{\frac{x^2}{x + 1}}\right]$.

62. $\ln \sqrt{\frac{x^4(x + 1)^3}{x + 2}}$.

In Problems 63 and 64, write each expression in terms of natural logarithms.

63. $\log(x + 8)$.

64. $\log_2 x$.

65. The cost c for a firm producing q units of a product is given by the cost function $c = (2q \ln q) + 20$. Evaluate the cost when $q = 6$. Give your answer to two decimal places.

66. A supply equation is $p = \log[10 + (q/2)]$, where q is the number of units supplied at a price p per unit. At what price will 1980 units be supplied?

67. If interest is compounded continuously at an annual rate of 0.05, how many years would it take for a principal P to triple? Give your answer to the nearest year.

68. If interest is compounded continuously, at what annual percentage rate will a principal of P double in ten years? Give your answer to the nearest percent.

69. The magnitude M of an earthquake and its energy E

are related in [39] by the equation

$$1.5M = \log\left(\frac{E}{2.5 \times 10^{11}}\right).$$

Here M is given in terms of Richter's preferred scale of 1958 and E is in ergs. Solve the equation for E.

70. For a certain population of cells, the number N of cells at time t is given by $N = N_0(2^{t/k})$, where N_0 is the number of cells at $t = 0$ and k is a positive constant.

a. Find N when $t = k$.

b. What is the significance of k?

c. Show that the time it takes to have population N_1 can be written $t = k \log_2(N_1/N_0)$.

71. Suppose that the daily output of units, q, of a new product on the tth day of a production run is given by $q = 500(1 - e^{-0.2t})$. Such an equation is called a *learning equation* and indicates that as time progresses output per day will increase. This may be due to the gain of the workers' proficiencies at their jobs.

a. Determine to the nearest complete unit the output on the first day and the tenth day after the start of a production run.

b. After how many days will a daily production run of 400 units be reached? Assume $\ln 0.2 = -1.6$.

72. In a discussion of market penetration by new products, Hurter and Rubenstein [40] refer to the function

$$F(t) = \frac{q - pe^{-(t+C)(p+q)}}{q[1 + e^{-(t+C)(p+q)}]},$$

where p, q, and C are constants. They claim that if $F(0) = 0$, then

$$C = -\frac{1}{p+q} \ln \frac{q}{p}.$$

Show that their claim is true.

73. On the surface of a glass slide is a grid that divides the surface into 225 equal squares. Suppose that a blood sample containing N red cells is spread on the slide and the cells are randomly distributed. Then the number of squares containing no cells is (approximately) given by $225e^{-N/225}$. If 100 of the squares contain no cells, estimate the number of cells the blood sample contained.

74. In a discussion [13] of the rate of cooling of isolated portions of the body when they are exposed to low temperatures, the following equation occurs:

$$T_t - T_e = (T_t - T_e)_o e^{-at},$$

where T_t is temperature of the portion at time t, T_e is environmental temperature, the subscript o refers to initial temperature difference, and a is a constant. Show that

$$a = \frac{1}{t} \ln \frac{(T_t - T_e)_o}{T_t - T_e}.$$

75. In an article concerning predators and prey, Holling[21] refers to an equation of the form $y = K(1 - e^{-ax})$, where x is prey density, y is the number of prey attacked, and K and a are constants. Verify his claim that

$$\ln \frac{K}{K - y} = ax.$$

76. In statistics the sample regression equation $y = ab^x$ is reduced to a linear form by taking logarithms of both sides. Find $\log y$.

77. In a study of rooted plants in a certain geographic region [25], it was determined that on plots of size A (in square meters), the average number of species that occurred was S. When $\log S$ was graphed as a function of $\log A$, the result was a straight line given by

$$\log S = \log 12.4 + 0.26 \log A.$$

Solve for S.

78. In an article, Taagepera and Hayes[31] refer to an equation of the form

$$\log T = 1.7 + 0.2068 \log P - 0.1334 \log^2 P.$$

Here T is the percentage of a country's gross national product (GNP) that corresponds to foreign trade (exports plus imports), and P is the country's population (in units of 100,000). The expression $\log^2 P$ means $(\log P)^2$. Verify the claim that

$$T = 50P^{(0.2068 - 0.1334 \log P)}.$$

You may assume that $\log 50 = 1.7$.

79. According to Richter[41], the magnitude M of an earthquake occurring 100 km from a certain type of seismometer is given by $M = \log(A) + 3$, where A is the recorded trace amplitude (in millimeters) of the quake.

a. Find the magnitude of an earthquake that records a trace amplitude of 1 mm.

b. If a particular earthquake has amplitude A_1 and magnitude M_1, determine the magnitude of a quake with amplitude $100A_1$. Express your answer in terms of M_1.

80. In a discussion of an inferior good, Persky[42] solves an equation of the form

$$u_0 = A \ln(x_1) + \frac{x_2^2}{2}$$

for x_1, where x_1 and x_2 are quantities of two products,

u_0 is a measure of utility, and A is a positive constant. Determine x_1.

81. In a study of military enlistments, Brown[43] considers total military compensation C as the sum of basic military compensation B (which includes the value of allowances, tax advantages, and base pay) and educational benefits E. Thus $C = B + E$. Brown states that

$$\ln C = \ln B + \ln\left(1 + \frac{E}{B}\right).$$

Verify this.

6.3 DERIVATIVES OF LOGARITHMIC FUNCTIONS

In this section the derivatives of logarithmic functions will be found. We begin with the derivative of $\ln x$. Let $f(x) = \ln x$, where x is positive. By the definition of the derivative,

$$\frac{d}{dx}(\ln x) = \lim_{h \to 0} \frac{f(x + h) - f(x)}{h} = \lim_{h \to 0} \frac{\ln(x + h) - \ln x}{h}.$$

Using the property that $\ln m - \ln n = \ln(m/n)$, we have

$$\frac{d}{dx}(\ln x) = \lim_{h \to 0} \frac{\ln\left(\dfrac{x + h}{x}\right)}{h}$$

$$= \lim_{h \to 0} \left[\frac{1}{h} \ln\left(\frac{x + h}{x}\right)\right] = \lim_{h \to 0} \left[\frac{1}{h} \ln\left(1 + \frac{h}{x}\right)\right].$$

Writing $\dfrac{1}{h}$ as $\dfrac{1}{x} \cdot \dfrac{x}{h}$ gives

$$\frac{d}{dx}(\ln x) = \lim_{h \to 0} \left[\frac{1}{x} \cdot \frac{x}{h} \ln\left(1 + \frac{h}{x}\right)\right]$$

$$= \lim_{h \to 0} \left[\frac{1}{x} \ln\left(1 + \frac{h}{x}\right)^{x/h}\right] \qquad \text{(since } r \ln m = \ln m^r)$$

$$= \frac{1}{x} \cdot \lim_{h \to 0} \left[\ln\left(1 + \frac{h}{x}\right)^{x/h}\right].$$

It can be shown that the limit of the logarithm is the logarithm of the limit $(\lim \ln u = \ln \lim u)$, so

$$\frac{d}{dx}(\ln x) = \frac{1}{x} \ln\left[\lim_{h \to 0}\left(1 + \frac{h}{x}\right)^{x/h}\right]. \qquad (1)$$

To evaluate $\lim\limits_{h \to 0} \left(1 + \dfrac{h}{x}\right)^{x/h}$, first note that as $h \to 0$, then $\dfrac{h}{x} \to 0$. Thus if we replace $\dfrac{h}{x}$ by k, the limit has the form

$$\lim_{k \to 0}(1 + k)^{1/k}.$$

As stated in Sec. 6.1, this limit is e. Thus Eq. (1) becomes

$$\frac{d}{dx}(\ln x) = \frac{1}{x}\ln e = \frac{1}{x}(1) = \frac{1}{x}.$$

Hence

$$\frac{d}{dx}(\ln x) = \frac{1}{x}. \tag{2}$$

EXAMPLE 1   *Find y' if $y = \dfrac{\ln x}{x}$.*

By the quotient rule and Eq. (2),

$$y' = \frac{x\,D_x(\ln x) - (\ln x)\,D_x(x)}{x^2}$$

$$= \frac{x\left(\dfrac{1}{x}\right) - (\ln x)\,(1)}{x^2} = \frac{1 - \ln x}{x^2}.$$

We now extend Eq. (2) to cover a broader class of functions. Let $y = \ln u$, where u is a positive differentiable function of x. By the chain rule,

$$\frac{d}{dx}(\ln u) = \frac{dy}{du}\cdot\frac{du}{dx} = \frac{d}{du}(\ln u)\cdot\frac{du}{dx} = \frac{1}{u}\cdot\frac{du}{dx}.$$

Thus

$$\frac{d}{dx}(\ln u) = \frac{1}{u}\cdot\frac{du}{dx}. \tag{3}$$

EXAMPLE 2   *Differentiate each of the following.*

a. $y = \ln(x^2 + 1)$.

This function has the form $\ln u$ with $u = x^2 + 1$. Using Eq. (3) gives

$$\frac{dy}{dx} = \frac{1}{x^2 + 1}\frac{d}{dx}(x^2 + 1) = \frac{1}{x^2 + 1}(2x) = \frac{2x}{x^2 + 1}.$$

b. $y = x^2 \ln(4x + 2)$.

Using the product rule and then Eq. (3) with $u = 4x + 2$, we obtain

$$\frac{dy}{dx} = x^2 D_x[\ln(4x + 2)] + [\ln(4x + 2)] D_x(x^2)$$

$$= x^2\left(\frac{1}{4x + 2}\right)(4) + [\ln(4x + 2)](2x)$$

$$= \frac{2x^2}{2x + 1} + 2x \ln(4x + 2).$$

c. $y = \ln(\ln x)$.

This has the form $y = \ln u$ where $u = \ln x$. Using Eqs. (3) and (2), we have

$$y' = \frac{1}{\ln x}\frac{d}{dx}(\ln x) = \frac{1}{\ln x}\left(\frac{1}{x}\right) = \frac{1}{x \ln x}.$$

Sometimes we can simplify the work involved in differentiating a function involving logarithms. We use properties of logarithms to rewrite the function *before* differentiating, as the next example shows.

EXAMPLE 3   *Differentiate each of the following.*

a. $f(p) = \ln[(p + 1)^2(p + 2)^3]$.

We rewrite the function using properties of logarithms.

$$f(p) = 2\ln(p + 1) + 3\ln(p + 2),$$

$$f'(p) = 2\left(\frac{1}{p + 1}\right)(1) + 3\left(\frac{1}{p + 2}\right)(1) = \frac{2}{p + 1} + \frac{3}{p + 2}.$$

Alternatively, differentiating the given function directly gives

$$f'(p) = \frac{1}{(p + 1)^2(p + 2)^3} D_p[(p + 1)^2(p + 2)^3].$$

Completing this involves the product and power rules. It is obvious that much work is saved by using properties of logarithms as originally done.

b. $f(w) = \ln \sqrt{\dfrac{1 + w^2}{w^2 - 1}}.$

Again, using properties of logarithms will simplify our work.

$$f(w) = \frac{1}{2}[\ln(1 + w^2) - \ln(w^2 - 1)].$$

$$f'(w) = \frac{1}{2}\left[\frac{1}{1 + w^2}(2w) - \frac{1}{w^2 - 1}(2w)\right]$$

$$= \frac{w}{1 + w^2} - \frac{w}{w^2 - 1} = -\frac{2w}{w^4 - 1}$$

$$= \frac{2w}{1 - w^4}.$$

c. $f(x) = \ln^3(2x + 1).$

The exponent 3 refers to the cubing of $\ln(2x + 1)$. That is,

$$f(x) = \ln^3(2x + 1) = [\ln(2x + 1)]^3.$$

By the power rule,

$$f'(x) = 3[\ln(2x + 1)]^2 D_x[\ln(2x + 1)]$$

$$= 3[\ln(2x + 1)]^2 \left[\frac{1}{2x + 1}(2)\right]$$

$$= \frac{6}{2x + 1}[\ln(2x + 1)]^2 = \frac{6}{2x + 1}\ln^2(2x + 1).$$

EXAMPLE 4   *If* $q - p = \ln q + \ln p$, *find* dq/dp.

We assume that q is a function of p and differentiate implicitly with respect to p.

$$D_p(q) - D_p(p) = D_p(\ln q) + D_p(\ln p),$$

$$\frac{dq}{dp} - 1 = \frac{1}{q}\frac{dq}{dp} + \frac{1}{p},$$

$$\frac{dq}{dp}\left(1 - \frac{1}{q}\right) = \frac{1}{p} + 1,$$

$$\frac{dq}{dp}\left(\frac{q - 1}{q}\right) = \frac{1 + p}{p},$$

$$\frac{dq}{dp} = \frac{(1 + p)q}{p(q - 1)}.$$

We can generalize Eq. (3) to any base b. By the change of base formula (Property 8 of Sec. 6.2), $\ln u = \dfrac{\log_b u}{\log_b e}$, so $\log_b u = (\log_b e) \ln u$. Hence

$$\frac{d}{dx}(\log_b u) = \frac{d}{dx}[(\log_b e) \ln u] = (\log_b e)\frac{d}{dx}(\ln u)$$

$$= (\log_b e)\left(\frac{1}{u}\frac{du}{dx}\right).$$

Thus

$$\frac{d}{dx}(\log_b u) = \frac{1}{u}(\log_b e)\frac{du}{dx}. \qquad (4)$$

Since the use of natural logarithms (that is, $b = e$) gives a value of 1 to the factor $\log_b e$ in Eq. (4), natural logarithms are used extensively in calculus.

EXAMPLE 5   *If $y = \log(2x + 1)$, find the rate of change of y with respect to x.*

We want to find dy/dx. By Eq. (4) with $u = 2x + 1$ and $b = 10$,

$$\frac{dy}{dx} = \frac{1}{2x + 1}(\log e)\,D_x(2x + 1) = \frac{1}{2x + 1}(\log e)(2) = \frac{2 \log e}{2x + 1}.$$

EXERCISE 6.3

*In Problems **1–34**, differentiate the functions. If possible, first use properties of logarithms to simplify the given function.*

1. $y = \ln(3x - 4)$.

2. $y = \ln(5x - 6)$.

3. $y = \ln x^2$.

4. $y = \ln(ax^2 + b)$.

5. $y = \ln(1 - x^2)$.

6. $y = \ln(-x^2 + 6x)$.

7. $f(p) = \ln(2p^3 + 3p)$.

8. $f(r) = \ln(2r^4 - 3r^2 + 2r + 1)$.

9. $f(t) = t \ln t$.

10. $y = x^2 \ln x$.

11. $y = \log_3(2x - 1)$.

12. $f(w) = \log(w^2 + w)$.

13. $f(z) = \dfrac{\ln z}{z}$.

14. $y = \dfrac{x^2 - 1}{\ln x}$.

15. $y = \ln(x^2 + 4x + 5)^3$.

16. $y = \ln x^{100}$.

17. $y = \ln\sqrt{1 + x^2}$.

18. $f(s) = \ln\left(\dfrac{s^2}{1 + s^2}\right)$.

19. $f(l) = \ln\left(\dfrac{1 + l}{1 - l}\right)$.

20. $y = \ln\left(\dfrac{2x + 3}{3x - 4}\right)$.

21. $y = \ln\sqrt[4]{\dfrac{1 + x^2}{1 - x^2}}$.

22. $y = \ln\sqrt{\dfrac{x^4 - 1}{x^4 + 1}}$.

23. $y = \ln[(x^2 + 2)^2(x^3 + x - 1)]$.

24. $y = \ln[(5x + 2)^4(8x - 3)^6]$.

25. $y = (x^2 + 1)\ln(2x + 1)$.

26. $y = (ax + b)\ln(ax)$.

27. $y = \ln x^3 + \ln^3 x$.

28. $y = x^{\ln 3}$.

29. $y = \ln^4(ax)$.

30. $y = \ln^2(2x + 3)$.

31. $y = x \ln\sqrt{x - 1}$.

32. $y = \ln(x^2\sqrt{3x - 2})$.

33. $y = \sqrt{4 + \ln x}$.

34. $y = \ln(x + \sqrt{1 + x^2})$.

35. If $y = x^2 \ln x$, find y''.

36. If $y = \ln(x^2 + x)$, find y''.

37. Find an equation of the tangent line to the curve $y = \ln(x^2 - 2x - 2)$ when $x = 3$.

38. Find the slope of the curve $y = \dfrac{x}{\ln x}$ when $x = 2$.

*In Problems **39** and **40**, determine when the function is increasing or decreasing and locate all relative extrema.*

39. $y = f(x) = x^2 - 2 \ln x$.

40. $y = f(x) = x \ln x$.

*In Problems **41** and **42**, find dy/dx by implicit differentiation.*

41. $y^2 + y = \ln x$.

42. $\ln(xy) + x = 4$.

43. Find the marginal revenue function if the demand function is $p = 25/\ln(q + 2)$.

44. A total cost function is $c = 25 \ln(q + 1) + 12$. Find the marginal cost when $q = 6$.

45. The magnitude M of an earthquake and its energy E are related by the equation [39]

$$1.5M = \log\left(\frac{E}{2.5 \times 10^{11}}\right).$$

Here M is given in terms of Richter's preferred scale of 1958 and E is in ergs. Determine the rate of change of energy with respect to magnitude.

46. According to Richter[41], the number N of earthquakes of magnitude M or greater per unit of time is given by $\log N = A - bM$, where A and b are constants. He claims that

$$\log\left(-\frac{dN}{dM}\right) = A + \log\left(\frac{b}{q}\right) - bM,$$

where $q = \log e$. Verify this statement.

47. In an experiment (adapted from [25]) on dispersal of a particular insect, a large number of insects are placed at a release point in an open field. Surrounding this point are traps that are placed in a concentric circular arrangement at a distance of 1 m, 2 m, 3 m, etc. from the release point. Twenty-four hours after the insects are released, the number of insects in each trap are counted. It is determined that at a distance of r meters from the release point, the average number of insects contained in a trap is n, where

$$n = f(r) = 0.1 \ln(r) + \frac{7}{r} - 0.8, \qquad 1 \le r \le 10.$$

a. Show that the graph of f is always falling and concave up.

b. Sketch the graph of f.

c. When $r = 5$, at what rate is the average number of insects in a trap decreasing with respect to distance?

48. Show that the relative rate of change of $y = f(x)$ with respect to x is equal to the derivative of $y = \ln[f(x)]$.

6.4   DERIVATIVES OF EXPONENTIAL FUNCTIONS

We now obtain a formula for the derivative of the exponential function e^u, where u is a differentiable function of x. By letting $y = e^u$, in logarithmic form we have $u = \ln y$. Differentiating both sides with respect to x gives

$$\frac{d}{dx}(u) = \frac{d}{dx}(\ln y),$$

$$\frac{du}{dx} = \frac{1}{y}\frac{dy}{dx}.$$

Solving for dy/dx and replacing y by e^u give

$$\frac{dy}{dx} = y\frac{du}{dx} = e^u\frac{du}{dx}.$$

Thus

$$\frac{d}{dx}(e^u) = e^u \frac{du}{dx}. \tag{1}$$

As a special case, let $u = x$. Then $du/dx = 1$ and

$$\frac{d}{dx}(e^x) = e^x. \tag{2}$$

Note that the function and its derivative are the same.

Pitfall
Do **not** use the power rule to find $D_x(e^x)$. That is, $D_x(e^x) \neq xe^{x-1}$.

EXAMPLE 1

a. *Find* $\dfrac{d}{dx}(3e^x)$.

Since 3 is a constant factor,

$$\frac{d}{dx}(3e^x) = 3\frac{d}{dx}(e^x)$$

$$= 3e^x \qquad \text{[by Eq. (2)]}.$$

b. *If* $y = \dfrac{x}{e^x}$, *find* y'.

We *first* use the quotient rule and then use Eq. (2).

$$\frac{dy}{dx} = \frac{e^x D_x(x) - x D_x(e^x)}{(e^x)^2} = \frac{e^x(1) - x(e^x)}{(e^x)^2} = \frac{e^x(1 - x)}{e^{2x}} = \frac{1 - x}{e^x}.$$

c. *If* $y = e^2 + e^x + \ln 3$, *find* y'.

Since e^2 and $\ln 3$ are constants, $y' = 0 + e^x + 0 = e^x$.

EXAMPLE 2

a. *Find* $\dfrac{d}{dx}(e^{x^3+3x})$.

The function has the form e^u with $u = x^3 + 3x$. From Eq. (1),

$$\frac{d}{dx}(e^{x^3+3x}) = e^{x^3+3x} D_x(x^3 + 3x) = e^{x^3+3x}(3x^2 + 3)$$

$$= 3(x^2 + 1)e^{x^3+3x}.$$

b. *Find $D_x[e^{x+1} \ln(x^2 + 1)]$.*

By the product rule,

$$D_x[e^{x+1} \ln(x^2 + 1)] = e^{x+1} D_x[\ln(x^2 + 1)] + [\ln(x^2 + 1)] D_x(e^{x+1})$$

$$= e^{x+1}\left(\frac{1}{x^2 + 1}\right)(2x) + [\ln(x^2 + 1)]e^{x+1}(1)$$

$$= e^{x+1}\left[\frac{2x}{x^2 + 1} + \ln(x^2 + 1)\right].$$

We can generalize Eq. (1) to other bases by considering the derivative of a^u where $a > 0$, $a \neq 1$, and u is a differentiable function of x. First we write a^u as an exponential function with base e. By Property 7 of Sec. 6.2, $a = e^{\ln a}$. Thus

$$D_x(a^u) = D_x[(e^{\ln a})^u] = D_x(e^{u \ln a})$$

$$= e^{u \ln a} D_x(u \ln a) \qquad\qquad \text{[by Eq. (1)]}$$

$$= e^{u \ln a}\left(\frac{du}{dx}\right) \ln a \qquad\qquad (\ln a \text{ is constant})$$

$$= a^u(\ln a)\frac{du}{dx} \qquad\qquad (\text{since } e^{u \ln a} = a^u).$$

Thus

$$\boxed{\frac{d}{dx}(a^u) = a^u (\ln a) \frac{du}{dx}.} \qquad\qquad (3)$$

Note that if $a = e$, then the factor $\ln a$ in Eq. (3) is 1. Thus if exponential functions with base e are used, we have a much simpler differentiation formula with which to work. This is the reason exponential functions with base e are used extensively in calculus.

EXAMPLE 3   *If $y = 8^{2x+3}$, find y'.*

By Eq. (3),

$$y' = 8^{2x+3}(\ln 8)\frac{d}{dx}(2x + 3)$$

$$= 8^{2x+3}(\ln 8)(2) = 2(\ln 8)8^{2x+3}.$$

EXAMPLE 4   *For a certain population of cells, the number N of cells at time t is given by*

$$N(t) = N_0(2^{t/k}),$$

where N_0 is the number of cells at $t = 0$ and k is a positive constant. Find the relative rate of change of $N(t)$.

The relative rate of change is $N'(t)/N(t)$. To find $N'(t)$ we use Eq. (3) with $a = 2$ and $u = t/k$.

$$N'(t) = N_0 2^{t/k}(\ln 2)D_t\left(\frac{t}{k}\right) = N_0 2^{t/k}(\ln 2)\left(\frac{1}{k}\right).$$

Thus

$$\frac{N'(t)}{N(t)} = \frac{N_0 2^{t/k}(\ln 2)\left(\frac{1}{k}\right)}{N_0 2^{t/k}} = \frac{\ln 2}{k}.$$

EXAMPLE 5   *Find $D_x y$ if $y = x^{100} + 100^x$.*

Note that this function involves a variable to a constant power and a constant raised to a variable power. Do not confuse these forms!

$$D_x y = 100x^{99} + 100^x(\ln 100)D_x(x) = 100x^{99} + 100^x \ln 100.$$

EXAMPLE 6   *An important function in the social sciences is the **normal density function***

$$y = f(x) = \frac{1}{\sigma\sqrt{2\pi}}\, e^{-(1/2)[(x-\mu)/\sigma]^2},$$

where σ (a Greek letter read "sigma") and μ (a Greek letter read "mu") are constants. Its graph, called the normal curve, is "bell-shaped" (see Fig. 6.8). Determine the rate of change of y with respect to x when $x = \mu$.

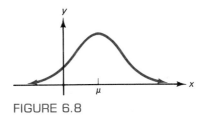

FIGURE 6.8

The rate of change of y with respect to x is dy/dx.

$$\frac{dy}{dx} = \frac{1}{\sigma\sqrt{2\pi}}\, [e^{-(1/2)[(x-\mu)/\sigma]^2}]\left[-\frac{1}{2}(2)\left(\frac{x-\mu}{\sigma}\right)\left(\frac{1}{\sigma}\right)\right].$$

Evaluating dy/dx when $x = \mu$, we obtain

$$\left. \frac{dy}{dx} \right|_{x=\mu} = 0.$$

Geometrically this means that the slope of the tangent line to the graph of $y = f(x)$ at $x = \mu$ is horizontal (see Fig. 6.8).

EXAMPLE 7　　*If $e^{xy} = x + y$, find y'.*

Differentiating implicitly gives

$$D_x(e^{xy}) = D_x(x) + D_x(y),$$
$$e^{xy} D_x(xy) = 1 + y',$$
$$e^{xy}(xy' + y) = 1 + y' \qquad \text{(product rule)},$$
$$xe^{xy}y' + ye^{xy} = 1 + y',$$
$$xe^{xy}y' - y' = 1 - ye^{xy},$$
$$y'(xe^{xy} - 1) = 1 - ye^{xy},$$
$$y' = \frac{1 - ye^{xy}}{xe^{xy} - 1}.$$

EXAMPLE 8　　*Test $f(x) = x^2 e^x$ for relative extrema.*

By the product rule,

$$f'(x) = x^2 e^x + e^x (2x) = xe^x(x + 2).$$

Since e^x is always positive, the critical values are 0 and -2. From the signs of $f'(x)$ given in Fig. 6.9, we conclude that there is a relative maximum when $x = -2$ and a relative minimum when $x = 0$.

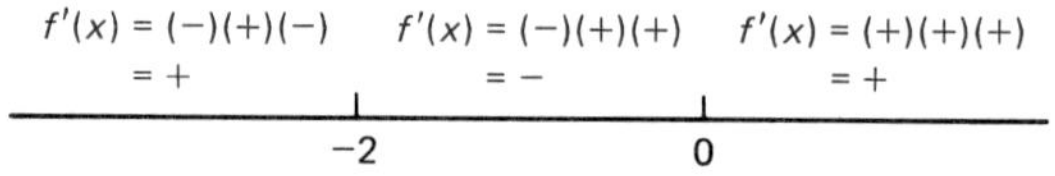

$$f'(x) = (-)(+)(-) \qquad f'(x) = (-)(+)(+) \qquad f'(x) = (+)(+)(+)$$
$$= + \qquad\qquad = - \qquad\qquad = +$$

$$\overline{\qquad\qquad\quad \underset{-2}{|} \qquad\qquad \underset{0}{|} \qquad\qquad\quad}$$

FIGURE 6.9

EXERCISE 6.4

In Problems 1–26, differentiate the functions.

1. $y = e^{x^2+1}$.

2. $y = e^{2x^2+5}$.

3. $y = e^{3-5x}$.

4. $f(q) = e^{-q^3+6q-1}$.

5. $f(r) = e^{3r^2+4r+4}$.

6. $y = e^{9x^2+5x^3-6}$.

7. $y = xe^x$.

8. $y = x^2 e^{-x}$.

9. $y = x^2 e^{-x^2}$.

10. $y = xe^{2x}$.

11. $y = \dfrac{e^x + e^{-x}}{2}$.

12. $y = \dfrac{e^x - e^{-x}}{2}$.

13. $y = 4^{3x^2}$.

14. $y = 2^x x^2$.

15. $f(w) = \dfrac{e^{2w}}{w^2}$.

16. $y = e^{x-\sqrt{x}}$.

17. $y = e^{1+\sqrt{x}}$.

18. $y = (e^{3x} + 1)^4$.

19. $y = x^3 - 3^x$.

20. $f(z) = e^{1/z}$.

21. $y = \dfrac{e^x - 1}{e^x + 1}$.

22. $y = e^{2x}(x + 1)$.

23. $y = e^{\ln x}$.

24. $y = e^{-x} \ln x$.

25. $y = e^{x \ln x}$.

26. $y = \ln e^{4x+1}$.

27. Find an equation of the tangent line to the graph of $y = e^x$ when $x = 2$.

28. Find the slope of the tangent line to the graph of $y = 2e^{-4x^2}$ when $x = 0$.

In Problems 29–32, find the indicated derivatives.

29. $y = x^3 + e^x$; $y^{(4)}$.

30. $y = e^{-4x^2}$; y''.

31. $f(z) = z^2 e^z$; $f''(z)$.

32. $y = \dfrac{x}{e^x}$; $\dfrac{d^2y}{dx^2}$.

In Problems 33 and 34, determine when the function is increasing or decreasing and locate all relative extrema.

33. $y = xe^x$.

34. $y = e^{-x^2}$.

In Problems 35 and 36, find dy/dx by implicit differentiation.

35. $xe^y + y = 4$.

36. $y \ln x = xe^y$.

For each of the demand equations in Problems 37 and 38, find the rate of change of price p with respect to quantity q. What is the rate of change for the indicated value of q?

37. $p = 15e^{-0.001q}$; $q = 500$.

38. $p = 8e^{-3q/800}$; $q = 400$.

39. For a manufacturer's product, the demand function is $q = 10,000e^{-0.02p}$. Find the value of p for which maximum revenue is obtained.

40. The population P of a city t years from now is given by $P = 20,000e^{0.03t}$. Show that $dP/dt = kP$ where k is a constant. This means that the rate of change of population at any time is proportional to the population at that time.

In Problems 41 and 42, $\bar{c}$ is the average cost of producing q units of a product. Find the marginal cost function and the marginal cost for the given values of q.

41. $\bar{c} = \dfrac{7000e^{q/700}}{q}$; $q = 350$, $q = 700$.

42. $\bar{c} = \dfrac{850}{q} + 4000 \dfrac{e^{(2q+6)/800}}{q}$; $q = 97$, $q = 197$.

43. For a firm the daily output q on the tth day of a production run is given by $q = 500(1 - e^{-0.2t})$. Find the rate of change of output q with respect to t on the tenth day.

44. After t years, the value S of a principal of P dollars which is invested at a nominal rate of r per year and compounded continuously is given by $S = Pe^{rt}$. Show that the relative rate of change of S with respect to t is r.

45. Sketch the graph of the normal density function

$$f(x) = \frac{1}{\sqrt{2\pi}} e^{-x^2/2}.$$

Include relative extrema and inflection points.

46. In an article concerning predators and prey, Holling[21] refers to an equation of the form

$$y = K(1 - e^{-ax}),$$

where x is the prey density, y is the number of prey attacked, and K and a are constants. Verify his statement that

$$\frac{dy}{dx} = a(K - y).$$

47. According to Richter[41], the number N of earthquakes of magnitude M or greater per unit of time is given by $N = 10^A 10^{-bM}$, where A and b are constants. Find dN/dM.

48. In a study of the effects of food deprivation on hunger [37], an insect was fed until its appetite was completely satisfied. Then it was deprived food for t hours (deprivation period). At the end of this period, the insect was refed until its appetite was again completely satisfied. The weight H (in grams) of the food that was consumed at this time was statistically found to be a function of t, where

$$H = 1.00[1 - e^{-(0.0464t + 0.0670)}].$$

Here H is a measure of hunger. Show that H is increasing with respect to t, but at a decreasing rate.

49. Short-term retention was studied by Peterson and Peterson[44]. They analyzed a procedure in which an experimenter verbally gave a subject a three-letter

consonant syllable, such as CHJ, followed by a three-digit number, such as 309. The subject then repeated the number and counted backwards by 3's, such as 309, 306, 303, ... After a period of time the subject was signaled by a light to recite the three-letter consonant syllable. The time interval between the experimenter's completion of the last consonant to the onset of the light was called the *recall interval*. The time between the onset of the light and the completion of a response was referred to as *latency*. After many trials, it was determined that for a recall interval of t seconds, the approximate proportion of correct recalls with latency below 2.83 sec was p, where

$$p = 0.89[0.01 + 0.99(0.85)^t].$$

a. Find dp/dt and interpret your result.

b. If you have a calculator, evaluate dp/dt when $t = 2$. Give your answer to two decimal places.

50. Hurter and Rubenstein[40] discuss the diffusion of a new process into a market. They refer to an equation of the form

$$Y = k\alpha^{\beta^t},$$

where Y is the cumulative level of diffusion of the new process at time t, and k, α, and β are positive constants. Verify their claim that

$$\frac{dY}{dt} = k\alpha^{\beta^t}(\beta^t \ln \alpha) \ln \beta.$$

51. When a deep-sea diver undergoes decompression, or a pilot climbs to a high altitude, nitrogen may bubble out of the blood, causing what is commonly called the *bends*. Suppose that the percentage P of people who suffer effects of the bends at an altitude of h thousand feet is given by (adapted from [7])

$$P = \frac{100}{1 + 100,000e^{-0.36h}}.$$

a. Determine $\lim_{h \to \infty} P$. What interpretation can you attach to this limit?

b. Find dP/dh.

c. Is P an increasing function of h?

52. In a psychological experiment involving conditioned response (adapted from [45]), subjects listened to four tones, denoted 0, 1, 2, and 3. Initially, the subjects

were conditioned to tone 0 by receiving a shock whenever this tone was heard. Later, when each of the four tones (stimuli) were heard without shocks, the subjects' responses were recorded by means of a tracking device that measures galvanic skin reaction. The average response to each stimulus (without shock) was determined, and the results were plotted on a coordinate plane where the x- and y-axes represent the stimuli (0, 1, 2, 3) and the average galvanic responses, respectively. It was determined that the points fit a curve that is approximated by the graph of $y = 12.5 + 5.8(0.42)^x$. Show that this function is decreasing and concave up.

53. Suppose a tracer, such as a colored dye, is injected instantly into the heart at time $t = 0$ and mixes uniformly with blood inside the heart. Let the initial concentration of the tracer in the heart be C_0 and assume the heart has constant volume V. As fresh blood flows into the heart, assume that the diluted mixture of blood and tracer flows out at the constant positive rate r. Then the concentration $C(t)$ of the tracer in the heart at time t is given by

$$C(t) = C_0 e^{-(r/V)t}.$$

a. Find $\lim_{t \to \infty} C(t)$.

b. Show that $dC/dt = (-r/V)C(t)$.

54. In Problem 53, suppose the tracer is injected at a constant rate R. Then the concentration at time t is

$$C(t) = \frac{R}{r}[1 - e^{-(r/V)t}].$$

a. Find $C(0)$.

b. Find $\lim_{t \to \infty} C(t)$.

c. Show that

$$\frac{dC}{dt} = \frac{R}{V} - \frac{r}{V} C(t).$$

55. Several models have been used to analyze the length of stay in a hospital. For a particular group of schizophrenics, one such model is (adapted from [46])

$$f(t) = 1 - e^{-0.008t},$$

where $f(t)$ is the proportion of the group that was discharged at the end of t days of hospitalization.

a. Find $\lim_{t \to \infty} f(t)$ and interpret your result.

b. Find the rate of discharge (proportion discharged per day) at the end of 100 days. Give your answer to four decimal places.

56. In a discussion of an inferior good, Persky[42] considers a function of the form

$$g(x) = e^{(U_0/A)} e^{-x^2/(2A)},$$

where x is a quantity of a good, U_0 is a constant that represents utility, and A is a positive constant. Persky claims that the graph of g is concave down when $x < \sqrt{A}$ and concave up when $x > \sqrt{A}$. Verify this.

6.5 LOGARITHMIC DIFFERENTIATION

There is a technique that often simplifies the differentiation of $y = f(x)$ when $f(x)$ involves products, quotients, or powers. We first take the natural logarithm of both sides of $y = f(x)$. After simplifying $\ln [f(x)]$ by using properties of logarithms, we then differentiate both sides with respect to x. The next example illustrates this method of **logarithmic differentiation**.

EXAMPLE 1   *Find y' if $y = \dfrac{(2x - 5)^3}{x^2\sqrt[4]{x^2 + 1}}$.*

Differentiating this function in the usual way is messy because it involves the quotient, power, and product rules. Logarithmic differentiation makes the work less of a chore. First, we take the natural logarithm of both sides and simplify.

$$\ln y = \ln \frac{(2x - 5)^3}{x^2\sqrt[4]{x^2 + 1}} = \ln(2x - 5)^3 - \ln[x^2\sqrt[4]{x^2 + 1}],$$

$$\ln y = 3 \ln(2x - 5) - 2 \ln x - \tfrac{1}{4} \ln(x^2 + 1).$$

Differentiating with respect to x gives

$$\frac{1}{y}y' = 3\left(\frac{1}{2x - 5}\right)(2) - 2\left(\frac{1}{x}\right) - \frac{1}{4}\left(\frac{1}{x^2 + 1}\right)(2x),$$

$$\frac{y'}{y} = \frac{6}{2x - 5} - \frac{2}{x} - \frac{x}{2(x^2 + 1)}.$$

We are not done yet! We must now solve for y'. Multiplying both sides by y and then substituting the original expression for y gives y' in terms of x only.

$$y' = y\left[\frac{6}{2x - 5} - \frac{2}{x} - \frac{x}{2(x^2 + 1)}\right],$$

$$y' = \frac{(2x - 5)^3}{x^2\sqrt[4]{x^2 + 1}}\left[\frac{6}{2x - 5} - \frac{2}{x} - \frac{x}{2(x^2 + 1)}\right].$$

Logarithmic differentiation can also be used to differentiate a function of the form $y = u^v$, where both u and v are differentiable functions of x.

Because both the base and exponent are not necessarily constants, the differentiation formulas for u^n and a^u do not apply here.

EXAMPLE 2   If $y = x^{4x}$, find y'.

This has the form $y = u^v$, where u and v are both functions of x. Taking the natural logarithm of both sides gives $\ln y = \ln x^{4x}$ or

$$\ln y = 4x \ln x.$$

Differentiating both sides with respect to x gives

$$\frac{1}{y}y' = 4x\left(\frac{1}{x}\right) + (\ln x)(4),$$

$$\frac{y'}{y} = 4(1 + \ln x).$$

Solving for y' and substituting x^{4x} for y give

$$y' = y[4(1 + \ln x)] = 4x^{4x}(1 + \ln x).$$

Pitfall
When using properties of logarithms, you must at all times be able to justify your steps. For example,

$$\text{if } y = x^x + x^5, \quad \text{then} \quad \ln y \neq \ln x^x + \ln x^5.$$

That is, the logarithm of a sum is *not* a sum of logarithms.

Be sure you understand how to differentiate each of the following forms:

$$y = \begin{cases} [f(x)]^n, & \text{(a)} \\ a^{f(x)}, & \text{(b)} \\ [f(x)]^{g(x)}. & \text{(c)} \end{cases}$$

For type (a) you may use the power rule; for type (b) use the differentiation formula for exponential functions; for type (c) use logarithmic differentiation. It would be nonsense to write $D_x(x^x) = x \cdot x^{x-1}$. Be sure that all steps are justified by the basic concepts that have been developed.

EXERCISE 6.5

In Problems 1–18, find y' by using logarithmic differentiation.

1. $y = (x + 1)^2(x - 1)(x^2 + 3)$.

2. $y = (3x + 4)(8x - 1)^2(3x^2 + 1)^4$.

3. $y = (3x^3 - 1)^2(2x + 5)^3$.

4. $y = (3x + 1)\sqrt{8x - 1}$.

5. $y = \sqrt{x + 1}\,\sqrt{x^2 - 2}\,\sqrt{x + 4}$.

6. $y = (x + 2)\sqrt{x^2 + 9}\,\sqrt[3]{2x + 1}$.

7. $y = \dfrac{\sqrt{1 - x^2}}{1 - 2x}$.

8. $y = \sqrt{\dfrac{x^2 + 5}{x + 9}}$.

9. $y = \dfrac{(2x^2 + 2)^2}{(x + 1)^2(3x + 2)}.$

10. $y = \sqrt{\dfrac{(x - 1)(x + 1)}{3x - 4}}.$

11. $y = \dfrac{(8x + 3)^{1/2}(x^2 + 2)^{1/3}}{(1 + 2x)^{1/4}}.$

12. $y = \dfrac{x(1 + x^2)^2}{\sqrt{2 + x^2}}.$

13. $y = x^{2x+1}.$

14. $y = x^{\sqrt{x}}.$

15. $y = x^{1/x}.$

16. $y = \left(\dfrac{2}{x}\right)^x.$

17. $y = (3x + 1)^{2x}.$

18. $y = x^{x^2}.$

19. Find an equation of the tangent line to

$$y = (x + 1)(x + 2)^2(x + 3)^2$$

at the point where $x = 0$.

20. Without using logarithmic differentiation, find the derivative of $y = x^x$. (*Hint:* First show that $y = x^x = e^{x \ln x}$.)

6.6 INTEGRALS INVOLVING EXPONENTIAL AND LOGARITHMIC FUNCTIONS

We now turn our attention to integrating exponential functions. If u is a differentiable function of x, then $D_x(e^u) = e^u\, du/dx$. Corresponding to this differentiation formula is the integration formula

$$\int e^u \frac{du}{dx}\, dx = e^u + C.$$

But $\dfrac{du}{dx}\, dx$ is the differential of u, namely du. Thus

$$\int e^u\, du = e^u + C. \tag{1}$$

In particular, if $u = x$, then $du = dx$ and

$$\int e^x\, dx = e^x + C. \tag{2}$$

EXAMPLE 1   *Find the following integrals.*

a. $\displaystyle\int 2xe^{x^2}\, dx.$

Let $u = x^2$. Then $du = 2x\, dx$ and by Eq. (1),

$$\int 2xe^{x^2}\, dx = \int e^{x^2}[2x\, dx] = \int e^u\, du$$
$$= e^u + C = e^{x^2} + C.$$

b. $\displaystyle\int (x^2 + 1)e^{x^3 + 3x} \, dx.$

If $u = x^3 + 3x$, then $du = (3x^2 + 3) \, dx = 3(x^2 + 1) \, dx$. If the integrand contained a factor of 3, the integral would have the form $\displaystyle\int e^u \, du$. Thus we write

$$\int (x^2 + 1)e^{x^3 + 3x} \, dx = \frac{1}{3}\int e^{x^3 + 3x}[3(x^2 + 1) \, dx]$$

$$= \frac{1}{3}\int e^u \, du = \frac{1}{3}e^u + C$$

$$= \frac{1}{3}e^{x^3 + 3x} + C.$$

c. $\displaystyle\int_0^1 e^{3t} \, dt.$

$$\int_0^1 e^{3t} \, dt = \frac{1}{3}\int_0^1 e^{3t}[3 \, dt] = \left(\frac{1}{3}\right) e^{3t}\bigg|_0^1 = \frac{1}{3}(e^3 - e^0) = \frac{1}{3}(e^3 - 1).$$

Pitfall

Do not apply the power rule formula for $\displaystyle\int u^n \, du$ to $\displaystyle\int e^u \, du$. For example,

$$\int e^x \, dx \neq \frac{e^{x+1}}{x + 1} + C.$$

Since $D_x\left(\dfrac{a^u}{\ln a}\right) = \dfrac{1}{\ln a}\left[a^u(\ln a)\dfrac{du}{dx}\right] = a^u \dfrac{du}{dx}$, it follows that

$$\int a^u \frac{du}{dx} \, dx = \frac{a^u}{\ln a} + C.$$

But $\dfrac{du}{dx} \, dx$ is du, so we have the integration formula

$$\boxed{\int a^u \, du = \frac{a^u}{\ln a} + C. \qquad\qquad (3)}$$

EXAMPLE 2   *Find* $\displaystyle\int 2^{3-x} \, dx.$

Let $u = 3 - x$. Then $du = -dx$, and from Eq. (3) with $a = 2$ we have

$$\int 2^{3-x} \, dx = -\int 2^{3-x}[-dx] = -\int a^u \, du$$

$$= -\frac{a^u}{\ln a} + C = -\frac{2^{3-x}}{\ln 2} + C.$$

Alternatively, we can avoid Eq. (3) by writing 2 in terms of e. Since $2 = e^{\ln 2}$ (by Property 7 of Sec. 6.2), we have

$$\int 2^{3-x}\, dx = \int (e^{\ln 2})^{3-x}\, dx = \int e^{(\ln 2)(3-x)}\, dx$$

$$= -\frac{1}{\ln 2} \int e^{(\ln 2)(3-x)}\, [(-\ln 2)\, dx] \qquad \left(\text{form: } \int e^u\, du\right)$$

$$= -\frac{1}{\ln 2} e^{(\ln 2)(3-x)} + C = -\frac{1}{\ln 2} 2^{3-x} + C.$$

As you know, the power rule formula

$$\int u^n\, du = \frac{u^{n+1}}{n+1} + C$$

assumes that $n \neq -1$. To find $\int u^{-1}\, du = \int \frac{1}{u}\, du$, we first recall that

$$\frac{d}{dx}(\ln u) = \frac{1}{u}\frac{du}{dx}.$$

It would seem that $\int \frac{1}{u}\frac{du}{dx}\, dx = \int \frac{1}{u}\, du = \ln u + C$. However, the logarithm of u is defined if and only if u is positive. If $u < 0$, then $\ln u$ is not defined. Thus $\int \frac{1}{u}\, du = \ln u + C$ as long as $u > 0$. On the other hand, if $u < 0$, then $-u > 0$ and $\ln(-u)$ *is* defined. Moreover,

$$\frac{d}{dx}[\ln(-u)] = \frac{1}{-u}(-1)\frac{du}{dx} = \frac{1}{u}\frac{du}{dx}.$$

In this case $(u < 0)$, $\int \frac{1}{u}\frac{du}{dx}\, dx = \int \frac{1}{u}\, du = \ln(-u) + C$. In summary, if $u > 0$, then $\int \frac{1}{u}\, du = \ln u + C$; if $u < 0$, then $\int \frac{1}{u}\, du = \ln(-u) + C$. Combining these cases, we have

$$\int \frac{1}{u}\, du = \ln|u| + C. \qquad (4)$$

In particular, if $u = x$, then $du = dx$ and

$$\int \frac{1}{x}\, dx = \ln|x| + C. \qquad (5)$$

EXAMPLE 3   *Find the following integrals.*

a. $\int \dfrac{7}{x}\, dx.$

From Eq. (5),

$$\int \frac{7}{x}\, dx = 7 \int \frac{1}{x}\, dx = 7\,\ln|x| + C.$$

Using properties of logarithms, we can write this answer another way:

$$\int \frac{7}{x}\, dx = \ln|x^7| + C.$$

b. $\int \dfrac{2x}{x^2 + 5}\, dx.$

Let $u = x^2 + 5$. Then $du = 2x\, dx$. From Eq. (4),

$$\int \frac{2x}{x^2 + 5}\, dx = \int \frac{1}{x^2 + 5}\,[2x\, dx] = \int \frac{1}{u}\, du$$

$$= \ln|u| + C = \ln|x^2 + 5| + C.$$

Since $x^2 + 5$ is always positive we can omit the absolute-value bars:

$$\int \frac{2x}{x^2 + 5}\, dx = \ln(x^2 + 5) + C.$$

c. $\int \dfrac{(2x^3 + 3x)\, dx}{x^4 + 3x^2 + 7}.$

If $u = x^4 + 3x^2 + 7$, then $du = (4x^3 + 6x)\, dx$, which is two times the numerator. We insert a factor of 2 and adjust for it with a factor of $\frac{1}{2}$.

$$\int \frac{2x^3 + 3x}{x^4 + 3x^2 + 7}\, dx = \frac{1}{2} \int \frac{2(2x^3 + 3x)}{x^4 + 3x^2 + 7}\, dx$$

$$= \frac{1}{2} \int \frac{1}{x^4 + 3x^2 + 7}\,[(4x^3 + 6x)\, dx]$$

$$= \frac{1}{2} \int \frac{1}{u}\, du = \frac{1}{2}\,\ln|u| + C$$

$$= \frac{1}{2}\,\ln|x^4 + 3x^2 + 7| + C$$

$$= \ln \sqrt{x^4 + 3x^2 + 7} + C.$$

EXAMPLE 4   *Find the following integrals.*

a. $\displaystyle\int\left[\frac{1}{(1-w)^2}+\frac{1}{w-1}\right]dw.$

$$\int\left[\frac{1}{(1-w)^2}+\frac{1}{w-1}\right]dw=\int(1-w)^{-2}\,dw+\int\frac{1}{w-1}\,dw$$

$$=-1\int(1-w)^{-2}[-dw]+\int\frac{1}{w-1}\,dw.$$

On the last line, the first integral has the form $\displaystyle\int u^{-2}\,du$, and the second

has the form $\displaystyle\int\frac{1}{v}\,dv$. Thus

$$\int\left[\frac{1}{(1-w)^2}+\frac{1}{w-1}\right]dw=-\frac{(1-w)^{-1}}{-1}+\ln|w-1|+C$$

$$=\frac{1}{1-w}+\ln|w-1|+C.$$

b. $\displaystyle\int_1^e\frac{1}{y}\,dy.$

$$\int_1^e\frac{1}{y}\,dy=\ln|y|\;\Big|_1^e=\ln e-\ln 1=1-0=1.$$

c. $\displaystyle\int\frac{1}{x\ln x}\,dx.$

If $u=\ln x$, then $du=\dfrac{1}{x}\,dx$ and

$$\int\frac{1}{x\ln x}\,dx=\int\frac{1}{\ln x}\left[\frac{1}{x}\,dx\right]=\int\frac{1}{u}\,du$$

$$=\ln|u|+C=\ln|\ln x|+C.$$

When you are integrating fractions, sometimes a preliminary division is needed to get familiar integration forms, as the next example shows.

EXAMPLE 5

a. $\displaystyle\int\frac{x^3+x-1}{x^2}\,dx.$

A familiar integration form is not apparent. However, we can break up the integrand into three fractions by dividing each term in the numerator by the denominator.

$$\int \frac{x^3 + x - 1}{x^2}\, dx = \int \left[\frac{x^3}{x^2} + \frac{x}{x^2} - \frac{1}{x^2} \right] dx = \int \left[x + \frac{1}{x} - \frac{1}{x^2} \right] dx$$

$$= \frac{x^2}{2} + \ln|x| - \int x^{-2}\, dx = \frac{x^2}{2} + \ln|x| + \frac{1}{x} + C.$$

b. $\displaystyle \int \frac{2x^3 + 3x^2 + x + 1}{2x + 1}\, dx.$

In order to integrate, we first use long division until the degree of the remainder is less than that of the divisor.

$$\int \frac{2x^3 + 3x^2 + x + 1}{2x + 1}\, dx = \int \left(x^2 + x + \frac{1}{2x + 1} \right) dx$$

$$= \frac{x^3}{3} + \frac{x^2}{2} + \int \frac{1}{2x + 1}\, dx$$

$$= \frac{x^3}{3} + \frac{x^2}{2} + \frac{1}{2} \int \frac{1}{2x + 1}\, [2\, dx]$$

$$= \frac{x^3}{3} + \frac{x^2}{2} + \frac{1}{2} \ln|2x + 1| + C.$$

Long division is useful whenever the integrand is a quotient of polynomials in which the degree of the numerator is greater than or equal to that of the denominator and the denominator has more than one term.

TABLE 6.2
Differentiation and Integration Formulas

1. $\dfrac{d}{dx} (e^x) = e^x,$ $\displaystyle \int e^x\, dx = e^x + C.$

2. $\dfrac{d}{dx} (e^u) = e^u \dfrac{du}{dx},$ $\displaystyle \int e^u\, du = e^u + C.$

3. $\dfrac{d}{dx} (a^u) = a^u (\ln a) \dfrac{du}{dx},$ $\displaystyle \int a^u\, du = \dfrac{a^u}{\ln a} + C.$

4. $\dfrac{d}{dx} (\ln x) = \dfrac{1}{x},$ $\displaystyle \int \dfrac{1}{x}\, dx = \ln|x| + C.$

5. $\dfrac{d}{dx} (\ln u) = \dfrac{1}{u} \dfrac{du}{dx},$ $\displaystyle \int \dfrac{1}{u}\, du = \ln|u| + C.$

6. $\dfrac{d}{dx} (\log_b u) = \dfrac{1}{u} (\log_b e) \dfrac{du}{dx}.$

For your convenience, Table 6.2 gives the differentiation and integration formulas discussed in this chapter. We assume that u is a function of x. For Formula 6, you may wonder why there is no corresponding integration rule. The reason is that Formula 5 is used to handle the integral $\int \frac{1}{u}(\log_b e)\, du$:

$$\int \frac{1}{u}(\log_b e)\, du = (\log_b e) \int \frac{1}{u}\, du$$

$$= (\log_b e) \ln|u| + C \qquad \text{(Formula 5)}.$$

EXERCISE 6.6

In Problems 1–50, determine the integrals.

1. $\int 3e^{3x}\, dx.$

2. $\int 2e^{2t+5}\, dt.$

3. $\int (2t + 1)e^{t^2+t}\, dt.$

4. $\int -3w^2 e^{-w^3}\, dw.$

5. $\int xe^{5x^2}\, dx.$

6. $\int x^3 e^{4x^4}\, dx.$

7. $\int v^2 e^{-2v^3+1}\, dv.$

8. $\int 2ye^{3y^2}\, dy.$

9. $\int (e^{-5x} + 2e^x)\, dx.$

10. $\int (e^x - e^{-x} + e^{2x})\, dx.$

11. $\int_0^2 x^2 e^{x^3}\, dx.$

12. $\int_0^1 (e^x - e^{-2x})\, dx.$

13. $\int 4^{7x}\, dx.$

14. $\int 3^x\, dx.$

15. $\int \frac{e^{7/x}}{x^2}\, dx.$

16. $\int_3^4 \frac{e^{\ln x}}{x}\, dx.$

17. $\int \frac{1}{x + 5}\, dx.$

18. $\int \frac{2x + 1}{x + x^2}\, dx.$

19. $\int \frac{3x^2 + 4x^3}{x^3 + x^4}\, dx.$

20. $\int \frac{3x^2 - 2x}{1 - x^2 + x^3}\, dx.$

21. $\int \frac{4}{x}\, dx.$

22. $\int \frac{3}{1 + 2y}\, dy.$

23. $\int \frac{s^2}{s^3 + 5}\, ds.$

24. $\int \frac{2x^2}{3 - 4x^3}\, dx.$

25. $\int \frac{7}{5 - 3x}\, dx.$

26. $\int_1^8 \frac{4}{y}\, dy.$

27. $\int \frac{6z}{(z^2 - 6)^5}\, dz.$

28. $\int \frac{1}{(8y - 3)^3}\, dy.$

29. $\int_0^{e-1} \frac{1}{x + 1}\, dx.$

30. $\int \frac{7}{3 - 2x}\, dx.$

31. $\int \frac{x^2 + 2}{x^3 + 6x}\, dx.$

32. $\int (e^{3.1})^2\, dx.$

33. $\int_1^e (x^{-1} + x^{-2} - x^{-3})\, dx.$

34. $\int_1^3 (x + 1)e^{x^2+2x}\, dx.$

35. $\int_0^1 \frac{2x^3 + x}{x^2 + x^4 + 1}\, dx.$

36. $\int \left[\frac{1}{x - 1} + \frac{1}{(x - 1)^2} \right] dx.$

37. $\int \left[\frac{x}{x^2 + 1} + \frac{x^5}{(x^6 + 1)^2} \right] dx.$

38. $\int \frac{8x^3 - 6x^2 - ex^4}{3x^3}\, dx.$

39. $\int \frac{3x^3 + x^2 - x}{x^2}\, dx.$

40. $\int \frac{3e^s}{6 + 5e^s}\, ds.$

41. $\int \frac{3e^{2x}}{e^{2x} + 1}\, dx.$

42. $\int \frac{2x^4 - 6x^3 + x - 2}{x - 2}\, dx.$

43. $\int \frac{\ln x}{x}\, dx.$

44. $\int \frac{x + 3}{x + 6}\, dx.$

45. $\int \frac{1}{(x + 3) \ln(x + 3)}\, dx.$

46. $\int \frac{e^x + e^{-x}}{e^x - e^{-x}}\, dx.$

47. $\int \frac{6x^2 - 11x + 5}{3x - 1}\, dx.$

48. $\int \frac{(2x - 1)(x + 3)}{x - 5}\, dx.$

49. $\int \dfrac{x}{x-1} \, dx.$

50. $\int \dfrac{x}{(x^2+1)\ln(x^2+1)} \, dx.$

*In Problems 51 and 52, dc/dq is a marginal cost function.
Find the total cost function if fixed costs are 2000.*

51. $\dfrac{dc}{dq} = \dfrac{20}{q+5}.$

52. $\dfrac{dc}{dq} = 2e^{0.001q}.$

*In Problems 53–56, find the area of the region bounded
by the given curves, the x-axis, and the given lines. Sketch
the region.*

53. $y = e^x,$ $x = 0,$ $x = 2.$

54. $y = \dfrac{1}{x},$ $x = 1,$ $x = e^2.$

55. $y = x + \dfrac{2}{x},$ $x = 1,$ $x = 2.$

56. $y = e^{-x},$ $x = 0,$ $x = 1.$

57. If $y' = \dfrac{x}{x^2+4}$ and $y(1) = 0,$ find $y.$

58. The total costs (in dollars) of a business over the next
five years is given by

$$\int_0^5 4000e^{0.05t} \, dt.$$

Evaluate the costs.

59. The present value (in dollars) of a continuous flow of
income of $2000 a year for five years at 6 percent
compounded continuously is given by

$$\int_0^5 2000e^{-0.06t} \, dt.$$

Evaluate the present value to the nearest dollar.

60. If c_0 is the yearly consumption of a mineral at time
$t = 0$, then under continuous consumption the total
amount of the mineral used in the interval $[0, t_1]$ is

$$\int_0^{t_1} c_0 e^{kt} \, dt,$$

where k is the consumption rate. Suppose that for a
rare-earth mineral, $c_0 = 3000$ units and $k = 0.05$.
Evaluate the foregoing integral for these data.

61. In biology, problems frequently arise involving transfer

of a substance between compartments. An example
would be transfer from the bloodstream to tissue. Evaluate the following integral which occurs in **[47]** in a
two-compartment diffusion problem:

$$\int_0^t (e^{-a\tau} - e^{-b\tau}) \, d\tau,$$

where τ (read "tau") is a Greek letter and a and b
are constants.

62. In a discussion of diffusion of oxygen from capillaries
[47], concentric cylinders of radius r are used as a
model for a capillary. The concentration C of oxygen
in the capillary is given by

$$C = \int \left(\dfrac{Rr}{2K} + \dfrac{B_1}{r} \right) dr,$$

where R is the constant rate at which oxygen diffuses
from the capillary, and K and B_1 are constants. Find
C. (Write the constant of integration as B_2.)

63. In a discussion of gene mutation **[48]**, the following
equation occurs:

$$\int_{q_0}^{q_n} \dfrac{dq}{q - \hat{q}} = -(u + v) \int_0^n dt,$$

where u and v are gene mutation rates, the q's are
gene frequencies, and n is the number of generations.
Assume that all letters represent constants except q
and t. Integrate both sides and then use your result
to show that

$$n = \dfrac{1}{u + v} \ln \left| \dfrac{q_0 - \hat{q}}{q_n - \hat{q}} \right|.$$

64. In statistics the function $f(x) = k/x$ is a density function
for $e \le x \le e^2$ if $\int_e^{e^2} f(x) = 1.$

a. Find k so that f is a density function.

b. The probability that $3 \le x \le 5$ is $\int_3^5 f(x) \, dx$. Evaluate
this integral by using the value of k found in part
(a).

c. The distribution function for f is given by $F(t) =$
$\int_e^t f(x) \, dx$, where $e \le t \le e^2$. Find $F(t)$ by using
the value of k found in part (a).

65. Taagepera**[49]** considers a special situation involving
a country's exports. The amount E of the exports is

given by

$$E = \int_{-R}^{R} \frac{i}{2} [e^{-k(R-x)} + e^{-k(R+x)}] \, dx,$$

where i and k are constants ($k \neq 0$). Evaluate E.

66. In a discussion of inventory, Barbosa and Friedman[50] refer to the function

$$g(x) = \frac{1}{k} \int_{1}^{1/x} k u^r \, du,$$

where k and r are constants, $k > 0$ and $r > -2$, and $x > 0$. They claim that

$$g'(x) = -\frac{1}{x^{r+2}}.$$

Show that their claim is indeed true. (*Hint:* Consider two cases: when $r \neq -1$, and when $r = -1$.)

6.7 REVIEW

Important Terms and Symbols

Section 6.1   exponential function, b^x e compound amount
interest compounded continuously   present value

Section 6.2   logarithmic function, $\log_b x$ common logarithm, $\log x$ natural logarithm, $\ln x$

Section 6.5   logarithmic differentiation

Summary

An exponential function has the form $f(x) = b^x$. A frequently used base in an exponential function is the irrational number e, where $e \approx 2.71828$. This base occurs in problems involving interest compounded continuously. The formula $S = Pe^{rt}$ gives the compound amount S of a principal of P dollars after t years at an annual interest rate r compounded continuously. The formula $P = Se^{-rt}$ gives the present value P of S.

A logarithmic function with base b is denoted $\log_b$, and $y = \log_b x$ if and only if $b^y = x$. Logarithms with base e are called natural logarithms and denoted by $\ln$; those with base 10, called common logarithms, are denoted by $\log$. Some important properties of logarithms are

$$\log_b(mn) = \log_b m + \log_b n,$$

$$\log_b \frac{m}{n} = \log_b m - \log_b n,$$

$$\log_b m^r = r \log_b m,$$

$$\log_b 1 = 0,$$

$$\log_b b = 1,$$

$$\log_b b^r = r,$$

$$b^{\log_b m} = m,$$

and $$\log_b m = \frac{\log_a m}{\log_a b}.$$

The derivative formulas for logarithmic and exponential functions are

$$D_x(\ln u) = \frac{1}{u} \frac{du}{dx},$$

$$D_x(\log_b u) = \frac{1}{u}(\log_b e)\frac{du}{dx},$$

$$D_x(e^u) = e^u \frac{du}{dx},$$

and $$D_x(a^u) = a^u(\ln a) \frac{du}{dx}.$$

Suppose that $f(x)$ consists of products, quotients, or powers. To differentiate $y = \log_b [f(x)]$, it may be helpful to use properties of logarithms to rewrite $\log_b [f(x)]$ in terms of simpler logarithms and then differentiate that form. To differentiate $y = f(x)$, the method of logarithmic differentiation may be used. In that method, we take the natural logarithm of both sides of $y = f(x)$ to obtain

$\ln y = \ln [f(x)]$. After simplifying $\ln [f(x)]$ by using properties of logarithms, we differentiate both sides of $\ln y = \ln [f(x)]$ with respect to x and then solve for y'. Logarithmic differentiation is also used to differentiate $y = u^v$, where both u and v are functions of x.

Integral formulas involving exponential and logarithmic functions are

$$\int e^u \, du = e^u + C,$$

$$\int a^u \, du = \frac{a^u}{\ln a} + C,$$

and $$\int \frac{1}{u} \, du = \ln|u| + C.$$

Review Problems

1. Convert $3^4 = 81$ to logarithmic form.

2. Convert $\log_5 \frac{1}{5} = -1$ to exponential form.

In Problems 3 and 4, find the limits.

3. $\lim\limits_{x \to \infty} \dfrac{7 + 3e^{-x}}{5}$.

4. $\lim\limits_{x \to -\infty} \dfrac{10}{4 + 2^{x+1}}$.

In Problems 5–12, find x.

5. $\log_5 125 = x$.

6. $\log_x \frac{1}{8} = -3$.

7. $\log x = -2$.

8. $\ln \dfrac{1}{e} = x$.

9. $\log_x(2x + 3) = 2$.

10. $\log(3x + 1) = 2$.

11. $e^{\ln(x+4)} = 7$.

12. $\log x + \log 2 = 1$.

13. Express $\ln \dfrac{x^2\sqrt{x+1}}{\sqrt[3]{x^2 + 2}}$ in terms of $\ln x$, $\ln(x+1)$, and $\ln(x^2 + 2)$.

14. If $\ln 3 = x$ and $\ln 4 = y$, express $\ln(16\sqrt{3})$ in terms of x and y.

15. Simplify $e^{\ln x} + \ln e^x + \ln 1$.

16. Simplify $\log 10^2 + \log 1000 - 5$.

17. If $\ln y = x^2 + 2$, find y.

18. Sketch the graphs of $y = 3^x$ and $y = \log_3 x$.

In Problems 19–46, differentiate.

19. $y = 2e^x + e^2 + e^{x^2}$.

20. $f(w) = we^w + w^2$.

21. $f(r) = \ln(r^2 + 5r)$.

22. $y = e^{\ln x}$.

23. $y = e^{x^2 + 4x + 5}$.

24. $f(t) = \log_6 \sqrt{t^2 + 1}$.

25. $y = e^x(x^2 + 2)$.

26. $y = 2^{7x^2}$.

27. $y = \sqrt{(x-6)(x+5)(9-x)}$.

28. $f(t) = e^{\sqrt{t}}$.

29. $y = \dfrac{\ln x}{e^x}$.

30. $y = \dfrac{e^x + e^{-x}}{x^2}$.

31. $f(q) = \ln[(q + 1)^2(q + 2)^3]$.

32. $y = (x - 6)^4(x + 4)^3(6 - x)^2$.

33. $y = 10^{2 - 7x}$.

34. $y = (e + e^2)^0$.

35. $y = \dfrac{4e^{3x}}{xe^{x-1}}$.

36. $y = \dfrac{e^x}{\ln x}$.

37. $y = \log_2(8x + 5)^2$.

38. $y = \ln(1/x)$.

39. $f(l) = \ln(1 + l + l^2 + l^3)$.

40. $y = x^{x^3}$.

41. $y = (x + 1)^{x+1}$.

42. $y = \dfrac{1 + e^x}{1 - e^x}$.

43. $f(t) = \ln(t^2\sqrt{1 - t})$.

44. $y = (x + 2)^{\ln x}$.

45. $y = \dfrac{(x^2 + 2)^{3/2}(x^2 + 9)^{4/9}}{(x^3 + 6x)^{4/11}}$.

46. $y = \dfrac{\ln x}{\sqrt{x}}$.

In Problems 47–50, find the indicated derivative at the given point.

47. $y = e^{x^2 - 4}$, y'', $(2, 1)$.

48. $y = x^2e^x$, y''', $(1, e)$.

49. $y = \ln(2x)$, y''', $(1, \ln 2)$.

50. $y = x \ln x$, y'', $(1, 0)$.

In Problems 51 and 52, find an equation of the tangent line to the curve at the point corresponding to the given value of x.

51. $y = e^x$, $x = \ln 2$.

52. $y = x + x^2 \ln x$, $x = 1$.

In Problems **53** and **54**, find the rate of change of y with respect to x at the given point.

53. $\ln(3y) = x$, $(0, \frac{1}{3})$.

54. $e^{x-y} = y$, $(1, 1)$.

In Problems **55** and **56**, find intervals on which the function is increasing, decreasing, concave up, concave down; find relative extrema and points of inflection. Sketch the graphs of the functions.

55. $f(x) = \dfrac{e^x + e^{-x}}{2}$.

56. $f(x) = 1 + \ln(x^2)$.

In Problems **57–74**, determine the integrals.

57. $\displaystyle\int 4e^{3x+5}\,dx$.

58. $\displaystyle\int_0^2 xe^{4-x^2}\,dx$.

59. $\displaystyle\int \frac{1}{2(x+1)}\,dx$.

60. $\displaystyle\int \frac{2}{5-3x}\,dx$.

61. $\displaystyle\int \frac{6x^2-12}{x^3-6x+1}\,dx$.

62. $\displaystyle\int \frac{2t}{5t^2+1}\,dt$.

63. $\displaystyle\int_1^2 \frac{t^2}{2+t^3}\,dt$.

64. $\displaystyle\int \frac{1}{e^{-3x}}\,dx$.

65. $\displaystyle\int_1^e \frac{p+8}{p}\,dp$.

66. $\displaystyle\int_0^1 \frac{e^{2x}}{1+e^{2x}}\,dx$.

67. $\displaystyle\int (e^{2y} - e^{-2y})\,dy$.

68. $\displaystyle\int x(5x^2+3)^{-1}\,dx$.

69. $\displaystyle\int \left(\frac{1}{x} + \frac{2}{x^2}\right)\,dx$.

70. $\displaystyle\int \frac{z^2}{z-1}\,dz$.

71. $\displaystyle\int 5^{2x}\,dx$.

72. $\displaystyle\int \ln(e^{4x})\,dx$.

73. $\displaystyle\int_{-1}^0 \frac{x^2+4x-1}{x+2}\,dx$.

74. $\displaystyle\int \frac{1}{\sqrt{x}\,(1+\sqrt{x})}\,dx$.

In Problems **75** and **76**, find the area of the region bounded by the given curves, the x-axis, and the given lines. Assume that the answer is in square units.

75. $y = \dfrac{1}{x} + 3$, $x = 1$, $x = 3$.

76. $y = 4e^{2x}$, $x = 0$, $x = 3$.

77. For a new product the yearly number of thousand packages sold, y, after t years from its introduction is predicted to be

$$y = f(t) = 150 - 76e^{-t}.$$

Show that $y = 150$ is a horizontal asymptote for the graph. This shows that after the product is established with consumers, the market tends to be constant.

78. Show that the graphs of $y = 6 - 3e^{-x}$ and $y = 6 + 3e^{-x}$ are both asymptotic to the same line. What is the equation of this line?

79. Due to ineffective advertising, the Kleer-Kut Razor Company finds its annual revenues have been cut sharply. Moreover, the annual revenue R at the end of t years of business satisfies the equation $R = 200{,}000e^{-0.2t}$. Find the annual revenue at the end of 2 years; at the end of 3 years.

80. In a study of water mites infesting a group of 589 adult flies **[36]**, a Poisson distribution was fitted to the data. If x is the number of mites per fly and $\mu = 0.44$, find the probability of 3 mites per fly. (See Sec. 6.1 for a discussion of the Poisson distribution.)

81. For a certain population of cells, the number N of cells at time t is given by $N = N_0(2^{t/k})$, where N_0 is the number of cells at $t = 0$ and k is a positive constant. Show that

$$k = \frac{t \ln 2}{\ln(N/N_0)}.$$

82. If $\overline{c} = (500/q)e^{q/300}$ is an average cost function, find the marginal cost function.

83. The demand function for a manufacturer's product is given by $p = 100e^{-0.1q}$, where p is the price per unit at which q units are demanded. For what value of q does the manufacturer maximize total revenue?

84. For a group of hospitalized individuals, suppose the discharge rate is given by $f(t) = 0.008e^{-0.008t}$, where $f(t)$ is the proportion discharged per day at the end of t days of hospitalization. What proportion of the group is discharged at the end of 100 days?

85. Several models have been used to analyze the length of stay in a hospital. For a particular group of schizophrenics, one such model is (adapted from **[46]**) $f(t) = 1 - (0.8e^{-0.01t} + 0.2e^{-0.0002t})$, where $f(t)$ is the proportion of the group that was discharged at the end of t days of hospitalization. Determine the discharge rate (proportion discharged per day) at the end of t days.

86. Another model for the group of schizophrenics of Problem 85 is

$$f(t) = 1 - e^{-0.01t^{0.9}}.$$

Show the rate of discharge is decreasing for $t > 0$.

87. New products or technologies often tend to replace old ones. For example, today most commercial airlines use jet engines rather than prop engines. In discussing the forecasting of technological substitution, Hurter and Rubenstein [40] refer to the equation

$$\ln \frac{f(t)}{1 - f(t)} + \sigma \frac{1}{1 - f(t)} = C_1 + C_2 t,$$

where $f(t)$ is the market share of the substitute over time t, and C_1, C_2, and σ (a Greek letter read "sigma") are constants. Verify their claim that the rate of substitution is

$$f'(t) = \frac{C_2 f(t)[1 - f(t)]^2}{\sigma f(t) + [1 - f(t)]}.$$

88. The "logit" function is employed in the study of the behavioral sciences. It is defined as follows:

$$\text{logit}(x) = \frac{1}{2} \ln \frac{1 - x}{x}, \qquad \text{where } 0 < x < 1.$$

If $\text{logit}(y) = a\, \text{logit}(x) + b$, show that

$$y = \frac{1}{1 + e^{2b} \left(\dfrac{1 - x}{x} \right)^a}.$$

CHAPTER 7

Topics on Integration

7.1 INTEGRATION BY TABLES

Certain forms of integrals that occur frequently may be found in standard tables of integration formulas.* A short table appears in Appendix F and its use will be illustrated in this section.

A given integral may have to be replaced by an equivalent form before it will fit a formula in the table. The equivalent form must match the formula *exactly*. Consequently, the steps that you perform should *not* be done mentally. *Write them down!* Failure to do this can easily lead to incorrect results. Before proceeding with the exercises, be sure you understand the illustrative examples *thoroughly*.

In the following examples, the formula numbers refer to the Table of Selected Integrals given in Appendix F.

EXAMPLE 1   *Find* $\int \dfrac{x\,dx}{(2 + 3x)^2}$.

Scanning the tables, we identify the integrand with Formula 7:

$$\int \frac{u\,du}{(a + bu)^2} = \frac{1}{b^2}\left(\ln|a + bu| + \frac{a}{a + bu}\right) + C.$$

Now we see if we can exactly match the given integrand with that in the formula. If we replace x by u, 2 by a, and 3 by b, then $du = dx$ and by

---

* See, for example, William A. Beyer (ed.), *CRC Standard Mathematical Tables*, 27th ed. (Boca Raton, Fla.: CRC Press), 1984.

substitution we have

$$\int \frac{x\ dx}{(2 + 3x)^2} = \int \frac{u\ du}{(a + bu)^2} = \frac{1}{b^2}\left(\ln|a + bu| + \frac{a}{a + bu}\right) + C.$$

Returning to the variable x and replacing a by 2 and b by 3, we obtain

$$\int \frac{x\ dx}{(2 + 3x)^2} = \frac{1}{9}\left(\ln|2 + 3x| + \frac{2}{2 + 3x}\right) + C.$$

EXAMPLE 2　　*Find* $\int x^2\sqrt{x^2 - 1}\ dx$.

This integral is identified with Formula 24:

$$\int u^2\sqrt{u^2 \pm a^2}\ du = \frac{u}{8}(2u^2 \pm a^2)\sqrt{u^2 \pm a^2} - \frac{a^4}{8}\ln\left|u + \sqrt{u^2 \pm a^2}\right| + C.$$

In this formula, if the bottommost sign in the dual symbol "$\pm$" on the left side is used, then the bottommost sign in the dual symbols on the right side must also be used. In the original integral, we let $u = x$ and $a = 1$. Then $du = dx$ and by substitution the integral becomes

$$\int x^2\sqrt{x^2 - 1}\ dx = \int u^2\sqrt{u^2 - a^2}\ du$$

$$= \frac{u}{8}(2u^2 - a^2)\sqrt{u^2 - a^2} - \frac{a^4}{8}\ln\left|u + \sqrt{u^2 - a^2}\right| + C.$$

Since $u = x$ and $a = 1$,

$$\int x^2\sqrt{x^2 - 1}\ dx = \frac{x}{8}(2x^2 - 1)\sqrt{x^2 - 1} - \frac{1}{8}\ln\left|x + \sqrt{x^2 - 1}\right| + C.$$

EXAMPLE 3　　*Find* $\int \dfrac{dx}{x\sqrt{16x^2 + 3}}$.

The integrand can be identified with Formula 28:

$$\int \frac{du}{u\sqrt{u^2 + a^2}} = \frac{1}{a}\ln\left|\frac{\sqrt{u^2 + a^2} - a}{u}\right| + C.$$

If we let $u = 4x$ and $a = \sqrt{3}$, then $du = 4\ dx$. Watch closely how, by inserting 4's in the numerator and denominator, we transform the given integral into an equivalent form that matches Formula 28.

$$\int \frac{dx}{x\sqrt{16x^2 + 3}} = \int \frac{(4\,dx)}{(4x)\sqrt{(4x)^2 + (\sqrt{3})^2}} = \int \frac{du}{u\sqrt{u^2 + a^2}}$$

$$= \frac{1}{a} \ln \left| \frac{\sqrt{u^2 + a^2} - a}{u} \right| + C$$

$$= \frac{1}{\sqrt{3}} \ln \left| \frac{\sqrt{16x^2 + 3} - \sqrt{3}}{4x} \right| + C.$$

Note that the answer was given in terms of x, the *original* variable of integration.

EXAMPLE 4   *Find* $\int \dfrac{dx}{x^2(2 - 3x^2)^{1/2}}.$

The integrand is identified with Formula 21:

$$\int \frac{du}{u^2\sqrt{a^2 - u^2}} = -\frac{\sqrt{a^2 - u^2}}{a^2 u} + C.$$

Letting $u = \sqrt{3}x$ and $a^2 = 2$, we have $du = \sqrt{3}\,dx$. Hence, by inserting two factors of $\sqrt{3}$ in both the numerator and denominator of the original integral, we have

$$\int \frac{dx}{x^2(2 - 3x^2)^{1/2}} = \sqrt{3} \int \frac{(\sqrt{3}\,dx)}{(\sqrt{3}x)^2[2 - (\sqrt{3}x)^2]^{1/2}} = \sqrt{3} \int \frac{du}{u^2(a^2 - u^2)^{1/2}}$$

$$= \sqrt{3} \left[-\frac{\sqrt{a^2 - u^2}}{a^2 u} \right] + C = \sqrt{3} \left[-\frac{\sqrt{2 - 3x^2}}{2(\sqrt{3}x)} \right] + C$$

$$= -\frac{\sqrt{2 - 3x^2}}{2x} + C.$$

EXAMPLE 5   *Find* $\int 7x^2 \ln(4x)\,dx.$

This is similar to Formula 42 with $n = 2$:

$$\int u^n \ln u\,du = \frac{u^{n+1} \ln u}{n + 1} - \frac{u^{n+1}}{(n + 1)^2} + C.$$

If we let $u = 4x$, then $du = 4\,dx$. Hence

$$\int 7x^2 \ln(4x)\,dx = \frac{7}{4^3} \int (4x)^2 \ln(4x)(4\,dx)$$

$$= \frac{7}{64} \int u^2 \ln u \, du = \frac{7}{64} \left(\frac{u^3 \ln u}{3} - \frac{u^3}{9} \right) + C$$

$$= \frac{7}{64} \left[\frac{(4x)^3 \ln(4x)}{3} - \frac{(4x)^3}{9} \right] + C$$

$$= 7x^3 \left[\frac{\ln(4x)}{3} - \frac{1}{9} \right] + C$$

$$= \frac{7x^3}{9} [3 \ln(4x) - 1] + C.$$

EXAMPLE 6   *Find* $\displaystyle \int \frac{e^{2x} \, dx}{7 + e^{2x}}.$

At first glance we do not identify the integrand with any form in the table. Perhaps rewriting the integral will help. Let $u = 7 + e^{2x}$; then $du = 2e^{2x} \, dx$.

$$\int \frac{e^{2x} \, dx}{7 + e^{2x}} = \frac{1}{2} \int \frac{(2e^{2x} \, dx)}{7 + e^{2x}} = \frac{1}{2} \int \frac{du}{u} = \frac{1}{2} \ln|u| + C$$

$$= \frac{1}{2} \ln|7 + e^{2x}| + C = \frac{1}{2} \ln(7 + e^{2x}) + C.$$

Thus we had only to use our knowledge of basic integration forms. Actually, this form appears as Formula 2 in the tables.

EXAMPLE 7   *Evaluate* $\displaystyle \int_1^4 \frac{dx}{(4x^2 + 2)^{3/2}}.$

We shall use Formula 32 to first get the indefinite integral:

$$\int \frac{du}{(u^2 \pm a^2)^{3/2}} = \frac{\pm u}{a^2 \sqrt{u^2 \pm a^2}} + C.$$

Letting $u = 2x$ and $a^2 = 2$, we have $du = 2 \, dx$. Thus

$$\int \frac{dx}{(4x^2 + 2)^{3/2}} = \frac{1}{2} \int \frac{(2 \, dx)}{[(2x)^2 + 2]^{3/2}} = \frac{1}{2} \int \frac{du}{(u^2 + 2)^{3/2}}$$

$$= \frac{1}{2} \left[\frac{u}{2\sqrt{u^2 + 2}} \right] + C.$$

Instead of substituting back to x and evaluating from $x = 1$ to $x = 4$, we can determine the corresponding limits of integration with respect to u and then evaluate the last expression between those limits. Since $u = 2x$, when $x =$

1 we have $u = 2$; when $x = 4$ we have $u = 8$. Thus

$$\int_1^4 \frac{dx}{(4x^2 + 2)^{3/2}} = \frac{1}{2}\int_2^8 \frac{du}{(u^2 + 2)^{3/2}}$$

$$= \frac{1}{2}\left(\frac{u}{2\sqrt{u^2 + 2}}\right)\bigg|_2^8 = \frac{2}{\sqrt{66}} - \frac{1}{2\sqrt{6}}.$$

Pitfall

When changing the variable of integration x to the variable of integration u, be certain to change the limits of integration so that they agree with u. That is,

$$\int_1^4 \frac{dx}{(4x^2 + 2)^{3/2}} \neq \frac{1}{2}\int_1^4 \frac{du}{(u^2 + 2)^{3/2}}.$$

Tables of integrals are useful when dealing with integrals associated with *annuities*. Suppose that you must pay out \$100 at the end of each year for the next two years. Any series of payments over a period of time, such as this, is called an **annuity**. If you were to pay off the debt now instead, you would pay the present value of the \$100 that is due at the end of the first year, plus the present value of the \$100 that is due at the end of the second year. (Present value is discussed in Sec. 6.1.) The sum of these present values is the present value of the annuity. We shall now consider the present value of payments made continuously over the time interval from $t = 0$ to $t = T$, t in years, when interest is compounded continuously at an annual rate of r.

Suppose a payment is made at time t such that on an annual basis this payment is $f(t)$. If we divide the interval $[0, T]$ into subintervals $[t_{i-1}, t_i]$ of length Δt (where Δt is small), then the total amount of all payments over such a subinterval is approximately $f(t_i) \, \Delta t$. [For example, if $f(t) = 2000$ and Δt were one day, then the total amount of the payments would be $2000(\frac{1}{365})$.] The present value of these payments is approximately $e^{-rt_i}f(t_i) \, \Delta t$. Over the interval $[0, T]$, the total of all such present values is

$$\sum e^{-rt_i}f(t_i) \, \Delta t.$$

This sum approximates the present value A of the annuity. The smaller Δt is, the better is the approximation. That is, as $\Delta t \to 0$ the limit of the sum *is* the present value. However, this limit is also a definite integral. That is,

$$A = \int_0^T f(t)e^{-rt} \, dt. \qquad (1)$$

This gives the **present value of a continuous annuity** at an annual rate r (compounded continuously) for T years if a payment at time t is at the rate of $f(t)$

per year. Sometimes we say that Eq. (1) gives the **present value of a continuous income stream**. Equation (1) may also be used to find the present value of future profits of a business. In this situation $f(t)$ is the annual rate of profit at time t.

We can also consider the *future* value of an annuity rather than its present value. If a payment is made at time t, then it has a certain value at the *end* of the period of the annuity, that is, $T - t$ years later. This value is

$$\left(\begin{matrix}\text{amount of} \\ \text{payment}\end{matrix}\right) + \left(\begin{matrix}\text{interest on this} \\ \text{payment for } T - t \text{ years}\end{matrix}\right).$$

If S is the total of such values for all payments, then S is called the **accumulated amount of a continuous annuity** and is given by

$$S = \int_0^T f(t)e^{r(T-t)}\, dt.$$

EXAMPLE 8   *Find the present value (to the nearest dollar) of a continuous annuity at an annual rate of 8 percent for 10 years if the payment at time t is at the rate of t^2 dollars per year.*

The present value is given by

$$A = \int_0^T f(t)e^{-rt}\, dt = \int_0^{10} t^2 e^{-0.08t}\, dt.$$

We shall use Formula 39,

$$\int u^n e^{au}\, du = \frac{u^n e^{au}}{a} - \frac{n}{a} \int u^{n-1} e^{au}\, du,$$

which is called a *reduction formula* since it reduces an integral into an expression that involves an integral which is easier to determine. If $u = t$, $n = 2$, and $a = -0.08$, then $du = dt$ and we have

$$A = \frac{t^2 e^{-0.08t}}{-0.08}\bigg|_0^{10} - \frac{2}{-0.08} \int_0^{10} t e^{-0.08t}\, dt.$$

In the new integral the exponent of t has been reduced to 1. We can match this integral with Formula 38,

$$\int u e^{au}\, du = \frac{e^{au}}{a^2}(au - 1) + C,$$

by letting $u = t$ and $a = -0.08$. Then $du = dt$ and

$$A = \int_0^{10} t^2 e^{-0.08t}\, dt = \frac{t^2 e^{-0.08t}}{-0.08}\bigg|_0^{10} - \frac{2}{-0.08}\left[\frac{e^{-0.08t}}{(-0.08)^2}(-0.08t - 1)\right]\bigg|_0^{10}$$

$$= \frac{100e^{-0.8}}{-0.08} - \frac{2}{-0.08}\left[\frac{e^{-0.8}}{(-0.08)^2}(-0.8 - 1) - \frac{1}{(-0.08)^2}(-1)\right]$$

$$= -1250e^{-0.8} + 25[-281.25e^{-0.8} + 156.25]$$

$$= -8281.25e^{-0.8} + 3906.25$$

$$\approx -8281.25(0.44933) + 3906.25 \approx 185.$$

The present value is $185.

EXERCISE 7.1

In Problems 1–33, find the integrals by using the table in Appendix F.

1. $\displaystyle\int \frac{dx}{x(6 + 7x)}.$

2. $\displaystyle\int \frac{x^2\,dx}{(1 + 2x)^2}.$

3. $\displaystyle\int \frac{dx}{x\sqrt{x^2 + 9}}.$

4. $\displaystyle\int \frac{dx}{(x^2 + 7)^{3/2}}.$

5. $\displaystyle\int \frac{x\,dx}{(2 + 3x)(4 + 5x)}.$

6. $\displaystyle\int 2^{5x}\,dx.$

7. $\displaystyle\int \frac{dx}{4 + 3e^{2x}}.$

8. $\displaystyle\int x^2\sqrt{1 + x}\,dx.$

9. $\displaystyle\int \frac{2\,dx}{x(1 + x)^2}.$

10. $\displaystyle\int \frac{dx}{x\sqrt{5 - 11x^2}}.$

11. $\displaystyle\int_0^1 \frac{x\,dx}{2 + x}.$

12. $\displaystyle\int \frac{x^2\,dx}{2 + 5x}.$

13. $\displaystyle\int \sqrt{x^2 - 3}\,dx.$

14. $\displaystyle\int \frac{dx}{(4 + 3x)(4x + 3)}.$

15. $\displaystyle\int_0^{1/12} xe^{12x}\,dx.$

16. $\displaystyle\int \sqrt{\frac{2 + 3x}{5 + 3x}}\,dx.$

17. $\displaystyle\int x^2e^x\,dx.$

18. $\displaystyle\int_1^2 \frac{dx}{x^2(1 + x)}.$

19. $\displaystyle\int \frac{\sqrt{4x^2 + 1}}{x^2}\,dx.$

20. $\displaystyle\int \frac{dx}{x\sqrt{2 - x}}.$

21. $\displaystyle\int \frac{x\,dx}{(1 + 3x)^2}.$

22. $\displaystyle\int \frac{dx}{\sqrt{(1 + 2x)(3 + 2x)}}.$

23. $\displaystyle\int \frac{dx}{7 - 5x^2}.$

24. $\displaystyle\int x^2\sqrt{2x^2 - 9}\,dx.$

25. $\displaystyle\int x^5 \ln(3x)\,dx.$

26. $\displaystyle\int \frac{dx}{x^2(1 + x)^2}.$

27. $\displaystyle\int 2x\sqrt{1 + 3x}\,dx.$

28. $\displaystyle\int x^2 \ln x\,dx.$

29. $\displaystyle\int \frac{dx}{\sqrt{4x^2 - 13}}.$

30. $\displaystyle\int \frac{dx}{x \ln(2x)}.$

31. $\displaystyle\int x \ln(2x)\,dx.$

32. $\displaystyle\int \frac{\sqrt{2 - 3x^2}}{x}\,dx.$

33. $\displaystyle\int \frac{dx}{x^2\sqrt{9 - 4x^2}}.$

In Problems 34–42, find the integrals by any method.

34. $\displaystyle\int \sqrt{x}\,e^{x3/2}\,dx.$

35. $\displaystyle\int \frac{x\,dx}{x^2 + 1}.$

36. $\displaystyle\int \frac{4x^2 - \sqrt{x}}{x}\,dx.$

37. $\displaystyle\int x\sqrt{2x^2 + 1}\,dx.$

38. $\displaystyle\int \frac{e^{2x}}{\sqrt{e^{2x} + 3}}\,dx.$

39. $\displaystyle\int \frac{dx}{x^2 - 5x + 6}.$

40. $\displaystyle\int_0^3 xe^{-x}\,dx.$

41. $\displaystyle\int_1^2 \frac{x\,dx}{\sqrt{4 - x}}.$

42. $\displaystyle\int_1^2 x^2\sqrt{3 + 2x}\,dx.$

43. In a discussion **[48]** about gene frequency, the following integral occurs:

$$\int_{q_0}^{q_n} \frac{dq}{q(1 - q)},$$

where the q's represent gene frequencies. Evaluate this integral.

44. Under certain conditions, the number n of generations required to change the frequency of a gene from 0.3 to 0.1 is given in [51] by

$$n = -\frac{1}{0.4} \int_{0.3}^{0.1} \frac{dq}{q^2(1-q)}.$$

Find n (to the nearest integer).

45. Find the present value, to the nearest dollar, of a continuous annuity at an annual rate of r for T years if the payment at time t is at an annual rate of $f(t)$ dollars given that

  a. $r = 0.06$, $T = 10$, $f(t) = 5000$.

  b. $r = 0.05$, $T = 8$, $f(t) = 200t$.

46. If $f(t) = k$, where k is a positive constant, show that the value of the integral in Eq. (1) of this section is

$$k\left(\frac{1 - e^{-rT}}{r}\right).$$

47. Find the accumulated amount, to the nearest dollar, of a continuous annuity at an annual rate of r for T years if the payment at time t is at an annual rate of $f(t)$ dollars given that

  a. $r = 0.06$, $T = 10$, $f(t) = 400$.

  b. $r = 0.04$, $T = 5$, $f(t) = 40t$.

48. Over the next five years, the profits of a business at time t are estimated to be $20,000t$ dollars per year. The business is to be sold at a price equal to the present value of these future profits. If interest is compounded continuously at the annual rate of 10 percent, to the nearest 10 dollars, at what price should the business be sold?

7.2 AVERAGE VALUE OF A FUNCTION

If we are given the three numbers 1, 2, and 9, then their average value, or *mean*, is their sum divided by 3. Denoting this average by $\bar{y}$, we have

$$\bar{y} = \frac{1 + 2 + 9}{3} = 4.$$

Similarly, suppose that we are given a function f defined on the interval $[a, b]$ and the points $x_1, x_2, \ldots, x_n$ are in the interval. Then the average value of the n corresponding function values $f(x_1), f(x_2), \ldots, f(x_n)$ is

$$\bar{y} = \frac{f(x_1) + f(x_2) + \cdots + f(x_n)}{n} = \frac{\sum\limits_{i=1}^{n} f(x_i)}{n}. \tag{1}$$

We can go a step further. Let us divide the interval $[a, b]$ into n subintervals of equal length. We shall choose x_1 to be in the first subinterval, x_2 to be in the second, etc. Since $[a, b]$ has length $b - a$, each subinterval has length $\frac{b - a}{n}$, which we shall call Δx. Thus (1) can be written*

$$\bar{y} = \frac{\sum\limits_{i=1}^{n} f(x_i)\left(\frac{\Delta x}{\Delta x}\right)}{n} = \frac{\frac{1}{\Delta x} \sum\limits_{i=1}^{n} f(x_i)\, \Delta x}{n} = \frac{1}{n\, \Delta x} \sum\limits_{i=1}^{n} f(x_i)\, \Delta x. \tag{2}$$

* Here we make use of the fact that a constant factor can be moved in front of a sigma: $\sum\limits_{i=1}^{n} cx_i = c \sum\limits_{i=1}^{n} x_i$. This follows from the distributive property:

$$\sum_{i=1}^{n} cx_i = cx_1 + cx_2 + \cdots + cx_n = c(x_1 + x_2 + \cdots + x_n) = c\sum_{i=1}^{n} x_i.$$

Since $\Delta x = \dfrac{b - a}{n}$, then $n \, \Delta x = b - a$, so the expression $\dfrac{1}{n \, \Delta x}$ in (2) can be replaced by $\dfrac{1}{b - a}$. Moreover, as $n \to \infty$ the number of function values used in computing $\bar{y}$ increases and we get the so-called *average value of the function f*, denoted $\bar{f}$:

$$\bar{f} = \lim_{n \to \infty} \left[\frac{1}{b - a} \sum_{i=1}^{n} f(x_i) \, \Delta x \right] = \frac{1}{b - a} \lim_{n \to \infty} \sum_{i=1}^{n} f(x_i) \, \Delta x.$$

The limit on the right is just the definite integral $\displaystyle\int_a^b f(x) \, dx$. Thus we have the following definition.

Definition
*The **average** (or **mean**) **value of a function** $y = f(x)$ over the interval $[a, b]$ is denoted $\bar{f}$ (or $\bar{y}$) and is given by*

$$\bar{f} = \frac{1}{b - a} \int_a^b f(x) \, dx.$$

EXAMPLE 1   *Find the average value of the function $f(x) = x^2$ over the interval $[1, 2]$.*

$$\bar{f} = \frac{1}{b - a} \int_a^b f(x) \, dx$$

$$= \frac{1}{2 - 1} \int_1^2 x^2 \, dx = \frac{x^3}{3} \bigg|_1^2 = \frac{7}{3}.$$

In Example 1 we found that the average value of $y = f(x) = x^2$ over $[1, 2]$ is $\frac{7}{3}$. We can interpret this value geometrically. Since

$$\frac{1}{2 - 1} \int_1^2 x^2 \, dx = \frac{7}{3},$$

by solving for the integral we have

$$\int_1^2 x^2 \, dx = \frac{7}{3}(2 - 1).$$

However, this integral gives the area of the region bounded by $f(x) = x^2$ and the x-axis from $x = 1$ to $x = 2$ (see Fig. 7.1). From the above equation, this area is $(\frac{7}{3})(2 - 1)$, which is the area of a rectangle whose height is the average value $\bar{f} = \frac{7}{3}$ and whose width is $b - a = 2 - 1 = 1$.

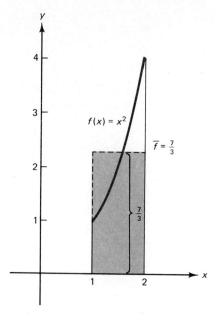

FIGURE 7.1

EXAMPLE 2   *Suppose that the flow of blood at time t in a system is given by*

$$F(t) = \frac{F_1}{(1 + \alpha t)^2}, \qquad 0 \le t \le T,$$

where F_1 and α (a Greek letter read "alpha") are constant [47]. *Find the average flow $\overline{F}$ on the interval* [0, T].

$$\overline{F} = \frac{1}{T - 0} \int_0^T F(t) \, dt$$

$$= \frac{1}{T} \int_0^T \frac{F_1}{(1 + \alpha t)^2} \, dt = \frac{F_1}{\alpha T} \int_0^T (1 + \alpha t)^{-2} (\alpha \, dt)$$

$$= \frac{F_1}{\alpha T} \left[\frac{(1 + \alpha t)^{-1}}{-1} \right] \Big|_0^T = \frac{F_1}{\alpha T} \left[-\frac{1}{1 + \alpha T} + 1 \right]$$

$$= \frac{F_1}{\alpha T} \left[\frac{-1 + 1 + \alpha T}{1 + \alpha T} \right] = \frac{F_1}{\alpha T} \left[\frac{\alpha T}{1 + \alpha T} \right] = \frac{F_1}{1 + \alpha T}.$$

EXERCISE 7.2

In Problems 1–8, find the average value of the function over the given interval.

1. $f(x) = x^2$; $[0, 4]$.

2. $f(x) = 3x - 1$; $[1, 2]$.

3. $f(x) = 2 - 3x^2$; $[-1, 2]$.

4. $f(x) = x^2 + x + 1$; $[1, 3]$.

5. $f(t) = 4t^3$; $[-2, 2]$.

6. $f(t) = t\sqrt{t^2 + 9}$; $[0, 4]$.

7. $f(x) = \sqrt{x}$; $[1, 9]$.   8. $f(x) = \dfrac{1}{x}$; $[2, 4]$.

9. The profit P (in dollars) of a business is given by

$$P = P(q) = 396q - 2.1q^2 - 400,$$

where q is the number of units of the product sold. Find the average profit on the interval from $q = 0$ to $q = 100$.

10. Suppose that the cost c (in dollars) of producing q units of a product is given by

$$c = 4000 + 10q + 0.1q^2.$$

Find the average cost on the interval from $q = 100$ to $q = 500$.

11. An investment of \$3000 earns interest at an annual rate of 10 percent compounded continuously. After t years, its value S (in dollars) is given by $S = 3000e^{0.10t}$. Find the average value of a two-year investment.

12. Suppose that colored dye is injected into the bloodstream at a constant rate R (see [47]). At time t, let $C(t)$ be the concentration of dye at a location distant (distal) from the point of injection, where

$$C(t) = \frac{R}{F(t)}$$

and $F(t)$ is given in Example 2. Show that the average concentration $\overline{C}$ on $[0, T]$ is

$$\overline{C} = \frac{R(1 + \alpha T + \frac{1}{3}\alpha^2 T^2)}{F_1}.$$

7.3 APPROXIMATE INTEGRATION ─────────────────────

When using the Fundamental Theorem to evaluate $\displaystyle\int_a^b f(x)\,dx$, you may find it extremely difficult, or perhaps impossible, to find an antiderivative of f, even using tables. Fortunately there are numerical methods that can be used to estimate a definite integral. These methods use values of $f(x)$ at various points and are especially suitable for computers or calculators. We shall consider two methods: the *trapezoidal rule* and *Simpson's rule*. In both cases we assume that f is continuous on $[a, b]$.

    In developing the trapezoidal rule, for convenience we shall also assume that $f(x) \geq 0$ on $[a, b]$ so that we can think in terms of area. Basically, this rule involves approximating the graph of f by straight-line segments.

    In Fig. 7.2 the interval $[a, b]$ is divided into n subintervals of equal length by the points $a = x_0, x_1, x_2, \ldots$, and $x_n = b$. Since the length of $[a, b]$ is $b - a$, the length of each subinterval is $(b - a)/n$, which we shall call h. Clearly, $x_1 = a + h$, $x_2 = a + 2h$, $\ldots$, $x_n = a + nh = b$. With each subinterval we can associate a trapezoid (a four-sided figure with two parallel sides). The area of the region bounded by the curve, the x-axis, and the lines

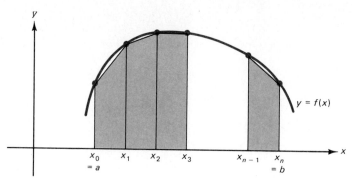

FIGURE 7.2

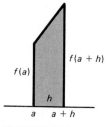

FIGURE 7.3

$x = a$ and $x = b$ is approximated by the sum of the areas of the trapezoids determined by the subintervals.

Consider the first trapezoid, which is redrawn in Fig. 7.3. Since the area of a trapezoid is equal to one-half the base times the sum of the lengths of the parallel sides, this trapezoid has area

$$\tfrac{1}{2}h[f(a) + f(a + h)].$$

Similarly, the second trapezoid has area

$$\tfrac{1}{2}h[f(a + h) + f(a + 2h)].$$

With n trapezoids the area A under the curve is approximated by

$$A \approx \tfrac{1}{2}h[f(a) + f(a + h)] + \tfrac{1}{2}h[f(a + h) + f(a + 2h)] +$$

$$\tfrac{1}{2}h[f(a + 2h) + f(a + 3h)] + \cdots + \tfrac{1}{2}h[f(a + (n - 1)h) + f(b)].$$

Since $A = \displaystyle\int_a^b f(x)\, dx$, by simplifying the above we have the trapezoidal rule:

THE TRAPEZOIDAL RULE

$$\int_a^b f(x)\, dx \approx \frac{h}{2}\{f(a) + 2f(a + h) + 2f(a + 2h) +$$

$$\cdots + 2f[a + (n - 1)h] + f(b)\},$$

where $h = (b - a)/n$.

The pattern of the coefficients inside the braces is 1, 2, 2, . . . , 2, 1. Usually, the more subintervals, the better is the approximation. In our development, we assumed for convenience that $f(x) \geq 0$ on $[a, b]$. However, the trapezoidal rule is valid without this restriction.

EXAMPLE 1   *Use the trapezoidal rule to estimate the value of*

$$\int_0^1 \frac{1}{1 + x^2} \, dx$$

by using n = 5. Compute each term to four decimal places and round off the answer to three decimal places.

Here $f(x) = 1/(1 + x^2)$, $n = 5$, $a = 0$, and $b = 1$. Thus

$$h = \frac{b - a}{n} = \frac{1 - 0}{5} = \frac{1}{5} = 0.2.$$

The terms to be added are

$$f(a) = f(0) = 1.0000$$
$$2f(a + h) = 2f(0.2) = 1.9231$$
$$2f(a + 2h) = 2f(0.4) = 1.7241$$
$$2f(a + 3h) = 2f(0.6) = 1.4706$$
$$2f(a + 4h) = 2f(0.8) = 1.2195$$
$$f(b) = f(1) = \underline{0.5000}$$
$$7.8373 = \text{sum.}$$

Thus our estimate for the integral is

$$\int_0^1 \frac{1}{1 + x^2} \, dx \approx \frac{0.2}{2}(7.8373) \approx 0.784.$$

The actual value of the integral is approximately 0.785.

Another method for estimating $\int_a^b f(x) \, dx$ is given by Simpson's rule, which involves approximating the graph of f by parabolic segments. We shall omit the derivation.

SIMPSON'S RULE

$$\int_a^b f(x) \, dx \approx \frac{h}{3}\{f(a) + 4f(a + h) + 2f(a + 2h) +$$
$$\cdots + 4f[a + (n - 1)h] + f(b)\},$$

where $h = (b - a)/n$ and n is even.

The pattern of coefficients inside the braces is 1, 4, 2, 4, 2, . . . , 2, 4, 1, and this requires that **n be even**. Let us use this rule for the integral in Example 1.

EXAMPLE 2   *Use Simpson's rule to estimate the value of* $\int_0^1 \frac{1}{1 + x^2} \, dx$
by using $n = 4$. *Compute each term to four decimal places and round off the answer to three decimal places.*

Here $f(x) = 1/(1 + x^2)$, $n = 4$, $a = 0$, and $b = 1$. Thus $h = (b - a)/n = 1/4 = 0.25$. The terms to be added are

$$
\begin{aligned}
f(a) = f(0) &= 1.0000 \\
4f(a + h) = 4f(0.25) &= 3.7647 \\
2f(a + 2h) = 2f(0.5) &= 1.6000 \\
4f(a + 3h) = 4f(0.75) &= 2.5600 \\
f(b) = f(1) &= \underline{0.5000} \\
9.4247 &= \text{sum.}
\end{aligned}
$$

Thus by Simpson's rule,

$$
\int_0^1 \frac{1}{1 + x^2} \, dx \approx \frac{0.25}{3}(9.4247) \approx 0.785.
$$

This is a better approximation than that which we obtained in Example 1 by using the trapezoidal rule.

___

Both Simpson's rule and the trapezoidal rule may be used if we know only $f(a)$, $f(a + h)$, etc.; we need not know f itself. Example 3 will illustrate.

EXAMPLE 3   *A function often used in demography (the study of births, marriages, mortality, etc., in a population) is the **life-table function**, denoted l. In a population having 100,000 births in any year of time, l(x) represents the number of persons who reach the age of x in any year of time. For example, if l(20) = 95,961, then the number of persons who attain age 20 in any year of time is 95,961. Suppose that the function l applies to all people born over an extended period of time. It can be shown that, at any time, the expected number of persons in the population between the exact ages of x and x + m inclusive is given by*

$$
\int_x^{x+m} l(t) \, dt.
$$

In Table 7.1 are values of l(x) for males and females in a certain population (adapted from [52]). Approximate the number of women in the 20–35 age group by using the trapezoidal rule with n = 3.

TABLE 7.1
Life Table

AGE, x	$l(x)$ MALES	$l(x)$ FEMALES
0	100,000	100,000
5	97,158	97,791
10	96,921	97,618
15	96,672	97,473
20	95,961	97,188
25	95,000	96,839
30	94,097	96,429
35	93,067	95,844
40	91,628	94,961
45	89,489	93,667
50	86,195	91,726
55	81,154	88,935
60	73,830	84,971
65	64,108	79,445
70	52,007	71,196
75	38,044	59,946
80	24,900	45,662

We want to estimate

$$\int_{20}^{35} l(t)\, dt.$$

We have $h = \dfrac{b-a}{n} = \dfrac{35-20}{3} = 5$. The terms to be added are

$$
\begin{aligned}
l(20) &= 97,188 \\
2l(25) = 2(96,839) &= 193,678 \\
2l(30) = 2(96,429) &= 192,858 \\
l(35) &= \underline{95,844} \\
& \ 579,568 = \text{sum.}
\end{aligned}
$$

By the trapezoidal rule,

$$\int_{20}^{35} l(t)\, dt \approx \frac{5}{2}(579,568) = 1,448,920.$$

There are formulas that are used to determine the accuracy of answers obtained by using the trapezoidal or Simpson's rule. They may be found in standard texts on numerical analysis.

EXERCISE 7.3

In each problem, compute each term to four decimal places and round off the answer to three decimal places.

*In Problems **1–6**, use the trapezoidal rule or Simpson's rule (as indicated) and the given value of n to estimate the integral. In Problems **1–4**, also find the answer by antidifferentiation (the Fundamental Theorem of Integral Calculus).*

1. $\int_0^1 x^2 \, dx$;   trapezoidal rule, $n = 5$.

2. $\int_0^1 x^2 \, dx$;   Simpson's rule, $n = 4$.

3. $\int_1^4 \dfrac{dx}{x}$;   Simpson's rule, $n = 6$.

4. $\int_1^4 \dfrac{dx}{x}$;   trapezoidal rule, $n = 6$.

5. $\int_0^2 \dfrac{x \, dx}{x + 1}$;   trapezoidal rule, $n = 4$.

6. $\int_2^4 \dfrac{dx}{x + x^2}$;   Simpson's rule, $n = 4$.

*In Problems **7** and **8**, use the life table (Table 7.1) in Example 3 to estimate the given integrals by the trapezoidal rule.*

7. $\int_{15}^{40} l(t) \, dt$,   males, $n = 5$.

8. $\int_{35}^{55} l(t) \, dt$,   females, $n = 4$.

*In Problems **9** and **10**, suppose the graph of a continuous function f, where f(x) ≥ 0, contains the given points. Use*

Simpson's rule and all of the points to approximate the area between the graph and the x-axis on the given interval.

9. (1, 0.4), (2, 0.6), (3, 1.2), (4, 0.8), (5, 0.5);   [1, 5].

10. (2, 0), (2.5, 3.6), (3, 10), (3.5, 19.9), (4, 34);   [2, 4].

*Problems **11** and **12** are designed for students with calculators having the square-root function. Estimate the given integrals.*

11. $\int_0^1 \sqrt{1 - x^2} \, dx$;   Simpson's rule, $n = 4$.

12. $\int_4^6 \dfrac{1}{\sqrt{1 + x}} \, dx$;   Simpson's rule, $n = 4$.   Also find the answer by the Fundamental Theorem of Integral Calculus.

13. Using all the information given in Fig. 7.4, estimate $\int_1^3 f(x) \, dx$ by Simpson's rule.

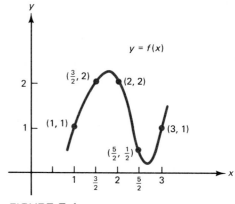

FIGURE 7.4

7.4 IMPROPER INTEGRALS

Suppose $f(x)$ is nonnegative for $a \le x < \infty$ (see Fig. 7.5). We know that the

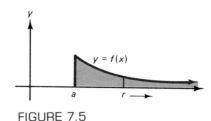

FIGURE 7.5

integral $\int_a^r f(x)\ dx$ is the area between the curve $y = f(x)$ and the x-axis from $x = a$ to $x = r$. As $r \to \infty$, we may think of

$$\lim_{r \to \infty} \int_a^r f(x)\ dx$$

as the area of the unbounded region that is shaded in Fig. 7.5. This limit is abbreviated by

$$\int_a^\infty f(x)\ dx, \tag{1}$$

called an **improper integral**. If this limit exists, $\int_a^\infty f(x)\ dx$ is said to be **convergent** or to *converge* to that limit. In this case the unbounded region is considered to have a finite area, and this area is represented by $\int_a^\infty f(x)\ dx$. If the limit does not exist, the improper integral is said to be **divergent** and the region does not have a finite area.

We can remove the restriction that $f(x) \geq 0$. In general, the improper integral $\int_a^\infty f(x)\ dx$ is defined by

$$\int_a^\infty f(x)\ dx = \lim_{r \to \infty} \int_a^r f(x)\ dx.$$

Other types of improper integrals are

$$\int_{-\infty}^b f(x)\ dx \tag{2}$$

$$\text{and} \quad \int_{-\infty}^\infty f(x)\ dx. \tag{3}$$

In each of these types of improper integrals [(1), (2), and (3)], the interval over which the integral is evaluated has infinite length. The improper integral in (2) is defined by

$$\int_{-\infty}^b f(x)\ dx = \lim_{r \to -\infty} \int_r^b f(x)\ dx.$$

If this limit exists, $\int_{-\infty}^b f(x)\ dx$ is said to be convergent. Otherwise, it is divergent. We shall define the improper integral in (3) after the following example.

EXAMPLE 1   *Determine whether the following improper integrals are convergent or divergent. If convergent, determine the value of the integral.*

a. $\int_1^\infty \dfrac{1}{x^3}\ dx.$

$$\int_1^\infty \frac{1}{x^3}\ dx = \lim_{r \to \infty} \int_1^r x^{-3}\ dx = \lim_{r \to \infty} \left. -\frac{x^{-2}}{2} \right|_1^r$$

$$= \lim_{r \to \infty} \left[-\frac{1}{2r^2} + \frac{1}{2} \right] = -0 + \frac{1}{2} = \frac{1}{2}.$$

Therefore, $\int_1^\infty \frac{1}{x^3} \, dx$ converges to $\frac{1}{2}$.

b. $\int_{-\infty}^0 e^x \, dx.$

$$\int_{-\infty}^0 e^x \, dx = \lim_{r \to -\infty} \int_r^0 e^x \, dx = \lim_{r \to -\infty} \left. e^x \right|_r^0$$

$$= \lim_{r \to -\infty} (1 - e^r) = 1 - 0 = 1.$$

(Here we used the fact that, as $r \to -\infty$, then $e^r \to 0$.) Therefore, $\int_{-\infty}^0 e^x \, dx$ converges to 1.

c. $\int_1^\infty \frac{1}{\sqrt{x}} \, dx.$

$$\int_1^\infty \frac{1}{\sqrt{x}} \, dx = \lim_{r \to \infty} \int_1^r x^{-1/2} \, dx = \lim_{r \to \infty} \left. 2x^{1/2} \right|_1^r$$

$$= \lim_{r \to \infty} 2(\sqrt{r} - 1) = \infty.$$

Therefore, the improper integral diverges.

The improper integral $\int_{-\infty}^\infty f(x) \, dx$ is defined in terms of improper integrals of the forms (1) and (2):

$$\int_{-\infty}^\infty f(x) \, dx = \int_{-\infty}^0 f(x) \, dx + \int_0^\infty f(x) \, dx. \tag{4}$$

If *both* integrals on the right side of (4) are convergent, then $\int_{-\infty}^\infty f(x) \, dx$ is said to be convergent; otherwise, it is divergent.

EXAMPLE 2   *Determine whether* $\int_{-\infty}^\infty e^x \, dx$ *is convergent or divergent.*

$$\int_{-\infty}^\infty e^x \, dx = \int_{-\infty}^0 e^x \, dx + \int_0^\infty e^x \, dx.$$

By Example 1(b), $\int_{-\infty}^0 e^x \, dx = 1$. On the other hand,

$$\int_0^\infty e^x \, dx = \lim_{r \to \infty} \int_0^r e^x \, dx = \lim_{r \to \infty} \left. e^x \right|_0^r = \lim_{r \to \infty} (e^r - 1) = \infty.$$

Since $\int_0^\infty e^x \, dx$ is divergent, $\int_{-\infty}^\infty e^x \, dx$ is also divergent.

EXAMPLE 3   *In statistics, a function f is called a **density function** provided f(x) ≥ 0 and*

$$\int_{-\infty}^\infty f(x) \, dx = 1.$$

Suppose that

$$f(x) = \begin{cases} ke^{-x}, & \text{for } x \geq 0, \\ 0, & \text{elsewhere} \end{cases}$$

is a density function. Find k.

We write the equation $\int_{-\infty}^\infty f(x) \, dx = 1$ as

$$\int_{-\infty}^0 f(x) \, dx + \int_0^\infty f(x) \, dx = 1.$$

Since $f(x) = 0$ for $x < 0$, $\int_{-\infty}^0 f(x) \, dx = 0$. Thus

$$\int_0^\infty ke^{-x} \, dx = 1,$$

$$\lim_{r \to \infty} \int_0^r ke^{-x} \, dx = 1,$$

$$\lim_{r \to \infty} -ke^{-x} \Big|_0^r = 1,$$

$$\lim_{r \to \infty} (-ke^{-r} + k) = 1,$$

$$0 + k = 1,$$

$$k = 1.$$

EXERCISE 7.4

In Problems 1–12, determine the integrals, if they exist. Indicate those that are divergent.

1. $\int_3^\infty \frac{1}{x^2} \, dx.$

2. $\int_2^\infty \frac{1}{(2x-1)^3} \, dx.$

3. $\int_1^\infty \frac{1}{x} \, dx.$

4. $\int_1^\infty \frac{1}{\sqrt[3]{x+1}} \, dx.$

5. $\int_1^\infty e^{-x} \, dx.$

6. $\int_0^\infty (5 + e^{-x}) \, dx.$

7. $\int_1^\infty \frac{1}{\sqrt{x}} \, dx.$

8. $\int_4^\infty \frac{x \, dx}{\sqrt{(x^2+9)^3}}.$

9. $\int_{-\infty}^{-2} \frac{1}{(x+1)^3} \, dx.$

10. $\int_{-\infty}^3 \frac{1}{\sqrt{7-x}} \, dx.$

11. $\int_{-\infty}^\infty xe^{-x^2} \, dx.$

12. $\int_{-\infty}^\infty (5 - 3x) \, dx.$

13. The density function for the life in hours x of an electronic component in a calculator is given by

$$f(x) = \begin{cases} \dfrac{k}{x^2}, & \text{for } x \geq 800 \\ 0, & \text{for } x < 800. \end{cases}$$

a. If k satisfies the condition that $\displaystyle\int_{800}^{\infty} f(x)\, dx = 1$, find k.

b. The probability that the component will last at least 1200 hours is given by $\displaystyle\int_{1200}^{\infty} f(x)\, dx$. Evaluate this integral.

14. Given the density function

$$f(x) = \begin{cases} ke^{-4x}, & \text{for } x \geq 0, \\ 0, & \text{elsewhere} \end{cases}$$

find k. (*Hint:* See Example 3.)

15. In a psychological model for signal detection [2], the probability α (a Greek letter read "alpha") of reporting a signal when no signal is present is given by

$$\alpha = \int_{x_c}^{\infty} e^{-x}\, dx, \qquad x \geq 0.$$

The probability β (a Greek letter read "beta") of detecting a signal when it is present is

$$\beta = \int_{x_c}^{\infty} ke^{-kx}\, dx, \qquad x \geq 0.$$

In both integrals x_c is a constant (called a criterion value in this model). Find α and β if $k = \frac{1}{8}$.

16. For a business the present value of all future profits at an annual interest rate r compounded continuously is given by

$$\int_0^{\infty} p(t)e^{-rt}\, dt,$$

where $p(t)$ is the profit per year in dollars at time t. If $p(t) = 240{,}000$ and $r = 0.06$, evaluate the above integral.

17. Find the area of the region in the first quadrant bounded by the curve $y = e^{-2x}$ and the x-axis.

18. In discussing entrance of a firm into an industry, Stigler[34] uses the equation

$$V = \pi_0 \int_0^{\infty} e^{\theta t} e^{-\rho t}\, dt,$$

where π_0, θ (a Greek letter read "theta"), and ρ (a Greek letter read "rho") are constants. Show that $V = \pi_0/(\rho - \theta)$ if $\theta < \rho$.

19. The predicted rate of growth per year of a population of a certain small city is given by $10{,}000/(t + 2)^2$, where t is the number of years from now. In the long run (that is, as $t \to \infty$), what is the expected change in population from today's level?

7.5 DIFFERENTIAL EQUATIONS

Occasionally, you may have to solve an equation that involves the derivative of an unknown function. Such an equation is called a **differential equation**. An example is

$$y' = xy^2. \tag{1}$$

More precisely, Eq. (1) is a **first-order differential equation** since it involves a derivative of the first order and none of higher order. A solution of Eq. (1) is any function $y = f(x)$ that is defined on an interval and that satisfies the equation for all x in the interval.

To solve $y' = xy^2$ or, equivalently,

$$\frac{dy}{dx} = xy^2, \tag{2}$$

we consider dy/dx to be a quotient of differentials and algebraically "separate variables" by rewriting the equation so that each side contains only one variable and a differential is not in a denominator:

$$\frac{dy}{y^2} = x \, dx.$$

Integrating both sides and combining the constants of integration, we obtain

$$\int \frac{1}{y^2} \, dy = \int x \, dx,$$

$$-\frac{1}{y} = \frac{x^2}{2} + C_1,$$

$$-\frac{1}{y} = \frac{x^2 + 2C_1}{2}.$$

Since $2C_1$ is an arbitrary constant, we can replace it by C.

$$-\frac{1}{y} = \frac{x^2 + C}{2}. \tag{3}$$

Solving Eq. (3) for y, we have

$$y = -\frac{2}{x^2 + C}. \tag{4}$$

We can verify by substitution that y is a solution to differential equation (2):

$$\frac{dy}{dx} = xy^2?$$

$$\frac{4x}{(x^2 + C)^2} = x \left[-\frac{2}{x^2 + C} \right]^2 ?$$

$$\frac{4x}{(x^2 + C)^2} = \frac{4x}{(x^2 + C)^2}.$$

Note in Eq. (4) that for *each* value of C, a different solution is obtained. We call Eq. (4) the **general solution** of the differential equation. The method that we used to find it is called **separation of variables**.

   In the example above, suppose we are given the condition that $y = -\frac{2}{3}$ when $x = 1$; that is, $y(1) = -\frac{2}{3}$. Then the *particular* function that satisfies Eq. (2) and this condition can be found by substituting the values $x = 1$ and $y = -\frac{2}{3}$ into Eq. (4) and solving for C:

$$-\frac{2}{3} = -\frac{2}{1^2 + C},$$

$$C = 2.$$

Therefore, the solution of $dy/dx = xy^2$ such that $y(1) = -\frac{2}{3}$ is

$$y = -\frac{2}{x^2 + 2}.$$ (5)

We call Eq. (5) a **particular solution** to the differential equation.

EXAMPLE 1   *Solve* $y' = -\dfrac{y}{x}$ *if* $x,y > 0$.

Writing y' as dy/dx, separating variables, and integrating, we have

$$\frac{dy}{dx} = -\frac{y}{x},$$

$$\frac{dy}{y} = -\frac{dx}{x},$$

$$\int \frac{1}{y}\, dy = -\int \frac{1}{x}\, dx,$$

$$\ln|y| = C_1 - \ln|x|.$$

Since $x,y > 0$,

$$\ln y = C_1 - \ln x.$$ (6)

The exponential form of Eq. (6) is

$$y = e^{C_1 - \ln x},$$

so

$$y = e^{C_1} e^{-\ln x} = \frac{e^{C_1}}{e^{\ln x}}.$$

Replacing e^{C_1} by C, where $C > 0$, and $e^{\ln x}$ by x gives

$$y = \frac{C}{x}, \qquad C, x > 0.$$

In Sec. 6.1 interest compounded continuously was developed. Let us now take a different approach to this topic. Suppose that P dollars are invested at an annual rate r compounded n times a year. Let the function $S = S(t)$ give the total amount S (or the compound amount) present after t years from the date of the initial investment. Then the initial principal is $S(0) = P$. Furthermore, since there are n interest periods per year, each period has length $1/n$ years, which we shall denote by Δt. At the end of the first period, the accrued interest for that period is added to the principal, and the sum acts as the principal for the second period, etc. Hence, if the beginning of an interest period occurs at time t, then the increase in the amount present at the end of a period of Δt is $S(t + \Delta t) - S(t)$, which we write as ΔS. This increase ΔS is also the interest earned for the period. Equivalently, the interest earned is principal times rate times time:

$$\Delta S = S \cdot r \cdot \Delta t.$$

Dividing both sides by Δt, we obtain

$$\frac{\Delta S}{\Delta t} = rS. \tag{7}$$

As $\Delta t \to 0$, then $n = \dfrac{1}{\Delta t} \to \infty$ and consequently interest is being *compounded continuously;* that is, the principal is subject to continuous growth at every instant. However, as $\Delta t \to 0$, then $\Delta S/\Delta t \to dS/dt$ and Eq. (7) takes the form

$$\frac{dS}{dt} = rS. \tag{8}$$

This differential equation means that *when interest is compounded continuously, the rate of change of the amount of money present at time t is proportional to the amount present at time t.* The constant of proportionality is r.

To determine the actual function S, we solve differential equation (8) by the method of separation of variables.

$$\frac{dS}{dt} = rS,$$

$$\frac{dS}{S} = r\,dt,$$

$$\int \frac{1}{S}\,dS = \int r\,dt,$$

$$\ln|S| = rt + C_1.$$

Since it can be assumed that $S > 0$, then $\ln|S| = \ln S$. Thus

$$\ln S = rt + C_1.$$

We can solve for S by converting to exponential form.

$$S = e^{rt + C_1} = e^{C_1} e^{rt}.$$

For simplicity, e^{C_1} can be replaced by C to obtain

$$S = Ce^{rt}.$$

Because of the condition that $S(0) = P$, we can determine the value of C:

$$P = Ce^{r(0)} = C \cdot 1.$$

Hence $C = P$ and

$$S = Pe^{rt}. \tag{9}$$

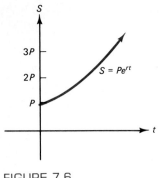

FIGURE 7.6

Equation (9) gives the total value after t years of an initial investment of P dollars, compounded continuously at an annual rate r (see Fig. 7.6).

In our compound interest discussion we saw from Eq. (8) that the rate of change in the amount present is proportional to the amount present. There are many natural quantities, such as population, whose rate of growth or decay at any time is considered proportional to the amount of that quantity present. If N denotes the amount of such a quantity at time t, then this rate of growth means that

$$\frac{dN}{dt} = kN,$$

where k is a constant. If we separate variables and solve for N as we did for Eq. (8), we get

$$N = N_0 e^{kt}, \tag{10}$$

where N_0 is a constant. Due to the form of Eq. (10), we say that the quantity follows an **exponential law of growth** if k is positive and an **exponential law of decay** if k is negative.

EXAMPLE 2   *In a certain city the rate at which the population grows at any time is proportional to the size of the population. If the population was 125,000 in 1960 and 140,000 in 1980, what is the expected population in 2000?*

Let N be the size of the population at time t. Since the exponential law of growth applies,

$$N = N_0 e^{kt}.$$

To find the population in 2000, we must first find the particular law of growth involved by determining the values of N_0 and k. Let the year 1960 correspond to $t = 0$. Then $t = 20$ for 1980 and $t = 40$ for 2000. To find N_0 we use the fact that $N = 125,000$ when $t = 0$:

$$N = N_0 e^{kt},$$
$$125,000 = N_0 e^0 = N_0.$$

Hence $N_0 = 125,000$ and

$$N = 125,000 e^{kt}.$$

To find k we use the fact that $N = 140,000$ when $t = 20$:

$$140,000 = 125,000 e^{20k}.$$

Thus

$$e^{20k} = \frac{140,000}{125,000} = 1.12,$$

$$20k = \ln(1.12) \quad \text{(logarithmic form)},$$

$$k = \tfrac{1}{20} \ln(1.12).$$

Therefore, the law of growth is

$$N = 125,000e^{(t/20)\,\ln\,1.12} \tag{11}$$

$$= 125,000[e^{\ln\,1.12}]^{t/20}.$$

$$N = 125,000(1.12)^{t/20}. \tag{12}$$

If $t = 40$,

$$N = 125,000(1.12)^2 = 156,800.$$

We can write Eq. (11) in a form different from Eq. (12). Because $\ln 1.12 \approx 0.11333$, we have $k \approx 0.11333/20 \approx 0.0057$. Thus

$$N \approx 125,000e^{0.0057t}.$$

The rate at which a radioactive element decays at any time is found to be proportional to the amount of that element present. If N is the amount of a radioactive substance at time t, then the rate of decay is given by

$$\frac{dN}{dt} = -\lambda N. \tag{13}$$

The positive constant λ (a Greek letter read "lambda") is called the **decay constant**, and the minus sign indicates that N is decreasing as t increases. Thus we have exponential decay. From Eq. (10), the solution of this differential equation is

$$N = N_0 e^{-\lambda t}. \tag{14}$$

If $t = 0$, then $N = N_0 \cdot 1 = N_0$, so N_0 represents the amount of the radioactive substance present when $t = 0$.

The time for one-half of the substance to decay is called the **half-life** of the substance. It is the value of t when $N = N_0/2$. From Eq. (14),

$$\frac{N_0}{2} = N_0 e^{-\lambda t},$$

$$\frac{1}{2} = e^{-\lambda t}.$$

In logarithmic form we have

$$-\lambda t = \ln \tfrac{1}{2} = \ln 1 - \ln 2 = -\ln 2,$$

$$t = \frac{\ln 2}{\lambda} \approx \frac{0.69315}{\lambda}. \tag{15}$$

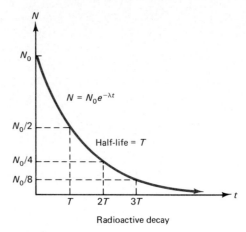

Radioactive decay

FIGURE 7.7

Note that the half-life depends on λ. Figure 7.7 shows the graph of radioactive decay.

EXAMPLE 3   *If* 60 *percent of a radioactive substance remains after* 50 *days, find the decay constant and the half-life of the element.*

From Eq. (14),

$$N = N_0 e^{-\lambda t},$$

where N_0 is the amount of the element present at $t = 0$. When $t = 50$, then $N = 0.6N_0$ and we have

$$0.6N_0 = N_0 e^{-50\lambda},$$

$$0.6 = e^{-50\lambda},$$

$$-50\lambda = \ln(0.6) \qquad \text{(logarithmic form*)},$$

$$\lambda = -\frac{1}{50}\ln(0.6) \approx -\frac{1}{50}(-0.51083)$$

$$\approx 0.01022.$$

The half-life, from Eq. (15), is approximately

$$\frac{0.69315}{\lambda} \approx \frac{0.69315}{0.01022} \approx 67.82 \text{ days.}$$

Radioactivity is useful in dating such things as fossil plant remains and archaeological remains made from organic material. Plants and other living organisms contain a small amount of radioactive carbon 14 (^{14}C) in addition to ordinary carbon (^{12}C). The ^{12}C atoms are stable, but the ^{14}C atoms are

* $\ln(0.6) = \ln(6/10) = \ln 6 - \ln 10 \approx 1.79176 - 2.30259 = -0.51083.$

decaying exponentially. However, ^{14}C is formed in the atmosphere due to the effect of cosmic rays. Eventually this ^{14}C is taken up by plants during photosynthesis and replaces what has decayed. As a result, the ratio of ^{14}C atoms to ^{12}C atoms is considered constant over a long period of time. When a plant dies, it stops absorbing ^{14}C and the remaining ^{14}C atoms decay. By comparing the proportion of ^{14}C to ^{12}C in a fossil plant to that of plants found today, we can estimate the age of the fossil. The half-life of ^{14}C is approximately 5600 years. Thus, for example, if a fossil is found to have a ^{14}C to ^{12}C ratio which is half that of a similar substance found today, we would estimate the fossil to be 5600 years old.

EXAMPLE 4   *A wooden tool found in a Middle East excavation site is found to have a ^{14}C to ^{12}C ratio which is 0.6 of the corresponding ratio in a present-day tree. Estimate the age of the tool to the nearest hundred years.*

Let N be the amount of ^{14}C present in the wood t years after the tool was made. Then $N = N_0 e^{-\lambda t}$ where N_0 is the amount of ^{14}C when $t = 0$. Since the ^{14}C to ^{12}C ratio is 0.6 of the corresponding ratio in a present-day tree, this means that we want to find the value of t for which $N = 0.6N_0$.

$$0.6N_0 = N_0 e^{-\lambda t},$$

$$0.6 = e^{-\lambda t},$$

$$-\lambda t = \ln(0.6),$$

$$t = -\frac{1}{\lambda} \ln(0.6).$$

From Eq. (15), the half-life is (approximately) $0.69315/\lambda$, which equals 5600, so $\lambda \approx 0.69315/5600$. Thus

$$t \approx -\frac{1}{0.69315/5600} \ln(0.6)$$

$$\approx -\frac{5600}{0.69315} (-0.51083)$$

$$\approx 4100 \text{ years.}$$

EXERCISE 7.5

In Problems 1–8, solve the differential equations.

1. $y' = 2xy^2$.

2. $y' = x^3 y^3$.

3. $\dfrac{dy}{dx} - x\sqrt{x^2 + 1} = 0$.

4. $\dfrac{dy}{dx} = \dfrac{x}{y}$.

5. $\dfrac{dy}{dx} = y, \quad y > 0$.

6. $y' = e^x y^2$.

7. $y' = \dfrac{y}{x}, \quad x, y > 0$.

8. $\dfrac{dy}{dx} + xe^x = 0$.

In Problems 9–14, solve each of the differential equations subject to the given conditions.

9. $y' = \dfrac{1}{y}; \quad y > 0, \quad y(2) = 2$.

10. $y' = e^{x-y}; \quad y(0) = 0.$ (*Hint:* $e^{x-y} = e^x/e^y$.)

11. $e^y y' - x^2 = 0$; $y = 0$ when $x = 0$.

12. $x^2 y' + \dfrac{1}{y^2} = 0$; $y(1) = 2$.

13. $(4x^2 + 3)^2 y' - 4xy^2 = 0$; $y(0) = \frac{3}{2}$.

14. $y' + x^2 y = 0$; $y > 0$, $y = 1$ when $x = 0$.

15. In a certain town the population at any time changes at a rate proportional to the population. If the population in 1975 was 20,000 and in 1985 it was 24,000, find an equation for the population at time t, where t is the number of years past 1975. Write your answer in two forms, one involving e. (You may assume ln 1.2 = 0.18.) What is the expected population in 1995?

16. The population of a town increases by natural growth at a rate which is proportional to the number N of persons present. If the population at time $t = 0$ is 10,000, find two expressions for the population N, t years later, if the population doubles in 50 years. Assume ln 2 = 0.69. Also, find N for $t = 100$.

17. Suppose that the population of the world in 1930 was 2 billion and in 1960 it was 3 billion. If the exponential law of growth is assumed, what is the expected population in 2000? Give your answer in terms of e.

18. If exponential growth is assumed, in approximately how many years will a population triple if it doubles in 50 years? (*Hint:* Let the population at $t = 0$ be N_0.)

19. If 30 percent of the initial amount of a radioactive sample remains after 100 seconds, find the decay constant and the half-life of the element.

20. If 30 percent of the initial amount of a radioactive sample *has decayed* after 100 seconds, find the decay constant and the half-life of the element.

21. An Egyptian scroll was found to have a ^{14}C to ^{12}C ratio that is 0.7 of the corresponding ratio in similar present-day material. Estimate the age of the scroll to the nearest hundred years.

22. A recently discovered archaeological specimen has a ^{14}C to ^{12}C ratio that is 0.2 of the corresponding ratio found in present-day organic material. Estimate the age of the specimen to the nearest hundred years.

23. Suppose that a population follows exponential growth given by $dN/dt = kN$ for $t \geq t_0$, and $N = N_0$ when $t = t_0$. Find N, the population size at time t.

24. Radon has a half-life of 3.82 days.

    a. Find the decay constant in terms of ln 2.

    b. What fraction of the original amount of it remains after $2(3.82) = 7.64$ days?

25. Radioactive isotopes are used in medical diagnoses as tracers to determine abnormalities that may exist in an organ. For example, if radioactive iodine is swallowed, after some time it is taken up by the thyroid gland. With the use of a detector, the rate at which it is taken up can be measured and a determination can be made as to whether the uptake is normal. Suppose that radioactive technetium-99m, which has a half-life of 6 hours, is to be used in a brain scan two hours from now. What should be its activity now if the activity when it is used is to be 10 units? Give your answer to one decimal place. (*Hint:* In Eq. (14), let $N =$ activity t hours from now, and $N_0 =$ activity now.)

26. A radioactive substance that has a half-life of 8 days is to be temporarily implanted in a hospital patient until there remains three-fifths of the amount originally present. How long should the implant remain in the patient?

27. Suppose that q is the amount of penicillin in the body at time t, and let q_0 be the amount at $t = 0$. Assume that the rate of change of q with respect to t is proportional to q and that q decreases as t increases. Then we have $dq/dt = -kq$, where $k > 0$. Solve for q and find $\lim_{t \to \infty} q$. What percentage of the original amount present is there when $t = 2/k$?

28. Suppose $A(t)$ is the amount of a product that is consumed at time t and A follows an exponential law of growth. If $t_1 < t_2$ and at time t_2 the amount consumed, $A(t_2)$, is double the amount consumed at time t_1, $A(t_1)$, then $t_2 - t_1$ is called a doubling period. In a discussion of exponential growth, Shonle[10] states that under exponential growth, ". . . the amount of a product consumed during one doubling period is equal to the total used for all time up to the beginning of the doubling period in question." To justify this statement, reproduce his argument as follows. The amount of the product used up to time t_1 is given by

$$\int_{-\infty}^{t_1} A_0 e^{kt}\, dt, \quad k > 0,$$

where A_0 is the amount when $t = 0$. Show that this is equal to $(A_0/k)e^{kt_1}$. Next, the amount used during the time interval from t_1 to t_2 is

$$\int_{t_1}^{t_2} A_0 e^{kt} \, dt.$$

Show that this is equal to

$$\frac{A_0}{k} e^{kt_1}[e^{k(t_2 - t_1)} - 1]. \qquad (16)$$

If the interval $[t_1, t_2]$ is a doubling period, then

$$A_0 e^{kt_2} = 2A_0 e^{kt_1}.$$

Show that this implies $e^{k(t_2 - t_1)} = 2$. Substitute this into (16); your result should be the same as the total used during all time up to t_1, namely, $(A_0/k)e^{kt_1}$.

29. In a forest natural litter occurs, such as fallen leaves and branches, dead animals, etc. [25]. Let $A = A(t)$ denote the amount of litter present at time t, where $A(t)$ is expressed in grams per square meter and t is in years. Suppose that there is no litter at $t = 0$. Thus $A(0) = 0$. Assume that

(1) Litter falls to the ground continuously at a constant rate of 200 grams per square meter per year.
(2) The accumulated litter decomposes continuously at the rate of 50 percent of the amount present per year (which is $0.50A$).

The difference of the two rates is the rate of change of the amount of litter present with respect to time:

$$\begin{pmatrix} \text{rate of change} \\ \text{of litter present} \end{pmatrix} = \\ \begin{pmatrix} \text{rate of falling} \\ \text{to ground} \end{pmatrix} - \begin{pmatrix} \text{rate of} \\ \text{decomposition} \end{pmatrix}.$$

Thus

$$\frac{dA}{dt} = 200 - 0.50A.$$

a. Solve for A.

b. To the nearest gram, determine the amount of litter per square meter after one year.

c. As $t \to \infty$, to what value does A tend?

7.6 MORE APPLICATIONS OF DIFFERENTIAL EQUATIONS _____

Suppose that the number N of individuals in a population at time t follows an exponential law of growth. From Sec. 7.5, $N = N_0 e^{kt}$, where $k > 0$ and N_0 is the population when $t = 0$. This law assumes that at time t the rate of growth, dN/dt, of the population is proportional to the number of individuals in the population. That is, $dN/dt = kN$.

Under exponential growth, a population would get infinitely large as time goes on. In reality, however, when the population gets large enough there are environmental factors that slow down the rate of growth. Examples are food supply, predators, overcrowding, etc. These factors cause dN/dt to eventually decrease. It is reasonable to assume that population size is limited to some maximum number M, where $0 < N < M$, and as $N \to M$, then $dN/dt \to 0$ and the population size tends to be stable.

In summary, we want a population model that has exponential growth initially but which also includes the effects of environmental resistance to large population growth. Such a model is obtained by multiplying the right side of $dN/dt = kN$ by the factor $(M - N)/M$:

$$\frac{dN}{dt} = kN \left(\frac{M - N}{M} \right).$$

Notice that if N is small, then $(M - N)/M$ is close to 1 and we have growth that is approximately exponential. As $N \to M$, then $M - N \to 0$ and

dN/dt approaches 0 as we wanted in our model. Because k/M is a constant, we can replace it by K. Thus

$$\frac{dN}{dt} = KN(M - N). \tag{1}$$

This states that the rate of growth is proportional to the product of the population size and the difference between the maximum size and the population size. We can solve differential equation (1) by the method of separation of variables.

$$\frac{dN}{N(M - N)} = K\, dt,$$

$$\int \frac{1}{N(M - N)}\, dN = \int K\, dt. \tag{2}$$

The integral on the left side can be found by using Formula (5) in the table of integrals. Thus Eq. (2) becomes

$$\frac{1}{M} \ln \left| \frac{N}{M - N} \right| = Kt + C,$$

$$\ln \left| \frac{N}{M - N} \right| = Mkt + MC.$$

Since $N > 0$ and $M - N > 0$, we can write

$$\ln \frac{N}{M - N} = Mkt + MC.$$

In exponential form we have

$$\frac{N}{M - N} = e^{MKt + MC} = e^{MKt}e^{MC}.$$

Replacing the positive constant e^{MC} by A gives

$$\frac{N}{M - N} = Ae^{MKt},$$

$$N = (M - N)Ae^{MKt},$$

$$N = MAe^{MKt} - NAe^{MKt},$$

$$NAe^{MKt} + N = MAe^{MKt},$$

$$N(Ae^{MKt} + 1) = MAe^{MKt},$$

$$N = \frac{MAe^{MKt}}{Ae^{MKt} + 1}.$$

Dividing numerator and denominator by Ae^{MKt}, we have

$$N = \frac{M}{1 + \dfrac{1}{Ae^{MKt}}} = \frac{M}{1 + \dfrac{1}{A}e^{-MKt}}.$$

Replacing $1/A$ by b and MK by c gives

$$N = \frac{M}{1 + be^{-ct}}. \qquad (3)$$

Equation (3) is called the **logistic function** or the **Verhulst-Pearl logistic function**. Its graph, called a *logistic curve*, is S-shaped and appears in Fig. 7.8. Notice in the graph that $N = M$ is a horizontal asymptote; that is,

$$\lim_{t \to \infty} \frac{M}{1 + be^{-ct}} = \frac{M}{1 + b(0)} = M.$$

Moreover, from Eq. (1) the rate of growth is

$$KN(M - N),$$

which can be considered as a function of N. To find when the maximum rate of growth occurs, we solve $\dfrac{d}{dN}[KN(M - N)] = 0$ for N.

$$\frac{d}{dN}[KN(M - N)] = \frac{d}{dN}[K(MN - N^2)]$$
$$= K[M - 2N] = 0.$$

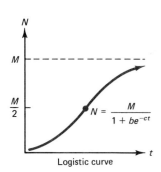

$N = \dfrac{M}{1 + be^{-ct}}$

Logistic curve

FIGURE 7.8

Thus $N = M/2$. The rate of growth increases until the population size is $M/2$ and decreases thereafter. The maximum rate of growth occurs when $N = M/2$ and corresponds to a point of inflection in the graph of N. To find the value of t for which this occurs, we substitute $M/2$ for N in Eq. (3) and solve for t.

$$\frac{M}{2} = \frac{M}{1 + be^{-ct}},$$
$$1 + be^{-ct} = 2,$$
$$e^{-ct} = \frac{1}{b},$$
$$e^{ct} = b,$$
$$ct = \ln b,$$
$$t = \frac{\ln b}{c}.$$

Thus the maximum rate of growth occurs at the point $([\ln b]/c, M/2)$. In Eq. (3) we may replace e^{-c} by C, and then the logistic function has the form

$$N = \frac{M}{1 + bC^t}.$$

EXAMPLE 1   *Suppose that the membership in a new country club is to be a maximum of 800 persons due to limitations of the physical plant. One year ago, the initial membership was 50 persons and now there are 200. Provided that enrollment follows a logistic function, how many members will there be three years from now?*

Let N be the number of members enrolled t years after the formation of the club. Then

$$N = \frac{M}{1 + be^{-ct}}.$$

Here $M = 800$, and when $t = 0$ we have $N = 50$.

$$50 = \frac{800}{1 + b},$$

$$1 + b = \frac{800}{50} = 16,$$

$$b = 15.$$

Thus

$$N = \frac{800}{1 + 15e^{-ct}}. \tag{4}$$

When $t = 1$, then $N = 200$.

$$200 = \frac{800}{1 + 15e^{-c}},$$

$$1 + 15e^{-c} = \frac{800}{200} = 4,$$

$$e^{-c} = \frac{3}{15} = \frac{1}{5}.$$

Hence $c = -\ln(\frac{1}{5}) = \ln 5$. Rather than substituting this value of c into Eq. (4), it is more convenient to substitute the value of e^{-c} there.

$$N = \frac{800}{1 + 15(\frac{1}{5})^t}.$$

Three years from now, $t = 4$. Thus

$$N = \frac{800}{1 + 15(\frac{1}{5})^4} \approx 781.$$

Let us now consider a simplified model of how a rumor spreads in a population of size M. A similar situation would be the spread of an epidemic or new fad.

Let $N = N(t)$ be the number of persons who know the rumor at time t. We shall assume that those who know the rumor spread it randomly in the population and that those who are told the rumor become spreaders of the rumor. Furthermore, we shall assume that each knower tells the rumor to k individuals per unit of time. (Some of these k individuals may already know the rumor.) We want an expression for the rate of increase of the knowers of the rumor. Over a unit of time, each of approximately N persons will tell the rumor to k persons. Thus the total number of persons who are told the rumor over the unit of time is (approximately) Nk. However, we are interested only in *new* knowers. The proportion of the population who do not know the rumor is $(M - N)/M$. Thus the total number of new knowers of the rumor is

$$Nk\left(\frac{M - N}{M}\right),$$

which can be written $(k/M)N(M - N)$. Therefore,

$$\frac{dN}{dt} = \frac{k}{M}N(M - N)$$

$$= KN(M - N), \quad \text{where } K = \frac{k}{M}.$$

This differential equation has the form of Eq. (1), so its solution, from Eq. (3), is a logistic function:

$$N = \frac{M}{1 + be^{-ct}}.$$

EXAMPLE 2   *In a large university of 45,000 students, a sociology major is researching the spread of a new campus rumor. When she begins her research, she determines that 300 students know the rumor. After one week she finds that 900 know it. Estimate the number who know it four weeks after the research begins by assuming logistic growth. Give the answer to the nearest thousand.*

Let N be the number of students who know the rumor after t weeks. Then

$$N = \frac{M}{1 + be^{-ct}}.$$

Here M, the size of the population, is 45,000, and when $t = 0$ we have $N = 300$.

$$300 = \frac{45,000}{1 + b},$$

$$1 + b = \frac{45,000}{300} = 150,$$

$$b = 149.$$

Thus

$$N = \frac{45,000}{1 + 149e^{-ct}}.$$

When $t = 1$, then $N = 900$.

$$900 = \frac{45,000}{1 + 149e^{-c}},$$

$$1 + 149e^{-c} = \frac{45,000}{900} = 50.$$

Therefore, $e^{-c} = \frac{49}{149}$, so

$$N = \frac{45,000}{1 + 149(\frac{49}{149})^t}.$$

When $t = 4$,

$$N = \frac{45,000}{1 + 149(\frac{49}{149})^4} \approx 16,000.$$

After four weeks, approximately 16,000 students know the rumor.

If a homicide is committed, the temperature of the victim's body will gradually decrease from 37°C (normal body temperature) to the temperature of the surroundings (ambient temperature). In general, the temperature of the cooling body changes at a rate proportional to the difference between the temperature of the body and the ambient temperature. This statement is known as **Newton's law of cooling**. Thus if $T(t)$ is the temperature of the body at time t and the ambient temperature is a, then

$$\frac{dT}{dt} = k(T - a),$$

where k is the constant of proportionality. Therefore, Newton's law of cooling is a differential equation. It can be applied to determine the time at which a homicide was committed, as the next example illustrates.

EXAMPLE 3　　*A wealthy industrialist was found murdered in his home. Police arrived on the scene at* 11:00 P.M. *The temperature of the body at that time was* 31°C, *and one hour later it was* 30°C. *The temperature of the room in which the body was found was* 22°C. *Estimate the time at which the murder occurred.*

Let t be the number of hours after the body was discovered, and $T(t)$ be the temperature (in degrees Celsius) of the body at time t. We want to find the value of t for which $T = 37$ (normal body temperature). This value of t will, of course, be negative. By Newton's law of cooling,

$$\frac{dT}{dt} = k(T - a),$$

where k is a constant and a (the ambient temperature) is 22. Thus

$$\frac{dT}{dt} = k(T - 22).$$

Separating variables, we have

$$\frac{dT}{T - 22} = k \, dt,$$

$$\int \frac{dT}{T - 22} = \int k \, dt,$$

$$\ln|T - 22| = kt + C.$$

Because $T - 22 > 0$,

$$\ln(T - 22) = kt + C.$$

When $t = 0$, then $T = 31$. Thus

$$\ln(31 - 22) = k \cdot 0 + C,$$

$$C = \ln 9.$$

Hence

$$\ln(T - 22) = kt + \ln 9,$$

$$\ln(T - 22) - \ln 9 = kt,$$

$$\ln \frac{T - 22}{9} = kt.$$

When $t = 1$, then $T = 30$, so

$$\ln \frac{30 - 22}{9} = k \cdot 1,$$

$$k = \ln \frac{8}{9} \approx -0.11778.$$

Thus

$$\ln \frac{T - 22}{9} \approx -0.11778t.$$

Now we find t when $T = 37$.

$$\ln \frac{37 - 22}{9} \approx -0.11778t,$$

$$t \approx -\frac{\ln(15/9)}{0.11778} \approx -\frac{0.51083}{0.11778},$$

$$\approx -4.34.$$

Thus the murder occurred about 4.34 hours *before* the time of discovery (11:00 P.M.). Since 4.34 hours is (approximately) 4 hours 20 minutes, the industrialist was murdered about 6:40 P.M.

EXERCISE 7.6

1. The population of a city follows logistic growth and is limited to 40,000. If the population in 1980 was 20,000 and in 1985 it was 25,000, what will be the population in 1990? Give your answer to the nearest hundred.

2. A company believes that the production of its product in present facilities will follow logistic growth. Presently, 200 units per day are produced, and production will increase to 300 units per day in one year. If production is limited to 500 units per day, what is the anticipated daily production in two years? Give your answer to the nearest unit.

3. In a country of 3,000,000 people the prime minister suffers a heart attack, which the government does not officially publicize. Initially 50 governmental personnel know of the attack but are spreading this information as a rumor. At the end of one week 5000 people know the rumor. Assuming logistic growth, find how many people know the rumor after two weeks. Give your answer to the nearest thousand.

4. A new fad is sweeping a college campus of 30,000 students. The college newspaper feels that its readers would be interested in a series on the fad. It assigns a reporter when the number of faddists is 400. One week later there are 1200 faddists. Assuming logistic growth, find a formula for the number N of faddists t weeks after the assignment of the reporter.

5. In a city whose population is 100,000 an outbreak of flu occurs. When the city health department begins its record-keeping, there are 500 infected persons. One week later there are 1000 infected persons. Assuming logistic growth, estimate the number of infected persons two weeks after record keeping begins.

6. The logistic curve for the U.S. population from 1790 to 1910 is estimated [53] to be

$$N = \frac{197.30}{1 + 35.60e^{-0.031186t}},$$

where N is the population in millions and t is in years counted from 1800. If this logistic function were valid for years after 1910, for what year would the point of inflection occur? Give your answer to one decimal place. Assume that $\ln 35.6 = 3.5723$.

7. In an experiment [54], five paramecia were placed in a test tube containing a nutritive medium. The number N of paramecia in the tube at the end of t days is approximately given by

$$N = \frac{375}{1 + e^{5.2-2.3t}}.$$

a. Show that this can be written as

$$N = \frac{375}{1 + 181.27e^{-2.3t}},$$

and hence is a logistic function.

b. Find $\lim_{t\to\infty} N$.

8. In the study of growth of a colony of unicellular organisms [16], the following equation was obtained:

$$N = \frac{0.2524}{e^{-2.128x} + 0.005125}, \qquad 0 \le x \le 5,$$

where N is the estimated area of the growth in square centimeters and x is the age of the colony in days after being first observed.

a. Put this equation in the form of a logistic function.

b. Find the area when the age of the colony is 0.

9. A waterfront murder was committed, and the victim's body was discovered at 3:15 A.M. by police. At that time the temperature of the body was 32°C. One hour later the body temperature was 30°C. After checking with the weather bureau, it was determined that the temperature at the waterfront was 10°C from 10:00 P.M. to 5:00 A.M. About what time did the murder occur?

10. An enzyme is a protein that acts as a catalyst for increasing the rate of a chemical reaction that occurs in cells. In a certain reaction an enzyme A is converted to another enzyme B. Enzyme B acts as a catalyst for its own formation. Let p be the amount of enzyme B at time t and I be the total amount of both enzymes when $t = 0$. Suppose the rate of formation of B is proportional to $p(I - p)$. Without directly using calculus, find the value of p for which the rate of formation will be a maximum.

11. A small town decides to conduct a fund-raising drive for a new fire engine whose cost is $70,000. The initial amount in the fund is $10,000. On the basis of past drives, it is determined that t months after the beginning of the drive, the rate dx/dt at which money is contributed to such a fund is proportional to the difference between the desired goal of $70,000 and the total amount x in

the fund at that time. After one month a total of $40,000 is in the fund. How much will be in the fund after three months?

12. In a discussion of unexpected properties of mathematical models of population, Bailey[55] considers the case in which the birth rate per *individual* is proportional to the population size N at time t. Since the growth rate per individual is $\dfrac{1}{N}\dfrac{dN}{dt}$, this means that

$$\frac{1}{N}\frac{dN}{dt} = kN$$

or $\dfrac{dN}{dt} = kN^2$,   (subject to $N = N_0$ at $t = 0$)

where $k > 0$. Show that

$$N = \frac{N_0}{1 - kN_0t}.$$

Use this result to show that

$$\lim N = \infty \quad \text{as} \quad t \rightarrow \left(\frac{1}{kN_0}\right)^-.$$

This means that over a finite interval of time there is an infinite amount of growth. Such a model might be useful only for rapid growth over a short interval of time.

13. Suppose that the rate of growth of a population is proportional to the difference between some maximum size M and the number N of individuals in the population at time t. Suppose that when $t = 0$ the population size is N_0. Find a formula for N.

7.7 REVIEW

Important Terms and Symbols

Section 7.1	present value of continuous annuity accumulated amount of continuous annuity
Section 7.2	average value of function
Section 7.3	trapezoidal rule Simpson's rule
Section 7.4	improper integral, $\displaystyle\int_a^\infty f(x)\,dx$, $\displaystyle\int_{-\infty}^b f(x)\,dx$, $\displaystyle\int_{-\infty}^\infty f(x)\,dx$
Section 7.5	first-order differential equation separation of variables exponential growth exponential decay decay constant half-life
Section 7.6	logistic function Newton's law of cooling

Summary

To determine an integral that does not have a familiar form, you may be able to match it with a formula in a table of integrals. However, it may be necessary to transform the given integral into an equivalent form before the matching can occur.

An annuity is a series of payments over a period of time. Suppose payments are made continuously for T years such that the payment at time t is at the rate of $f(t)$ per year. If the rate of interest is r compounded continuously, then the present value A of the continuous annuity is given by

$$A = \int_0^T f(t)e^{-rt}\,dt,$$

and the accumulated amount S is given by

$$S = \int_0^T f(t)e^{r(T-t)}\,dt.$$

The average value $\bar{f}$ of a function f over the interval $[a, b]$ is given by

$$\bar{f} = \frac{1}{b-a}\int_a^b f(x)\,dx.$$

There are formulas that allow us to approximate the value of a definite integral. One formula is the trapezoidal rule:

The Trapezoidal Rule

$$\int_a^b f(x)\,dx \approx \frac{h}{2}\{f(a) + 2f(a+h) + 2f(a+2h) +$$

$$\cdots + 2f[a+(n-1)h] + f(b)\},$$

where $h = (b-a)/n$.

Another formula is Simpson's rule:

Simpson's Rule

$$\int_a^b f(x)\,dx \approx \frac{h}{3}\{f(a) + 4f(a+h) + 2f(a+2h) +$$

$$\cdots + 4f[a+(n-1)h] + f(b)\},$$

where $h = (b-a)/n$ and n is even.

An integral of the form

$$\int_a^\infty f(x)\,dx, \qquad \int_{-\infty}^b f(x)\,dx \quad \text{or} \quad \int_{-\infty}^\infty f(x)\,dx$$

is called an improper integral. The first two integrals are defined as follows.

$$\int_a^\infty f(x)\,dx = \lim_{r\to\infty} \int_a^r f(x)\,dx$$

and

$$\int_{-\infty}^b f(x)\,dx = \lim_{r\to-\infty} \int_r^b f(x)\,dx.$$

If $\int_a^\infty f(x)\,dx$ (or $\int_{-\infty}^b f(x)\,dx$) is a finite number, we say that the integral is convergent; otherwise, it is divergent. The improper integral $\int_{-\infty}^\infty f(x)\,dx$ is defined by

$$\int_{-\infty}^\infty f(x)\,dx = \int_{-\infty}^0 f(x)\,dx + \int_0^\infty f(x)\,dx.$$

If both integrals on the right side are convergent, then $\int_{-\infty}^\infty f(x)\,dx$ is said to be convergent; otherwise, it is divergent.

An equation that involves the derivative of an unknown function is called a differential equation. If the highest-order derivative that occurs is the first, the equation is called a first-order differential equation. Some first-order differential equations may be solved by the method of separation of variables. In that method, by considering the derivative to be a quotient of differentials, we rewrite the equation so that each side contains only one variable and

a differential is not in a denominator. Integrating both sides of the resulting equation gives the solution. This solution involves a constant of integration and is called the general solution of the differential equation. If the unknown function must satisfy the condition that it has a specific function value for a given value of the independent variable, then a particular solution may be found.

Differential equations arise when we know a relation involving the rate of change of a function. For example, if a quantity N at time t is such that it changes at a rate proportional to the amount present, then

$$\frac{dN}{dt} = kN, \qquad \text{where } k \text{ is a constant.}$$

The solution of this differential equation is

$$N = N_0 e^{kt},$$

where N_0 is the quantity present at $t = 0$. The value of k may be determined when the value of N is known for a given value of t (other than $t = 0$). If k is positive, N follows an exponential law of growth; if k is negative, N follows an exponential law of decay. When N represents a quantity of a radioactive element, then

$$\frac{dN}{dt} = -\lambda N, \qquad \text{where } \lambda \text{ is a positive constant.}$$

Thus N follows an exponential law of decay, and hence

$$N = N_0 e^{-\lambda t}.$$

The constant λ is called the decay constant. The time for one half of the element to decay is the half-life of the element and

$$\text{half-life} = \frac{\ln 2}{\lambda} \approx \frac{0.69315}{\lambda}.$$

At times a quantity N will follow a rate of growth given by

$$\frac{dN}{dt} = KN(M - N), \qquad \text{where } K, M \text{ are constants.}$$

Solving this differential equation gives a function of the form

$$N = \frac{M}{1 + be^{-ct}}, \qquad \text{where } b, c \text{ are constants,}$$

which is called a logistic function. Many population sizes can be described by logistic functions. In this case, M

represents the limit of the size of a population. A logistic function is also used in analyzing the spread of a rumor.

Newton's law of cooling states that the temperature T of a cooling body at time t changes at a rate proportional to the difference $T - a$, where a is the ambient temperature. Thus

$$\frac{dT}{dt} = k(T - a), \qquad \text{where } k, a \text{ are constants.}$$

The solution of this differential equation can be used to determine, for example, the time at which a homicide was committed.

Review Problems

In Problems 1–18, determine the integrals.

1. $\int x \ln x \, dx.$

2. $\int \frac{1}{\sqrt{4x^2 + 1}} \, dx.$

3. $\int_0^2 \sqrt{4x^2 + 9} \, dx.$

4. $\int \frac{2x}{3 - 4x} \, dx.$

5. $\int \frac{x \, dx}{(2 + 3x)(3 + x)}.$

6. $\int_e^{e^2} \frac{1}{x \ln x} \, dx.$

7. $\int \frac{dx}{x(x + 2)^2}.$

8. $\int \frac{dx}{x^2 - 1}.$

9. $\int \frac{dx}{x^2\sqrt{9 - 16x^2}}.$

10. $\int x^2 \ln(4x) \, dx.$

11. $\int \frac{9 \, dx}{x^2 - 9}.$

12. $\int \frac{3x}{\sqrt{1 + 3x}} \, dx.$

13. $\int xe^{7x} \, dx.$

14. $\int \frac{dx}{2 + 3e^{4x}}.$

15. $\int \frac{dx}{2x \ln 2x}.$

16. $\int \frac{dx}{x(2 + x)}.$

17. $\int \frac{2x}{3 + 2x} \, dx.$

18. $\int \frac{dx}{\sqrt{4x^2 - 9}}.$

19. Find the average value of $f(x) = 3x^2 + 2x$ over the interval $[2, 4]$.

20. Find the average value of $f(t) = te^{t^2}$ over the interval $[2, 5]$.

*In Problems **21** and **22**, use (a) the trapezoidal rule and (b) Simpson's rule to estimate the integral. Use the given value of n. Give your answer to three decimal places.*

21. $\int_0^3 \frac{1}{x + 1} \, dx, \quad n = 6.$

22. $\int_0^1 \frac{1}{2 - x^2} \, dx, \quad n = 4.$

*In Problems **23–26**, determine the improper integrals, if they exist. Indicate those that are divergent.*

23. $\int_3^\infty \frac{1}{x^3} \, dx.$

24. $\int_{-\infty}^0 e^{3x} \, dx.$

25. $\int_1^\infty \frac{1}{2x} \, dx.$

26. $\int_{-\infty}^\infty xe^{1 - x^2} \, dx.$

*In Problems **27** and **28**, solve the differential equations.*

27. $y' = 3x^2y + 2xy, \quad y > 0.$

28. $y' - 2xe^{x^2 - y + 3} = 0, \quad y(0) = 3.$

29. The population of a city in 1965 was 100,000 and in 1980 it was 120,000. Assuming exponential growth, project the population in 1995.

30. The population of a city doubles every 10 years due to exponential growth. At a certain time the population is 10,000. Find an expression for the number of people N at time t years later. Assume that $\ln 2 = 0.69$.

31. If 95 percent of a radioactive substance remains after 100 years, find the decay constant and, to the nearest percentage, give the percentage of the original amount present after 200 years.

32. For a group of hospitalized individuals, suppose that

$$\int_0^t f(x) \, dx, \text{ where}$$

$$f(x) = 0.008e^{-0.01x} + 0.00004e^{-0.0002x},$$

gives the proportion that has been discharged at the end of t days. Evaluate $\int_0^\infty f(x) \, dx$.

33. Two organisms are initially placed in a medium and begin to multiply. The number N of organisms that are present after t days is recorded on a graph with the horizontal axis labeled t and the vertical axis labeled N. It is observed that the points lie on a logistic curve. The number of organisms present after 6 days is 300, and beyond 10 days the number approaches a limit of 450. Find the logistic equation.

34. A college believes that enrollment follows logistic growth. Last year, enrollment was 1000, and this year it is 1100. If the college can accommodate a maximum of 2000 students, what is the anticipated enrollment next year? Give your answer to the nearest hundred.

35. A coroner is called in on a murder case. He arrives at 6:00 P.M. and finds that the victim's temperature is 35°C. One hour later the body temperature is 34°C.

The temperature of the room is 25°C. About what time was the murder committed? (Assume that normal body temperature is 37°C.)

36. Find the present value, to the nearest dollar, of a continuous annuity at an annual rate of 5 percent for 10 years if the payment at time t is at the annual rate of $f(t) = 40t$ dollars.

Multivariable Calculus

8.1 FUNCTIONS OF SEVERAL VARIABLES _____

Suppose that a manufacturer produces two products, X and Y. Then the total cost depends on the levels of production of *both* X and Y. Table 8.1 is a

TABLE 8.1

UNITS OF X PRODUCED, x	UNITS OF Y PRODUCED, y	TOTAL COST OF PRODUCTION, c
5	6	17
5	7	19
6	6	18
6	7	20

schedule that indicates total cost at various levels. For example, when 5 units of X and 6 units of Y are produced, the total cost c is 17. In this situation, it seems natural to associate the number 17 with the *ordered pair* (5, 6):

$$(5, 6) \rightarrow 17.$$

The first element of the ordered pair, 5, represents the number of units of X produced, while the second element, 6, represents the number of units of Y produced. Corresponding to the other production situations, we have

$$(5, 7) \rightarrow 19,$$
$$(6, 6) \rightarrow 18,$$
and $$(6, 7) \rightarrow 20.$$

This correspondence can be considered an input-output relation where the inputs are ordered pairs. With each input we associate exactly one output. Thus the correspondence defines a function f where

the domain consists of (5, 6), (5, 7), (6, 6), (6, 7),

and   the range consists of 17, 19, 18, 20.

In function notation,

$$f(5, 6) = 17, \qquad f(5, 7) = 19,$$
$$f(6, 6) = 18, \qquad f(6, 7) = 20.$$

We say that the total cost schedule can be described by $c = f(x, y)$, a function of the two independent variables x and y. The letter c is the dependent variable.

Turning to another function of two variables, we see that the equation

$$z = \frac{2}{x^2 + y^2}$$

defines z as a function of x and y:

$$z = f(x, y) = \frac{2}{x^2 + y^2}.$$

The domain of f is all ordered pairs of real numbers (x, y) for which the equation has meaning when the first and second elements of (x, y) are substituted for x and y, respectively, in the equation. Thus the domain of f is all ordered pairs except $(0, 0)$. To find $f(2, 3)$, for example, we substitute $x = 2$ and $y = 3$ into $2/(x^2 + y^2)$. Hence $f(2, 3) = 2/(2^2 + 3^2) = 2/13$.

EXAMPLE 1

a. $f(x, y) = \dfrac{x + 3}{y - 2}$ is a function of two variables. Because the denominator is zero when $y = 2$, the domain of f is all (x, y) such that $y \neq 2$. Some function values are

$$f(0, 3) = \frac{0 + 3}{3 - 2} = 3,$$

$$f(3, 0) = \frac{3 + 3}{0 - 2} = -3.$$

Note that $f(0, 3) \neq f(3, 0)$.

b. $h(x, y) = 4x$ defines h as a function of x and y. The domain is all ordered pairs of real numbers. Some function values are

$$h(2, 5) = 4(2) = 8,$$
$$h(2, 6) = 4(2) = 8.$$

Note that the function values are independent of the choice of y.

c. If $z^2 = x^2 + y^2$ and $x = 3$ and $y = 4$, then $z^2 = 3^2 + 4^2 = 25$. Consequently, $z = \pm 5$. Thus with the ordered pair $(3, 4)$ we *cannot* associate exactly one output number. Hence z is *not* a function of x and y.

EXAMPLE 2   *Under certain conditions, if two brown-eyed parents have exactly k children, the probability $P = P(r, k)$ that there will be exactly r blue-eyed children is given by*

$$P(r, k) = \frac{k! \left(\frac{1}{4}\right)^r \left(\frac{3}{4}\right)^{k-r}}{r!\,(k-r)!}, \qquad r = 0, 1, 2, \ldots, k.$$

Find the probability that out of a total of four children exactly three will be blue-eyed.

We want to find $P(3, 4)$:

$$P(3, 4) = \frac{4! \left(\frac{1}{4}\right)^3 \left(\frac{3}{4}\right)^1}{3!\,(1)!} = \frac{3}{64}.$$

If $y = f(x)$ is a function of one variable, the domain of f can be geometrically represented by points on the real number line. The function itself can be represented by its graph in a coordinate plane, sometimes called a two-dimensional coordinate system. However, for a function of two variables, $z = f(x, y)$, its domain (consisting of ordered pairs of real numbers) can be geometrically represented by a *region* in the plane. The function itself can be geometrically represented in a **three-dimensional coordinate system**. Such a system is formed when three mutually perpendicular real number lines in space intersect at the origin of each line as in Fig. 8.1. The three number lines are called the x-, y-, and z-axes, and their point of intersection is called the origin of the system.

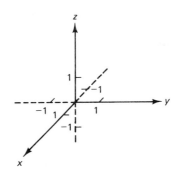

FIGURE 8.1

To each point P in space we can assign a unique ordered triple of numbers, called the *coordinates* of P. To do this [see Fig. 8.2(a)], from P we construct a line perpendicular to the x,y-plane, that is, the plane determined

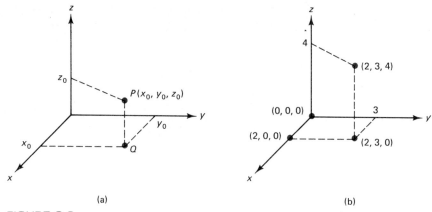

(a)   (b)

FIGURE 8.2

by the x- and y-axes. Let Q be the point where the line intersects this plane. From Q we construct perpendiculars to the x- and y-axes. These lines intersect the x- and y-axes at x_0 and y_0, respectively. From P a perpendicular to the z-axis is constructed which intersects it at z_0. Thus to P we assign the ordered triple (x_0, y_0, z_0). It should also be evident that with each ordered triple of numbers we can assign a unique point in space. Due to this one-to-one correspondence between points in space and ordered triples, an ordered triple may be called a point. In Fig. 8.2(b) points $(2, 0, 0)$, $(2, 3, 0)$, and $(2, 3, 4)$ are shown. Note that the origin corresponds to $(0, 0, 0)$.

There is a way to represent geometrically a function of two variables, $z = f(x, y)$. To each ordered pair (x, y) in the domain of f we assign the point $(x, y, f(x, y))$. The set of all such points is called the *graph* of f. Such a graph appears in Fig. 8.3. You can consider $z = f(x, y)$ as representing a surface in space.*

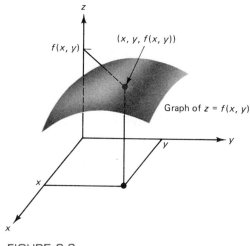

FIGURE 8.3

* We shall freely use the term "surface" in the intuitive sense.

In Chapter 1, continuity of a function of one variable was discussed. If $y = f(x)$ is continuous, then the graph of f has no break in it. Extending this concept, we say that a function of two variables is continuous if its graph is an "unbroken surface."

Until now we have considered only functions of either one or two variables. In general, a **function of n variables** is one whose domain consists of ordered n-tuples $(x_1, x_2, \ldots, x_n)$. For example, $f(x, y, z) = 2x + 3y + 4z$ is a function of three variables with a domain consisting of all ordered triples. The function $g(x_1, x_2, x_3, x_4) = x_1 x_2 x_3 x_4$ is a function of four variables with a domain consisting of all ordered 4-tuples. Although functions of several variables are extremely important and useful, we cannot geometrically represent functions of more than two variables.

We now give a brief discussion of sketching surfaces in space. We begin with planes that are parallel to a coordinate plane. By a "coordinate plane" we mean a plane containing two coordinate axes. For example, the plane determined by the x- and y-axes is the **x,y-plane**. Similarly, we speak of the **x,z-plane** and the **y,z-plane**. The coordinate planes divide space into eight parts, called *octants*. In particular, the part containing all points (x, y, z) such that $x > 0$, $y > 0$, and $z > 0$ is called the **first octant**.

Suppose S is a plane that is parallel to the x,y-plane and passes through the point $(0, 0, 5)$ [see Fig. 8.4(a)]. Then the point (x, y, z) will lie on S if and only if $z = 5$; that is, x and y can be any real numbers, but z must equal

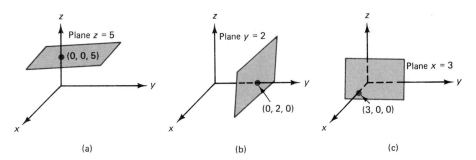

(a)   (b)   (c)

FIGURE 8.4

5. For this reason we say that $z = 5$ is an equation of S. Similarly, an equation of the plane parallel to the x,z-plane and passing through the point $(0, 2, 0)$ is $y = 2$ [Fig. 8.4(b)]. The equation $x = 3$ is an equation of the plane passing through $(3, 0, 0)$ and parallel to the y,z-plane [Fig. 8.4(c)]. Now let us look at planes in general.

In space the graph of an equation of the form

$$Ax + By + Cz + D = 0,$$

where D is a constant and A, B and C are constants that are not all zero, is a plane. Since three distinct points (not lying on the same line) determine a plane, a convenient way to sketch a plane is to first determine the points, if any, where the plane intersects the x-, y-, or z-axes. These points are called *intercepts*.

EXAMPLE 3   *Sketch the plane $2x + 3y + z = 6$.*

The plane intersects the x-axis when $y = 0$ and $z = 0$. Thus $2x = 6$, which gives $x = 3$. Similarly, if $x = z = 0$, then $y = 2$; if $x = y = 0$, then $z = 6$. Thus the intercepts are $(3, 0, 0)$, $(0, 2, 0)$, and $(0, 0, 6)$. The portion of the plane in the first octant is shown in Fig. 8.5(a); however, you should realize that the plane extends indefinitely in space.

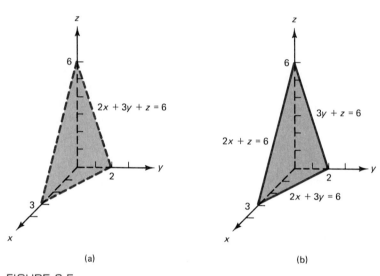

(a)   (b)

FIGURE 8.5

A surface can be sketched with the aid of its **traces**. These are the intersections of the surface with the coordinate planes. For the plane $2x + 3y + z = 6$ in Example 3, the trace in the x,y-plane is obtained by setting $z = 0$. This gives $2x + 3y = 6$, which is an equation of a *line* in the x,y-plane. Similarly, setting $x = 0$ gives the trace in the y,z-plane: the line $3y + z = 6$. The x,z-trace is the line $2x + z = 6$ [see Fig. 8.5(b)].

EXAMPLE 4   *Sketch the surface $2x + z = 4$.*

This equation has the form of a plane. The x- and z-intercepts are $(2, 0, 0)$ and $(0, 0, 4)$, and there is no y-intercept since x and z cannot both be zero. Setting $y = 0$ gives the x,z-trace $2x + z = 4$, which is a line in the x,z-plane. In fact, the intersection of the surface with *any* plane $y = k$ is also $2x + z = 4$. Hence the plane appears as in Fig. 8.6.

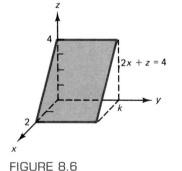

FIGURE 8.6

Our final examples deal with surfaces that are not planes but whose graphs can be easily obtained.

EXAMPLE 5   *Sketch the surface $z = x^2$.*

The x,z-trace is the curve $z = x^2$, which is a parabola. In fact, for *any* fixed value of y we get $z = x^2$. Thus the graph appears as in Fig. 8.7.

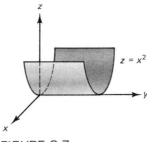

FIGURE 8.7

EXAMPLE 6   *Sketch the surface $x^2 + y^2 + z^2 = 25$.*

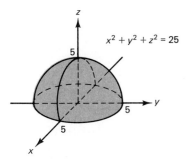

FIGURE 8.8

Setting $z = 0$ gives the x,y-trace $x^2 + y^2 = 25$, which is a circle of radius 5. Similarly, the y,z- and x,z-traces are the circles $y^2 + z^2 = 25$ and $x^2 + z^2 = 25$, respectively. Note also that since $x^2 + y^2 = 25 - z^2$, the intersection of the surface with the plane $z = k$, where $-5 \le k \le 5$, is a circle. For example, if $z = 3$ the intersection is the circle $x^2 + y^2 = 16$. If $z = 4$, the intersection is $x^2 + y^2 = 9$. That is, cross sections of the surface that are parallel to the x,y-plane are circles. A portion of the surface appears in Fig. 8.8. The entire surface is a sphere.

EXERCISE 8.1

In Problems 1–12, determine the indicated function values for the given functions.

1. $f(x, y) = 3x + y - 1$; $f(0, 4)$.

2. $f(x, y) = xy^2 + 2$; $f(1, -4)$.

3. $g(x, y, z) = ze^{x+y}$; $g(-2, 2, 6)$.

4. $g(x, y, z) = xy + xz + yz$; $g(1, 2, -3)$.

5. $h(r, s, t, u) = \dfrac{r + s^2}{t - u}$; $h(-3, 3, 5, 4)$.

6. $h(r, s, t, u) = \ln(ru)$; $h(1, 5, 3, 1)$.

7. $g(p_A, p_B) = 2p_A(p_A^2 - 5)$; $g(4, 8)$.

8. $g(p_A, p_B) = p_A \sqrt{p_B} + 10$; $g(8, 4)$.

9. $F(x, y, z) = 3$; $F(2, 0, -1)$.

10. $F(x, y, z) = \dfrac{x}{yz}$; $F(0, 0, 3)$.

11. $f(x, y) = 2x - 5y + 4$; $f(x_0 + h, y_0)$.

12. $f(x, y) = x^2y - 3y^3$; $f(r + t, r)$.

In Problems 13–16, find equations of the planes that satisfy the given conditions.

13. Parallel to the x,z-plane and passes through the point $(0, -4, 0)$.

14. Parallel to the y,z-plane and passes through the point $(8, 0, 0)$.

15. Parallel to the x,y-plane and passes through the point $(2, 7, 6)$.

16. Parallel to the y,z-plane and passes through the point $(-4, -2, 7)$.

In Problems 17–26, sketch the given surfaces.

17. $x + y + z = 1$.

18. $2x + y + 2z = 6$.

19. $3x + 6y + 2z = 12$.

20. $x + 2y + 3z = 4$.

21. $x + 2y = 2$.

22. $y + z = 1$.

23. $z = 4 - x^2$.

24. $y = x^2$.

25. $x^2 + y^2 + z^2 = 1$.

26. $x^2 + y^2 = 1$.

27. A method of ecological sampling to determine animal populations in a given area involves first marking all the animals obtained in a sample of R animals from the area and then releasing them so that they can mix with unmarked animals. At a later date a second sample is taken of M animals and the number of these which are marked, S, is noted. Based on R, M, and S an estimate of the total population of animals, N, in the sample area is given by

$$N = f(R, M, S) = \frac{RM}{S}.$$

Find $f(400, 400, 80)$. This method is called the *mark and recapture procedure* **[56]**.

28. On hot and humid days, many people tend to feel uncomfortable. The degree of discomfort is numerically given by the temperature-humidity index, THI, which is a function of two variables:

$$\text{THI} = f(t_d, t_w) = 15 + 0.4(t_d + t_w),$$

where t_d is the dry-bulb temperature (in degrees Fahrenheit) and t_w is the wet-bulb temperature (in degrees Fahrenheit) of the air. When the THI is greater than 75, most people are uncomfortable. In fact, the THI was once called the "discomfort index." Many electric utilities closely follow this index so that they can anticipate the demand of air conditioning on their systems. Evaluate the THI when $t_d = 90$ and $t_w = 80$.

29. In Example 2, find a function which gives the probability that out of a total of five children, exactly r will be blue-eyed.

8.2 PARTIAL DERIVATIVES

Figure 8.9 shows the surface $z = f(x, y)$ and a plane that is parallel to the x,z-plane and that passes through the point $(x_0, y_0, f(x_0, y_0))$ on the surface. An equation of this plane is $y = y_0$. Hence any point on the curve that is the

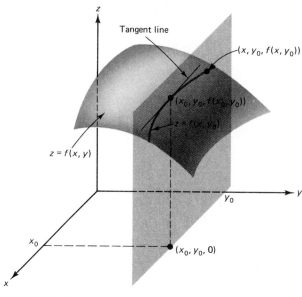

FIGURE 8.9

intersection of the surface with the plane must have the form $(x, y_0, f(x, y_0))$. Thus the curve can be described by $z = f(x, y_0)$. Since y_0 is constant, $z = f(x, y_0)$ can be considered a function of one variable, x. When the derivative of this function is evaluated at x_0, it gives the slope of the tangent line to this curve at $(x_0, y_0, f(x_0, y_0))$ (see Fig. 8.9). This slope is called the *partial derivative of f with respect to x* at (x_0, y_0) and is denoted $f_x(x_0, y_0)$. In terms of limits,

$$f_x(x_0, y_0) = \lim_{h \to 0} \frac{f(x_0 + h, y_0) - f(x_0, y_0)}{h} \tag{1}$$

On the other hand, in Fig. 8.10 the plane $x = x_0$ is parallel to the y, z-plane and cuts the surface $z = f(x, y)$ in a curve given by $z = f(x_0, y)$, a

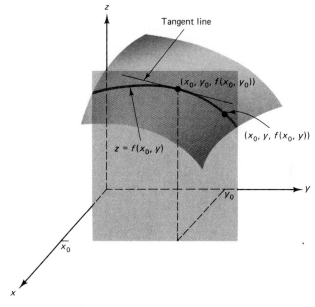

FIGURE 8.10

function of y. When the derivative of this function is evaluated at y_0, it gives the slope of the tangent line to this curve at the point $(x_0, y_0, f(x_0, y_0))$. This slope is called the *partial derivative of f with respect to y* at (x_0, y_0) and is denoted $f_y(x_0, y_0)$. In terms of limits,

$$f_y(x_0, y_0) = \lim_{h \to 0} \frac{f(x_0, y_0 + h) - f(x_0, y_0)}{h}. \tag{2}$$

Sometimes $f_x(x_0, y_0)$ is said to be the slope of the tangent line to the graph of f at $(x_0, y_0, f(x_0, y_0))$ *in the x-direction;* similarly $f_y(x_0, y_0)$ is the slope of the tangent line *in the y-direction.*

For generality, by replacing x_0 and y_0 in Eqs. (1) and (2) by x and y, respectively, we get the following definition.

Definition
*If $z = f(x, y)$, the **partial derivative of f with respect to x**, denoted f_x, is the function given by*

$$f_x(x, y) = \lim_{h \to 0} \frac{f(x + h, y) - f(x, y)}{h},$$

provided that this limit exists.
   *The **partial derivative of f with respect to y**, denoted f_y, is the function given by*

$$f_y(x, y) = \lim_{h \to 0} \frac{f(x, y + h) - f(x, y)}{h},$$

provided that this limit exists.

From the definition, we see that to find f_x we treat y as a constant and differentiate f with respect to x in the usual way. Similarly, to find f_y we treat x as a constant and differentiate f with respect to y.

EXAMPLE 1   *If $f(x, y) = xy^2 + x^2y$, find $f_x(x, y)$ and $f_y(x, y)$. Also find $f_x(3, 4)$ and $f_y(3, 4)$.*

To find $f_x(x, y)$, we treat y as a constant and differentiate f with respect to x:

$$f_x(x, y) = (1)y^2 + (2x)y = y^2 + 2xy.$$

To find $f_y(x, y)$, we treat x as a constant and differentiate with respect to y.

$$f_y(x, y) = x(2y) + x^2(1) = 2xy + x^2.$$

Note that $f_x(x, y)$ and $f_y(x, y)$ are each functions of the two variables x and y. To find $f_x(3, 4)$ we evaluate $f_x(x, y)$ when $x = 3$ and $y = 4$.

$$f_x(3, 4) = 4^2 + 2(3)(4) = 40.$$

Similarly,

$$f_y(3, 4) = 2(3)(4) + 3^2 = 33.$$

Notations for partial derivatives of $z = f(x, y)$ are in Table 8.2. Table 8.3 gives notations for partial derivatives evaluated at (x_0, y_0).

TABLE 8.2

PARTIAL DERIVATIVE OF f (OR z) WITH RESPECT TO x	PARTIAL DERIVATIVE OF f (OR z) WITH RESPECT TO y
$f_x(x, y)$	$f_y(x, y)$
$\dfrac{\partial}{\partial x}[f(x, y)]$	$\dfrac{\partial}{\partial y}[f(x, y)]$
$\dfrac{\partial z}{\partial x}$	$\dfrac{\partial z}{\partial y}$

TABLE 8.3

PARTIAL DERIVATIVE OF f (OR z) WITH RESPECT TO x EVALUATED AT (x_0, y_0)	PARTIAL DERIVATIVE OF f (OR z) WITH RESPECT TO y EVALUATED AT (x_0, y_0)
$f_x(x_0, y_0)$	$f_y(x_0, y_0)$
$\dfrac{\partial z}{\partial x}\bigg|_{(x_0, y_0)}$	$\dfrac{\partial z}{\partial y}\bigg|_{(x_0, y_0)}$
$\dfrac{\partial z}{\partial x}\bigg|_{\substack{x=x_0 \\ y=y_0}}$	$\dfrac{\partial z}{\partial y}\bigg|_{\substack{x=x_0 \\ y=y_0}}$

EXAMPLE 2

a. *If* $z = 3x^3y^3 - 9x^2y + xy^2 + 4y$, *find* $\dfrac{\partial z}{\partial x}, \dfrac{\partial z}{\partial x}\bigg|_{(1,0)}, \dfrac{\partial z}{\partial y}$, *and* $\dfrac{\partial z}{\partial y}\bigg|_{(1,0)}$.

To find $\partial z/\partial x$ we differentiate z with respect to x while treating y as a constant.

$$\frac{\partial z}{\partial x} = 3(3x^2)y^3 - 9(2x)y + (1)y^2 + 0$$
$$= 9x^2y^3 - 18xy + y^2.$$

Evaluating at $(1,0)$, we obtain

$$\frac{\partial z}{\partial x}\bigg|_{(1,0)} = 9(1)^2(0)^3 - 18(1)(0) + 0^2 = 0.$$

To find $\partial z/\partial y$ we differentiate z with respect to y while treating x as a constant.

$$\frac{\partial z}{\partial y} = 3x^3(3y^2) - 9x^2(1) + x(2y) + 4(1)$$
$$= 9x^3y^2 - 9x^2 + 2xy + 4.$$

Thus

$$\frac{\partial z}{\partial y}\bigg|_{(1,0)} = 9(1)^3(0)^2 - 9(1)^2 + 2(1)(0) + 4 = -5.$$

b. *If* $w = x^2e^{2x+3y}$, *find* $\partial w/\partial x$ *and* $\partial w/\partial y$.

To find $\partial w/\partial x$, we treat y as a constant and differentiate with respect to x. Since x^2e^{2x+3y} is a product of two functions, each involving x, we use the product rule.

$$\frac{\partial w}{\partial x} = x^2 \frac{\partial}{\partial x}(e^{2x+3y}) + e^{2x+3y}\frac{\partial}{\partial x}(x^2)$$
$$= x^2(2e^{2x+3y}) + e^{2x+3y}(2x)$$
$$= 2x(x + 1)e^{2x+3y}.$$

To find $\partial w/\partial y$, we treat x as a constant and differentiate with respect to y.

$$\frac{\partial w}{\partial y} = x^2 \frac{\partial}{\partial y}(e^{2x+3y}) = 3x^2 e^{2x+3y}.$$

We have seen that for a function of two variables, two partial derivatives can be considered. Actually the concept of partial derivatives can be extended to functions of more than two variables. For example, with $w = f(x, y, z)$ we have three partial derivatives:

the partial with respect to x, denoted $f_x(x, y, z)$, $\partial w/\partial x$, etc.;

the partial with respect to y, denoted $f_y(x, y, z)$, $\partial w/\partial y$, etc.;

and   the partial with respect to z, denoted $f_z(x, y, z)$, $\partial w/\partial z$, etc.

To determine $\partial w/\partial x$, treat y and z as constants and differentiate w with respect to x. For $\partial w/\partial y$, treat x and z as constants and differentiate with respect to y. For $\partial w/\partial z$, treat x and y as constants and differentiate with respect to z. With a function of n variables we have n partial derivatives, which are determined in the obvious way.

EXAMPLE 3

a. *If $f(x, y, z) = x^2 + y^2 z + z^3$, find $f_x(x, y, z)$, $f_y(x, y, z)$, and $f_z(x, y, z)$.*

To find $f_x(x, y, z)$, we treat y and z as constants and differentiate f with respect to x.

$$f_x(x, y, z) = 2x.$$

Treating x and z as constants and differentiating with respect to y, we have

$$f_y(x, y, z) = 2yz.$$

Treating x and y as constants and differentiating with respect to z, we have

$$f_z(x, y, z) = y^2 + 3z^2.$$

b. *If $p = g(r, s, t, u) = \dfrac{rsu}{rt^2 + s^2 t}$, find $\dfrac{\partial p}{\partial s}$, $\dfrac{\partial p}{\partial t}$, and $\dfrac{\partial p}{\partial t}\bigg|_{(0,1,1,1)}$*

To find $\partial p/\partial s$, first note that p is a quotient of two functions, each involving the variable s. Thus we use the quotient rule and treat r, t, and u as constants.

$$\frac{\partial p}{\partial s} = \frac{(rt^2 + s^2 t)\dfrac{\partial}{\partial s}(rsu) - rsu\dfrac{\partial}{\partial s}(rt^2 + s^2 t)}{(rt^2 + s^2 t)^2}$$

$$= \frac{(rt^2 + s^2 t)(ru) - (rsu)(2st)}{(rt^2 + s^2 t)^2}.$$

Simplifying gives

$$\frac{\partial p}{\partial s} = \frac{ru(rt - s^2)}{t(rt + s^2)^2}.$$

To find $\partial p/\partial t$ we can first write p as

$$p = rsu(rt^2 + s^2 t)^{-1}.$$

Next we use the power rule and treat r, s, and u as constants.

$$\frac{\partial p}{\partial t} = rsu(-1)(rt^2 + s^2 t)^{-2}\frac{\partial}{\partial t}(rt^2 + s^2 t)$$

$$= -rsu(rt^2 + s^2 t)^{-2}(2rt + s^2).$$

$$\frac{\partial p}{\partial t} = -\frac{rsu(2rt + s^2)}{(rt^2 + s^2 t)^2}.$$

Letting $r = 0$, $s = 1$, $t = 1$ and $u = 1$ gives

$$\left.\frac{\partial p}{\partial t}\right|_{(0,1,1,1)} = -\frac{0(1)(1)[2(0)(1) + (1)^2]}{[0(1)^2 + (1)^2(1)]^2} = 0.$$

EXERCISE 8.2

In each of Problems 1–26, a function of two or more variables is given. Find the partial derivative of the function with respect to each of the variables.

1. $f(x, y) = x - 5y + 3$.

2. $f(x, y) = 4 - 5x^2 + 6y^3$.

3. $f(x, y) = 3x - 4$.

4. $f(x, y) = \sqrt{7}$.

5. $g(x, y) = x^5 y^4 - 3x^4 y^3 + 7x^3 + 2y^2 - 3xy + 4$.

6. $g(x, y) = x^8 - 2x^6 y^5 + 3x^5 y^3 + x^3 y^3 + 3x - 4$.

7. $g(p, q) = \sqrt{pq}$.

8. $g(w, z) = \sqrt[3]{w^2 + z^2}$.

9. $h(s, t) = \dfrac{s^2 + 4}{t - 3}$.

10. $h(u, v) = \dfrac{4uv^2}{u^2 + v^2}$.

11. $u(q_1, q_2) = \frac{3}{4} \ln q_1 + \frac{1}{4} \ln q_2$.

12. $Q(l, k) = 3l^{0.41} k^{0.59}$.

13. $h(x, y) = \dfrac{x^2 + 3xy + y^2}{\sqrt{x^2 + y^2}}$.

14. $h(x, y) = \dfrac{\sqrt{x + 4}}{x^2 y + y^2 x}$.

15. $z = e^{5xy}$.

16. $z = (x^2 + y)e^{3x + 4y}$.

17. $z = 5x \ln(x^2 + y)$.

18. $z = \ln(3x^2 + 4y^4)$.

19. $f(r, s) = \sqrt{r + 2s}\,(r^3 - 2rs + s^2)$.

20. $f(r, s) = \sqrt{rs}\, e^{2 + r}$.

21. $f(r, s) = e^{3 - r} \ln(7 - s)$.

22. $f(r, s) = (5r^2 + 3s^3)(2r - 5s)$.

23. $g(x, y, z) = 3x^2 y + 2xy^2 z + 3z^3$.

24. $g(x, y, z) = x^2 y^3 z^5 - 3x^2 y^4 z^3 + 5xz$.

25. $g(r, s, t) = e^{s + t}(r^2 + 7s^3)$.

26. $g(r, s, t, u) = rs \ln(2t + 5u)$.

In Problems 27–32, evaluate the given partial derivatives.

27. $f(x, y) = x^3 y + 7x^2 y^2$; $f_x(1, -2)$.

28. $z = \sqrt{5x^2 + 3xy + 2y}$; $\left.\dfrac{\partial z}{\partial x}\right|_{\substack{x=0 \\ y=2}}$.

29. $g(x, y, z) = e^x\sqrt{y + 2z}$; $g_z(0, 1, 4)$.

30. $g(x, y, z) = \dfrac{3x^2 + 2y}{xy + xz}$; $g_y(1, 1, 1)$.

31. $h(r, s, t, u) = (s^2 + tu) \ln(2r + 7st)$; $h_s(1, 0, 0, 1)$.

32. $h(r, s, t, u) = \dfrac{7r + 3s^2u^2}{s}$; $h_t(4, 3, 2, 1)$.

33. In a discussion of inventory theory of money demand, Swanson[57] considers the function

$$F(b, C, T, i) = \frac{bT}{C} + \frac{iC}{2}$$

and determines that $\dfrac{\partial F}{\partial C} = -\dfrac{bT}{C^2} + \dfrac{i}{2}$. Verify this partial derivative.

34. In a discussion of stock prices of a dividend cycle, Palmon and Yaari [58] consider the function f given by

$$u = f(t, r, z) = \frac{(1 + r)^{1-z} \ln(1 + r)}{(1 + r)^{1-z} - t},$$

where u is the instantaneous rate of ask-price appreciation, r is an annual opportunity rate of return, z is the fraction of a dividend cycle over which a share of stock is held by a midcycle seller, and t is the effective rate of capital gains tax. They claim that

$$\frac{\partial u}{\partial z} = \frac{t(1 + r)^{1-z} \ln^2(1 + r)}{[(1 + r)^{1-z} - t]^2}.$$

Verify this.

35. In an analysis of advertising and profitability, Swales

[59] considers a function f given by

$$R = f(r, a, n) = \frac{r}{1 + a\dfrac{n-1}{2}},$$

where R is adjusted rate of profit, r is accounting rate of profit, a is a measure of advertising expenditures, and n is the number of years that advertising fully depreciates. In the analysis Swales determines $\partial R/\partial n$. Find this partial derivative.

36. In an article on interest rate deregulation, Christofi and Agapos [11] arrive at the equation

$$r_L = r + D\frac{\partial r}{\partial D} + \frac{dC}{dD}, \tag{3}$$

where r is the deposit rate paid by commercial banks, r_L is the rate earned by commercial banks, C is the administrative cost of transforming deposits into return-earning assets, and D is the savings deposits level. Christofi and Agapos state that

$$r_L = r\left[\frac{1 + \eta}{\eta}\right] + \frac{dC}{dD}, \tag{4}$$

where η is the deposit elasticity with respect to the deposit rate and is given by $\eta = \dfrac{r/D}{\partial r/\partial D}$. Express Eq. (3) in terms of η to verify Eq. (4).

8.3 APPLICATIONS OF PARTIAL DERIVATIVES

Suppose a manufacturer produces x units of product X and y units of product Y. Then the total cost c of these units is a function of x and y and is called a **joint-cost function**. If such a function is $c = f(x, y)$, then $\partial c/\partial x$ is called the **(partial) marginal cost with respect to x**. It is the rate of change of c with respect to x when y is held fixed. Similarly, $\partial c/\partial y$ is the **(partial) marginal cost with respect to y**. It is the rate of change of c with respect to y when x is held fixed.

For example, if c is expressed in dollars and $\partial c/\partial y = 2$, then the cost of producing an extra unit of Y when the level of production of X is fixed is approximately two dollars.

If a manufacturer produces n products, the joint-cost function is a function of n variables and there are n (partial) marginal cost functions.

EXAMPLE 1   *A company manufactures two types of skis, the Lightning and the Alpine models. Suppose the joint-cost function for producing x pairs* of the Lightning model and y pairs of the Alpine model per week is

$$c = f(x, y) = 0.06x^2 + 7x + 15y + 1000,$$

where c is expressed in dollars. Determine the marginal costs $\partial c/\partial x$ and $\partial c/\partial y$ when $x = 100$ and $y = 50$ and interpret the results.

The marginal costs are

$$\frac{\partial c}{\partial x} = 0.12x + 7 \quad \text{and} \quad \frac{\partial c}{\partial y} = 15.$$

Thus

$$\left.\frac{\partial c}{\partial x}\right|_{(100,50)} = 0.12(100) + 7 = 19 \tag{1}$$

and

$$\left.\frac{\partial c}{\partial y}\right|_{(100,50)} = 15. \tag{2}$$

Equation (1) means that increasing the output of the Lightning model from 100 to 101, while maintaining production of the Alpine model at 50, increases costs by approximately $19. Equation (2) means that increasing output of the Alpine model from 50 to 51 and holding production of the Lightning model at 100 will increase costs by approximately $15. In fact, since $\partial c/\partial y$ is a constant function, the marginal cost with respect to y is $15 at all levels of production.

EXAMPLE 2   *On a cold day a person may feel colder when the wind is blowing than when the wind is calm because the rate of heat loss is a function of both temperature and wind speed. The equation*

$$H = (10.45 + \sqrt{100w} - w)(33 - t)$$

indicates the rate of heat loss H (in kilocalories per square meter per hour) when the air temperature is t (in degrees Celsius) and the wind speed is w (in meters per second). For H = 2000, exposed flesh will freeze in one minute [7].
a. *Evaluate H when t = 0 and w = 4.*
b. *Evaluate $\partial H/\partial w$ and $\partial H/\partial t$ when t = 0 and w = 4, and interpret the results.*
c. *When t = 0 and w = 4, which has a greater effect on H: a change in wind speed of 1 m/sec or a change in temperature of 1°C?*

a. Note that $\sqrt{100w} = 10\sqrt{w}$ and thus H can be written

$$H = (10.45 + 10\sqrt{w} - w)(33 - t).$$

When $t = 0$ and $w = 4$, then

$$H = (10.45 + 10\sqrt{4} - 4)(33 - 0) = 872.85.$$

b. $\dfrac{\partial H}{\partial w} = \left(\dfrac{5}{\sqrt{w}} - 1\right)(33 - t),$ $\dfrac{\partial H}{\partial w}\bigg|_{\substack{t=0 \\ w=4}} = 49.5;$

$\dfrac{\partial H}{\partial t} = (10.45 + 10\sqrt{w} - w)(-1),$ $\dfrac{\partial H}{\partial t}\bigg|_{\substack{t=0 \\ w=4}} = -26.45.$

This means that when $t = 0$ and $w = 4$, then increasing w by a small amount while keeping t fixed will make H increase approximately 49.5 times as much as w increases. Increasing t by a small amount while keeping w fixed will make H *decrease* approximately 26.45 times as much as t increases.

c. Since the partial derivative of H with respect to w is greater in magnitude than the partial with respect to t when $t = 0$ and $w = 4$, a change in wind speed of 1 m/sec has a greater effect on H.

Output of a product depends on many factors of production. Among these may be labor, capital, land, machinery, etc. For simplicity, let us suppose that output depends only on labor and capital. If the function $P = f(l, k)$ gives the output P when the producer uses l units of labor and k units of capital, then this function is called a **production function**. We define the **marginal productivity with respect to l** to be $\partial P/\partial l$. This is the rate of change of P with respect to l when k is held fixed. Likewise, the **marginal productivity with respect to k** is $\partial P/\partial k$. It is the rate of change of P with respect to k when l is held fixed.

EXAMPLE 3   *A manufacturer of a popular toy has determined that the production function is $P = \sqrt{lk}$, where l is the number of labor-hours per week and k is the capital (expressed in hundreds of dollars per week) required for a weekly production of P gross of the toy. (One gross is 144 units.) Determine the marginal productivity functions and evaluate them when l = 400 and k = 16. Interpret the results.*

Since $P = (lk)^{1/2}$,

$$\frac{\partial P}{\partial l} = \frac{1}{2}(lk)^{-1/2}k = \frac{k}{2\sqrt{lk}}$$

and $$\frac{\partial P}{\partial k} = \frac{1}{2}(lk)^{-1/2}l = \frac{l}{2\sqrt{lk}}.$$

Evaluating when $l = 400$ and $k = 16$, we obtain

$$\frac{\partial P}{\partial l}\bigg|_{\substack{l=400 \\ k=16}} = \frac{16}{2\sqrt{400(16)}} = \frac{1}{10}$$

$$\text{and} \qquad \left.\frac{\partial P}{\partial k}\right|_{\substack{l=400 \\ k=16}} = \frac{400}{2\sqrt{400(16)}} = \frac{5}{2}.$$

Thus if $l = 400$ and $k = 16$, increasing l to 401 and holding k at 16 will increase output by approximately $\frac{1}{10}$ gross. But if k is increased to 17 while l is held at 400, the output increases by approximately $\frac{5}{2}$ gross.

Sometimes two products may be related so that changes in the price of one of them can affect the demand for the other. A typical example is that of butter and margarine. If such a relationship exists between products A and B, then the demand for each product is dependent on the prices of both. Suppose q_A and q_B are the quantities demanded for A and B, respectively, and p_A and p_B are their respective prices. Then both q_A and q_B are functions of p_A and p_B:

$$q_A = f(p_A, p_B), \quad \text{demand function for A}$$

$$q_B = g(p_A, p_B), \quad \text{demand function for B.}$$

We can find four partial derivatives:

$$\frac{\partial q_A}{\partial p_A}, \quad \textit{the marginal demand for A with respect to } p_A;$$

$$\frac{\partial q_A}{\partial p_B}, \quad \textit{the marginal demand for A with respect to } p_B;$$

$$\frac{\partial q_B}{\partial p_A}, \quad \textit{the marginal demand for B with respect to } p_A;$$

$$\frac{\partial q_B}{\partial p_B}, \quad \textit{the marginal demand for B with respect to } p_B.$$

Under typical conditions, if the price of B is fixed and the price of A increases, the quantity of A demanded will decrease. Thus $\partial q_A/\partial p_A < 0$. Similarly, $\partial q_B/\partial p_B < 0$. However, $\partial q_A/\partial p_B$ and $\partial q_B/\partial p_A$ may be either positive or negative. If

$$\frac{\partial q_A}{\partial p_B} > 0 \quad \text{and} \quad \frac{\partial q_B}{\partial p_A} > 0,$$

then A and B are said to be **competitive products** or **substitutes**. In this situation, an increase in the price of B causes an increase in the demand for A, if it is assumed that the price of A does not change. Similarly, an increase in the price of A causes an increase in the demand for B when the price of B is held fixed. Butter and margarine are examples of substitutes.

Proceeding to a different situation, we say that if

$$\frac{\partial q_A}{\partial p_B} < 0 \quad \text{and} \quad \frac{\partial q_B}{\partial p_A} < 0,$$

then A and B are **complementary products**. In this case an increase in the price of B causes a decrease in the demand for A if the price of A does not change. Similarly, an increase in the price of A causes a decrease in the demand for B when the price of B is held fixed. For example, cameras and film are complementary products. An increase in the price of film will make picture-taking more expensive. Hence the demand for cameras will decrease.

EXAMPLE 4   *The demand functions for products* A *and* B *are each a function of the prices of* A *and* B *and are given by*

$$q_A = \frac{50\sqrt[3]{p_B}}{\sqrt{p_A}} \quad and \quad q_B = \frac{75p_A}{\sqrt[3]{p_B^2}},$$

respectively. Find the four marginal demand functions and also determine whether A *and* B *are competitive products, complementary products, or neither.*

Writing $q_A = 50p_A^{-1/2}p_B^{1/3}$ and $q_B = 75p_Ap_B^{-2/3}$, we have

$$\frac{\partial q_A}{\partial p_A} = 50\left(-\frac{1}{2}\right)p_A^{-3/2}p_B^{1/3} = -25p_A^{-3/2}p_B^{1/3},$$

$$\frac{\partial q_A}{\partial p_B} = 50p_A^{-1/2}\left(\frac{1}{3}\right)p_B^{-2/3} = \frac{50}{3}p_A^{-1/2}p_B^{-2/3},$$

$$\frac{\partial q_B}{\partial p_A} = 75(1)p_B^{-2/3} = 75p_B^{-2/3},$$

$$\frac{\partial q_B}{\partial p_B} = 75p_A\left(-\frac{2}{3}\right)p_B^{-5/3} = -50p_Ap_B^{-5/3}.$$

Since p_A and p_B represent prices, they are both positive. Thus $\partial q_A/\partial p_B > 0$ and $\partial q_B/\partial p_A > 0$. We conclude that A and B are competitive products.

EXERCISE 8.3

*For the joint-cost functions in Problems **1–3**, find the indicated marginal cost at the given production level.*

1. $c = 4x + 0.3y^2 + 2y + 500$; $\dfrac{\partial c}{\partial y}$, $x = 20$, $y = 30$.

2. $c = x\sqrt{x+y} + 1000$; $\dfrac{\partial c}{\partial x}$, $x = 40$, $y = 60$.

3. $c = 0.03(x+y)^3 - 0.6(x+y)^2 + 4.5(x+y) + 7700$; $\dfrac{\partial c}{\partial x}$, $x = 50$, $y = 50$.

*For the production functions in Problems **4** and **5**, find the marginal production functions $\partial P/\partial k$ and $\partial P/\partial l$.*

4. $P = 20lk - 2l^2 - 4k^2 + 800$.

5. $P = 1.582l^{0.192}k^{0.764}$.

6. A Cobb-Douglas production function is a production function of the form $P = Al^\alpha k^\beta$, where A, α, and β are constants and $\alpha + \beta = 1$. For such a function, show that

a. $\partial P/\partial l = \alpha P/l$.

b. $\partial P/\partial k = \beta P/k$.

c. $l\dfrac{\partial P}{\partial l} + k\dfrac{\partial P}{\partial k} = P$. This means that summing the products of the marginal productivity of each factor and the amount of that factor results in the total product P.

In Problems 7–9, q_A and q_B are demand functions for products A and B, respectively. In each case find $\partial q_A/\partial p_A$, $\partial q_A/\partial p_B$, $\partial q_B/\partial p_A$, $\partial q_B/\partial p_B$ and determine whether A and B are competitive, complementary, or neither.

7. $q_A = 1000 - 50p_A + 2p_B$; $q_B = 500 + 4p_A - 20p_B$.

8. $q_A = 20 - p_A - 2p_B$; $q_B = 50 - 2p_A - 3p_B$.

9. $q_A = \dfrac{100}{p_A\sqrt{p_B}}$; $q_B = \dfrac{500}{p_B\sqrt[3]{p_A}}$.

10. The production function for the Canadian manufacturing industries for 1927 is estimated in **[60]** by the equation $P = 33.0l^{0.46}k^{0.52}$, where P is product, l is labor, and k is capital. Find the marginal productivities for labor and capital and evaluate when $l = 1$ and $k = 1$.

11. An estimate of the production function for dairy farming in Iowa (1939) is given in **[61]** by

$$P = A^{0.27}B^{0.01}C^{0.01}D^{0.23}E^{0.09}F^{0.27},$$

where P is product, A is land, B is labor, C is improvements, D is liquid assets, E is working assets, and F is cash operating expenses. Find the marginal productivities for labor and improvements.

12. A person's general status, S_g, is believed to be a function of status attributable to education, S_e, and status attributable to income, S_i, where S_g, S_e, and S_i are represented numerically. If $S_g = 7\sqrt[3]{S_e}\sqrt{S_i}$, determine $\partial S_g/\partial S_e$ and $\partial S_g/\partial S_i$ when $S_e = 125$ and $S_i = 100$, and interpret your results (adapted from **[4]**).

13. In a study of success among master of business administration (MBA) graduates published in 1974 **[62]**, it was estimated that for staff managers (which includes accountants, analysts, etc.) current annual compensation z (in dollars) was given by

$$z = 10{,}990 + 1120x + 873y,$$

where x and y are the number of years of work experience before and after receiving the MBA degree, respectively. Find $\partial z/\partial x$ and interpret your result.

14. The study of frequency of vibrations of a taut wire is useful in considering such things as a person's voice. Suppose that **[28]**

$$\omega = \frac{1}{bL}\sqrt{\frac{\tau}{\pi\rho}},$$

where ω (a Greek letter read "omega") is frequency, b is diameter, L is length, ρ (a Greek letter read "rho")

is density, and τ (a Greek letter read "tau") is tension. Find $\partial\omega/\partial b$, $\partial\omega/\partial L$, $\partial\omega/\partial\rho$, and $\partial\omega/\partial\tau$.

15. Sometimes we want to evaluate the degree of readability of a piece of writing. Rudolf Flesch **[63]** developed a function of two variables that will do this:

$$R = f(w, s) = 206.835 - (1.015w + 0.846s),$$

where R is called the *reading ease score*, w is the average number of words per sentence in 100-word samples, and s is the average number of syllables in such samples. Flesch says that an article for which $R = 0$ is "practically unreadable," but one with $R = 100$ is "easy for any literate person."

a. Find $\partial R/\partial w$ and $\partial R/\partial s$.

b. Which is "easier" to read: an article for which $w = w_0$ and $s = s_0$, or one for which $w = w_0 + 1$ and $s = s_0$?

16. Consider the following traffic-flow situation. On a highway where two lanes of traffic flow in the same direction, there is a maintenance vehicle blocking the left lane (see Fig. 8.11). Two vehicles (*lead* and *following*) are

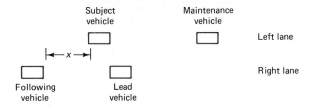

FIGURE 8.11

in the right lane with a gap between them. The *subject* vehicle can choose either to fill or not to fill the gap. That decision may be based not only on the distance x shown in the diagram, but on other factors (such as the velocity of the *following* vehicle). A *gap index*, g, has been used in analyzing such a decision **[64, 65]**. The greater the g-value, the greater is the propensity for the *subject* vehicle to fill the gap. Suppose

$$g = \frac{x}{V_F} - \left(0.75 + \frac{V_F - V_S}{19.2}\right),$$

where x (in feet) is as before, V_F is the velocity of the *following* vehicle (in feet per second), and V_S is the velocity of the *subject* vehicle (in feet per second). From the diagram it seems reasonable that if both V_F and V_S are fixed and x increases, then g should increase. Show that this is true by applying calculus to the function g given above. Assume that x, V_F, and V_S are positive.

17. For the congressional elections of 1974, the Republican percentage, R, of the Republican-Democratic vote in a district is given (approximately) in **[66]** by

$$R = f(E_r, E_d, I_r, I_d, N)$$

$$= 15.4725 + 2.5945E_r - 0.0804E_r^2 - 2.3648E_d +$$

$$0.0687E_d^2 + 2.1914I_r - 0.0912I_r^2 - 0.8096I_d +$$

$$0.0081I_d^2 - 0.0277E_rI_r + 0.0493E_dI_d +$$

$$0.8579N - 0.0061N^2.$$

Here E_r and E_d are the campaign expenditures (in units of $10,000) by Republicans and Democrats, respectively; I_r and I_d are the number of terms served in Congress, *plus* one, for the Republican and Democratic candidates,

respectively; and N is the percentage of the two-party presidential vote that Richard Nixon received in the district for 1968. The variable N gives a measure of Republican strength in the district.

a. In the Federal Election Campaign Act of 1974, Congress set a limit of $188,000 on campaign expenditures. By analyzing $\partial R/\partial E_r$, would you have advised a Republican candidate who served nine terms in Congress to spend $188,000 on his campaign?

b. Determine the percentage above which the Nixon vote had a negative effect on R; that is, determine when $\partial R/\partial N < 0$. Give your answer to the nearest percentage.

8.4 HIGHER-ORDER PARTIAL DERIVATIVES

If $z = f(x, y)$, then not only is z a function of x and y, but also f_x and f_y are each functions of x and y. Hence we may differentiate f_x and f_y to obtain **second-order partial derivatives** of f. Symbolically,

$$f_{xx} \quad \text{means} \quad (f_x)_x, \qquad f_{xy} \quad \text{means} \quad (f_x)_y,$$

$$f_{yx} \quad \text{means} \quad (f_y)_x, \qquad f_{yy} \quad \text{means} \quad (f_y)_y.$$

In terms of ∂-notation,

$$\frac{\partial^2 z}{\partial x^2} \quad \text{means} \quad \frac{\partial}{\partial x}\left[\frac{\partial z}{\partial x}\right], \qquad \frac{\partial^2 z}{\partial y \partial x} \quad \text{means} \quad \frac{\partial}{\partial y}\left[\frac{\partial z}{\partial x}\right],$$

$$\frac{\partial^2 z}{\partial x \partial y} \quad \text{means} \quad \frac{\partial}{\partial x}\left[\frac{\partial z}{\partial y}\right], \qquad \frac{\partial^2 z}{\partial y^2} \quad \text{means} \quad \frac{\partial}{\partial y}\left[\frac{\partial z}{\partial y}\right].$$

Note that to find f_{xy}, first differentiate f with respect to x. For $\partial^2 z/\partial x \partial y$, first differentiate with respect to y.

We can extend our notation beyond second-order partial derivatives. For example, f_{xxy} (or $\partial^3 z/\partial y \partial x^2$) is a third-order partial derivative of f. It is the partial derivative of f_{xx} (or $\partial^2 z/\partial x^2$) with respect to y. A generalization of higher-order partial derivatives to functions of more than two variables should be obvious.

EXAMPLE 1   *Find the four second-order partial derivatives of*

$$f(x, y) = x^2y + x^2y^2.$$

Since

$$f_x(x, y) = 2xy + 2xy^2,$$

we have

$$f_{xx}(x, y) = \frac{\partial}{\partial x}(2xy + 2xy^2) = 2y + 2y^2$$

and $\quad f_{xy}(x, y) = \frac{\partial}{\partial y}(2xy + 2xy^2) = 2x + 4xy.$

Since

$$f_y(x, y) = x^2 + 2x^2y,$$

we have

$$f_{yy}(x, y) = \frac{\partial}{\partial y}(x^2 + 2x^2y) = 2x^2$$

and $\quad f_{yx}(x, y) = \frac{\partial}{\partial x}(x^2 + 2x^2y) = 2x + 4xy.$

The derivatives f_{xy} and f_{yx} are called **mixed partial derivatives**. Observe in Example 1 that $f_{xy}(x, y) = f_{yx}(x, y)$. Under suitable conditions, mixed partial derivatives for a function are equal; that is, the order of differentiation is of no concern. You may assume this is the case for all the functions that we consider.

EXAMPLE 2   *Determine the value of*

$$\frac{\partial^3 w}{\partial z\, \partial y\, \partial x}\bigg|_{(1,2,3)}$$

if $w = (2x + 3y + 4z)^3.$

$$\frac{\partial w}{\partial x} = 3(2x + 3y + 4z)^2 \frac{\partial}{\partial x}(2x + 3y + 4z)$$

$$= 6(2x + 3y + 4z)^2.$$

$$\frac{\partial^2 w}{\partial y\, \partial x} = 6 \cdot 2(2x + 3y + 4z)\frac{\partial}{\partial y}(2x + 3y + 4z)$$

$$= 36(2x + 3y + 4z).$$

$$\frac{\partial^3 w}{\partial z\, \partial y\, \partial x} = 36 \cdot 4 = 144.$$

Thus

$$\frac{\partial^3 w}{\partial z\, \partial y\, \partial x}\bigg|_{(1,2,3)} = 144.$$

EXERCISE 8.4

In Problems 1–10, find the indicated partial derivatives.

1. $f(x, y) = 3x^2y^2$; $f_x(x, y)$, $f_{xy}(x, y)$.

2. $f(x, y) = 3x^2y + 2xy^2 - 7y$; $f_x(x, y)$, $f_{xx}(x, y)$.

3. $f(x, y) = e^{3xy} + 4x^2y$; $f_y(x, y)$, $f_{yx}(x, y)$, $f_{yxy}(x, y)$.

4. $f(x, y) = 7x^2 + 3y$; $f_y(x, y)$, $f_{yy}(x, y)$, $f_{yyx}(x, y)$.

5. $f(x, y) = (x^2 + xy + y^2)(x^2 + xy + 1)$; $f_x(x, y)$, $f_{xy}(x, y)$.

6. $f(x, y) = \ln(x^2 + y^2) + 2$; $f_x(x, y)$, $f_{xx}(x, y)$, $f_{xy}(x, y)$.

7. $f(x, y) = (x + y)^2(xy)$; $f_x(x, y)$, $f_y(x, y)$, $f_{xx}(x, y)$, $f_{yy}(x, y)$.

8. $f(x, y, z) = xy^2z^3$; $f_x(x, y, z)$, $f_{xz}(x, y, z)$, $f_{xy}(x, y, z)$.

9. $z = \sqrt{x^2 + y^2}$; $\dfrac{\partial z}{\partial x}$, $\dfrac{\partial^2 z}{\partial x^2}$.

10. $z = \dfrac{\ln(x^2 + 5)}{y}$; $\dfrac{\partial z}{\partial x}$, $\dfrac{\partial^2 z}{\partial y \partial x}$.

11. If $f(x, y, z) = 7$, find $f_{yxx}(4, 3, -2)$.

12. If $f(x, y, z) = z^2(3x^2 - 4xy^3)$, find $f_{xyz}(1, 2, 3)$.

13. If $f(l, k) = 5l^3k^6 - lk^7$, find $f_{kkl}(2, 1)$.

14. If $f(x, y) = 2x^2y + xy^2 - x^2y^2$, find $f_{xxy}(0, 1)$.

15. If $f(x, y) = y^2e^x + \ln(xy)$, find $f_{xyy}(1, 1)$.

16. If $f(x, y) = x^3 - 3xy^2 + x^2 - y^3$, find $f_{xy}(1, -1)$.

17. If $f(x, y) = 8x^3 + 2x^2y^2 + 5y^4$, show that $f_{xy}(x, y) = f_{yx}(x, y)$.

18. If $f(x, y) = x^4y^4 + 3x^3y^2 - 7x + 4$, show that $f_{xyx}(x, y) = f_{xxy}(x, y)$.

19. If $z = \ln(x^2 + y^2)$, show that $\dfrac{\partial^2 z}{\partial x^2} + \dfrac{\partial^2 z}{\partial y^2} = 0$.

8.5 MAXIMA AND MINIMA FOR FUNCTIONS OF TWO VARIABLES

We now extend to functions of two variables the notion of relative maxima and minima (or relative extrema).

Definition
*A function $z = f(x, y)$ has a **relative maximum** at the point (x_0, y_0), that is, when $x = x_0$ and $y = y_0$, if for all points (x, y) in the plane that are sufficiently close to (x_0, y_0) we have*

$$f(x_0, y_0) \geq f(x, y). \tag{1}$$

*For a **relative minimum**, in (1) we replace $\geq$ by $\leq$.*

To say that $z = f(x, y)$ has a relative maximum at (x_0, y_0) means geometrically that the point (x_0, y_0, z_0) on the graph of f is higher than (or is as high as) all other points on the surface that are "near" (x_0, y_0, z_0). In Fig. 8.12(a), f has a relative maximum at (x_1, y_1). Similarly, the function f in Fig. 8.12(b) has a relative minimum when $x = y = 0$, which corresponds to a *low* point on the surface.

Recall that in locating extrema for a function $y = f(x)$ of one variable, we examine those values of x in the domain of f for which $f'(x) = 0$ or $f'(x)$ does not exist. For functions of two (or more) variables, a similar procedure is followed. However, for the functions that concern us, extrema will not

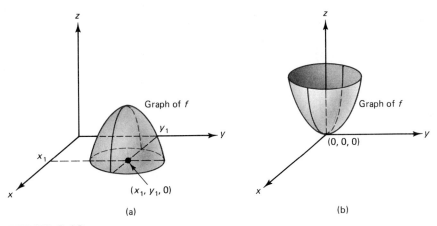

FIGURE 8.12

occur where a derivative does not exist, and such situations will be excluded from consideration.

Suppose that $z = f(x, y)$ has a relative maximum at (x_0, y_0), as indicated in Fig. 8.13(a). Then the curve where the plane $y = y_0$ intersects the surface must have a relative maximum when $x = x_0$. Hence the slope of the tangent line to the surface in the x-direction must be 0 at (x_0, y_0). Equivalently, $f_x(x, y) = 0$ at (x_0, y_0). Similarly, on the curve where the plane $x = x_0$ intersects

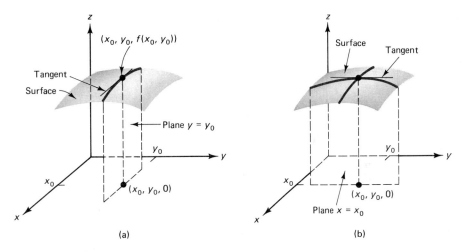

FIGURE 8.13

the surface [Fig. 8.13(b)], there must be a relative maximum when $y = y_0$. Thus in the y-direction, the slope of the tangent to the surface must be 0 at (x_0, y_0). Equivalently, $f_y(x, y) = 0$ at (x_0, y_0). Since a similar discussion can be given for a relative minimum, we can combine these results as follows.

> ### Rule 1
>
> If $z = f(x, y)$ has a relative maximum or minimum at (x_0, y_0) and if both f_x and f_y are defined for all points close to (x_0, y_0), it is necessary that (x_0, y_0) be a solution of the system
>
> $$\begin{cases} f_x(x, y) = 0, \\ f_y(x, y) = 0. \end{cases}$$

A point (x_0, y_0) for which $f_x(x, y) = f_y(x, y) = 0$ is called a **critical point** of f. Thus from Rule 1 we infer that to locate relative extrema for a function we should examine its critical points.

Pitfall

Rule 1 does not imply that there must be an extremum at a critical point. Just as in the case of functions of one variable, a critical point can give rise to a relative maximum, a relative minimum, or neither.

Two additional comments: First, Rule 1, as well as the notion of a critical point, can be extended to functions of more than two variables. Thus to locate possible extrema for $w = f(x, y, z)$, we would examine those points for which $w_x = w_y = w_z = 0$. Second, for a function whose domain is restricted, a thorough examination for absolute extrema would include consideration of boundary points.

EXAMPLE 1   *Find the critical points of the following functions.*

a. $f(x, y) = 2x^2 + y^2 - 2xy + 5x - 3y + 1$.

Since $f_x(x, y) = 4x - 2y + 5$ and $f_y(x, y) = 2y - 2x - 3$, we solve the system

$$\begin{cases} 4x - 2y + 5 = 0, \\ -2x + 2y - 3 = 0. \end{cases}$$

This gives $x = -1$ and $y = \frac{1}{2}$. Thus $(-1, \frac{1}{2})$ is the only critical point.

b. $f(l, k) = l^3 + k^3 - lk$.

$$\begin{cases} f_l(l, k) = 3l^2 - k = 0, & (2) \\ f_k(l, k) = 3k^2 - l = 0. & (3) \end{cases}$$

From Eq. (2), $k = 3l^2$. Substituting for k in Eq. (3) gives the equation $0 = 27l^4 - l = l(27l^3 - 1)$. Hence $l = 0$ or $l = \frac{1}{3}$. If $l = 0$, then $k = 0$; if $l = \frac{1}{3}$, then $k = \frac{1}{3}$. The critical points are thus $(0, 0)$ and $(\frac{1}{3}, \frac{1}{3})$.

c. $f(x, y, z) = 2x^2 + xy + y^2 + 100 - z(x + y - 100)$.

Solving the system

$$\begin{cases} f_x(x, y, z) = 4x + y - z = 0, \\ f_y(x, y, z) = x + 2y - z = 0, \\ f_z(x, y, z) = -x - y + 100 = 0 \end{cases}$$

gives the critical point $(25, 75, 175)$, as you may verify.

EXAMPLE 2   *Find the critical points of*

$$f(x, y) = x^2 - 4x + 2y^2 + 4y + 7.$$

We have $f_x(x, y) = 2x - 4$ and $f_y(x, y) = 4y + 4$. The system

$$\begin{cases} 2x - 4 = 0, \\ 4y + 4 = 0 \end{cases}$$

gives the critical point $(2, -1)$. Observe that the given function can be written

$$\begin{aligned} f(x, y) &= x^2 - 4x + 4 + 2(y^2 + 2y + 1) + 1 \\ &= (x - 2)^2 + 2(y + 1)^2 + 1, \end{aligned}$$

and $f(2, -1) = 1$. Clearly, if $(x, y) \neq (2, -1)$, then $f(x, y) > 1$. Hence a relative minimum occurs at $(2, -1)$. Moreover, there is an *absolute minimum* at $(2, -1)$ since $f(x, y) > f(2, -1)$ for *all* $(x, y) \neq (2, -1)$.

Although in Example 2 we were able to show that the critical point gave rise to a relative extremum, in many cases this is not so easy to do. There is, however, a second-derivative test that gives conditions under which a critical point will be a relative maximum or minimum. We state it now, omitting the proof.

Rule 2

Second-Derivative Test for Functions of Two Variables. Suppose $z = f(x, y)$ has continuous partial derivatives f_{xx}, f_{yy}, and f_{xy} at all points (x, y) near the critical point (x_0, y_0). Let D be the function defined by

$$D(x, y) = f_{xx}(x, y) f_{yy}(x, y) - [f_{xy}(x, y)]^2.$$

Then

a. If $D(x_0, y_0) > 0$ and $f_{xx}(x_0, y_0) < 0$, f has a relative maximum at (x_0, y_0);
b. if $D(x_0, y_0) > 0$ and $f_{xx}(x_0, y_0) > 0$, f has a relative minimum at (x_0, y_0);
c. if $D(x_0, y_0) < 0$, f has neither a relative maximum nor a relative minimum at (x_0, y_0);
d. if $D(x_0, y_0) = 0$, no conclusion about extrema at (x_0, y_0) can yet be drawn and further analysis is required.

EXAMPLE 3   *Examine $f(x, y) = x^3 + y^3 - xy$ for relative maxima or minima by using the second-derivative test.*

First we find critical points.

$$f_x(x, y) = 3x^2 - y, \qquad f_y(x, y) = 3y^2 - x.$$

In the same manner as in Example 1(b), solving $f_x(x, y) = f_y(x, y) = 0$ gives the critical points $(0, 0)$ and $(\frac{1}{3}, \frac{1}{3})$.
   Now,

$$f_{xx}(x, y) = 6x, \qquad f_{yy}(x, y) = 6y, \qquad f_{xy}(x, y) = -1.$$

Thus

$$D(x, y) = (6x)(6y) - (-1)^2 = 36xy - 1.$$

Since $D(0, 0) = 36(0)(0) - 1 = -1 < 0$, there is no relative extremum at $(0, 0)$.

Since $D(\frac{1}{3}, \frac{1}{3}) = 36(\frac{1}{3})(\frac{1}{3}) - 1 = 3 > 0$ and $f_{xx}(\frac{1}{3}, \frac{1}{3}) = 6(\frac{1}{3}) = 2 > 0$, there is a relative minimum at $(\frac{1}{3}, \frac{1}{3})$. At this point, the value of the function is

$$f(\tfrac{1}{3}, \tfrac{1}{3}) = (\tfrac{1}{3})^3 + (\tfrac{1}{3})^3 - (\tfrac{1}{3})(\tfrac{1}{3}) = -\tfrac{1}{27}.$$

EXAMPLE 4   *Examine $f(x, y) = y^2 - x^2$ for relative extrema.*

Solving

$$f_x(x, y) = -2x = 0 \qquad \text{and} \qquad f_y(x, y) = 2y = 0,$$

we get the critical point $(0, 0)$. Moreover, at $(0, 0)$, and indeed at any point,

$$f_{xx}(x, y) = -2, \qquad f_{yy}(x, y) = 2, \qquad f_{xy}(x, y) = 0.$$

Hence $D(0, 0) = (-2)(2) - (0)^2 = -4 < 0$ and no relative extremum exists. A sketch of $z = f(x, y) = y^2 - x^2$ appears in Fig. 8.14. Note that for the surface curve cut by the plane $y = 0$ there is a *maximum* at $(0, 0)$; but for the surface curve cut by the plane $x = 0$ there is a *minimum* at $(0, 0)$. Thus

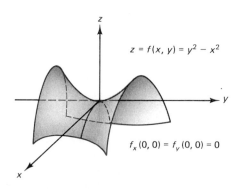

$z = f(x, y) = y^2 - x^2$

$f_x(0, 0) = f_y(0, 0) = 0$

FIGURE 8.14

on the *surface* no relative extremum can exist at the origin, although $(0, 0)$ is a critical point. Around the origin the curve is saddle-shaped and $(0, 0)$ is called a *saddle point* of f.

EXAMPLE 5   *Examine $f(x, y) = x^4 + (x - y)^4$ for relative extrema.*

If we set

$$f_x(x, y) = 4x^3 + 4(x - y)^3 = 0 \tag{4}$$

$$\text{and} \quad f_y(x, y) = -4(x - y)^3 = 0, \tag{5}$$

then from Eq. (5) we have $x - y = 0$, or $x = y$. Substituting into Eq . (4) gives $4x^3 = 0$, or $x = 0$. Thus $x = y = 0$ and $(0, 0)$ is the only critical point. At $(0, 0)$, $f_{xx}(x, y) = 12x^2 + 12(x - y)^2 = 0$, $f_{yy}(x, y) = 12(x - y)^2 = 0$, and $f_{xy}(x, y) = -12(x - y)^2 = 0$. Hence $D(0, 0) = 0$ and the second-derivative test gives no information. However, note that for all $(x, y) \neq (0, 0)$ we have $f(x, y) > 0$, whereas $f(0, 0) = 0$. Hence at $(0, 0)$ the graph of f has a low point and we conclude that f has a relative (and absolute) minimum at $(0, 0)$.

In many situations involving functions of two variables, and especially in their applications, the nature of the given problem is an indicator of whether a critical point is in fact a relative (or absolute) maximum or a relative (or absolute) minimum. In such cases the second-derivative test is not needed. Often, in mathematical studies of applied problems, the appropriate second-order conditions are assumed to hold.

EXAMPLE 6   *Let P be a production function given by*

$$P = f(l, k) = 0.54l^2 - 0.02l^3 + 1.89k^2 - 0.09k^3,$$

where l and k are the amounts of labor and capital, respectively, and P is the quantity of output produced. Find the values of l and k that maximize P.

$$P_l = 1.08l - 0.06l^2 \qquad\qquad P_k = 3.78k - 0.27k^2$$
$$= 0.06l(18 - l) = 0. \qquad\qquad = 0.27k(14 - k) = 0.$$
$$l = 0, l = 18. \qquad\qquad k = 0, k = 14.$$

The critical points are $(0, 0)$, $(0, 14)$, $(18, 0)$, and $(18, 14)$.

Now,

$$P_{ll} = 1.08 - 0.12l, \qquad P_{kk} = 3.78 - 0.54k, \qquad P_{lk} = 0.$$

Thus

$$D(l, k) = P_{ll} P_{kk} - [P_{lk}]^2$$
$$= (1.08 - 0.12l)(3.78 - 0.54k).$$

At $(0, 0)$,

$$D(0, 0) = 1.08(3.78) > 0.$$

Since $D(0, 0) > 0$ and $P_{ll} = 1.08 > 0$, there is a relative minimum at $(0, 0)$.

At $(0, 14)$,

$$D(0, 14) = 1.08(-3.78) < 0.$$

Since $D(0, 14) < 0$, there is no relative extremum at $(0, 14)$.

At $(18, 0)$,

$$D(18, 0) = (-1.08)(3.78) < 0.$$

Since $D(18, 0) < 0$, there is no relative extremum at $(18, 0)$.

At $(18, 14)$,

$$D(18, 14) = (-1.08)(-3.78) > 0.$$

Since $D(18, 14) > 0$ and $P_{ll} = -1.08 < 0$, there is a relative maximum at $(18, 14)$. The maximum output is obtained when $l = 18$ and $k = 14$.

EXAMPLE 7   *A food manufacturer produces two types of candy, A and B, for which the average costs of production are constant at 70 and 80 cents per pound, respectively. The quantities q_A, q_B (in pounds) of A and B that can be sold each week are given by the joint-demand functions*

$$q_A = 240(p_B - p_A)$$

and

$$q_B = 240(150 + p_A - 2p_B),$$

where p_A and p_B are the selling prices (in cents per pound) of A and B, respectively. Determine the selling prices that will maximize the manufacturer's profit P.

For A and B the profits per pound are $p_A - 70$ and $p_B - 80$, respectively. Hence total profit P is given by

$$P = \left(\begin{array}{c} \text{profit} \\ \text{per pound} \\ \text{of A} \end{array}\right)\left(\begin{array}{c} \text{pounds} \\ \text{of A} \\ \text{sold} \end{array}\right) + \left(\begin{array}{c} \text{profit} \\ \text{per pound} \\ \text{of B} \end{array}\right)\left(\begin{array}{c} \text{pounds} \\ \text{of B} \\ \text{sold} \end{array}\right)$$

or

$$P = (p_A - 70)q_A + (p_B - 80)q_B$$
$$= (p_A - 70)[240(p_B - p_A)] + (p_B - 80)[240(150 + p_A - 2p_B)].$$

To maximize P we set its partial derivatives equal to 0:

$$\frac{\partial P}{\partial p_A} = (p_A - 70)[240(-1)] + [240(p_B - p_A)](1) + (p_B - 80)[(240)(1)] = 0,$$

$$\frac{\partial P}{\partial p_B} = (p_A - 70)[240(1)] + (p_B - 80)[240(-2)] + [240(150 + p_A - 2p_B)](1) = 0$$

Simplifying the preceding two equations gives

$$\begin{cases} p_B - p_A - 5 = 0, \\ -2p_B + p_A + 120 = 0, \end{cases}$$

whose solution is $p_A = 110$ and $p_B = 115$ (both in cents). Moreover,

$$\frac{\partial^2 P}{\partial p_A^2} = -480 < 0, \qquad \frac{\partial^2 P}{\partial p_B^2} = -960, \qquad \frac{\partial^2 P}{\partial p_B \, \partial p_A} = 480.$$

Thus $D(110, 115) = (-480)(-960) - (480)^2 > 0$, Since $\partial^2 P / \partial p_A^2 < 0$, we indeed have a maximum. The manufacturer should sell candy A at \$1.10 per pound and B at \$1.15 per pound.

EXERCISE 8.5

In Problems **1–6,** *find the critical points of the functions.*

1. $f(x, y) = x^2 + y^2 - 5x + 4y + xy$.

2. $f(x, y) = x^2 + 4y^2 - 6x + 16y$.

3. $f(x, y) = 2x^3 + y^3 - 3x^2 + 1.5y^2 - 12x - 90y$.

4. $f(x, y) = xy - \dfrac{1}{x} - \dfrac{1}{y}$.

5. $f(x, y, z) = 2x^2 + xy + y^2 + 100 - z(x + y - 200)$.

6. $f(x, y, z, w) = x^2 + y^2 + z^2 - w(x - y + 2z - 6)$.

In Problems **7–18,** *find the critical points of the functions. Determine, by the second-derivative test, whether each point corresponds to a relative maximum, to a relative minimum, to neither, or whether the test gives no information.*

7. $f(x, y) = x^2 + 3y^2 + 4x - 9y + 3$.

8. $f(x, y) = -2x^2 + 8x - 3y^2 + 24y + 7$.

9. $f(x, y) = y - y^2 - 3x - 6x^2$.

10. $f(x, y) = x^2 + y^2 + xy - 9x + 1$.

11. $f(x, y) = x^3 - 3xy + y^2 + y - 5$.

12. $f(x, y) = \dfrac{x^3}{3} + y^2 - 2x + 2y - 2xy$.

13. $f(x, y) = \frac{1}{3}(x^3 + 8y^3) - 2(x^2 + y^2) + 1$.

14. $f(x, y) = x^2 + y^2 - xy + x^3$.

15. $f(l, k) = 2lk - l^2 + 264k - 10l - 2k^2$.

16. $f(l, k) = l^3 + k^3 - 3lk$.

17. $f(p, q) = pq - \dfrac{1}{p} - \dfrac{1}{q}$.

18. $f(x, y) = (x - 3)(y - 3)(x + y - 3)$.

In Problems **19–26,** *unless otherwise indicated the variables* p_A *and* p_B *denote selling prices of products A and B, respectively. Similarly,* q_A *and* q_B *denote quantities of A and B which are produced and sold during some time period. In all cases, the variables employed will be assumed to be units of output, input, money, etc.*

19. Suppose that

$$P = f(l, k) = 1.08l^2 - 0.03l^3 + 1.68k^2 - 0.08k^3$$

is a production function for a firm. Find the quantities of input, l and k, so as to maximize output P.

20. In a certain automated manufacturing process, machines M and N are utilized for m and n hours, respectively. If daily output Q is a function of m and n, namely $Q = 4.5m + 5n - 0.5m^2 - n^2 - 0.25mn$, find the values of m and n that maximize Q.

21. A candy company produces two varieties of candy, A and B, for which the constant average costs of production are 60 and 70 (cents per pound), respectively. The demand functions for A and B are respectively $q_A = 5(p_B - p_A)$ and $q_B = 500 + 5(p_A - 2p_B)$. Find the selling prices p_A and p_B that maximize the company's profit.

22. Repeat Problem 21 if the constant costs of production of A and B are a and b (cents per pound), respectively.

23. Suppose that a monopolist is practicing price discrimination in the sale of a product by charging different prices in two separate markets. In market A the demand function is $p_A = 100 - q_A$ and in B it is $p_B = 84 - q_B$, where q_A and q_B are the quantities sold per week in A and B, and p_A and p_B are the respective prices per unit. If the monopolist's cost function is given by $c = 600 + 4(q_A + q_B)$, how much should be sold in each market to maximize profit? What selling prices give this maximum profit? Find the maximum profit.

24. A monopolist sells two competitive products, A and B, for which the demand functions are given by $q_A = 1 - 2p_A + 4p_B$ and $q_B = 11 + 2p_A - 6p_B$. If the constant average cost of producing a unit of A is 4 and for B it is 1, how many units of A and B should be sold to maximize the monopolist's profit?

25. For products A and B, the joint-cost function for a manufacturer is $c = 1.5q_A^2 + 4.5q_B^2$ and the demand functions are $p_A = 36 - q_A^2$ and $p_B = 30 - q_B^2$. Find the level of production that maximizes profit.

26. For a monopolist's products, A and B, the joint-cost function is $c = (q_A + q_B)^2$ and the demand functions are $q_A = 26 - p_A$ and $q_B = 10 - 0.25p_B$. Find the values of p_A and p_B that maximize profit. What are the quantities of A and B that correspond to these prices? What is the total profit?

27. An open-top rectangular box is to have a volume of 6 cubic feet. The cost per square foot of materials is $3 for the bottom, $1 for the front and back, and $0.50 for the other two sides. Find the dimensions of the box so that the cost of materials is minimized (see Fig. 8.15).

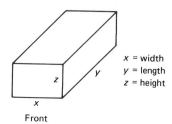

x = width
y = length
z = height

Front

FIGURE 8.15

28. Suppose that A and B are the only two firms in the market selling the same product (we say they are *duopolists*). The industry demand function for the product is $p = 92 - q_A - q_B$, where q_A and q_B denote the output produced and sold by A and B, respectively. For A the cost function is $c_A = 10q_A$ and for B it is $c_B = 0.5q_B^2$. Suppose the firms decide to enter into an agreement on output and price control by jointly acting as a monopoly. In this case we say they enter into *collusion*. Show that the profit function for the monopoly is given by

$$P = pq_A - c_A + pq_B - c_B.$$

Express P as a function of q_A and q_B and determine how output should be allocated so as to maximize the profit of the monopoly.

29. Suppose that $f(x, y) = -2x^2 + 5y^2 + 7$, where x and y must satisfy the equation $3x - 2y = 7$. Find the relative extrema of f subject to the given condition on x and y by first solving the second equation for y. Substitute the result for y in the given equation. Thus f is expressed as a function of one variable for which extrema may be found in the usual way.

30. Repeat Problem 29 if $f(x, y) = x^2 + 4y^2 + 6$ subject to the condition that $2x - 8y = 20$.

8.6　LAGRANGE MULTIPLIERS

We shall now find relative maxima and minima for a function on which certain *constraints* are imposed. Such a situation could arise if a manufacturer wished to minimize the total cost of factors of input and yet obtain a particular level of output.

Suppose that we want to find the relative extrema of

$$w = x^2 + y^2 + z^2 \tag{1}$$

subject to the constraint that x, y, and z must satisfy

$$x - y + 2z = 6. \tag{2}$$

We can transform w, which is a function of three variables, into a function of two variables such that the new function reflects the constraint (2). Solving Eq. (2) for x, we get

$$x = y - 2z + 6, \tag{3}$$

which when substituted for x in Eq. (1) gives

$$w = (y - 2z + 6)^2 + y^2 + z^2. \tag{4}$$

Since w is now expressed as a function of two variables, to find relative extrema we follow the usual procedure of setting its partial derivatives equal to 0:

$$\frac{\partial w}{\partial y} = 2(y - 2z + 6) + 2y = 4y - 4z + 12 = 0, \tag{5}$$

$$\frac{\partial w}{\partial z} = -4(y - 2z + 6) + 2z = -4y + 10z - 24 = 0. \tag{6}$$

Solving Eqs. (5) and (6) simultaneously gives $y = -1$ and $z = 2$. Substituting in Eq. (3), we get $x = 1$. Hence the only critical point of (1) subject to constraint (2) is $(1, -1, 2)$. By using the second-derivative test on (4) when $y = -1$ and $z = 2$, we have

$$\frac{\partial^2 w}{\partial y^2} = 4, \qquad \frac{\partial^2 w}{\partial z^2} = 10, \qquad \frac{\partial^2 w}{\partial z \, \partial y} = -4,$$

$$D(-1, 2) = 4(10) - (-4)^2 = 24 > 0.$$

Thus w, subject to the constraint, has a relative minimum at $(1, -1, 2)$.

This solution was found by using the constraint to express one of the variables in the original function in terms of the other variables. Often this is not practical, but there is another technique, called the method of **Lagrange multipliers**,* that avoids this step and yet allows us to obtain critical points.

The method is as follows. Suppose that we have a function $f(x, y, z)$ subject to the constraint $g(x, y, z) = 0$. We construct a new function F of four variables defined by the following (where λ is a Greek letter read "lambda"):

$$F(x, y, z, \lambda) = f(x, y, z) - \lambda g(x, y, z).$$

It can be shown that if (x_0, y_0, z_0) is a critical point of f subject to the constraint $g(x, y, z) = 0$, there exists a value of λ, say λ_0, such that $(x_0, y_0, z_0, \lambda_0)$ is a critical point of F. The number λ_0 is called a **Lagrange multiplier**. Also, if

* After the French mathematician, Joseph-Louis Lagrange (1736–1813).

$(x_0, y_0, z_0, \lambda_0)$ is a critical point of F, then (x_0, y_0, z_0) is a critical point of f subject to the constraint. Thus to find all the critical points of f subject to $g(x, y, z) = 0$, we instead find all the critical points of F. These are obtained by solving the simultaneous equations

$$\begin{cases} F_x(x, y, z, \lambda) = 0, \\ F_y(x, y, z, \lambda) = 0, \\ F_z(x, y, z, \lambda) = 0, \\ F_\lambda(x, y, z, \lambda) = 0. \end{cases}$$

At times, ingenuity must be used to do this. Once we obtain a critical point $(x_0, y_0, z_0, \lambda_0)$ of F, we can conclude that (x_0, y_0, z_0) is a critical point of f subject to the constraint $g(x, y, z) = 0$. Although here f and g are functions of three variables, the method of Lagrange multipliers can be extended to n variables.

Let us illustrate the method of Lagrange multipliers for the original situation:

$$f(x, y, z) = x^2 + y^2 + z^2 \quad \text{subject to} \quad x - y + 2z = 6.$$

First, we write the constraint as $g(x, y, z) = x - y + 2z - 6 = 0$. Second, we form the function

$$\begin{aligned} F(x, y, z, \lambda) &= f(x, y, z) - \lambda g(x, y, z) \\ &= x^2 + y^2 + z^2 - \lambda(x - y + 2z - 6). \end{aligned}$$

Next we set each partial derivative of F equal to 0. For convenience we shall write $F_x(x, y, z, \lambda)$ as F_x, etc.

$$\begin{cases} F_x = 2x - \lambda = 0, & (7) \\ F_y = 2y + \lambda = 0, & (8) \\ F_z = 2z - 2\lambda = 0, & (9) \\ F_\lambda = -x + y - 2z + 6 = 0. & (10) \end{cases}$$

From Eqs. (7)–(9) we see immediately that

$$x = \frac{\lambda}{2}, \qquad y = -\frac{\lambda}{2}, \qquad \text{and} \qquad z = \lambda. \tag{11}$$

Substituting these values in Eq. (10), we obtain

$$-\frac{\lambda}{2} - \frac{\lambda}{2} - 2\lambda + 6 = 0,$$

$$-3\lambda + 6 = 0,$$

$$\lambda = 2.$$

Thus from Eq. (11), $x = 1$, $y = -1$, and $z = 2$. Hence the only critical point of f subject to the constraint is $(1, -1, 2)$, at which there may exist a relative maximum, a relative minimum, or neither of these. The method of Lagrange multipliers does not directly indicate which of these possibilities occurs, although from our previous work we saw that it is indeed a relative minimum. In applied problems the nature of the problem itself may give a clue as to how a critical point is to be regarded. Often the existence of either a relative minimum or a relative maximum is assumed and a critical point is treated accordingly. Actually, sufficient second-order conditions for relative extrema are available, but we shall not consider them.

EXAMPLE 1   *Find the critical points for* $z = f(x, y) = 3x - y + 6$ *subject to the constraint* $x^2 + y^2 = 4$.

We write the constraint as $g(x, y) = x^2 + y^2 - 4 = 0$ and construct the function

$$F(x, y, \lambda) = f(x, y) - \lambda g(x, y) = 3x - y + 6 - \lambda(x^2 + y^2 - 4).$$

Setting $F_x = F_y = F_\lambda = 0$ yields

$$\begin{cases} 3 - 2x\lambda = 0, & (12) \\ -1 - 2y\lambda = 0, & (13) \\ -x^2 - y^2 + 4 = 0. & (14) \end{cases}$$

From Eqs. (12) and (13),

$$x = \frac{3}{2\lambda} \quad \text{and} \quad y = -\frac{1}{2\lambda}.$$

Substituting in Eq. (14), we obtain

$$-\frac{9}{4\lambda^2} - \frac{1}{4\lambda^2} + 4 = 0,$$

$$-\frac{10}{4\lambda^2} + 4 = 0,$$

$$\lambda = \pm \frac{\sqrt{10}}{4}.$$

If $\lambda = \sqrt{10}/4$, then

$$x = \frac{3}{2\left(\dfrac{\sqrt{10}}{4}\right)} = \frac{3\sqrt{10}}{5}, \qquad y = -\frac{1}{2\left(\dfrac{\sqrt{10}}{4}\right)} = -\frac{\sqrt{10}}{5}.$$

Similarly, if $\lambda = -\sqrt{10}/4$, then

$$x = -\frac{3\sqrt{10}}{5}, \qquad y = \frac{\sqrt{10}}{5}.$$

Thus the critical points of f subject to the constraint are $(3\sqrt{10}/5, -\sqrt{10}/5)$ and $(-3\sqrt{10}/5, \sqrt{10}/5)$. Note that the values of λ do not appear in the answer. They are simply a means to obtain it.

EXAMPLE 2   *Find critical points for $f(x, y, z) = xyz$, where $xyz \neq 0$, subject to the constraint $x + 2y + 3z = 36$.*

Set

$$F(x, y, z, \lambda) = xyz - \lambda(x + 2y + 3z - 36).$$

Then

$$\begin{cases} F_x = yz - \lambda = 0, \\ F_y = xz - 2\lambda = 0, \\ F_z = xy - 3\lambda = 0, \\ F_\lambda = -x - 2y - 3z + 36 = 0. \end{cases}$$

We can write the system as

$$\begin{cases} yz = \lambda, & (15) \\ xz = 2\lambda, & (16) \\ xy = 3\lambda, & (17) \\ x + 2y + 3z - 36 = 0. & (18) \end{cases}$$

Dividing each side of Eq. (15) by the corresponding side of Eq. (16), we get

$$\frac{yz}{xz} = \frac{\lambda}{2\lambda} \qquad \text{or} \qquad y = \frac{x}{2}.$$

This division is valid since $xyz \neq 0$. Similarly, from Eqs. (15) and (17) we get

$$z = \frac{x}{3}.$$

Substituting into Eq. (18) gives

$$x + 2\left(\frac{x}{2}\right) + 3\left(\frac{x}{3}\right) - 36 = 0,$$

$$x = 12.$$

Thus $y = 6$ and $z = 4$. Hence $(12, 6, 4)$ is the only critical point satisfying the given conditions.

EXAMPLE 3   *Suppose that a firm has an order for 200 units of its product and wishes to distribute their manufacture between two of its plants, plant 1 and plant 2. Let q_1 and q_2 denote the outputs of plants 1 and 2, respectively, and suppose the total cost function is given by*

$$c = f(q_1, q_2) = 2q_1^2 + q_1 q_2 + q_2^2 + 200.$$

How should the output be distributed in order to minimize costs?

We must minimize $c = f(q_1, q_2)$ subject to the constraint $q_1 + q_2 = 200$.

$$F(q_1, q_2, \lambda) = 2q_1^2 + q_1 q_2 + q_2^2 + 200 - \lambda(q_1 + q_2 - 200).$$

$$
\begin{cases}
\dfrac{\partial F}{\partial q_1} = 4q_1 + q_2 - \lambda = 0, & \text{(19)} \\[2mm]
\dfrac{\partial F}{\partial q_2} = q_1 + 2q_2 - \lambda = 0, & \text{(20)} \\[2mm]
\dfrac{\partial F}{\partial \lambda} = -q_1 - q_2 + 200 = 0. & \text{(21)}
\end{cases}
$$

Solving Eqs. (19) and (20) simultaneously for q_1 and q_2 gives

$$
q_1 = \frac{\lambda}{7}, \qquad q_2 = \frac{3\lambda}{7}.
$$

Substituting these values in Eq. (21) gives $\lambda = 350$. Thus $q_1 = 50$ and $q_2 = 150$. Plant 1 should produce 50 units and plant 2 should produce 150 units in order to minimize costs.

An interesting observation can be made concerning Example 3. From Eq. (19), $\lambda = 4q_1 + q_2 = \partial c/\partial q_1$, the marginal cost of plant 1. From Eq. (20), $\lambda = q_1 + 2q_2 = \partial c/\partial q_2$, the marginal cost of plant 2. Hence $\partial c/\partial q_1 = \partial c/\partial q_2$, and we conclude that to minimize cost it is necessary that the marginal costs of each plant be equal to each other.

EXAMPLE 4   *Suppose that a firm must produce a given quantity P_0 of output in the cheapest possible manner. If there are two input factors l and k, and their prices per unit are fixed at p_l and p_k respectively, discuss the economic significance of combining input to achieve least cost. That is, describe the least-cost input combination.*

Let $P = f(l, k)$ be the production function. Then we must minimize the cost function

$$
c = lp_l + kp_k
$$

subject to

$$
P_0 = f(l, k).
$$

We construct

$$
F(l, k, \lambda) = lp_l + kp_k - \lambda[f(l, k) - P_0].
$$

We have

$$
\begin{cases}
\dfrac{\partial F}{\partial l} = p_l - \lambda \dfrac{\partial}{\partial l}[f(l, k)] = 0, & \text{(22)} \\[2mm]
\dfrac{\partial F}{\partial k} = p_k - \lambda \dfrac{\partial}{\partial k}[f(l, k)] = 0, & \text{(23)} \\[2mm]
\dfrac{\partial F}{\partial \lambda} = -f(l, k) + P_0 = 0.
\end{cases}
$$

From Eqs. (22) and (23),

$$\lambda = \frac{p_l}{\dfrac{\partial}{\partial l}[f(l, k)]} = \frac{p_k}{\dfrac{\partial}{\partial k}[f(l, k)]}. \tag{24}$$

Hence

$$\frac{p_l}{p_k} = \frac{\dfrac{\partial}{\partial l}[f(l, k)]}{\dfrac{\partial}{\partial k}[f(l, k)]}.$$

We conclude that when the least-cost combination of factors is used, the ratio of the marginal products of the input factors must be equal to the ratio of their corresponding prices.

The method of Lagrange multipliers is by no means restricted to problems involving a single constraint. For example, suppose $f(x, y, z, w)$ were subject to constraints $g_1(x, y, z, w) = 0$ and $g_2(x, y, z, w) = 0$. Then there would be two lambdas, λ_1 and λ_2 (one for each constraint), and we would construct the function $F = f - \lambda_1 g_1 - \lambda_2 g_2$. We would then solve the system $F_x = F_y = F_z = F_w = F_{\lambda_1} = F_{\lambda_2} = 0$.

EXAMPLE 5   *Find critical points for $f(x, y, z) = xy + yz$ subject to the constraints $x^2 + y^2 = 8$ and $yz = 8$.*

Set

$$F(x, y, z, \lambda_1, \lambda_2) = xy + yz - \lambda_1(x^2 + y^2 - 8) - \lambda_2(yz - 8).$$

Then

$$\begin{cases} F_x = y - 2x\lambda_1 = 0, \\ F_y = x + z - 2y\lambda_1 - z\lambda_2 = 0, \\ F_z = y - y\lambda_2 = 0, \\ F_{\lambda_1} = -x^2 - y^2 + 8 = 0, \\ F_{\lambda_2} = -yz + 8 = 0. \end{cases}$$

We can write the system as

$$\begin{cases} \dfrac{y}{2x} = \lambda_1, & (25) \\[2mm] x + z - 2y\lambda_1 - z\lambda_2 = 0, & (26) \\[2mm] \lambda_2 = 1, & (27) \\[2mm] x^2 + y^2 = 8, & (28) \\[2mm] z = \dfrac{8}{y}. & (29) \end{cases}$$

Substituting $\lambda_2 = 1$ from Eq. (27) into Eq. (26) and simplifying gives $x - 2y\lambda_1 = 0$, so

$$\lambda_1 = \frac{x}{2y}.$$

Substituting into Eq. (25) gives

$$\frac{y}{2x} = \frac{x}{2y},$$
$$y^2 = x^2. \tag{30}$$

Substituting into Eq. (28) gives $x^2 + x^2 = 8$ from which $x = \pm 2$. If $x = 2$, then from Eq. (30) we have $y = \pm 2$. Similarly, if $x = -2$, then $y = \pm 2$. Thus if $x = 2$ and $y = 2$, then from Eq. (29) we obtain $z = 4$. Continuing in this manner, we obtain four critical points:

$$(2, 2, 4), \quad (2, -2, -4), \quad (-2, 2, 4), \quad (-2, -2, -4).$$

EXERCISE 8.6

In Problems 1–12 find, by the method of Lagrange multipliers, the critical points of the functions subject to the given constraints.

1. $f(x, y) = x^2 + 4y^2 + 6$; $\quad 2x - 8y = 20$.

2. $f(x, y) = -2x^2 + 5y^2 + 7$; $\quad 3x - 2y = 7$.

3. $f(x, y, z) = x^2 + y^2 + z^2$; $\quad 2x + y - z = 9$.

4. $f(x, y, z) = x + y + z$; $\quad xyz = 27$.

5. $f(x, y, z) = x^2 + xy + 2y^2 + z^2$; $\quad x - 3y - 4z = 16$.

6. $f(x, y, z) = xyz^2$; $\quad x - y + z = 20$, $\quad (xyz^2 \neq 0)$.

7. $f(x, y, z) = xyz$; $\quad x + 2y + 3z = 18$, $\quad (xyz \neq 0)$.

8. $f(x, y, z) = x^2 + y^2 + z^2$, $\quad x + y + z = 1$.

9. $f(x, y, z) = x^2 + 2y - z^2$; $\quad 2x - y = 0$, $\quad y + z = 0$.

10. $f(x, y, z) = x^2 + y^2 + z^2$; $\quad x + y + z = 1$, $\quad x = y - z + 1$.

11. $f(x, y, z) = xyz$; $\quad x + y + z = 12$, $\quad x + y - z = 0$, $\quad (xyz \neq 0)$.

12. $f(x, y, z, w) = 2x^2 + 2y^2 + 3z^2 - 4w^2$; $\quad 4x - 8y + 6z + 16w = 6$.

13. To fill an order for 100 units of its product, a firm wishes to distribute the production between its two plants, plant 1 and plant 2. The total cost function is given by $c = f(q_1, q_2) = 0.1q_1^2 + 7q_1 + 15q_2 + 1000$, where q_1 and q_2 are the number of units produced at plants 1 and 2, respectively. How should the output be distributed in order to minimize costs?

14. Repeat Problem 13 if the cost function is $c = 3q_1^2 + q_1q_2 + 2q_2^2$ and a total of 200 units are to be produced.

15. The production function for a firm is $f(l, k) = 12l + 20k - l^2 - 2k^2$. The cost to the firm of l and k is 4 and 8 per unit, respectively. If the firm wants the total cost of input to be 88, find the greatest output possible subject to this budget constraint.

16. Repeat Problem 15 given that $f(l, k) = 60l + 30k - 2l^2 - 3k^2$ and the budget constraint is $2l + 3k = 30$.

*Problems 17–20 refer to the following definition. A **utility function** is a function that attaches a measure to the satisfaction or utility a consumer gets from the consumption of products per unit of time. Suppose that $U = f(x, y)$ is such a function, where x and y are the amounts of two products, X and Y. The **marginal utility** of X is $\partial U/\partial x$ and approximately represents the change in total utility resulting from a one-unit change in consumption of product X per unit of time. We define the marginal utility of Y in similar fashion. If the prices of X and Y are p_x and p_y, respectively, and the consumer has an income or budget of I to spend, then the budget constraint is $xp_x + yp_y = I$. In Problems 17–19, find the quantities of each product that the consumer* *should buy, subject to the budget, that will allow maximum satisfaction. That is, in Problems **17** and **18**, find values of x and y that maximize $U = f(x, y)$ subject to $xp_x + yp_y = I$. Perform a similar procedure for Problem **19**. Assume such a maximum exists.*

17. $U = x^3y^3$; $p_x = 2$, $p_y = 3$, $I = 48$, $(x^3y^3 \neq 0)$.

18. $U = 46x - (5x^2/2) + 34y - 2y^2$; $p_x = 5$, $p_y = 2$, $I = 30$.

19. $U = f(x, y, z) = xyz$; $p_x = 2$, $p_y = 1$, $p_z = 4$, $I = 60$, $(xyz \neq 0)$.

20. Let $U = f(x, y)$ be a utility function subject to the budget constraint $xp_x + yp_y = I$, where p_x, p_y, and

I are constants. Show that to maximize satisfaction it is necessary that

$$\lambda = \frac{f_x(x, y)}{p_x} = \frac{f_y(x, y)}{p_y},$$

where $f_x(x, y)$ and $f_y(x, y)$ are the marginal utilities of X and Y, respectively. Deduce that $f_x(x, y)/p_x$ is the marginal utility of one dollar's worth of X. Hence, maximum satisfaction is obtained when the consumer allocates the budget so that the marginal utility of a dollar's worth of X is equal to the marginal utility per dollar's worth of Y. Performing the same procedure as above, verify that this is true for $U = f(x, y, z, w)$ subject to the corresponding budget equation. In each case, λ is called the *marginal utility of income*.

8.7 LINES OF REGRESSION

To study the influence of advertising on sales, a firm compiled the data in Table 8.4. The variable x denotes advertising expenditures in hundreds of dollars, and the variable y denotes the resulting sales revenue in thousands

TABLE 8.4

EXPENDITURES, x	2	3	4.5	5.5	7
REVENUE, y	3	6	8	10	11

of dollars. If each pair (x, y) of data is plotted, the result is called a **scatter diagram** [Fig. 8.16(a)].

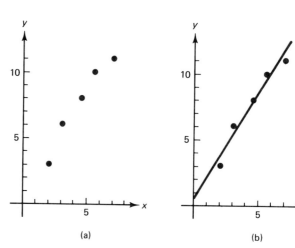

(a)   (b)

FIGURE 8.16

From an observation of the distribution of the points, it is reasonable to assume that a relationship exists between x and y and that it is approximately linear. On this basis we may fit "by eye" a straight line that approximates the given data [Fig. 8.16(b)], and from this line predict a value of y for a given value of x. This line seems consistent with the trend of the data, although other lines could be drawn as well. Unfortunately, determining a line "by eye" is not very objective. We want to apply criteria in specifying what we shall call a line of "best fit." A frequently used technique is called the **method of least squares**.

To apply the method of least squares to the data in Table 8.4, we first assume that x and y are approximately linearly related and that we can fit a straight line

$$\hat{y} = \hat{a} + \hat{b}x \qquad (1)$$

that approximates the given points by a suitable objective choice of the constants $\hat{a}$ and $\hat{b}$ (read "a hat" and "b hat," respectively). For a given value of x in Eq. (1), $\hat{y}$ is the corresponding predicted value of y, and $(x, \hat{y})$ will be on the line. Our aim is that $\hat{y}$ be near y.

When $x = 2$, the observed value of y is 3. Our predicted value of y is obtained by substituting $x = 2$ in Eq. (1), which gives $\hat{y} = \hat{a} + 2\hat{b}$. The error of estimation, or vertical deviation of the point $(2, 3)$ from the line, is $\hat{y} - y$, or

$$\hat{a} + 2\hat{b} - 3.$$

This vertical deviation is indicated (although exaggerated for clarity) in Fig. 8.17. Similarly, the vertical deviation of $(3, 6)$ from the line is $\hat{a} + 3\hat{b} - 6$,

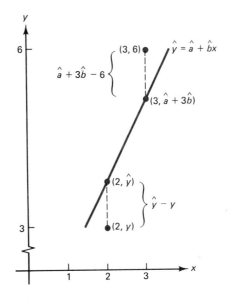

FIGURE 8.17

as is also illustrated. To avoid possible difficulties associated with positive and negative deviations, we shall consider the squares of the deviations and shall form the sum S of all such squares for the given data.

$$S = (\hat{a} + 2\hat{b} - 3)^2 + (\hat{a} + 3\hat{b} - 6)^2 + (\hat{a} + 4.5\hat{b} - 8)^2 +$$
$$(\hat{a} + 5.5\hat{b} - 10)^2 + (\hat{a} + 7\hat{b} - 11)^2.$$

The method of least squares requires that we choose as the line of "best fit" the one obtained by selecting $\hat{a}$ and $\hat{b}$ so as to minimize S. We can minimize S with respect to $\hat{a}$ and $\hat{b}$ by solving the system

$$\begin{cases} \dfrac{\partial S}{\partial \hat{a}} = 0, \\[2ex] \dfrac{\partial S}{\partial \hat{b}} = 0. \end{cases}$$

We have

$$\frac{\partial S}{\partial \hat{a}} = 2(\hat{a} + 2\hat{b} - 3) + 2(\hat{a} + 3\hat{b} - 6) + 2(\hat{a} + 4.5\hat{b} - 8) +$$
$$2(\hat{a} + 5.5\hat{b} - 10) + 2(\hat{a} + 7\hat{b} - 11) = 0,$$

$$\frac{\partial S}{\partial \hat{b}} = 4(\hat{a} + 2\hat{b} - 3) + 6(\hat{a} + 3\hat{b} - 6) + 9(\hat{a} + 4.5\hat{b} - 8) +$$
$$11(\hat{a} + 5.5\hat{b} - 10) + 14(\hat{a} + 7\hat{b} - 11) = 0,$$

which when simplified gives

$$\begin{cases} 5\hat{a} + 22\hat{b} = 38, \\ 44\hat{a} + 225\hat{b} = 384. \end{cases}$$

Solving for $\hat{a}$ and $\hat{b}$, we obtain

$$\hat{a} = \frac{102}{157} \approx 0.65, \qquad \hat{b} = \frac{248}{157} \approx 1.58.$$

It can be shown that these values of $\hat{a}$ and $\hat{b}$ lead to a minimum value of S. Hence, in the sense of least squares, the line of best fit is

$$\hat{y} = 0.65 + 1.58x. \tag{2}$$

This is, in fact, the line indicated in Fig. 8.16(b). It is called the **least squares line of y on x** or the **linear regression line of y on x**. The constants $\hat{a}$ and $\hat{b}$ are called **linear regression coefficients**. With Eq. (2) we would predict that when $x = 5$, the corresponding value of y is $\hat{y} = 0.65 + 1.58(5) = 8.55$.

More generally, suppose that we are given the following n pairs of observations:

$$(x_1, y_1), (x_2, y_2), \ldots, (x_n, y_n).$$

If we assume that x and y are approximately linearly related and that we can fit a straight line $\hat{y} = \hat{a} + \hat{b}x$ that approximates the data, the sum of the squares of the errors $\hat{y} - y$ is

$$S = (\hat{a} + \hat{b}x_1 - y_1)^2 + (\hat{a} + \hat{b}x_2 - y_2)^2 + \ldots + (\hat{a} + \hat{b}x_n - y_n)^2.$$

Since S must be minimized with respect to $\hat{a}$ and $\hat{b}$,

$$\begin{cases} \dfrac{\partial S}{\partial \hat{a}} = 2(\hat{a} + \hat{b}x_1 - y_1) + 2(\hat{a} + \hat{b}x_2 - y_2) + \ldots + 2(\hat{a} + \hat{b}x_n - y_n) = 0, \\[2mm] \dfrac{\partial S}{\partial \hat{b}} = 2x_1(\hat{a} + \hat{b}x_1 - y_1) + 2x_2(\hat{a} + \hat{b}x_2 - y_2) + \ldots \\[2mm] \hspace{6cm} + 2x_n(\hat{a} + \hat{b}x_n - y_n) = 0. \end{cases}$$

Dividing both equations by 2 and using sigma notation, we have

$$\begin{cases} \hat{a}n + \hat{b}\displaystyle\sum_{i=1}^{n} x_i - \sum_{i=1}^{n} y_i = 0, \\[3mm] \hat{a}\displaystyle\sum_{i=1}^{n} x_i + \hat{b}\sum_{i=1}^{n} x_i^2 - \sum_{i=1}^{n} x_i y_i = 0. \end{cases}$$

Equivalently, we have the system of so-called *normal equations*:

$$\sum_{i=1}^{n} y_i = \hat{a}n + \hat{b}\sum_{i=1}^{n} x_i, \tag{3}$$

$$\sum_{i=1}^{n} x_i y_i = \hat{a}\sum_{i=1}^{n} x_i + \hat{b}\sum_{i=1}^{n} x_i^2. \tag{4}$$

To solve for $\hat{b}$ we first multiply Eq. (3) by $\displaystyle\sum_{i=1}^{n} x_i$ and Eq. (4) by n:

$$\left(\sum_{i=1}^{n} x_i\right)\left(\sum_{i=1}^{n} y_i\right) = \hat{a}n\sum_{i=1}^{n} x_i + \hat{b}\left(\sum_{i=1}^{n} x_i\right)^2, \tag{5}$$

$$n\sum_{i=1}^{n} x_i y_i = \hat{a}n\sum_{i=1}^{n} x_i + \hat{b}n\sum_{i=1}^{n} x_i^2. \tag{6}$$

Subtracting Eq. (5) from Eq. (6), we obtain

$$n\sum_{i=1}^{n} x_i y_i - \left(\sum_{i=1}^{n} x_i\right)\left(\sum_{i=1}^{n} y_i\right) = \hat{b}n\sum_{i=1}^{n} x_i^2 - \hat{b}\left(\sum_{i=1}^{n} x_i\right)^2$$

$$= \hat{b}\left[n\sum_{i=1}^{n} x_i^2 - \left(\sum_{i=1}^{n} x_i\right)^2\right].$$

Thus

$$\hat{b} = \frac{n \sum\limits_{i=1}^{n} x_i y_i - \left(\sum\limits_{i=1}^{n} x_i\right)\left(\sum\limits_{i=1}^{n} y_i\right)}{n \sum\limits_{i=1}^{n} x_i^2 - \left(\sum\limits_{i=1}^{n} x_i\right)^2}. \tag{7}$$

Solving Eqs. (3) and (4) for $\hat{a}$ gives

$$\hat{a} = \frac{\left(\sum\limits_{i=1}^{n} x_i^2\right)\left(\sum\limits_{i=1}^{n} y_i\right) - \left(\sum\limits_{i=1}^{n} x_i\right)\left(\sum\limits_{i=1}^{n} x_i y_i\right)}{n \sum\limits_{i=1}^{n} x_i^2 - \left(\sum\limits_{i=1}^{n} x_i\right)^2}. \tag{8}$$

It can be shown that these values of $\hat{a}$ and $\hat{b}$ minimize S.

Computing the linear regression coefficients $\hat{a}$ and $\hat{b}$ by the formulas of Eqs. (7) and (8) gives the linear regression line of y on x, namely, $\hat{y} = \hat{a} + \hat{b}x$, which can be used to estimate y for a given value of x.

In the next example, as well as in the exercises, you will encounter **index numbers**. They are used to relate a variable in one period of time to the same variable in another period, this latter period called the *base period*. An index number is a *relative* number that describes data that are changing over time. Such data are referred to as *time series*.

For example, consider the time series data of total production of widgets in the United States for 1983–1987 indicated in Table 8.5. If we choose 1984 as the base year and assign to it the index number 100, then the other index

TABLE 8.5

YEAR	PRODUCTION (THOUSANDS)	INDEX (1984 = 100)
1983	828	92
1984	900	100
1985	936	104
1986	891	99
1987	954	106

numbers are obtained by dividing each year's production by the 1984 production of 900 and multiplying the result by 100. We can, for example, interpret the index 106 for 1987 as meaning that production for that year was 106 percent of the production in 1984.

In time series analysis, index numbers are obviously of great advantage if the data involve numbers of great magnitude. But regardless of the magnitude of the data, index numbers simplify the task of comparing changes in data over periods of time.

EXAMPLE 1   *By means of the linear regression line, represent the trend for the Index of Industrial Production from 1963 to 1968 (Table 8.6).*

TABLE 8.6
Index of Industrial Production (1967 = 100)

YEAR	1963	1964	1965	1966	1967	1968
INDEX	79	84	91	99	100	105

Source: Economic Report of the President, 1971 (Washington, D.C.: U.S. Government Printing Office), 1971.

We shall let x denote time, y denote the index, and treat y as a linear function of x. Also, we shall designate 1963 by $x = 1$, 1964 by $x = 2$, etc. There are $n = 6$ pairs of measurements. To determine the linear regression coefficients by using Eqs. (7) and (8), we first perform the arithmetic shown in Table 8.7.

TABLE 8.7

YEAR	x_i	y_i	$x_i y_i$	x_i^2
1963	1	79	79	1
1964	2	84	168	4
1965	3	91	273	9
1966	4	99	396	16
1967	5	100	500	25
1968	6	105	630	36
Total	21	558	2046	91
	$= \sum_{i=1}^{6} x_i$	$= \sum_{i=1}^{6} y_i$	$= \sum_{i=1}^{6} x_i y_i$	$= \sum_{i=1}^{6} x_i^2$

Hence by Eq. (8),

$$\hat{a} = \frac{91(558) - 21(2046)}{6(91) - (21)^2} = \frac{7812}{105} = 74.4,$$

and by Eq. (7),

$$\hat{b} = \frac{6(2046) - 21(558)}{6(91) - (21)^2} = \frac{558}{105} = 5.31 \text{ (approximately)}.$$

Thus the regression line of y on x is

$$\hat{y} = 74.4 + 5.31x,$$

whose graph, as well as a scatter diagram, appears in Fig. 8.18.

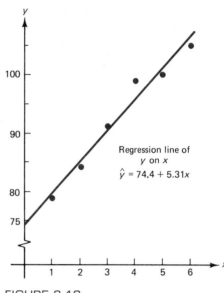

Regression line of
y on x
$\hat{y} = 74.4 + 5.31x$

FIGURE 8.18

EXERCISE 8.7

In Problems 1–4, find an equation of the least squares linear regression line of y on x for the given data and sketch both the line and the data. Predict the value of y corresponding to x = 3.5.

1.

x	1	2	3	4	5	6
y	1.5	2.3	2.6	3.7	4.0	4.5

2.

x	1	2	3	4	5	6	7
y	1	1.8	2	4	4.5	7	9

3.

x	2	3	4.5	5.5	7
y	3	5	8	10	11

4.

x	2	3	4	5	6	7
y	2.4	2.9	3.3	3.8	4.3	4.9

5. A firm finds that when the price of its product is p dollars per unit, the number of units sold is q, as indicated below. Find an equation of the regression line of q on p.

Price, p	10	30	40	50	60	70
Demand, q	70	68	63	50	46	32

6. On a farm an agronomist finds that the amount of water applied (in inches) and the corresponding yield of a certain crop (in tons per acre) are as given below. Find an equation of the regression line of y on x. Predict y when $x = 12$.

Water, x	8	16	24	32
Yield, y	4.1	4.5	5.1	6.1

7. A rabbit was injected with a virus, and x hours after the injection the temperature y (in degrees Fahrenheit) of the rabbit was measured [36]. The data are given below. Find an equation of the regression line of y on x, and estimate the rabbit's temperature 40 hours after the injection.

Elapsed time, x	24	32	48	56
Temperature, y	102.8	104.5	106.5	107.0

8. In a psychological experiment, four persons were subjected to a stimulus. Both before and after the stimulus, the systolic blood pressure (in millimeters of mercury) of each subject was measured. The data are given below. Find an equation of the regression line of y on x where x and y are defined below.

Before stimulus, x	blood pressure			
	130	132	136	140
After stimulus, y	139	140	144	146

For the time series in Problems 9 and 10 fit a linear regression line by least squares; that is, find an equation of the linear regression line of y on x. In each case let the first year in the table correspond to $x = 1$.

9. Production of Product A, 1983–1987 (thousands of units)

YEAR	PRODUCTION
1983	10
1984	15
1985	16
1986	18
1987	21

10. In the following, let 1961 correspond to $x = 1$, 1963 correspond to $x = 3$, etc.

Wholesale Price Index—Lumber and Wood Products (1967 = 100)

YEAR	INDEX
1961	91
1963	94
1965	96
1967	100

Source: *Economic Report of the President, 1971* (Washington, D.C., U.S. Government Printing Office), 1971.

11. a. Find an equation of the least squares line of y on x for the following data. Refer to 1981 as year $x = 1$, etc.

Overseas Shipments of Computers by Acme Computer Co. (in thousands)

YEAR	QUANTITY
1981	35
1982	31
1983	26
1984	24
1985	26

b. For the data in part (a), refer to 1981 as year $x = -2$, 1982 as year $x = -1$, 1983 as year $x = 0$, etc. Then $\sum_{i=1}^{5} x_i = 0$. Fit a least squares line and observe how the calculation is simplified.

12. For the following time series, find an equation of the linear regression line that best fits the data. Refer to 1965 as year $x = -2$, 1966 as year $x = -1$, etc.

Consumer Price Index—Medical Care 1965–1969 (1967 = 100)

YEAR	INDEX
1965	90
1966	93
1967	100
1968	106
1969	113

Source: *Economic Report of the President, 1971* (Washington, D.C.: U.S. Government Printing Office), 1971.

8.8 MULTIPLE INTEGRALS

Recall that the definite integral of a function of one variable is concerned with integration over an *interval*. There are also definite integrals of functions of two variables, called (definite) **double integrals.** These involve integration over a *region* in the plane.

For example, the symbol

$$\int_0^2 \int_3^4 xy \, dx \, dy, \quad \text{or equivalently} \quad \int_0^2 \left[\int_3^4 xy \, dx \right] dy,$$

is the double integral of $f(x, y) = xy$ over a region determined by the limits of integration. That region consists of all points (x, y) in the x,y-plane such that $3 \le x \le 4$ and $0 \le y \le 2$, which is shown in Fig. 8.19. Essentially, a double integral is a limit of a sum of the form $\Sigma f(x, y) \, \Delta x \, \Delta y$, where in our

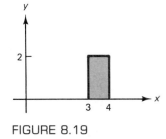

FIGURE 8.19

case the points (x, y) are in the shaded region. A geometric interpretation of a double integral will be given later.

To evaluate

$$\int_0^2 \int_3^4 xy \, dx \, dy \qquad \text{or} \qquad \int_0^2 \left[\int_3^4 xy \, dx \right] dy,$$

we use successive integrations starting with the innermost integral. First, we evaluate

$$\int_3^4 xy \, dx$$

by treating y as a constant and integrating with respect to x between the limits 3 and 4:

$$\int_3^4 xy \, dx = \frac{x^2 y}{2} \Big|_3^4.$$

Substituting the limits for the variable x, we have

$$\frac{4^2 \cdot y}{2} - \frac{3^2 \cdot y}{2} = \frac{16y}{2} - \frac{9y}{2} = \frac{7}{2} y.$$

Now we integrate this result with respect to y between the limits of 0 and 2.

$$\int_0^2 \frac{7}{2} y \, dy = \frac{7y^2}{4} \Big|_0^2 = \frac{7 \cdot 2^2}{4} - 0 = 7.$$

Thus

$$\int_0^2 \int_3^4 xy \, dx \, dy = 7.$$

Now consider the double integral

$$\int_0^1 \int_{x^3}^{x^2} (x^3 - xy) \, dy \, dx \qquad \text{or} \qquad \int_0^1 \left[\int_{x^3}^{x^2} (x^3 - xy) \, dy \right] dx.$$

Here we integrate first with respect to y and then with respect to x. The region over which the integration takes place is all points (x, y) where $x^3 \le y \le x^2$ and $0 \le x \le 1$ (see Fig. 8.20). This double integral is evaluated by first treating

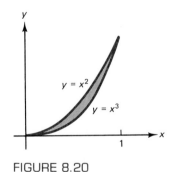

FIGURE 8.20

x as a constant and integrating $x^3 - xy$ with respect to y between x^3 and x^2. Then we integrate the result with respect to x between 0 and 1.

$$\int_0^1 \int_{x^3}^{x^2} (x^3 - xy)\, dy\, dx$$

$$= \int_0^1 \left[\int_{x^3}^{x^2} (x^3 - xy)\, dy \right] dx = \int_0^1 \left(x^3 y - \frac{xy^2}{2} \right) \Big|_{x^3}^{x^2} dx$$

$$= \int_0^1 \left(\left[x^3(x^2) - \frac{x(x^2)^2}{2} \right] - \left[x^3(x^3) - \frac{x(x^3)^2}{2} \right] \right) dx$$

$$= \int_0^1 \left(x^5 - \frac{x^5}{2} - x^6 + \frac{x^7}{2} \right) dx = \int_0^1 \left(\frac{x^5}{2} - x^6 + \frac{x^7}{2} \right) dx$$

$$= \left(\frac{x^6}{12} - \frac{x^7}{7} + \frac{x^8}{16} \right) \Big|_0^1 = \left(\frac{1}{12} - \frac{1}{7} + \frac{1}{16} \right) - 0 = \frac{1}{336}.$$

EXAMPLE 1   *Evaluate* $\displaystyle\int_{-1}^1 \int_0^{1-x} (2x + 1)\, dy\, dx.$

Here we first integrate with respect to y.

$$\int_{-1}^1 \int_0^{1-x} (2x + 1)\, dy\, dx$$

$$= \int_{-1}^1 \left[\int_0^{1-x} (2x + 1)\, dy \right] dx$$

$$= \int_{-1}^1 (2xy + y) \Big|_0^{1-x} dx = \int_{-1}^1 \{[2x(1 - x) + (1 - x)] - 0\}\, dx$$

$$= \int_{-1}^1 (-2x^2 + x + 1)\, dx = \left(-\frac{2x^3}{3} + \frac{x^2}{2} + x \right) \Big|_{-1}^1$$

$$= \left(-\frac{2}{3} + \frac{1}{2} + 1 \right) - \left(\frac{2}{3} + \frac{1}{2} - 1 \right) = \frac{2}{3}.$$

EXAMPLE 2   *Evaluate* $\displaystyle\int_1^{\ln 2} \int_{e^y}^2 dx\, dy.$

Here we first integrate with respect to x.

$$\int_1^{\ln 2} \int_{e^y}^2 dx\, dy = \int_1^{\ln 2} \left[\int_{e^y}^2 dx \right] dy = \int_1^{\ln 2} x \Big|_{e^y}^2 dy$$

$$= \int_1^{\ln 2} (2 - e^y)\, dy = (2y - e^y) \Big|_1^{\ln 2}$$

$$= (2 \ln 2 - 2) - (2 - e) = 2 \ln 2 - 4 + e$$

$$= \ln 4 - 4 + e.$$

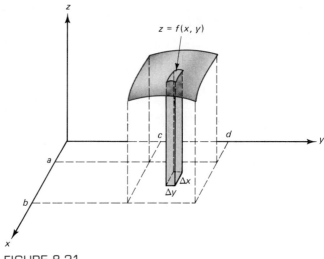

FIGURE 8.21

A double integral can be interpreted in terms of the volume of a region between the x,y-plane and a surface $z = f(x, y)$ if $z \geq 0$. In Fig. 8.21 is a region whose volume we shall consider. The element of volume for this region is a vertical column. It has height $z = f(x, y)$ and a base area of $\Delta y \, \Delta x$. Thus its volume is $f(x, y) \, \Delta y \, \Delta x$. The volume of the entire region can be found by summing the volumes of all such elements for $a \leq x \leq b$ and $c \leq y \leq d$ via a double integral. Thus

$$\text{volume} = \int_a^b \int_c^d f(x, y) \, dy \, dx.$$

Triple integrals are handled by successively evaluating three integrals, as the next example shows.

EXAMPLE 3   *Evaluate* $\displaystyle\int_0^1 \int_0^x \int_0^{x-y} x \, dz \, dy \, dx.$

$$\int_0^1 \int_0^x \int_0^{x-y} x \, dz \, dy \, dx = \int_0^1 \int_0^x \left[\int_0^{x-y} x \, dz \right] dy \, dx$$

$$= \int_0^1 \int_0^x (xz) \Big|_0^{x-y} dy \, dx = \int_0^1 \int_0^x [x(x - y) - 0] \, dy \, dx$$

$$= \int_0^1 \int_0^x (x^2 - xy) \, dy \, dx = \int_0^1 \left[\int_0^x (x^2 - xy) \, dy \right] dx$$

$$= \int_0^1 \left(x^2 y - \frac{xy^2}{2} \right) \Big|_0^x dx = \int_0^1 \left[\left(x^3 - \frac{x^3}{2} \right) - 0 \right] dx$$

$$= \int_0^1 \frac{x^3}{2} \, dx = \frac{x^4}{8} \Big|_0^1 = \frac{1}{8}.$$

EXERCISE 8.8

In Problems **1–22,** *evaluate the multiple integrals.*

1. $\int_0^3 \int_0^4 x \, dy \, dx.$ **2.** $\int_0^2 \int_1^2 y \, dy \, dx.$

3. $\int_0^1 \int_0^1 xy \, dx \, dy.$ **4.** $\int_0^2 \int_0^3 x^2 \, dy \, dx.$

5. $\int_1^3 \int_1^2 (x^2 - y) \, dx \, dy.$

6. $\int_{-1}^2 \int_1^4 (x^2 - 2xy) \, dy \, dx.$

7. $\int_0^1 \int_0^2 (x + y) \, dy \, dx.$ **8.** $\int_0^3 \int_0^x (x^2 + y^2) \, dy \, dx.$

9. $\int_0^6 \int_0^{3x} y \, dy \, dx.$ **10.** $\int_1^2 \int_0^{x-1} y \, dy \, dx.$

11. $\int_0^1 \int_{3x}^{x^2} 2x^2 y \, dy \, dx.$ **12.** $\int_0^2 \int_0^{x^2} xy \, dy \, dx.$

13. $\int_0^2 \int_0^{\sqrt{4-y^2}} x \, dx \, dy.$ **14.** $\int_0^1 \int_y^{\sqrt{y}} y \, dx \, dy.$

15. $\int_{-1}^1 \int_x^{1-x} (x + y) \, dy \, dx.$

16. $\int_0^3 \int_{y^2}^{3y} x \, dx \, dy.$

17. $\int_0^1 \int_0^y e^{x+y} \, dx \, dy.$ **18.** $\int_2^3 \int_0^2 e^{x-y} \, dx \, dy.$

19. $\int_{-1}^0 \int_{-1}^2 \int_1^2 6xy^2 z^3 \, dx \, dy \, dz.$

20. $\int_0^1 \int_0^x \int_0^{x-y} x \, dz \, dy \, dx.$

21. $\int_0^1 \int_{x^2}^x \int_0^{xy} dz \, dy \, dx.$ **22.** $\int_0^2 \int_{y^2}^{3y} \int_0^x dz \, dx \, dy.$

23. In statistics a joint density function $z = f(x, y)$ defined on a region in the x,y-plane is represented by a surface in space. The probability that $a \le x \le b$ and $c \le y \le d$ is given by

$$P(a \le x \le b, c \le y \le d) = \int_c^d \int_a^b f(x, y) \, dx \, dy,$$

and is represented by the volume between the graph of f and the rectangular region given by $a \le x \le b$ and $c \le y \le d$. If $f(x, y) = e^{-(x+y)}$ is a joint density function, where $x \ge 0$ and $y \ge 0$, find

$$P(0 \le x \le 2, 1 \le y \le 2)$$

and give your answer in terms of e.

24. In Problem 23, let $f(x, y) = 12e^{-4x-3y}$ for $x, y \ge 0$. Find $P(3 \le x \le 4, 2 \le y \le 6)$ and give your answer in terms of e.

25. In Problem 23, let $f(x, y) = x/8$ where $0 \le x \le 2$ and $0 \le y \le 4$. Find $P(x \ge 1, y \ge 2)$.

26. In Problem 23, let f be the uniform density function $f(x, y) = 1$ defined over the unit square $0 \le x \le 1$, $0 \le y \le 1$. Find the probability that $0 \le x \le \frac{1}{3}$ and $\frac{1}{4} \le y \le \frac{3}{4}$.

8.9 REVIEW _____

Important Terms and Symbols

Section 8.1	three-dimensional coordinate system $f(x_1, x_2, \ldots, x_n)$ function of n variables	
	x,y-plane x,z-plane y,z-plane octant traces	

Section 8.2   partial derivative $\dfrac{\partial z}{\partial x}$ $f_x(x, y)$ $\dfrac{\partial z}{\partial x}\Big|_{(x_0, y_0)}$ $f_x(x_0, y_0)$

Section 8.3   joint-cost function   production function   marginal productivity   competitive products   complementary products

Section 8.4 $\dfrac{\partial^2 z}{\partial y \, \partial x}$ $\dfrac{\partial^2 z}{\partial x \, \partial y}$ $\dfrac{\partial^2 z}{\partial x^2}$ $\dfrac{\partial^2 z}{\partial y^2}$ f_{xy} f_{yx} f_{xx} f_{yy}

Summary

We can extend the concept of a function of one variable to functions of several variables. The inputs for functions of n variables are n-tuples. Generally, the graph of a function of two variables is a surface in a three-dimensional coordinate system. Functions of more than two variables cannot be geometrically represented.

For a function of n variables, we can consider n partial derivatives. For example, if $w = f(x, y, z)$, we have the partial derivatives of f with respect to x, with respect to y, and with respect to z, denoted f_x, f_y, and f_z, or $\partial f/\partial x$, $\partial f/\partial y$, and $\partial f/\partial z$, respectively. To find $f_x(x, y, z)$, treat y and z as constants and differentiate f with respect to x in the usual way. The other partial derivatives are found in a similar fashion. We can interpret $f_x(x, y, z)$ as the approximate change in w that results from a one-unit change in x when y and z are held fixed. There are similar interpretations for the other partial derivatives.

Functions of several variables occur frequently in business and economic analysis. If a manufacturer produces x units of product X and y units of product Y, then the total cost c of these units is a function of x and y and is called a joint-cost function. The partial derivatives $\partial c/\partial x$ and $\partial c/\partial y$ are called the marginal costs with respect to x and y, respectively. We can interpret, for examaple, $\partial c/\partial x$ as the approximate cost of producing an extra unit of X while the level of production of Y is held fixed.

If l units of labor and k units of capital are used to produce P units of a product, the function $P = f(l, k)$ is called a production function. The partial derivatives of P are called marginal productivity functions.

Suppose two products, A and B, are such that the quantity demanded of each is dependent on the prices of both. If q_A and q_B are the quantities of A and B demanded when the prices of A and B are p_A and p_B, respectively, then q_A and q_B are each functions of p_A and p_B. When $\partial q_A/\partial p_B > 0$ and $\partial q_B/\partial p_A > 0$, then A and B are called competitive products (or substitutes). When $\partial q_A/\partial p_B < 0$ and $\partial q_B/\partial p_A < 0$, then A and B are called complementary products.

A partial derivative of a function of n variables is itself a function of n variables. By successively taking partial derivatives of partial derivatives, we obtain higher-order partial derivatives. For example, if f is a function of x and y, then f_{xy} denotes the partial derivative of f_x with respect to y; f_{xy} is called the second partial derivative of f, first with respect to x and then with respect to y.

If the function $f(x, y)$ has a relative extremum at (x_0, y_0), then (x_0, y_0) must be a solution of the system

$$f_x(x_0, y_0) = 0, \qquad f_y(x_0, y_0) = 0.$$

Any solution of this system is called a critical point of f. Thus critical points are the candidates at which a relative extremum may occur. The second-derivative test for functions of two variables gives conditions under which a critical point corresponds to a relative maximum or a relative minimum. It states that if (x_0, y_0) is a critical point of f and

$$D(x, y) = f_{xx}(x, y) f_{yy}(x, y) - [f_{xy}(x, y)]^2,$$

then

a. if $D(x_0, y_0) > 0$ and $f_{rr}(x_0, y_0) < 0$, f has a relative maximum at (x_0, y_0);

b. if $D(x_0, y_0) > 0$ and $f_{xx}(x_0, y_0) > 0$, f has a relative minimum at (x_0, y_0);

c. if $D(x_0, y_0) < 0$, f has neither a relative maximum nor a relative minimum at (x_0, y_0);

d. if $D(x_0, y_0) = 0$, no conclusion about extrema at (x_0, y_0) can yet be drawn and further analysis is required.

To find critical points of a function of several variables subject to a constraint, we may use the method of Lagrange multipliers. For example, to find the critical points of $f(x, y, z)$ subject to the constraint $g(x, y, z) = 0$, we first form the function

$$F(x, y, z, \lambda) = f(x, y, z) - \lambda g(x, y, z).$$

By solving the system

$$F_x = 0, \quad F_y = 0, \quad F_z = 0, \quad F_\lambda = 0,$$

we obtain the critical points of F. If $(x_0, y_0, z_0, \lambda_0)$ is such a critical point, then (x_0, y_0, z_0) is a critical point of f subject to the constraint. It is important to write the constraint in the form $g(x, y, z) = 0$. For example, if the constraint is $2x + 3y - z = 4$, then $g(x, y, z) = 2x + 3y - z - 4$ [or $g(x, y, z) = 4 - 2x - 3y + z$]. If $f(x, y, z)$ is subject to two constraints, $g_1(x, y, z) = 0$ and $g_2(x, y, z) = 0$, then we would form the function $F = f - \lambda_1 g_1 - \lambda_2 g_2$ and solve the system

$$F_x = 0, \quad F_y = 0, \quad F_z = 0, \quad F_{\lambda_1} = 0, \quad F_{\lambda_2} = 0.$$

Sometimes two variables, say x and y, may be related in such a way that the relationship is approximately linear. When the data points (x_i, y_i), where $i = 1, 2, 3, \ldots, n$, are given to us, we can fit a straight line that approximates them. Such a line is the linear regression line (or least squares line) of y on x and is given by $\hat{y} = \hat{a} + \hat{b}x$ where

$$\hat{a} = \frac{\left(\sum_{i=1}^{n} x_i^2 \right)\left(\sum_{i=1}^{n} y_i \right) - \left(\sum_{i=1}^{n} x_i \right)\left(\sum_{i=1}^{n} x_i y_i \right)}{n \sum_{i=1}^{n} x_i^2 - \left(\sum_{i=1}^{n} x_i \right)^2}$$

and

$$\hat{b} = \frac{n \sum_{i=1}^{n} x_i y_i - \left(\sum_{i=1}^{n} x_i \right)\left(\sum_{i=1}^{n} y_i \right)}{n \sum_{i=1}^{n} x_i^2 - \left(\sum_{i=1}^{n} x_i \right)^2}$$

The $\hat{y}$-values can be used to predict y-values for given values of x.

When working with functions of several variables, we can consider their multiple integrals. These are determined by successive integration. For example, the double integral

$$\int_1^2 \int_0^y (x + y) \, dx \, dy$$

is determined by first treating y as a constant and integrating $x + y$ with respect to x. After evaluating between the limits 0 and y, we integrate that result with respect to y from $x = 1$ to $x = 2$. Thus

$$\int_1^2 \int_0^y (x + y) \, dx \, dy = \int_1^2 \left[\int_0^y (x + y) \, dx \right] dy.$$

Triple integrals involve functions of three variables and are also evaluated by successive integration.

Review Problems

In Problems 1–4, sketch the given surfaces.

1. $2x + 3y + z = 9$.

2. $z = x$.

3. $z = y^2$.

4. $x^2 + z^2 = 1$.

In Problems 5–16, find the indicated partial derivatives.

5. $f(x, y) = 2x^2 + 3xy + y^2 - 1$; $f_x(x, y)$, $f_y(x, y)$.

6. $P = l^3 + k^3 - lk$; $\partial P/\partial l$, $\partial P/\partial k$.

7. $z = \dfrac{x}{x + y}$; $\partial z/\partial x$, $\partial z/\partial y$.

8. $w = \dfrac{\sqrt{x^2 + y^2}}{y}$; $\partial w/\partial x$.

9. $w = e^{x^2 yz}$; $w_{xy}(x, y)$.

10. $f(x, y) = xy \ln(xy)$; $f_{xy}(x, y)$.

11. $f(x, y) = \ln\sqrt{x^2 + y^2}$; $\dfrac{\partial}{\partial y}[f(x, y)]$.

12. $f(p_A, p_B) = 2(p_A - 20) + 3(p_B - 30)$; $f_{p_A}(p_A, p_B)$.

13. $f(x, y, z) = (x + y)(y + z^2)$; $\dfrac{\partial^2}{\partial z^2}[f(x, y, z)]$.

14. $z = (x^2 - y)(y^2 - 2xy)$; $\partial^2 z/\partial y^2$.

15. $w = xe^{yz} \ln z$; $\partial w/\partial y$, $\partial^2 w/\partial x \partial z$.

16. $P = 2.4l^{0.11}k^{0.89}$; $\partial P/\partial k$.

17. If a manufacturer's production function is defined by $P = 20l^{0.7}k^{0.3}$, determine the marginal productivity functions.

18. A manufacturer's cost for producing x units of product X and y units of product Y is given by $c = 5x + 0.03xy + 7y + 200$. Determine the (partial) marginal cost with respect to x when $x = 100$ and $y = 200$.

19. If $q_A = 200 - 3p_A + p_B$ and $q_B = 50 - 5p_B + p_A$,

where q_A and q_B are the number of units demanded of products A and B, respectively, and p_A and p_B are their respective prices per unit, determine whether A and B are competitive or complementary products.

20. For industry there is a model that describes the rate α (a Greek letter read "alpha") at which a new innovation substitutes for an established process [40]. It is given by

$$\alpha = Z + 0.530P - 0.027S,$$

where Z is a constant that depends on the particular industry, P is an index of profitability of the new innovation, and S is an index of the extent of the investment necessary to make use of the innovation. Find $\partial\alpha/\partial P$ and $\partial\alpha/\partial S$.

21. Examine $f(x, y) = x^2 + 2y^2 - 2xy - 4y + 3$ for relative extrema.

22. Examine $f(w, z) = 2w^3 + 2z^3 - 6wz + 7$ for relative extrema.

23. An open-top rectangular cardboard box is to have a volume of 32 cubic feet. Find the dimensions of the box so that the amount of cardboard used is minimized.

24. Find all critical points of $f(x, y, z) = xyz$ subject to the condition that $3x + 2y + 4z - 120 = 0$ $(xyz \neq 0)$.

25. Find all critical points of $f(x, y, z) = x^2 + y^2 + z^2$ subject to the constraint $3x + 2y + z = 14$.

26. Find the least squares linear regression line of y on x for the data given below. Refer to year 1982 as year $x = 1$, etc.

Equipment Expenditures of Allied Computer Company, 1982–1987 (millions of dollars)

YEAR	EXPENDITURES
1982	15
1983	22
1984	21
1985	27
1986	26
1987	34

27. In an experiment [67] a group of fish were injected with living bacteria. Of those fish maintained at 28° C, the percentage p that survived the infection t hours after the injection is given below. Find the linear regression line of p on t.

t	8	10	18	20	48
p	82	79	78	78	64

In Problems 28–31, evaluate the double integrals.

28. $\displaystyle\int_0^4 \int_{y/2}^2 xy \, dx \, dy.$

29. $\displaystyle\int_1^2 \int_0^y x^2 y^2 \, dx \, dy.$

30. $\displaystyle\int_0^3 \int_{y^2}^{3y} x \, dx \, dy.$

31. $\displaystyle\int_0^1 \int_{\sqrt{x}}^{x^2} (x^2 + 2xy - 3y^2) \, dy \, dx.$

Trigonometric Functions

There are many situations in which cyclic, or wave, patterns are evident. For example, you may have heard of business cycles, breathing cycles, or brain waves. Such cycles or waves may be described by trigonometric functions. In this chapter we consider the calculus of these functions. Because the inputs for trigonometric functions are angles, we begin with a discussion of radian measure of an angle.

9.1 RADIAN MEASURE

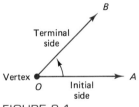

FIGURE 9.1

An angle is generated when a ray, or half-line, is rotated about its endpoint. For example, the angle in Fig. 9.1 is generated by rotating the ray *OA* about its endpoint *O* to the position of *OB*. We call *OA* the **initial side** of the angle, *OB* the **terminal side**, and *O* the **vertex**. An angle generated by a counterclockwise rotation is taken to be *positive,* and an angle is *negative* if the rotation is clockwise. The direction of rotation is indicated by an arrow, as in Fig. 9.2.

An angle is said to be in **standard position** when its vertex is the origin of a rectangular coordinate system and its initial side lies on the positive

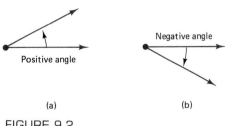

(a)   (b)

FIGURE 9.2

343

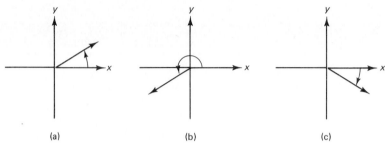

Angles in standard position

FIGURE 9.3

horizontal axis. Figure 9.3 shows some angles in standard position. If the terminal side lies in Quadrant I, then the angle is called a **first-quadrant angle**. Similarly, there are **second-**, **third-**, and **fourth-quadrant angles**. An angle whose terminal side lies on an axis is called a **quadrantal angle**.

EXAMPLE 1

a. Figure 9.3 shows a first-quadrant angle in part (a), a third-quadrant angle in part (b), and a fourth-quadrant angle in part (c).

b. The angles in Fig. 9.4 are quadrantal angles. For the first angle, the initial and terminal sides coincide.

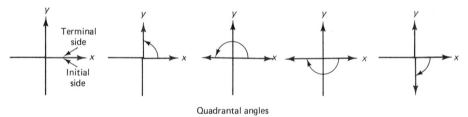

Quadrantal angles

FIGURE 9.4

One common unit for measuring an angle is the *degree*. One **degree**, denoted 1°, is the measure of an angle generated by $\frac{1}{360}$ of a counterclockwise revolution. Thus an angle generated by one revolution has a measure of 360°. Figure 9.5 shows the degree measure of several angles. In part (a), the angle has measure 45°. If θ (the Greek lowercase letter theta) represents this angle, then we write $\theta = 45°$. In part (c), notice that the measure of a negative angle includes a minus sign.

Another unit of angular measurement is the *radian*. Our focus will be on the radian because it is more useful than the degree in obtaining differentiation and integration formulas for trigonometric functions. To define a radian, consider

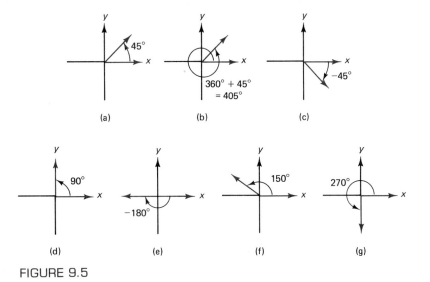

FIGURE 9.5

a circle of radius r. A central angle that subtends (cuts off) an arc of length r on the circle [see Fig. 9.6(a)] is defined to be an angle of one **radian**, abbreviated 1 rad. Thus an arc of length $2r$ corresponds to a central angle

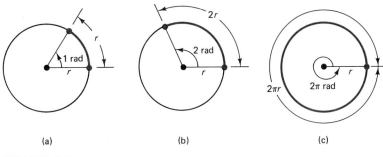

FIGURE 9.6

with measure 2 rad [see Fig. 9.6(b)]. Because the circumference of the circle is $2\pi r$, an angle of one revolution is 2π rad [see Fig. 9.6(c)]. Thus

$$2\pi \text{ rad} = 360°.$$

Dividing both sides by 2 gives

$$\pi \text{ rad} = 180°,$$

which is a useful formula for converting between radians and degrees. Examples 2 and 3 will illustrate.

EXAMPLE 2

a. *Convert 2 rad to degrees.*

Dividing both sides by of π rad $= 180°$ by π rad gives $1 = \dfrac{180°}{\pi \text{ rad}}$. Thus

$$2 \text{ rad} = 2 \text{ rad} \cdot (1) = 2 \text{ rad} \cdot \dfrac{180°}{\pi \text{ rad}}.$$

On the right side the radian units cancel and we are left with degrees.

$$2 \text{ rad} = 2 \text{ \cancel{rad}} \cdot \dfrac{180°}{\pi \text{ \cancel{rad}}} = \left(\dfrac{360}{\pi}\right)^{\circ} \approx 114.59°.$$

b. *Convert $\pi/4$ rad to degrees.*

$$\dfrac{\pi}{4} \text{ rad} = \dfrac{\cancel{\pi}}{\cancel{4}} \text{ \cancel{rad}} \cdot \dfrac{\overset{45°}{\cancel{180°}}}{\cancel{\pi} \text{ \cancel{rad}}} = 45°.$$

c. *Convert $-2\pi/3$ rad to degrees.*

$$-\dfrac{2\pi}{3} \text{ rad} = -\dfrac{2\pi}{3} \text{ \cancel{rad}} \cdot \dfrac{180°}{\pi \text{ \cancel{rad}}} = -120°.$$

EXAMPLE 3

a. *Convert 90° to radians.*

Since π rad $= 180°$, then $\dfrac{\pi \text{ rad}}{180°} = 1$. Thus

$$90° = 90° \cdot (1) = 90° \cdot \dfrac{\pi \text{ rad}}{180°}.$$

On the right side the degree units cancel and we are left with

$$90° = 90^{\cancel{\circ}} \cdot \dfrac{\pi \text{ rad}}{180^{\cancel{\circ}}} = \dfrac{\pi}{2} \text{ rad}.$$

b. *Convert 30° to radians.*

$$30° = 30° \cdot \dfrac{\pi \text{ rad}}{180°} = \dfrac{\pi}{6} \text{ rad}.$$

Usually, we omit the symbol *rad* on radian measure. Thus an angle of 2 means 2 rad and an angle of π means π rad. Figure 9.7 shows some angles measured in degrees and radians.

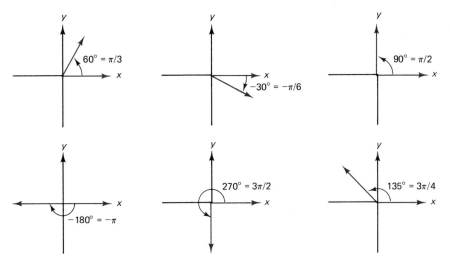

FIGURE 9.7

There are special names for certain types of angles. An angle t between 0 and $\pi/2$ (that is, $0 < t < \pi/2$) is called an **acute angle**; one between $\pi/2$ and π is an **obtuse angle**; an angle of $\pi/2$ is called a **right angle**. These are shown in Fig. 9.8.

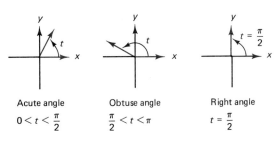

Acute angle
$0 < t < \dfrac{\pi}{2}$

Obtuse angle
$\dfrac{\pi}{2} < t < \pi$

Right angle
$t = \dfrac{\pi}{2}$

FIGURE 9.8

In Fig. 9.9, notice that the angles $\pi/3$ and $2\pi + (\pi/3) = 7\pi/3$ have the same terminal side. Such angles are called **coterminal angles**. In general, adding (or subtracting) a multiple of 2π to (or from) an angle yields an angle coterminal with it.

EXAMPLE 4   *Find the angle t that is coterminal with $16\pi/3$ and such that $0 \le t < 2\pi$.*

By adding or subtracting a multiple of 2π to or from any angle, we obtain a coterminal angle. Since $16\pi/3 \ge 2\pi$, we shall successively subtract 2π from $16\pi/3$ until we reach an angle between 0 and 2π.

$$\frac{16\pi}{3} - 2\pi = \frac{16\pi}{3} - \frac{6\pi}{3} = \frac{10\pi}{3}.$$

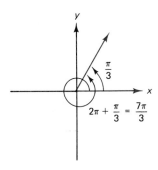

FIGURE 9.9

Since $10\pi/3 \geq 2\pi$, we continue.

$$\frac{10\pi}{3} - 2\pi = \frac{10\pi}{3} - \frac{6\pi}{3} = \frac{4\pi}{3}.$$

Now, $0 \leq 4\pi/3 < 2\pi$ and $4\pi/3$ is coterminal with $16\pi/3$. Thus $t = 4\pi/3$.

EXERCISE 9.1

*In Problems **1–8**, convert each angle to degrees.*

1. $\dfrac{\pi}{6}$.

2. $\dfrac{\pi}{3}$.

3. $\dfrac{\pi}{2}$.

4. $-\dfrac{3\pi}{4}$.

5. $-\dfrac{7\pi}{3}$.

6. $\dfrac{5\pi}{3}$.

7. $\dfrac{5\pi}{4}$.

8. -4π.

*In Problems **9–16**, convert each angle to radians.*

9. $0°$.

10. $180°$.

11. $45°$.

12. $720°$.

13. $-120°$.

14. $150°$.

15. $315°$.

16. $-270°$.

*In Problems **17–24**, determine whether the angle is quadrantal, or a first-, second-, third-, or fourth-quadrant angle.*

17. $\dfrac{2\pi}{3}$.

18. $\dfrac{\pi}{3}$.

19. $-\dfrac{\pi}{6}$.

20. 0.

21. $\dfrac{5\pi}{6}$.

22. 4π.

23. $-\pi$.

24. $-\dfrac{\pi}{4}$.

*In Problems **25** and **26**, find two positive and two negative angles that are coterminal with the given angle.*

25. $\dfrac{\pi}{6}$.

26. $\dfrac{2\pi}{3}$.

*In Problems **27–32**, find the angle t that is coterminal with the given angle and such that $0 \leq t < 2\pi$.*

27. $\dfrac{13\pi}{4}$.

28. $\dfrac{14\pi}{3}$.

29. $\dfrac{7\pi}{3}$.

30. 6π.

31. $-\dfrac{17\pi}{2}$.

32. $-\dfrac{5\pi}{6}$.

33. Suppose that a pendulum of length 20 cm swings through an arc of length 12 cm. Through what angle, in radians, does the pendulum swing?

34. A portion of a cloverleaf of an exit on a certain interstate road can be considered to be an arc of a circle of radius 300 m. The central angle of this arc is $2\pi/3$. Find the length of this portion of the cloverleaf.

35. The minute hand of a clock is 6 cm long. Over a time period of 50 minutes, through what distance does the tip of the minute hand move?

9.2 THE SINE AND COSINE FUNCTIONS

There are six special functions, called *trigonometric functions*, that we shall consider in this chapter. As you will see, their inputs are angles and their outputs are real numbers. In this section we shall discuss two of these functions: the *sine* and *cosine* functions. We shall assume that all angles are in radians.

To define them we begin with any angle t in standard position (see Fig. 9.10). Let $P = (x, y)$ be a point (other than the vertex) on the terminal side

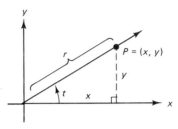

FIGURE 9.10

of t, and let r be the distance of P from the origin. The **sine** and **cosine** of t, abbreviated sin t and cos t, are defined by the following ratios:

$$\sin t = \frac{y}{r}, \qquad \cos t = \frac{x}{r}.$$

If we were to choose any other point on the terminal side, it can be shown, by using geometry, that we would obtain the same values as above for sin t and cos t. From the right triangle in Fig. 9.10 and the Pythagorean theorem,* we have

$$r = \sqrt{x^2 + y^2}.$$

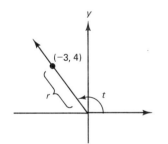

FIGURE 9.11

EXAMPLE 1   *For angle t in Fig. 9.11, find* sin t *and* cos t.

Here the point $(x, y) = (-3, 4)$ lies on the terminal side of t and

$$r = \sqrt{x^2 + y^2} = \sqrt{(-3)^2 + 4^2} = \sqrt{9 + 16} = \sqrt{25} = 5.$$

Thus

$$\sin t = \frac{y}{r} = \frac{4}{5}$$

and

$$\cos t = \frac{x}{r} = \frac{-3}{5} = -\frac{3}{5}.$$

Because

$$\sin t = \frac{y}{r} \quad \text{and} \quad \cos t = \frac{x}{r},$$

---

* The Pythagorean theorem states that the square of the length of the hypotenuse of a right triangle is equal to the sum of the squares of the lengths of the other two sides.

the signs of sin t and cos t depend on the signs of x, y, and r. If t is a first-quadrant angle, then x, y, and r are positive (r is always positive), so sin t and cos t are positive. If t is a second-quadrant angle, then $x < 0$ and $y > 0$. Thus sin t is positive and cos t is negative. Similarly, we can determine the signs of the sine and cosine functions in the other quadrants. Figure 9.12 gives a summary.

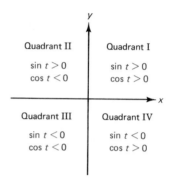

FIGURE 9.12

Figure 9.13(a) shows an *acute* angle t in standard position and a right triangle formed by dropping a perpendicular from (x, y) to the x-axis. Figure 9.13(b) shows the triangle by itself. Note that y is the length of the **opposite**

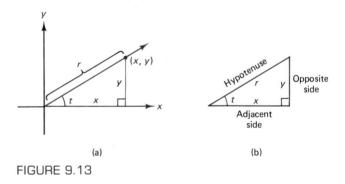

FIGURE 9.13

side of angle t and x is the length of the **adjacent side**. The **hypotenuse** is r. The values of sin t and cos t can be given in terms of these sides:

$$\sin t = \frac{y}{r} = \frac{\text{opposite side}}{\text{hypotenuse}}$$

and

$$\cos t = \frac{x}{r} = \frac{\text{adjacent side}}{\text{hypotenuse}}.$$

More simply, if t is an acute angle of a right triangle, then

$$\sin t = \frac{\text{opp}}{\text{hyp}} \quad \text{and} \quad \cos t = \frac{\text{adj}}{\text{hyp}}.$$

By using right triangles we can easily determine values of the sine and cosine for some *special* angles, namely $\pi/4$, $\pi/6$, and $\pi/3$. Figure 9.14 shows two right triangles: one with an acute angle of $\pi/4$ (45°) and the other with acute angles of $\pi/6$ (30°) and $\pi/3$ (60°). In both triangles the vertical side has

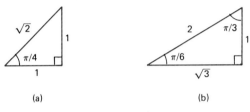

(a)   (b)

FIGURE 9.14

length 1. The lengths of the other sides can be established by geometric considerations. From Fig. 9.14(a),

$$\sin \frac{\pi}{4} = \frac{\text{opp}}{\text{hyp}} = \frac{1}{\sqrt{2}}$$

$$= \frac{1}{\sqrt{2}} \cdot \frac{\sqrt{2}}{\sqrt{2}} = \frac{\sqrt{2}}{2} \quad \text{(rationalizing the denominator)}$$

and $\quad \cos \dfrac{\pi}{4} = \dfrac{\text{adj}}{\text{hyp}} = \dfrac{1}{\sqrt{2}} = \dfrac{\sqrt{2}}{2}.$

From Fig. 9.14(b),

$$\sin \frac{\pi}{6} = \frac{\text{opp}}{\text{hyp}} = \frac{1}{2}$$

$$\text{and} \quad \cos \frac{\pi}{6} = \frac{\text{adj}}{\text{hyp}} = \frac{\sqrt{3}}{2}.$$

For the angle $\pi/3$ [see Fig. 9.4(b)], its opposite side is $\sqrt{3}$ and its adjacent side is 1. Thus

$$\sin \frac{\pi}{3} = \frac{\text{opp}}{\text{hyp}} = \frac{\sqrt{3}}{2}$$

$$\text{and} \quad \cos \frac{\pi}{3} = \frac{\text{adj}}{\text{hyp}} = \frac{1}{2}.$$

Table 9.1 gives a summary of the trigonometric values of the special angles. Rather than memorize the table, you should memorize the special triangles from which it came.

TABLE 9.1

	$\pi/6$ (30°)	$\pi/4$ (45°)	$\pi/3$ (60°)
sin	$\dfrac{1}{2}$	$\dfrac{\sqrt{2}}{2}$	$\dfrac{\sqrt{3}}{2}$
cos	$\dfrac{\sqrt{3}}{2}$	$\dfrac{\sqrt{2}}{2}$	$\dfrac{1}{2}$

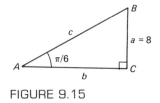

FIGURE 9.15

EXAMPLE 2   *For the right triangle shown in Fig. 9.15, determine the lengths of side b and the hypotenuse c.*

The vertices of the triangle are A, B, and C. It is typical to label the side opposite vertex A as a, and similarly for the other sides. The angle with vertex A is called angle A, and similarly there are angles B and C. Typically, angle C is the right angle. From the triangle,

$$\sin A = \sin \frac{\pi}{6} = \frac{\text{opp}}{\text{hyp}} = \frac{8}{c}.$$

Thus

$$\sin \frac{\pi}{6} = \frac{8}{c}.$$

Since $\sin(\pi/6) = \frac{1}{2}$,

$$\frac{1}{2} = \frac{8}{c}.$$

Solving for c gives

$$c = 8 \cdot 2 = 16.$$

To find b we consider

$$\cos \frac{\pi}{6} = \frac{\text{adj}}{\text{hyp}} = \frac{b}{c} = \frac{b}{16}$$

or

$$\cos \frac{\pi}{6} = \frac{b}{16},$$
$$\frac{\sqrt{3}}{2} = \frac{b}{16},$$
$$b = \frac{16\sqrt{3}}{2} = 8\sqrt{3}.$$

In Appendix G is a table of values of trigonometric functions for angles between 0 and 1.60 (radians). Most entries are approximate. Many calculators have keys for determining these values.

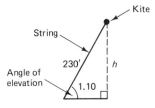

FIGURE 9.16

EXAMPLE 3 *The length of a kite string is 230 ft and the angle between the string and the horizontal is 1.10 (see Fig. 9.16). This angle is called the **angle of elevation** of the kite. Find, to the nearest foot, the height of the kite.*

If h is the height in feet, then from the right triangle constructed in Fig. 9.16,

$$\sin 1.10 = \frac{\text{opp}}{\text{hyp}} = \frac{h}{230}.$$
$$h = 230 \sin 1.10$$
$$= 230(0.8912) \qquad \text{(Appendix G)}$$
$$= 204.976.$$

Thus the height (to the nearest foot) is 205 ft.

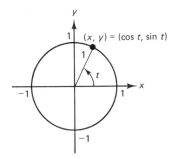

FIGURE 9.17

There is an alternative way of defining the sine and the cosine. Consider the unit circle (the circle with radius 1 centered at the origin) and let t be an angle in standard position, as shown in Fig. 9.17. Let (x, y) be the point where the terminal side of t intersects the unit circle. Because $r = 1$, we find that $\sin t = y/1 = y$ and $\cos t = x/1 = x$:

$$\sin t = y$$
$$\text{and} \qquad \cos t = x.$$

Thus the point (x, y) has first coordinate $\cos t$ and second coordinate $\sin t$.

With this interpretation of the sine and cosine functions, we can easily determine function values for quadrantal angles.

EXAMPLE 4 *Determine the values of* $\sin 0$, $\sin(\pi/2)$, $\cos \pi$, *and* $\cos(3\pi/2)$.

Each point on the unit circle in Fig. 9.18 has the form $(\cos t, \sin t)$. Because the angle 0 corresponds to the point $(1, 0)$, we have $\sin 0 = 0$. The angle $\pi/2$

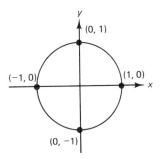

FIGURE 9.18

corresponds to $(0, 1)$, so $\sin(\pi/2) = 1$. The point $(-1, 0)$ gives $\cos \pi = -1$, and the point $(0, -1)$ gives $\cos(3\pi/2) = 0$.

Table 9.2 gives values of the sine and cosine of some quadrantal angles.

TABLE 9.2

	0	$\pi/2$	π	$3\pi/2$
sin	0	1	0	-1
cos	1	0	-1	0

Because the coterminal angles t and $t + 2\pi$ correspond to the same point on the unit circle (see Fig. 9.19), we have

$$\cos(t + 2\pi) = \cos t$$

and $$\sin(t + 2\pi) = \sin t.$$

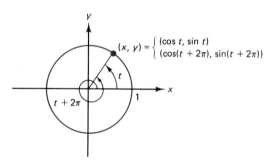

FIGURE 9.19

These equations are true for any angle t and are called **trigonometric identities**. Similarly,

$$\cos(t - 2\pi) = \cos t$$

and $$\sin(t - 2\pi) = \sin t.$$

We can easily establish some other identities. Since an equation of the unit circle is $x^2 + y^2 = 1$, and $x = \cos t$ and $y = \sin t$, by substitution we have

$$(\cos t)^2 + (\sin t)^2 = 1.$$

Usually, $(\sin t)^2$ is written $\sin^2 t$ and similarly for $(\cos t)^2$. Thus

$$\sin^2 t + \cos^2 t = 1.$$

In Fig. 9.20 you can see that if (x, y) is the point on the unit circle that corresponds to angle t, then $(x, -y)$ corresponds to angle $-t$. Thus

$$\cos(-t) = \cos t \quad \text{and} \quad \sin(-t) = -\sin t.$$

Table 9.3 summarizes these identities and gives others for angles s and t.

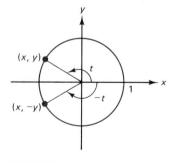

FIGURE 9.20

TABLE 9.3
Trigonometric Identities

1. $\sin^2 t + \cos^2 t = 1$.
2. $\sin(t \pm 2\pi) = \sin t$.
3. $\cos(t \pm 2\pi) = \cos t$.
4. $\sin(-t) = -\sin t$.
5. $\cos(-t) = \cos t$.
6. $\sin(s \pm t) = \sin s \cos t \pm \cos s \sin t$.
7. $\cos(s \pm t) = \cos s \cos t \mp \sin s \sin t$.

EXAMPLE 5

a. *Determine* $\sin(-\pi/6)$.

Since $\sin(\pi/6) = 1/2$, by Identity 4,

$$\sin\left(-\frac{\pi}{6}\right) = -\sin\frac{\pi}{6} = -\frac{1}{2}.$$

b. *If* $\sin(7\pi/8) = 0.3827$, *find* $\cos(7\pi/8)$.

By Identity 1,

$$\sin^2\frac{7\pi}{8} + \cos^2\frac{7\pi}{8} = 1,$$

$$(0.3827)^2 + \cos^2\frac{7\pi}{8} = 1,$$

$$\cos^2\frac{7\pi}{8} = 0.8535.$$

Thus $\cos(7\pi/8)$ is $\sqrt{0.8535}$ or $-\sqrt{0.8535}$. Because $7\pi/8$ is a second-quadrant angle and the cosine of such an angle is negative,

$$\cos\frac{7\pi}{8} = -\sqrt{0.8535} = -0.9239.$$

EXAMPLE 6   *If* $\sin t = a$, *find* $\cos\left(t + \frac{\pi}{2}\right)$.

We use Identity 7. Since we must use the upper sign $(+)$ on the left side of the identity, we must also use the upper sign $(-)$ on the right side.

$$\cos\left(t + \frac{\pi}{2}\right) = \cos t \cos\frac{\pi}{2} - \sin t \sin\frac{\pi}{2}$$

$$= (\cos t)(0) - (a)(1)$$

$$= -a.$$

With each nonquadrantal angle t, we can associate another angle called the **reference angle** of t, denoted t_R. It is the *acute* angle between the terminal side of t and the horizontal axis. Figure 9.21 shows a typical angle in each quadrant.

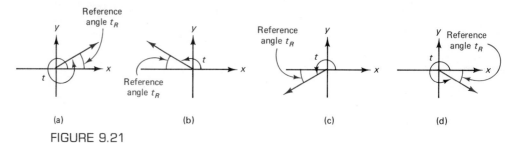

(a)   (b)   (c)   (d)

FIGURE 9.21

EXAMPLE 7   *Find the reference angle of each of the following angles.*

a. $t = 9\pi/4$ [*see Fig. 9.22(a)*].

Since $9\pi/4 = 2\pi + t_R$, we have

$$t_R = \frac{9\pi}{4} - 2\pi = \frac{\pi}{4}.$$

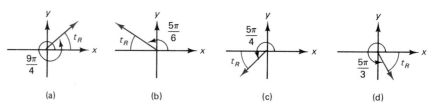

(a)   (b)   (c)   (d)

FIGURE 9.22

b. $t = 5\pi/6$ [*see Fig. 9.22(b)*].

$$\frac{5\pi}{6} + t_R = \pi,$$

$$t_R = \pi - \frac{5\pi}{6} = \frac{\pi}{6}.$$

c. $t = 5\pi/4$ [*see Fig. 9.22(c)*].

$$\pi + t_R = \frac{5\pi}{4},$$

$$t_R = \frac{5\pi}{4} - \pi = \frac{\pi}{4}.$$

d. $t = 5\pi/3$ [*see Fig. 9.22(d)*].

$$\frac{5\pi}{3} + t_R = 2\pi,$$

$$t_R = 2\pi - \frac{5\pi}{3} = \frac{\pi}{3}.$$

If an angle is nonquadrantal, we can find its sine and cosine from its reference angle. For example, consider $5\pi/6$ [see Fig. 9.23(a)]. Its reference angle is $\pi/6$. To locate a point on the terminal side of $5\pi/6$, we shall use our

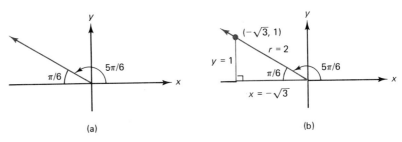

(a)   (b)

FIGURE 9.23

special right triangle with an acute angle of $\pi/6$. From Fig. 9.23(b), $(-\sqrt{3}, 1)$ lies on the terminal side and $r = 2$. Thus

$$\sin\frac{5\pi}{6} = \frac{y}{r} = \frac{1}{2}$$

and $\quad \cos\frac{5\pi}{6} = \frac{x}{r} = \frac{-\sqrt{3}}{2} = -\frac{\sqrt{3}}{2}.$

Now, consider the reference angle. We have

$$\sin\frac{\pi}{6} = \frac{1}{2} \quad \text{and} \quad \cos\frac{\pi}{6} = \frac{\sqrt{3}}{2}.$$

Notice that the trigonometric values of $5\pi/6$ are the same as those of its reference angle except, perhaps, for sign. Thus to find $\cos 5\pi/6$, we can find $\cos \pi/6$ and place a minus sign in front of it. The minus sign is needed because the cosine of a second-quadrant angle ($5\pi/6$) is negative.

In general, to find the value of a trigonometric function of an angle t, find the value of the same function of the reference angle and attach the proper sign. This sign depends on the function involved and the quadrant of t.

$$\sin t = \pm\sin t_R \quad \text{and} \quad \cos t = \pm\cos t_R$$

sign depends on
quadrant of t

EXAMPLE 8

a. *Find* $\sin(3\pi/4)$.

Since $3\pi/4$ is a second-quadrant angle [see Fig. 9.24(a)], $\sin(3\pi/4)$ is positive. The reference angle is $\pi - (3\pi/4) = \pi/4$. Thus from the right triangle with acute angle $\pi/4$ in Fig. 9.24(b),

$$\sin\frac{3\pi}{4} = +\sin\frac{\pi}{4} = +\frac{\text{opp}}{\text{hyp}} = +\frac{1}{\sqrt{2}} = \frac{\sqrt{2}}{2}.$$

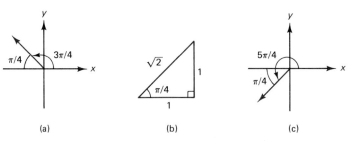

(a)   (b)   (c)

FIGURE 9.24

b. *Find* $\cos(5\pi/4)$.

Since $5\pi/4$ is a third-quadrant angle [see Fig. 9.24(c)], $\cos(5\pi/4)$ is negative. The reference angle is $(5\pi/4) - \pi = \pi/4$. From Fig. 9.24(b),

$$\cos\frac{5\pi}{4} = -\cos\frac{\pi}{4} = -\frac{\text{adj}}{\text{hyp}} = -\frac{1}{\sqrt{2}} = -\frac{\sqrt{2}}{2}.$$

EXERCISE 9.2

In Problems **1–6**, *the given point lies on the terminal side of an angle t in standard position. Find (a)* $\sin t$ *and (b)* $\cos t$.

1. (6, 8).

2. (12, 5).

3. (4, −3).

4. (2, −3).

5. (−1, −1).

6. (−3, −3).

In Problems **7–14**, *give the sign of each value.*

7. $\sin\dfrac{\pi}{4}$.

8. $\sin\dfrac{7\pi}{6}$.

9. $\cos\dfrac{4\pi}{3}$.

10. $\cos\dfrac{5\pi}{6}$.

11. $\sin\dfrac{11\pi}{6}$.

12. $\cos\dfrac{7\pi}{4}$.

13. $\sin\dfrac{8\pi}{3}$.

14. $\cos\dfrac{25\pi}{6}$.

In Problems **15** *and* **16**, *determine the sines and cosines of angles A and B in the given figure.*

15. Figure 9.25(a).

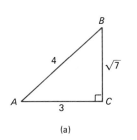

FIGURE 9.25

16. Figure 9.25(b).

*In Problems **17–22**, use special triangles to determine each value. Do not use Appendix G or a calculator.*

17. $\sin \dfrac{\pi}{6}$.

18. $\sin \dfrac{\pi}{3}$.

19. $\cos \dfrac{\pi}{4}$.

20. $\cos \dfrac{\pi}{6}$.

21. $\cos \dfrac{\pi}{3}$.

22. $\sin \dfrac{\pi}{4}$.

*In Problems **23–26**, ABC is a right triangle with right angle C. Solve for the remaining sides of the triangle in each case. Do not use Appendix G or a calculator.*

23. $A = \dfrac{\pi}{4}$, $c = 6$.

24. $A = \dfrac{\pi}{6}$, $b = 3$.

25. $B = \dfrac{\pi}{3}$, $a = 3$.

26. $B = \dfrac{\pi}{3}$, $c = 1$.

*In Problems **27–32**, ABC is a right triangle with right angle C. Determine the remaining sides to one decimal place by using Appendix G or a calculator.*

27. $c = 4$, $A = 0.47$.

28. $a = 10$, $A = 0.30$.

29. $b = 5$, $B = 1.00$.

30. $c = 10$, $B = 0.70$.

31. $b = 100$, $A = 0.25$.

32. $a = 4$, $B = 1.20$.

33. If t is an angle near 0, then $\sin t \approx t$, where t is in radians. Confirm this by finding $\sin t$ if t is (a) 0.05, (b) 0.04, (c) 0.03, (d) 0.02, and (e) 0.01.

34. The points (2, 1) and (6, 3) both lie on the terminal side of an angle t in standard position.

a. Choose (2, 1) as (x, y) and find $\sin t$ and $\cos t$.

b. Do the same for (6, 3).

Your answers in parts (a) and (b) should be the same. This illustrates that it is immaterial as to the point you use on the terminal side of t, except (0, 0), to find $\sin t$ and $\cos t$.

35. The length of a kite string is 600 ft, and the angle of elevation of the kite is 0.60. How high is the kite? Give your answer to the nearest foot.

36. After passing a toll gate, a car travels on a road that rises until it reaches a bridge. If the angle of elevation of the road is 0.06 and the length of the road is 170 m, how high above ground level is the bridge?

37. A 24-ft ladder leans against a vertical wall. It is recommended that the acute angle between ground level and the ladder be 1.32. If this recommendation is followed, how far, to the nearest half foot, should the bottom of the ladder be placed from the wall?

38. An airplane pilot wants to increase his altitude by 3 miles. He plans to climb at a constant angle of 0.15 and a constant rate of 200 mph. To the nearest minute, how long will it take to reach that altitude?

39. An escalator inclined at an angle of 0.75 moves at a constant speed of 0.6 m/s. A person steps on the escalator and reaches the top in 42 seconds. Through what vertical height has the person been lifted? Give your answer to the nearest tenth of a meter.

40. A wire bracing an antenna is 50 ft long. One end is attached to the top of the antenna, and the other end is attached to level ground at a distance of 40 ft from the base of the antenna. Find the angle that the wire makes with the ground. Give your answer to the nearest hundredth radian. (*Hint:* Find the value of the cosine of the angle and then use Appendix G to find an angle whose cosine is that value.)

*In Problems **41–46**, find the sine and cosine of the angle. Do not use Appendix G or a calculator.*

41. $\dfrac{\pi}{2}$.

42. π.

43. $\dfrac{3\pi}{2}$.

44. 2π.

45. -3π.

46. $-\dfrac{3\pi}{2}$.

*In Problems **47–52**, use trigonometric identities.*

47. If $\sin t = a$, find $\sin(-t)$.

48. If $\cos t = a$, find $\cos(-t)$.

49. If $\cos t = a$, find $\cos(t + \pi)$.

50. If $\cos t = a$, find $\sin\left(t - \dfrac{\pi}{2}\right)$.

51. If $\cos \dfrac{6\pi}{5} = -0.8090$, find $\sin \dfrac{6\pi}{5}$.

52. If $\sin \dfrac{15\pi}{8} = -0.3827$, find $\cos \dfrac{15\pi}{8}$.

*In Problems **53–60**, find the reference angle of the given angle.*

53. $\dfrac{7\pi}{4}$.

54. $\dfrac{2\pi}{3}$.

55. $\dfrac{7\pi}{6}$.

56. $\dfrac{5\pi}{4}$.

57. $-\dfrac{\pi}{3}$.

58. $\dfrac{5\pi}{6}$.

59. $\dfrac{13\pi}{6}$.

60. $-\dfrac{3\pi}{4}$.

*In Problems **61–75**, use reference angles and special angles to determine the values.*

61. $\sin \dfrac{2\pi}{3}$.

62. $\sin \dfrac{4\pi}{3}$.

63. $\cos \dfrac{7\pi}{6}$.

64. $\cos \dfrac{5\pi}{3}$.

65. $\sin \dfrac{5\pi}{4}$.

66. $\cos \dfrac{3\pi}{4}$.

67. $\cos \dfrac{2\pi}{3}$.

68. $\sin \dfrac{11\pi}{6}$.

69. $\cos \dfrac{5\pi}{6}$.

70. $\cos \dfrac{7\pi}{4}$.

71. $\sin \dfrac{13\pi}{6}$.

72. $\cos \dfrac{8\pi}{3}$.

73. $\cos\left(-\dfrac{11\pi}{6}\right)$.

74. $\sin\left(-\dfrac{2\pi}{3}\right)$.

75. $\sin\left(-\dfrac{5\pi}{4}\right)$.

9.3 GRAPHS AND OTHER TRIGONOMETRIC FUNCTIONS

To understand the behavior of the sine and cosine functions, an analysis of their graphs is essential. You will see that these graphs have unique features. We begin with the graph of the basic sine function

$$y = f(t) = \sin t,$$

where t is measured in radians. Table 9.4 lists some values of t between 0 and 2π along with the corresponding y-values. The points (t, y) are plotted in Fig. 9.26 and connected by a smooth curve.

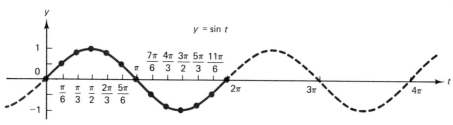

FIGURE 9.26

Because $\sin(t + 2\pi) = \sin t$, the graph of $y = \sin t$ repeats itself every 2π radians. This is indicated by the dashed curve in Fig. 9.26 and gives the graph a characteristic wavelike appearance. To describe the repetition, we

TABLE 9.4

t (rad)	$y = \sin t$	t (rad)	$y = \sin t$
0	0	$\dfrac{7\pi}{6}$	$-\dfrac{1}{2} = -0.5$
$\dfrac{\pi}{6}$	$\dfrac{1}{2} = 0.5$	$\dfrac{4\pi}{3}$	$-\dfrac{\sqrt{3}}{2} \approx -0.87$
$\dfrac{\pi}{3}$	$\dfrac{\sqrt{3}}{2} \approx 0.87$	$\dfrac{3\pi}{2}$	-1
$\dfrac{\pi}{2}$	1	$\dfrac{5\pi}{3}$	$-\dfrac{\sqrt{3}}{2} \approx -0.87$
$\dfrac{2\pi}{3}$	$\dfrac{\sqrt{3}}{2} \approx 0.87$	$\dfrac{11\pi}{6}$	$-\dfrac{1}{2} = -0.5$
$\dfrac{5\pi}{6}$	$\dfrac{1}{2} = 0.5$	2π	0
π	0		

say that the sine function is *periodic*. In general, any function f is **periodic** if there is a positive number c such that $f(t + c) = f(t)$ for all t in the domain of f. The least such value of c is called the **period** of f. Thus the sine function has period 2π.

The graph of $y = \sin t$ over an interval of one period (2π) is called a **cycle** of the sine curve. Observe in Fig. 9.26 that the cycle from $t = 0$ to $t = 2\pi$ has four basic parts:

1. From $t = 0$ to $t = \dfrac{\pi}{2}$ the curve *rises* from $y = 0$ to $y = 1$.

2. From $t = \dfrac{\pi}{2}$ to $t = \pi$ it *falls* from $y = 1$ to $y = 0$.

3. From $t = \pi$ to $t = \dfrac{3\pi}{2}$ it continues to *fall* from $y = 0$ to $y = -1$.

4. From $t = \dfrac{3\pi}{2}$ to $t = 2\pi$ it *rises* from $y = -1$ to $y = 0$.

For any periodic function, one-half of the difference of the maximum and minimum values is called the **amplitude** of the function. You can see that the maximum value of $\sin t$ is 1 and the minimum value is -1. Thus the amplitude of $y = \sin t$ is 1, because $\frac{1}{2}[1 - (-1)] = \frac{1}{2}[2] = 1$. In summary we have:

The sine function $y = \sin t$ has period 2π and amplitude 1.

The graph of the basic cosine function $y = \cos t$ appears in Fig. 9.27. Like the sine function, the cosine function has period 2π and amplitude 1.

> The cosine function $y = \cos t$ has period 2π and amplitude 1.

t	0	$\dfrac{\pi}{6}$	$\dfrac{\pi}{3}$	$\dfrac{\pi}{2}$	$\dfrac{2\pi}{3}$	$\dfrac{5\pi}{6}$	π	$\dfrac{7\pi}{6}$	$\dfrac{4\pi}{3}$	$\dfrac{3\pi}{2}$	$\dfrac{5\pi}{3}$	$\dfrac{11\pi}{6}$	2π
$y = \cos t$	1	0.87	0.5	0	−0.5	−0.87	−1	−0.87	−0.5	0	0.5	0.87	1

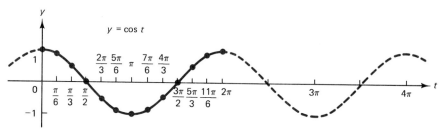

FIGURE 9.27

The cycle from $t = 0$ to $t = 2\pi$ has four basic parts:

1. From $t = 0$ to $t = \dfrac{\pi}{2}$ the curve *falls* from $y = 1$ to $y = 0$.

2. From $t = \dfrac{\pi}{2}$ to $t = \pi$ it continues to *fall* from $y = 0$ to $y = -1$.

3. From $t = \pi$ to $t = \dfrac{3\pi}{2}$ it *rises* from $y = -1$ to $y = 0$.

4. From $t = \dfrac{3\pi}{2}$ to $t = 2\pi$ it continues to *rise* from $y = 0$ to $y = 1$.

Comparing the cosine curve in Fig. 9.27 to the sine curve in Fig. 9.26, you can see that the cosine curve is simply the sine curve shifted $\pi/2$ to the left. This means that $\sin\left(t + \dfrac{\pi}{2}\right) = \cos t$.

One simple variation of the sine function is a function of the form

$$y = a \sin t, \qquad \text{where } a \text{ is a constant.}$$

As $\sin t$ varies between 1 and -1, the values of $a \sin t$ vary between a and $-a$. In fact, the y-value of each point on the curve $y = a \sin t$ is just a times the corresponding y-value on the curve $y = \sin t$. Thus the graphs of $y =$

$a \sin t$ and $y = \sin t$ may differ in a vertical sense. The period is 2π in both cases. If a is positive, the graph of $y = a \sin t$ resembles the basic sine curve. If a is negative, then when $\sin t$ is positive, $a \sin t$ is negative; when $\sin t$ is negative, $a \sin t$ is positive. As a result, the graph of $y = a \sin t$ appears like an *inverted* sine curve. But, regardless of a, the amplitude of $y = a \sin t$ is $|a|$. We remark that the above discussion on period, inversion, and amplitude applies equally well to the graph of $y = a \cos t$.

EXAMPLE 1   *Draw the graph of $y = 2 \sin t$.*

Here $a = 2 > 0$. Thus the graph is a sine curve with amplitude 2 and period 2π. To draw the graph we sketch a basic sine curve and label the axes to indicate the amplitude and period (see Fig. 9.28).

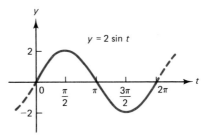

FIGURE 9.28

EXAMPLE 2   *Draw the graph of $y = -4 \cos t$.*

The amplitude is $|a| = |-4| = 4$ and the period is 2π. Since $a < 0$, the basic shape is an *inverted* cosine curve (see Fig. 9.29).

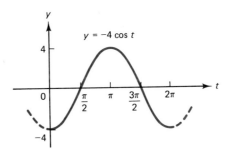

FIGURE 9.29

Another variation of the sine function has the form

$$y = a \sin bt, \qquad a, b \text{ constants}, b > 0.$$

As before, the amplitude is $|a|$. The basic shape is still a sine curve if $a > 0$, or an inverted sine curve if $a < 0$. However, the graph may be affected in a horizontal way, because the value of b affects the period. We obtain one cycle as bt ranges from 0 to 2π. This means that t ranges from 0 to $2\pi/b$. Thus the period of $y = a \sin bt$ is $2\pi/b$. In summary:

The amplitude of $y = a \sin bt$ is $|a|$ and the period is $2\pi/b$. The basic shape is that of a sine curve if $a > 0$, or an inverted sine curve if $a < 0$. These facts are also true if *sine* is replaced by *cosine*.

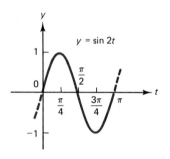

FIGURE 9.30

EXAMPLE 3   *Draw the graph of* $y = \sin 2t$.

Here $a = 1 > 0$, so the graph is a basic sine curve with amplitude 1 and period $2\pi/b = 2\pi/2 = \pi$. Because a cycle of a sine curve has four basic parts, to draw one cycle of $y = \sin 2t$ that begins at $t = 0$ and ends at $t = \pi$, we first divide the interval $[0, \pi]$ into four subintervals by the equally spaced points 0, $\pi/4$, $\pi/2$, $3\pi/4$, and π. Then we draw a sine curve that has a value of 0 at $t = 0$ and then rises to a maximum value of 1 at $t = \pi/4$, then decreases to 0 at $t = \pi/2$, then continues to decrease to the minimum value of -1 at $t = 3\pi/4$, and then rises to the value of 0 at $t = \pi$ (see Fig. 9.30).

Using the sine and cosine functions, we can define four other trigonometric functions. Their names (and abbreviations) are tangent function (tan), cotangent function (cot), secant function (sec), and cosecant function (csc). They are defined as follows:

$$\tan t = \frac{\sin t}{\cos t},$$

$$\cot t = \frac{\cos t}{\sin t},$$

$$\sec t = \frac{1}{\cos t},$$

$$\csc t = \frac{1}{\sin t}.$$

Alternatively, if angle t is in standard position, these functions can be defined in terms of a point (x, y) on the terminal side of t [see Fig. 9.31(a)].

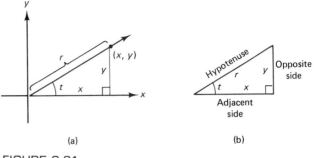

(a)   (b)

FIGURE 9.31

For example, because $\sin t = y/r$ and $\cos t = x/r$,

$$\tan t = \frac{\sin t}{\cos t} = \frac{y/r}{x/r} = \frac{y}{x}.$$

Similarly, we can define $\cot t$, $\sec t$, and $\csc t$ in terms of x, y, and r.

$$\tan t = \frac{y}{x}, \qquad \cot t = \frac{x}{y},$$

$$\sec t = \frac{r}{x}, \qquad \csc t = \frac{r}{y}.$$

Looking at the denominators in the above box, you can see that $\tan t$ and $\sec t$ are not defined when $x = 0$, that is, when the terminal side is on the y-axis (such as for $t = \pi/2$ or $3\pi/2$). Also, $\cot t$ and $\csc t$ are not defined when $y = 0$; that is, when the terminal side is on the x-axis (such as for $t = 0$ or π).

Furthermore, if t is a positive acute angle, these functions can be defined in terms of ratios of the sides of a right triangle. Using Fig. 9.31(b) we have:

$$\tan t = \frac{\text{opposite side}}{\text{adjacent side}}, \qquad \cot t = \frac{\text{adjacent side}}{\text{opposite side}},$$

$$\sec t = \frac{\text{hypotenuse}}{\text{adjacent side}}, \qquad \csc t = \frac{\text{hypotenuse}}{\text{opposite side}}.$$

Some useful trigonometric identities are given in Table 9.5. For example, to prove Identity 2 we have

$$1 + \tan^2 t = 1 + \left(\frac{y}{x}\right)^2 = 1 + \frac{y^2}{x^2}$$

$$= \frac{x^2 + y^2}{x^2} = \frac{r^2}{x^2} = \left(\frac{r}{x}\right)^2$$

$$= \sec^2 t.$$

TABLE 9.5

1. $\cot t = \dfrac{1}{\tan t}$,
2. $1 + \tan^2 t = \sec^2 t$,
3. $1 + \cot^2 t = \csc^2 t$,
4. $\tan(-t) = -\tan t$.

EXAMPLE 4   *Find the six trigonometric functions of* $t = \dfrac{\pi}{6}$.

Since

$$\sin \frac{\pi}{6} = \frac{1}{2} \quad \text{and} \quad \cos \frac{\pi}{6} = \frac{\sqrt{3}}{2},$$

we have

$$\tan \frac{\pi}{6} = \frac{\sin \dfrac{\pi}{6}}{\cos \dfrac{\pi}{6}} = \frac{1/2}{\sqrt{3}/2} = \frac{1}{\sqrt{3}} = \frac{\sqrt{3}}{3},$$

$$\cot \frac{\pi}{6} = \frac{\cos \dfrac{\pi}{6}}{\sin \dfrac{\pi}{6}} = \frac{\sqrt{3}/2}{1/2} = \sqrt{3},$$

$$\sec \frac{\pi}{6} = \frac{1}{\cos \dfrac{\pi}{6}} = \frac{1}{\sqrt{3}/2} = \frac{2}{\sqrt{3}} = \frac{2\sqrt{3}}{3},$$

$$\csc \frac{\pi}{6} = \frac{1}{\sin \dfrac{\pi}{6}} = \frac{1}{1/2} = 2.$$

You should verify these results for this special angle by using the right-triangle definitions of the trigonometric functions.

EXAMPLE 5   *If the point* $(-3, 4)$ *lies on the terminal side of angle t,* *find* tan *t,* cot *t,* sec *t, and* csc *t (see Example 1·of Sec. 9.2).*

Here $x = -3$, $y = 4$, and $r = 5$. Thus

$$\tan t = \frac{y}{x} = \frac{4}{-3} = -\frac{4}{3},$$

$$\cot t = \frac{x}{y} = \frac{-3}{4} = -\frac{3}{4},$$

$$\sec t = \frac{r}{x} = \frac{5}{-3} = -\frac{5}{3},$$

$$\csc t = \frac{r}{y} = \frac{5}{4}.$$

EXAMPLE 6   *Find the trigonometric functions of* $t = \pi$.

The angle π corresponds to the point $(-1, 0)$ on the unit circle. Thus

$$\sin \pi = y = 0, \qquad\qquad \cos \pi = x = -1,$$

$$\tan \pi = \frac{\sin \pi}{\cos \pi} = \frac{0}{-1} = 0, \qquad \cot \pi = \frac{\cos \pi}{\sin \pi} = \frac{-1}{0} = \text{undefined},$$

$$\sec \pi = \frac{1}{\cos \pi} = \frac{1}{-1} = -1, \qquad \csc \pi = \frac{1}{\sin \pi} = \frac{1}{0} = \text{undefined}.$$

Figure 9.32 gives the trigonometric functions that are positive for angles in the various quadrants.

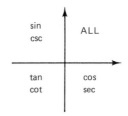

Positive Trigonometric Functions

FIGURE 9.32

If you recall, the sine of angle t can be expressed in terms of the sine of its reference angle, and a similar statement is true for the cosine of t. Therefore, since the tangent function can be expressed in terms of the sine and cosine, we can find tan t by making use of the reference angle of t. We can do the same for the cotangent, secant, and cosecant functions. In general,

$$\left.\begin{array}{c}\text{trigonometric}\\\text{function of } t\end{array}\right\} = \pm \left\{\begin{array}{l}\text{trigonometric function}\\\text{of reference angle of } t\end{array}\right.$$

↑

sign depends on trigonometric
function and quadrant of t

EXAMPLE 7   *Find* $\cot \dfrac{11\pi}{6}$.

Since $\dfrac{11\pi}{6}$ is a fourth-quadrant angle, $\cot \dfrac{11\pi}{6}$ is negative (see Fig. 9.32). The reference angle is $\dfrac{\pi}{6}$ (or 30°). Thus

$$\cot \frac{11\pi}{6} = -\cot \frac{\pi}{6}.$$

From the special triangle in Fig. 9.14(b),

$$\cot \frac{\pi}{6} = \frac{\text{adj}}{\text{opp}} = \frac{\sqrt{3}}{1} = \sqrt{3}.$$

Hence

$$\cot \frac{11\pi}{6} = -\sqrt{3}.$$

Figure 9.33 shows the graphs of the tangent, cotangent, secant, and cosecant functions. These functions do not have amplitudes because there are no maximum or minimum values. Note that their graphs have asymptotes. Also observe that the secant and cosecant functions have period 2π, whereas the tangent and cotangent functions have period π.

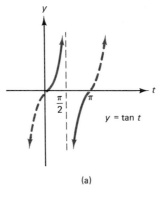

(a)

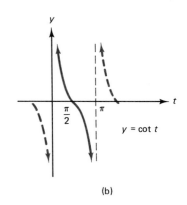

(b)

FIGURE 9.33

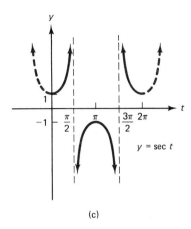

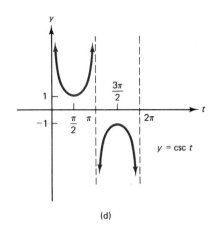

FIGURE 9.33 (Cont.)

You have seen that the trigonometric functions are functions of angles and their graphs exhibit a cyclical pattern. There are practical situations that also behave in a cyclical fashion, such as business cycles. However, a business cycle depends on time, not an angle. For this reason there are advantages to thinking of trigonometric functions of *numbers*. In particular, by the sine of the number 0.3 we shall mean the sine of 0.3 radian. As a result, the domain of sin t can be considered as consisting of real numbers and similarly for the other trigonometric functions.

EXERCISE 9.3

In Problems **1–14**, *draw one cycle of each curve. Give the amplitude A and the period p.*

1. $y = 3 \sin t$.

2. $y = \frac{1}{2} \sin t$.

3. $y = -4 \sin t$.

4. $y = -\sin t$.

5. $y = \frac{1}{2} \cos t$.

6. $y = 5 \cos t$.

7. $y = -\cos t$.

8. $y = -\frac{2}{3} \cos t$.

9. $y = \sin 2t$.

10. $y = \sin \frac{t}{2}$.

11. $y = -4 \cos 3t$.

12. $y = \frac{1}{2} \cos 5t$.

13. $y = 2 \sin \frac{\pi}{3} t$.

14. $y = -2 \sin 4\pi t$.

15. Draw the graphs of $y = \sin t$ and $y = \sin 2t$ on the same coordinate plane. How many cycles of the curve $y = \sin 2t$ are there for each cycle of $y = \sin t$?

16. Draw the graphs of $y = \cos t$ and $y = \cos \frac{t}{2}$ on the same coordinate plane. How many cycles of the curve $y = \cos t$ are there for each cycle of $y = \cos \frac{t}{2}$?

17. Draw the graph of $y = 2 \sin \frac{t}{4}$ from $t = -2\pi$ to $t = 4\pi$.

18. Draw the graph of $y = \frac{1}{4} \cos 4t$ from $t = -\frac{\pi}{2}$ to $t = \pi$.

In Problems **19–24**, *the given point lies on the terminal side of an angle t in standard position. Find the values of the six trigonometric functions of t.*

19. (6, 8).

20. (12, 5).

21. (−4, 3).

22. (2, −3).

23. (−1, −1).

24. (−3, −3).

In Problems **25–30**, *use special triangles to determine each value.*

25. $\tan \dfrac{\pi}{4}$.

26. $\tan \dfrac{\pi}{3}$.

27. $\cot \dfrac{\pi}{6}$.

28. $\cot \dfrac{\pi}{4}$.

29. $\sec \dfrac{\pi}{3}$.

30. $\csc \dfrac{\pi}{4}$.

*In Problems **31–34**, two trigonometric values of t are given. Find the other four trigonometric values.*

31. $\sin t = \dfrac{1}{4}, \quad \cos t = \dfrac{\sqrt{15}}{4}$.

32. $\sin t = -\dfrac{\sqrt{21}}{5}, \quad \cos t = \dfrac{2}{5}$.

33. $\sec t = -\dfrac{5}{4}, \quad \csc t = -\dfrac{5}{3}$.

34. $\cos t = -\dfrac{\sqrt{5}}{3}, \quad \csc t = \dfrac{3}{2}$.

*In Problems **35–44**, find the values.*

35. $\tan 0$.

36. $\sec 0$.

37. $\cot \dfrac{\pi}{2}$.

38. $\csc \dfrac{\pi}{2}$.

39. $\csc 0$.

40. $\sec \dfrac{3\pi}{2}$.

41. $\tan \dfrac{7\pi}{4}$.

42. $\tan \dfrac{11\pi}{6}$.

43. $\sec \dfrac{4\pi}{3}$.

44. $\cot \dfrac{3\pi}{4}$.

45. If $\tan t = 2$, what are the possible values for $\sec t$?

46. Show that $\tan(-t) = -\tan t$.

47. The theory of *biorhythms* states that there are cyclical highs and lows in the physical, emotional, and intellectual aspects of a person's life. These three cycles, or biorhythms, can be represented by the following equations.

$$\text{Physical:} \quad y = \sin \dfrac{2\pi}{23}t,$$

$$\text{Emotional:} \quad y = \sin \dfrac{2\pi}{28}t,$$

$$\text{Intellectual:} \quad y = \sin \dfrac{2\pi}{33}t,$$

where t is the number of days after birth. Positive y-values indicate high levels of cycles, or good days, whereas negative y-values indicate low levels, or bad days.

a. Find the period (in days) of each cycle. Because these cycles have different lengths, a person can have a day in which there is a high in one cycle and lows in the other two cycles. At times, all three cycles are high (or *in phase*).

b. If you have a calculator, find the biorhythm levels on the 5000th day of your life.

9.4 DERIVATIVES OF TRIGONOMETRIC FUNCTIONS

We now state the derivatives of the sine and cosine functions:

$$\dfrac{d}{dt}(\sin t) = \cos t \tag{1}$$

and

$$\dfrac{d}{dt}(\cos t) = -\sin t. \tag{2}$$

Proofs of these rules appear in Appendix B.

EXAMPLE 1   *If $y = t^2 \cos t$, find y'.*

By the product rule,

$$y' = t^2 \cdot D_t(\cos t) + (\cos t) \cdot D_t(t^2)$$
$$= t^2(-\sin t) + (\cos t)(2t)$$
$$= t(2 \cos t - t \sin t).$$

Equations (1) and (2) will now be extended to cover a broader class of functions. Suppose that $y = \sin u$, where u is a differentiable function of t. By the chain rule,

$$\frac{d}{dt}(\sin u) = \frac{dy}{du} \cdot \frac{du}{dt} = \frac{d}{du}(\sin u) \cdot \frac{du}{dt} = \cos u \frac{du}{dt}.$$

Similarly, if $y = \cos u$, then

$$\frac{d}{dt}(\cos u) = \frac{dy}{du} \cdot \frac{du}{dt} = \frac{d}{du}(\cos u) \cdot \frac{du}{dt} = -\sin u \frac{du}{dt}.$$

Thus we have the rules:

> **1.** $\dfrac{d}{dt}(\sin u) = \cos u \dfrac{du}{dt},$
>
> **2.** $\dfrac{d}{dt}(\cos u) = -\sin u \dfrac{du}{dt}.$

EXAMPLE 2   *Differentiate each of the following.*

a. $y = \sin(t^2 + 3t + 2)$.

By applying Rule 1 with $u = t^2 + 3t + 2$, we have

$$y' = \cos(t^2 + 3t + 2) \cdot D_t(t^2 + 3t + 2)$$
$$= \cos(t^2 + 3t + 2) \cdot [2t + 3]$$
$$= (2t + 3)\cos(t^2 + 3t + 2).$$

b. $y = \cos \sqrt{t + 5}$.

Applying Rule 2 with $u = (t + 5)^{1/2}$ gives

$$y' = -\sin \sqrt{t + 5} \cdot D_t[(t + 5)^{1/2}]$$
$$= -\sin \sqrt{t + 5} \cdot \left[\frac{1}{2}(t + 5)^{-1/2}\right] = -\frac{\sin \sqrt{t + 5}}{2\sqrt{t + 5}}.$$

c. $y = \sin^3(4t - 1)$.

Here y is the third power of the function $\sin(4t - 1)$. By the power rule,

$$\frac{d}{dt}[\sin^3(4t - 1)] = \frac{d}{dt}\{[\sin(4t - 1)]^3\}$$

$$= 3[\sin(4t - 1)]^2 \frac{d}{dt}[\sin(4t - 1)]$$

$$= 3\sin^2(4t - 1) \cdot \cos(4t - 1) \cdot D_t(4t - 1)$$
$$= 3\sin^2(4t - 1) \cdot \cos(4t - 1) \cdot [4]$$
$$= 12\sin^2(4t - 1)\cos(4t - 1).$$

d. $y = \dfrac{\sin(1 - t)}{1 + \cos t}$.

By the quotient rule,

$$y' = \frac{(1 + \cos t) \cdot D_t[\sin(1 - t)] - \sin(1 - t) \cdot D_t(1 + \cos t)}{(1 + \cos t)^2}$$

$$= \frac{(1 + \cos t) \cdot \cos(1 - t) \cdot D_t(1 - t) - \sin(1 - t) \cdot [-\sin t]}{(1 + \cos t)^2}$$

$$= \frac{(1 + \cos t) \cdot \cos(1 - t) \cdot [-1] + \sin(1 - t) \cdot \sin t}{(1 + \cos t)^2}$$

$$= \frac{\sin(1 - t) \cdot \sin t - \cos(1 - t) - \cos t \cdot \cos(1 - t)}{(1 + \cos t)^2}.$$

To differentiate the other trigonometric functions, we first express them in terms of the sine and cosine. For example, $\tan u = (\sin u)/(\cos u)$ and, by the quotient rule,

$$\frac{d}{dt}(\tan u) = \frac{d}{dt}\left(\frac{\sin u}{\cos u}\right) = \frac{\cos u \cdot D_t(\sin u) - \sin u \cdot D_t(\cos u)}{\cos^2 u}$$

$$= \frac{\cos u \cdot \left(\cos u \dfrac{du}{dt}\right) - \sin u \cdot \left(-\sin u \dfrac{du}{dt}\right)}{\cos^2 u}$$

$$= \frac{1}{\cos^2 u}(\cos^2 u + \sin^2 u)\frac{du}{dt}.$$

Because $\cos^2 u + \sin^2 u = 1$ and $1/\cos^2 u = \sec^2 u$, we have

$$\frac{d}{dt}(\tan u) = \sec^2 u \frac{du}{dt}.$$

Table 9.6 gives the differentiation rules for all six trigonometric functions.

TABLE 9.6

1. $\dfrac{d}{dt}(\sin u) = \cos u \dfrac{du}{dt}$.   **2.** $\dfrac{d}{dt}(\cos u) = -\sin u \dfrac{du}{dt}$.

3. $\dfrac{d}{dt}(\tan u) = \sec^2 u \dfrac{du}{dt}$.   **4.** $\dfrac{d}{dt}(\cot u) = -\csc^2 u \dfrac{du}{dt}$.

5. $\dfrac{d}{dt}(\sec u) = \sec u \tan u \dfrac{du}{dt}$.   **6.** $\dfrac{d}{dt}(\csc u) = -\csc u \cot u \dfrac{du}{dt}$.

EXAMPLE 3   *Differentiate each of the following.*

a. $y = \tan \dfrac{t}{t + 1}$.

By Rule 3 in Table 9.6,

$$y' = \sec^2 \frac{t}{t + 1} \cdot D_t\left(\frac{t}{t + 1}\right)$$

$$= \sec^2 \frac{t}{t + 1} \cdot \left[\frac{(t + 1)(1) - t(1)}{(t + 1)^2}\right] \qquad \text{(quotient rule)}$$

$$= \frac{1}{(t + 1)^2} \sec^2 \frac{t}{t + 1}.$$

b. $y = \sec(1 - t) \cot t$.

By the product rule and Rules 5 and 4,

$$y' = \sec(1 - t) \cdot D_t(\cot t) + \cot t \cdot D_t[\sec(1 - t)]$$

$$= \sec(1 - t) \cdot [-\csc^2 t] \cdot D_t(t) +$$
$$\cot t \cdot [\sec(1 - t) \tan(1 - t) \cdot D_t(1 - t)]$$

$$= -\sec(1 - t) \csc^2 t - \cot t \sec(1 - t) \tan(1 - t)$$

$$= -\sec(1 - t) \cdot [\csc^2 t + \cot t \tan(1 - t)].$$

c. $y = \dfrac{\csc(\pi t)}{t}$.

By the quotient rule and Rule 6,

$$y' = \frac{t \cdot D_t[\csc(\pi t)] - \csc(\pi t) \cdot D_t(t)}{t^2}$$

$$= \frac{t[-\csc(\pi t) \cdot \cot(\pi t) \cdot D_t(\pi t)] - \csc(\pi t) \cdot [1]}{t^2}$$

$$= \frac{-\pi t \csc(\pi t) \cot(\pi t) - \csc(\pi t)}{t^2}$$

$$= \frac{-\csc(\pi t) \cdot [\pi t \cot(\pi t) + 1]}{t^2}$$

$$= -\frac{\pi t \cot(\pi t) + 1}{t^2 \sin(\pi t)} \qquad \left[\text{since } \csc(\pi t) = \frac{1}{\sin(\pi t)}\right].$$

EXAMPLE 4   *In a breathing cycle, a person inhales and then exhales air. Normally this takes approximately 5 seconds for a person at rest. Let v be the volume, in cubic centimeters, of air in the lungs that is attributable to the breathing cycle at time t. Suppose that*

$$v = 250\left[1 + \sin\frac{2\pi(t - 1.25)}{5}\right],$$

where time is in seconds. Figure 9.34 shows the graph of this equation for $0 \le t \le 10$. *Determine the rate of change of volume at* $t = 3.125$.

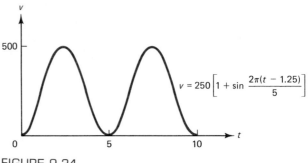

FIGURE 9.34

The rate of change at time t is given by dv/dt.

$$\frac{dv}{dt} = 250\cos\frac{2\pi(t - 1.25)}{5} D_t\left[\frac{2\pi(t - 1.25)}{5}\right]$$

$$= 250\cos\frac{2\pi(t - 1.25)}{5}\left[\frac{2\pi}{5}\right]$$

$$= 100\pi\cos\frac{2\pi(t - 1.25)}{5}.$$

Evaluating at $t = 3.125$ gives

$$\frac{dv}{dt}\bigg|_{t=3.125} = 100\pi\cos\frac{2\pi(3.125 - 1.25)}{5}$$

$$= 100\pi\cos\frac{3\pi}{4} = 100\pi\left(-\frac{\sqrt{2}}{2}\right)$$

$$= -50\sqrt{2}\pi \approx -222.$$

The minus sign indicates that the volume of air is *decreasing* at the rate of 222 cc per minute. Thus the person is exhaling.

EXAMPLE 5   *A toy manufacturer has sales given by*

$$q = 3000\left[1.05 - \cos\frac{2\pi(t - 65)}{365}\right], \qquad 0 \le t \le 365,$$

where q is the number of toys sold per day after t days from January 1. Determine the date of maximum daily sales.

First we find dq/dt.

$$\frac{dq}{dt} = 3000\left[-\left(-\sin\frac{2\pi(t-65)}{365} \right)D_t\left(\frac{2\pi(t-65)}{365} \right) \right]$$

$$= 3000\left[\frac{2\pi}{365}\sin\frac{2\pi(t-65)}{365} \right].$$

The critical values of q occur when $dq/dt = 0$; that is, when

$$\sin\frac{2\pi(t-65)}{365} = 0.$$

Now, $\sin u$ is 0 when u is 0 or π. Thus critical values occur when $\frac{2\pi(t-65)}{365}$ is 0 or π. If

$$\frac{2\pi(t-65)}{365} = 0,$$

then $t = 65$. If

$$\frac{2\pi(t-65)}{365} = \pi,$$

then

$$t = \frac{365}{2} + 65 = 247.5.$$

To test the values 65 and 247.5 for relative extrema, we shall use the second-derivative test.

$$\frac{d^2q}{dt^2} = 3000\left[\frac{2\pi}{365}\cos\frac{2\pi(t-65)}{365} \cdot D_t\left(\frac{2\pi(t-65)}{365} \right) \right]$$

$$= 3000\left[\left(\frac{2\pi}{365} \right)^2 \cos\frac{2\pi(t-65)}{365} \right].$$

$$\left.\frac{d^2q}{dt^2}\right|_{t=65} = 3000\left[\left(\frac{2\pi}{365} \right)^2 \cos 0 \right] = 3000\left[\left(\frac{2\pi}{365} \right)^2 \cdot 1 \right] > 0.$$

$$\left.\frac{d^2q}{dt^2}\right|_{t=247.5} = 3000\left[\left(\frac{2\pi}{365} \right)^2 \cos \pi \right] = 3000\left[\left(\frac{2\pi}{365} \right)^2 (-1) \right] < 0.$$

When $t = 247.5$, $d^2q/dt^2 < 0$ and thus a relative maximum occurs there. It can be shown that the endpoints 0 and 365 yield no absolute extrema. Hence maximum sales occur when $t = 247.5$. From a practical point of view, this value of t corresponds to the 248th day of the year, namely September 5.

Figure 9.35 shows a graph of the function. Although you may think that maximum sales should occur nearer to Christmas, toy retailers usually fill their stock well in advance of that time.

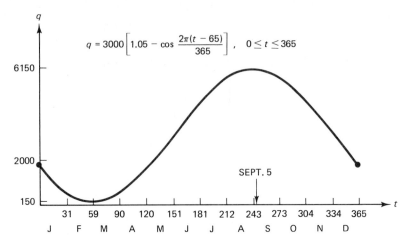

$$q = 3000\left[1.05 - \cos\frac{2\pi(t - 65)}{365}\right], \quad 0 \le t \le 365$$

FIGURE 9.35

EXERCISE 9.4

In Problems 1–30, differentiate.

1. $y = \sin \pi t$.

2. $y = \cos 3\pi t$.

3. $y = \sin(t^2 + 3t + 5)$.

4. $y = \cos(5 - t)$.

5. $y = \cos \dfrac{t}{2}$.

6. $y = \sin \sqrt{t}$.

7. $y = \tan(t + 1)^2$.

8. $y = \cot \dfrac{1}{t}$.

9. $y = \csc e^t$.

10. $y = \sec \dfrac{t}{t + 1}$.

11. $y = \sin^2(2\pi t + 1)$.

12. $y = t^2 \cos t$.

13. $y = \dfrac{\cos t^2}{t^2}$.

14. $y = \tan t - t$.

15. $y = \dfrac{\sqrt{\csc t}}{\sin t}$.

16. $y = \sin^2(2t - 3) + \cos^2(2t - 3)$.

17. $y = \sqrt{t} \cot t$.

18. $y = \sqrt{t} + \cot(3 - t)$.

19. $y = (1 + \sec 2t)^3$.

20. $y = \sin t \cos t$.

21. $y = \sin^2 t \cos^3 t$.

22. $y = e^{\cos t}$.

23. $y = \ln(\sin t)$.

24. $y = \sec(\tan t)$.

25. $y = \ln(\sec t + \tan t)$.

26. $y = \ln(\csc t - \cot t)$.

27. $y = \dfrac{1}{1 + \cos \pi t}$.

28. $y = \dfrac{\sin t + \cos t}{\cos t}$.

29. $y = \dfrac{1 + \sin t}{1 - \cos t}$.

30. $y = \dfrac{\csc t}{1 + \sin t}$.

31. Derive Rule 4 in Table 9.6.

32. Derive Rule 5 in Table 9.6.

33. Find an equation of the tangent line to $y = \sin t$ at the point where $t = 5\pi/3$.

34. Find an equation of the tangent line to $y = t \tan t$ at the point where $t = \pi/4$.

35. If $y = \tan t$, find the relative rate of change of y with respect to t when $t = 3\pi/4$.

36. If $y = (1 + \sin t)^2 \cos t$, find the relative rate of change of y with respect to t when $t = 0$.

37. Show that $y = c_1 \sin t + c_2 \cos t$, where c_1 and c_2 are constants, satisfies the differential equation $y'' + y = 0$.

38. Show that $y = e^t \sin t$ satisfies the differential equation $y'' - 2y' + 2y = 0$.

39. For a person at rest, suppose that the blood pressure P after t seconds is given by

$$P = 100 + 20 \cos \frac{7\pi t}{3},$$

where P is in millimeters of mercury. At what rate is the pressure changing after 5 seconds?

40. For a person at rest, suppose that the volume v, in cubic centimeters, of air in the lungs that is attributable to a breathing cycle is given by

$$v = 250 \left[1 + \sin \frac{2\pi(t - 1.25)}{5} \right],$$

where t is in seconds and $0 \le t \le 5$. Determine the time at which the rate of change of volume is maximum.

41. A manufacturer of swimming pool filters sells q filters a month after t months from the beginning of the year. Suppose that

$$q = 2000 \left[1.03 - \cos \frac{\pi t}{6} \right], \qquad 0 \le t \le 12.$$

Determine the maximum monthly number of filters sold and the month in which these sales occur.

42. After t months, a manufacturer has monthly sales of q units, where $q = 4t + \sin(\pi t / 6)$. Show that sales are increasing for $t > 0$.

43. Over a period of several days in a certain city, the average temperature T (in degrees Fahrenheit) at time t is given by

$$T = 71 - 9 \sin \frac{\pi(t + 2)}{12}, \qquad 0 \le t \le 24,$$

where t is the number of hours past midnight. At what time does the maximum temperature occur?

44. In Problem 43, at what rate is the temperature changing at noon?

45. In a discussion of vascular branching, Batschelet[68] assumes that the resistance R of blood along a certain path is given by

$$R = k \left[\frac{l_0 - s \cot \theta}{r_1^4} + \frac{s}{r_2^4 \sin \theta} \right].$$

Suppose that k, l_0, s, r_1, and r_2 are positive constants and θ is acute. To minimize R, Batschelet finds $dR/d\theta$ and claims that it is 0 when $\cos \theta = r_2^4 / r_1^4$. Verify this.

46. In a model of the flight of a homing pigeon across a lake [68], the total energy E involved in flying between a point over the lake to a point over land is given by

$$E = e_2 s + e_2 r \left[\frac{c}{\sin \theta} - \cot \theta \right],$$

where e_2, s, r, and c are positive constants and θ is the acute angle between the line of flight over water and a straight shore line. If $c = 2$, determine the angle θ that minimizes E. (*Hint:* Express $dE/d\theta$ in terms of sines and cosines.)

In Problems **47–50**, find $f_s(s, t)$ and $f_t(s, t)$.

47. $f(s, t) = \sin(s^2 + 2st)$.

48. $f(s, t) = \sin s \cos t$.

49. $f(s, t) = \dfrac{\tan^2 s}{\sin^2 t}$.

50. $f(s, t) = t \sin s + t \cos t$.

9.5 INTEGRALS OF TRIGONOMETRIC FUNCTIONS _____

For each differentiation formula given in Sec. 9.4, there is a corresponding integration formula. Table 9.7 gives these basic trigonometric integrals, which

TABLE 9.7

If u is a function of t, then:

1. $\displaystyle\int \sin u \; du = -\cos u + C.$

2. $\displaystyle\int \cos u \; du = \sin u + C.$

3. $\displaystyle\int \sec^2 u \; du = \tan u + C.$

4. $\displaystyle\int \csc^2 u \; du = -\cot u + C.$

5. $\displaystyle\int \sec u \tan u \; du = \sec u + C.$

6. $\displaystyle\int \csc u \cot u \; du = -\csc u + C.$

can be easily verified. For example, the differentiation formula

$$\frac{d}{dt}(\cos u) = -\sin u \, \frac{du}{dt}$$

implies that

$$\int -\sin u \, \frac{du}{dt} \, dt = \cos u + C_1,$$

from which

$$\int \sin u \, \frac{du}{dt} \, dt = -\cos u + C,$$

where $C = -C_1$. Since $\dfrac{du}{dt} \, dt$ is du, we have Formula 1:

$$\int \sin u \; du = -\cos u + C.$$

In particular, if $u = t$, then $du = dt$ and

$$\int \sin t \; dt = -\cos t + C.$$

Similarly, from Formula 2 it follows that

$$\int \cos t \; dt = \sin t + C.$$

EXAMPLE 1   *Find the following integrals.*

a. $\int \cos 7t \, dt.$

If we let $u = 7t$, then $du = 7 \, dt$. By adjusting for the constant 7, we can get an integral of the form $\int \cos u \, du$ and then use Formula 2.

$$\int \cos 7t \, dt = \frac{1}{7} \int \cos 7t \, [7 \, dt]$$

$$= \frac{1}{7} \int \cos u \, du = \frac{1}{7} \sin 7t + C.$$

b. $\int 2x \sin x^2 \, dx.$

If we let $u = x^2$, then $du = 2x \, dx$. Thus we have the form $\int \sin u \, du$ and we can use Formula 1.

$$\int 2x \sin x^2 \, dx = \int \sin x^2 \, [2x \, dx]$$

$$= \int \sin u \, du = -\cos x^2 + C.$$

EXAMPLE 2   *Find* $\int \sin^3 t \cos t \, dt.$

This integral does not match any formula in Table 9.7. However, if we let $u = \sin t$, then $du = \cos t \, dt$ and the integral has the form $\int u^3 \, du$. Thus the power rule for integration applies.

$$\int \sin^3 t \cos t \, dt = \int (\sin t)^3 [\cos t \, dt]$$

$$= \int u^3 \, du = \frac{\sin^4 t}{4} + C.$$

EXAMPLE 3   *Find the following integrals.*

a. $\int (\sec^2 x + \csc^2 x) \, dx.$

From Formulas 3 and 4 with $u = x$ and $du = dx$, we have

$$\int (\sec^2 x + \csc^2 x) \, dx = \tan x - \cot x + C.$$

b. $\int e^t \sec e^t \tan e^t \, dt.$

If $u = e^t$, then $du = e^t \, dt$ and by Formula 5,

$$\int \sec e^t \tan e^t \, [e^t \, dt] = \int \sec u \tan u \, du = \sec e^t + C.$$

EXAMPLE 4   *Find the area under one arch of the sine curve y = sin t.*

A sketch of the arch on the interval $[0, \pi]$ is given in Fig. 9.36.

$$\text{Area} = \int_0^\pi y \, dt = \int_0^\pi \sin t \, dt = -\cos t \Big|_0^\pi$$

$$= (-\cos \pi) - (-\cos 0) = [-(-1)] - [-(1)]$$

$$= 2 \text{ square units.}$$

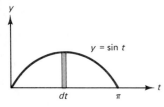

FIGURE 9.36

EXERCISE 9.5

In Problems 1–22, find the indefinite integrals.

1. $\int 3 \sin t \, dt.$

2. $\int \frac{\sin x}{4} \, dx.$

3. $\int \sin 2t \, dt.$

4. $\int \sin \frac{t}{4} \, dt.$

5. $\int t \cos t^2 \, dt.$

6. $\int \cos 4x \, dx.$

7. $\int \sin 3\pi t \, dt.$

8. $\int \sin 2\pi \, dt.$

9. $\int \cos \frac{x}{5} \, dx.$

10. $\int t^5 \cos t^6 \, dt.$

11. $\int x \sin 6x^2 \, dx.$

12. $\int e^t \cos e^t \, dt.$

13. $\int \sec^2 3t \, dt.$

14. $\int \csc^2 8w \, dw.$

15. $\int \sec 5x \tan 5x \, dx.$

16. $\int x \csc x^2 \cot x^2 \, dx.$

17. $\int \frac{\csc^2 \sqrt{t}}{\sqrt{t}} \, dt.$

18. $\int \frac{\sec^2 e^{-x}}{e^x} \, dx.$

19. $\int \sin^4 x \cos x \, dx.$

20. $\int \cot^2 x \csc^2 x \, dx.$

21. $\int \sqrt{\cos t} \sin t \, dt.$

22. $\int \sin t \cos t \, dt.$

In Problems 23–28, find the definite integrals.

23. $\int_0^{\pi/3} \sin t \, dt.$

24. $\int_0^\pi \sin 3t \, dt.$

25. $\int_0^{\pi/4} \sin t \cos^3 t \, dt.$

26. $\int_0^{\pi/2} \cos 2x \, dx.$

27. $\int_{0.1}^{0.2} \cos t \, dt.$

28. $\int_{\pi/2}^\pi (\sin t + \cos t) \, dt.$

29. Find the area of the region between $y = \sin t$ and the t-axis from $t = \pi/3$ to $t = \pi/2$.

30. Find the area of the region between $y = \sin 2t$ and the t-axis from $t = 0$ to $t = \pi/3$.

31. Find the area of the region between $y = \cos t$ and the t-axis from $t = -\pi/2$ to $t = 3\pi/4$.

32. Find the area of the region between the curve $y = \sin \frac{t}{3}$ and the t-axis from $t = 0$ to $t = 4\pi$.

33. Find $\int \cot x \, dx.$ $\left(Hint: \cot x = \frac{\cos x}{\sin x}. \right)$

34. Find $\int \tan x \, dx.$ $\left(Hint: \tan x = \frac{\sin x}{\cos x}. \right)$

35. The marginal cost (in dollars per unit) for a manufacturer to produce q units of a product is given by

$$\frac{dc}{dq} = 10 + 3 \cos\left(\frac{\pi q}{200}\right).$$

If fixed costs are $1000, find the cost of producing 100 units.

36. The temperature T, in degrees Fahrenheit, of a patient taking an antibiotic is given at the end of the first t days by

$$T = 101 + 1.5 \cos \frac{\pi t}{4}.$$

Find the average temperature over the interval $[0, 2]$.

37. The blood pressure P of a person performing a task is given by

$$P = 90 - 10 \cos \frac{5\pi}{2}t,$$

where t is time. Find the average value of P over the interval $[0, 10]$.

38. An electrical current i varies with time t according to the function

$$i = 5 \sin 2t.$$

Find the average value of i during one period of the function.

9.6 INVERSE TRIGONOMETRIC FUNCTIONS

The equation $w = \sin t$ defines w as a function of t. However, it does not define t as a function of w. To see why, let the input w be $1/2$. Then the output is a value of t such that

$$\frac{1}{2} = \sin t.$$

Because the sine function is periodic, this equation has many solutions. Some are $\pi/6$, $5\pi/6$, $13\pi/6$, and $-7\pi/6$, as shown in Fig. 9.37.

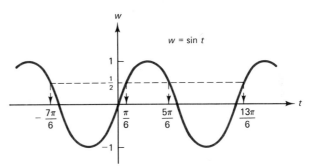

FIGURE 9.37

Because there is more than one output for a single input, the equation $w = \sin t$ does *not* define t as a function of w. However, if we restrict the domain of $\sin t$ so that

$$-\frac{\pi}{2} \le t \le \frac{\pi}{2},$$

then to each input w, where $-1 \le w \le 1$, there corresponds exactly one output t (see Fig. 9.38) and we *do* have a function of w. Thus if the input is $w = 1/2$, then the output t must be $\pi/6$ because

$$\sin \frac{\pi}{6} = \frac{1}{2} \quad \text{and} \quad -\frac{\pi}{2} \le \frac{\pi}{6} \le \frac{\pi}{2}.$$

Similarly, if $w = -1$, then $t = -\pi/2$ because

$$\sin\left(-\frac{\pi}{2}\right) = -1 \quad \text{and} \quad -\frac{\pi}{2} \le -\frac{\pi}{2} \le \frac{\pi}{2}.$$

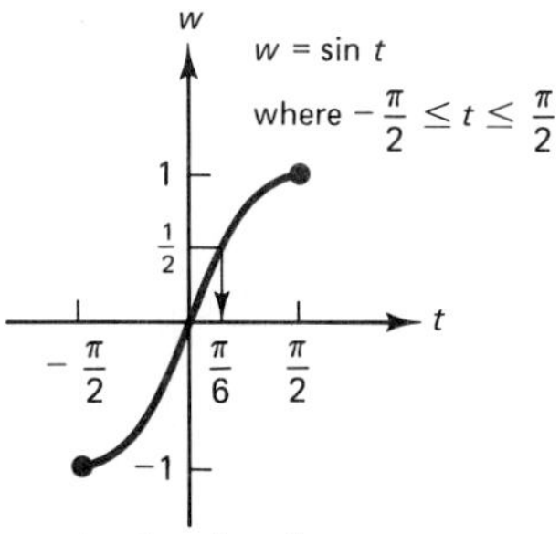

$w = \sin t$
where $-\frac{\pi}{2} \le t \le \frac{\pi}{2}$

t is a function of w
for $-\frac{\pi}{2} \le t \le \frac{\pi}{2}$

FIGURE 9.38

This function of w that reverses the action of the restricted sine function is called the *inverse sine function* (or *arcsine function*) and is denoted $\sin^{-1}$ (or arcsin). Thus $t = \sin^{-1} w$. For example, we have shown that $\sin^{-1}(\frac{1}{2}) = \pi/6$ and $\sin^{-1}(-1) = -\pi/2$.

In the equation $t = \sin^{-1} w$, let's replace the input w by x and the output t by y. Then to say that y is the inverse sine of x, that is,

$$y = \sin^{-1} x,$$

means that $x = \sin y$, where $-\pi/2 \le y \le \pi/2$. Note that the domain of $\sin^{-1}$ is the interval $[-1, 1]$. In summary we have:

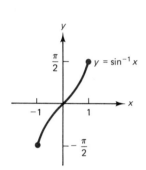

$$y = \sin^{-1} x \text{ if and only if } \sin y = x,$$
$$\text{where } -1 \le x \le 1 \quad \text{and} \quad -\frac{\pi}{2} \le y \le \frac{\pi}{2}.$$

The graph of $y = \sin^{-1} x$ is shown in Fig. 9.39.

Restricting the domains of the cosine and tangent functions allows us to define inverse functions for them as follows.

FIGURE 9.39

$$y = \cos^{-1} x \text{ if and only if } \cos y = x,$$
$$\text{where } -1 \le x \le 1 \quad \text{and} \quad 0 \le y \le \pi.$$

$$y = \tan^{-1} x \text{ if and only if } \tan y = x,$$
$$\text{where } x \text{ is any real number and } -\frac{\pi}{2} < y < \frac{\pi}{2}.$$

EXAMPLE 1   *Find each of the following.*

a. $\sin^{-1} \dfrac{\sqrt{2}}{2}$.

$\sin^{-1}(\sqrt{2}/2)$ is the value of y between $-\pi/2$ and $\pi/2$ such that $\sin y = \sqrt{2}/2$. Thus $y = \pi/4$, so $\sin^{-1}(\sqrt{2}/2) = \pi/4$.

b. $\cos^{-1} \dfrac{1}{2}$.

$\cos^{-1}(1/2)$ is the value of y between 0 and π such that $\cos y = 1/2$. Thus $y = \pi/3$, so $\cos^{-1}(1/2) = \pi/3$.

EXAMPLE 2   *Find each of the following.*

a. $\cos^{-1}\left(-\dfrac{\sqrt{3}}{2}\right)$.

$\cos^{-1}(-\sqrt{3}/2)$ is the value of y between 0 and π such that $\cos y = -\sqrt{3}/2$. Since $\cos y$ is negative, y must lie between $\pi/2$ and π. If the reference angle is y_R, then $\cos y_R = \sqrt{3}/2$, so $y_R = \pi/6$. Thus $y = \pi - (\pi/6) = 5\pi/6$. Hence $\cos^{-1}(-\sqrt{3}/2) = 5\pi/6$.

b. $\tan^{-1}(-1)$.

$\tan^{-1}(-1)$ is the value of y between $-\pi/2$ and $\pi/2$ such that $\tan y = -1$. Since $\tan y$ is negative, y must lie between $-\pi/2$ and 0. If its reference angle is y_R, then $\tan y_R = 1$, so $y_R = \pi/4$. Thus $y = \tan^{-1}(-1) = -\pi/4$.

EXAMPLE 3   *Find each of the following.*

a. $\sin\left(\sin^{-1}\dfrac{1}{4}\right)$.

$\sin^{-1}(1/4)$ is a value of y such that $\sin y = 1/4$. Thus

$$\sin\left(\sin^{-1}\frac{1}{4}\right) = \sin(y) = \frac{1}{4}.$$

As a general rule,

$$\boxed{\sin(\sin^{-1} x) = x.}$$

b. $\cos(\sin^{-1} 0)$.

Since $\sin^{-1} 0 = 0$, then

$$\cos(\sin^{-1} 0) = \cos(0) = 1.$$

c. $\sin^{-1}\left(\sin\dfrac{5\pi}{6}\right)$.

$$\sin^{-1}\left(\sin\frac{5\pi}{6}\right) = \sin^{-1}\left(\frac{1}{2}\right) = \frac{\pi}{6}.$$

Note that $\sin^{-1}\left(\sin\dfrac{5\pi}{6}\right) \neq \dfrac{5\pi}{6}$. Compare this with the rule in part (a).

The remaining inverse trigonometric functions are defined as follows:

$y = \cot^{-1} x$ if and only if $\cot y = x$, where
x is any real number and $0 < y < \pi$.

$y = \sec^{-1} x$ if and only if $\sec y = x$, where
$x \leq -1$ or $x \geq 1$, and $-\pi \leq y < -\dfrac{\pi}{2}$ or $0 \leq y < \dfrac{\pi}{2}$.

$y = \csc^{-1} x$ if and only if $\csc y = x$, where
$x \leq -1$ or $x \geq 1$, and $-\pi < y \leq -\dfrac{\pi}{2}$ or $0 < y \leq \dfrac{\pi}{2}$.

Let us now consider the derivatives of the inverse trigonometric functions. To differentiate $y = \sin^{-1} u$, where u is a differentiable function of x and $-\pi/2 \leq y \leq \pi/2$, we use the equivalent form

$$\sin y = u. \tag{1}$$

Differentiating both sides with respect to x gives

$$\cos y \cdot \frac{dy}{dx} = \frac{du}{dx}.$$

Solving for dy/dx, we have

$$\frac{dy}{dx} = \frac{1}{\cos y} \frac{du}{dx}. \tag{2}$$

We can express $\cos y$ in terms of u. From the identity $\sin^2 y + \cos^2 y = 1$, it follows that

$$\cos y = \pm\sqrt{1 - \sin^2 y}$$

$$= \pm\sqrt{1 - u^2} \qquad \text{[from Eq. (1)]}.$$

Because $-\pi/2 \leq y \leq \pi/2$, $\cos y$ cannot be negative, and thus we can delete the $\pm$ sign. Substituting for $\cos y$ in Eq. (2) gives

$$\frac{dy}{dx} = \frac{1}{\sqrt{1 - u^2}} \frac{du}{dx}.$$

Thus we have the formula

$$\frac{d}{dx}(\sin^{-1} u) = \frac{1}{\sqrt{1 - u^2}} \frac{du}{dx}. \tag{3}$$

EXAMPLE 4 *Differentiate $y = \sin^{-1} x^3$.*

Applying Formula (3) with $u = x^3$ gives

$$\frac{dy}{dx} = \frac{1}{\sqrt{1 - (x^3)^2}}(3x^2) = \frac{3x^2}{\sqrt{1 - x^6}}.$$

To obtain the derivative of the inverse tangent function, we let $y = \tan^{-1} u$ or

$$\tan y = u. \tag{4}$$

By differentiating with respect to x, we have

$$\sec^2 y \cdot \frac{dy}{dx} = \frac{du}{dx}.$$

Solving for dy/dx and using the identity $1 + \tan^2 y = \sec^2 y$ give

$$\frac{dy}{dx} = \frac{1}{\sec^2 y} \frac{du}{dx}$$

$$= \frac{1}{1 + \tan^2 y} \frac{du}{dx}$$

$$= \frac{1}{1 + u^2} \frac{du}{dx} \qquad \text{[from Eq. (4)]}.$$

Thus

$$\frac{d}{dx}(\tan^{-1} u) = \frac{1}{1 + u^2} \frac{du}{dx}. \tag{5}$$

EXAMPLE 5 *Differentiate $f(x) = 2x \tan^{-1} 3x$.*

By the product rule,

$$f'(x) = (2x) D_x(\tan^{-1} 3x) + (\tan^{-1} 3x) D_x(2x).$$

By Formula (5),

$$f'(x) = (2x)\left[\frac{1}{1 + (3x)^2}\right] D_x(3x) + (\tan^{-1} 3x)(2)$$

$$= \frac{6x}{1 + 9x^2} + 2 \tan^{-1} 3x.$$

Derivative formulas for the remaining inverse trigonometric functions can be obtained by methods similar to those shown for $\sin^{-1}$ and $\tan^{-1}$. Table 9.8 gives a complete list. Note that the derivatives for $\sin^{-1} u$ and $\cos^{-1} u$

TABLE 9.8

1. $\dfrac{d}{dx}(\sin^{-1} u) = \dfrac{1}{\sqrt{1 - u^2}}\dfrac{du}{dx}.$

2. $\dfrac{d}{dx}(\cos^{-1} u) = -\dfrac{1}{\sqrt{1 - u^2}}\dfrac{du}{dx}.$

3. $\dfrac{d}{dx}(\tan^{-1} u) = \dfrac{1}{1 + u^2}\dfrac{du}{dx}.$

4. $\dfrac{d}{dx}(\cot^{-1} u) = -\dfrac{1}{1 + u^2}\dfrac{du}{dx}.$

5. $\dfrac{d}{dx}(\sec^{-1} u) = \dfrac{1}{u\sqrt{u^2 - 1}}\dfrac{du}{dx}.$

6. $\dfrac{d}{dx}(\csc^{-1} u) = -\dfrac{1}{u\sqrt{u^2 - 1}}\dfrac{du}{dx}.$

differ only in sign. The same is true for the derivatives of $\tan^{-1} u$ and $\cot^{-1} u$, and for $\sec^{-1} u$ and $\csc^{-1} u$.

EXAMPLE 6   *Differentiate* $y = \cos^{-1}(x + 1)^2.$

Applying Formula 2 from Table 9.8 with $u = (x + 1)^2$, we have

$$\frac{dy}{dx} = -\frac{1}{\sqrt{1 - [(x + 1)^2]^2}}[2(x + 1)]$$

$$= -\frac{2(x + 1)}{\sqrt{1 - (x + 1)^4}}.$$

EXAMPLE 7   *Differentiate* $y = \sec^{-1} x^3.$

Applying Formula 5 from Table 9.8 with $u = x^3$, we have

$$\frac{dy}{dx} = \frac{1}{x^3\sqrt{(x^3)^2 - 1}}(3x^2) = \frac{3}{x\sqrt{x^6 - 1}}.$$

EXERCISE 9.6

In Problems 1–22, find the value.

1. $\sin^{-1} 1.$

2. $\sin^{-1} \dfrac{\sqrt{3}}{2}.$

3. $\cos^{-1} \dfrac{\sqrt{3}}{2}.$

4. $\cos^{-1}(-1).$

5. $\tan^{-1} 0.$

6. $\sin^{-1}\left(-\dfrac{\sqrt{2}}{2}\right).$

7. $\sin^{-1}\left(-\dfrac{1}{2}\right).$

8. $\cos^{-1} 0.$

9. $\tan^{-1} 1.$

10. $\tan^{-1}(-\sqrt{3}).$

11. $\cos^{-1}\left(-\dfrac{\sqrt{2}}{2}\right)$.

12. $\cos^{-1}\left(-\dfrac{1}{2}\right)$.

13. $\sin\left(\sin^{-1}\dfrac{1}{3}\right)$.

14. $\cos\left[\cos^{-1}\left(-\dfrac{1}{5}\right)\right]$.

15. $\cos\left(\sin^{-1}\dfrac{1}{2}\right)$.

16. $\sin[\cos^{-1}(-1)]$.

17. $\cot(\cos^{-1}0)$.

18. $\tan(\cos^{-1}1)$.

19. $\sin^{-1}\left(\sin\dfrac{2\pi}{3}\right)$.

20. $\sin^{-1}\left(\sin\dfrac{\pi}{3}\right)$.

In Problems 21–38, find y'.

21. $y = \sin^{-1}3x$.

22. $y = \cos^{-1}x^2$.

23. $y = \tan^{-1}x^3$.

24. $y = \tan^{-1}\dfrac{x}{3}$.

25. $y = \cos^{-1}(t + 2)$.

26. $y = t\cot^{-1}t$.

27. $y = 3x\sin^{-1}3x$.

28. $y = \sin^{-1}(t - 1)$.

29. $y = t^2\tan^{-1}t$.

30. $y = \sin^{-1}x + \cos^{-1}x$.

31. $y = \cos^{-1}e^x$.

32. $y = \cot^{-1}\sqrt{x}$.

33. $y = (\sin^{-1}x)^2$.

34. $y = \sin x\sin^{-1}x$.

35. $y = \sec^{-1}x^2$.

36. $y = \csc^{-1}3x$.

37. $y = \sin(\sin^{-1}x)$.

38. $y = \sec^{-1}(e^{2x})$.

39. If $f(x) = \tan^{-1}x$, find $f''(x)$.

40. If $f(x) = \sin^{-1}x$, find $f''(x)$.

41. Find the slope of the curve $y = \tan^{-1}x^2$ when $x = \sqrt{2}/2$.

42. Find an equation of the tangent line to the curve $y = \sin^{-1}x$ when $x = 1/2$.

43. If $f(x, y) = \sin^{-1}(xy)$, find $f_x(x, y)$.

44. If $f(x, y) = \tan^{-1}(x + y)$, find $f_{yx}(x, y)$.

45. A lighthouse is 1000 ft off a straight shore [Fig. 9.40(a)]. Find the angle θ (the Greek letter read "theta") when the light beam shines along a sea wall at a point x feet down the coast. (Give answer in terms of x.)

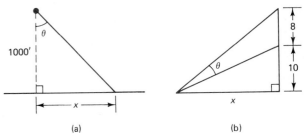

(a)   (b)

FIGURE 9.40

46. In Fig. 9.40(b), show that $\theta = \tan^{-1}\dfrac{18}{x} - \tan^{-1}\dfrac{10}{x}$.

47. A lighthouse is 1000 ft off a straight shore. If the light revolves at 2 revolutions per minute, how fast is the light beam moving along a sea wall at a point 500 ft down the coast? [*Hint:* Differentiate the result in Problem 45 implicitly with respect to time t. Using the fact that $d\theta/dt = 4\pi$ (why?), solve for dx/dt when $x = 500$.]

9.7 INTEGRALS INVOLVING INVERSE TRIGONOMETRIC FUNCTIONS

Because $\dfrac{d}{dt}(\sin^{-1}u) = \dfrac{1}{\sqrt{1 - u^2}}\dfrac{du}{dt}$, we have

$$\int \frac{1}{\sqrt{1 - u^2}}\,du = \sin^{-1}u + C. \qquad (1)$$

Similarly, the differentiation formulas for $\tan^{-1}u$ and $\sec^{-1}u$, give rise to the integration formulas

$$\int \frac{1}{1 + u^2}\, du = \tan^{-1} u + C \tag{2}$$

and $\quad\displaystyle\int \frac{1}{u\sqrt{u^2 - 1}}\, du = \sec^{-1} u + C. \tag{3}$

EXAMPLE 1   *Find the following integrals.*

a. $\displaystyle\int \frac{1}{\sqrt{1 - t^2}}\, dt.$

We use Formula (1) with $u = t$. Then $du = dt$ and

$$\int \frac{1}{\sqrt{1 - t^2}}\, dt = \sin^{-1} t + C.$$

b. $\displaystyle\int \frac{dt}{\sqrt{1 - 16t^2}}.$

Because the radical is $\sqrt{1 - (4t)^2}$, we are motivated to use Formula (1). If $u = 4t$, then $du = 4\, dt$. Thus

$$\int \frac{dt}{\sqrt{1 - 16t^2}} = \frac{1}{4} \int \frac{1}{\sqrt{1 - (4t)^2}} [4\, dt]$$

$$= \frac{1}{4} \int \frac{1}{\sqrt{1 - u^2}}\, du = \sin^{-1} u + C$$

$$= \frac{1}{4} \sin^{-1}(4t) + C.$$

EXAMPLE 2   *Find the following integrals.*

a. $\displaystyle\int \frac{t}{1 + t^4}\, dt.$

The denominator is $1 + (t^2)^2$. If $u = t^2$, then $du = 2t\, dt$ and, by Formula (2),

$$\int \frac{t}{1 + t^4}\, dt = \frac{1}{2} \int \frac{1}{1 + (t^2)^2} [2t\, dt]$$

$$= \frac{1}{2} \int \frac{1}{1 + u^2}\, du = \frac{1}{2} \tan^{-1} u + C$$

$$= \frac{1}{2} \tan^{-1} t^2 + C.$$

b. $\int_{1/2}^{1} \dfrac{dt}{t\sqrt{4t^2 - 1}}$.

We shall use Formula (3). If $u = 2t$, then $du = 2\ dt$.

$$\int_{1/2}^{1} \frac{dt}{t\sqrt{4t^2 - 1}} = \int_{1/2}^{1} \frac{1}{(2t)\sqrt{(2t)^2 - 1}}[2\ dt].$$

Note that 2 was inserted in the numerator to obtain du, and 2 was inserted in the denominator to obtain u. Thus no factor is required in front of the integral. To complete the integration, we have

$$\int_{1/2}^{1} \frac{1}{(2t)\sqrt{(2t)^2 - 1}}[2\ dt] = \sec^{-1} 2t \ \Big|_{1/2}^{1}$$

$$= \sec^{-1} 2 - \sec^{-1} 1$$

$$= \frac{\pi}{3} - 0 = \frac{\pi}{3}.$$

In order to use Formulas (1)–(3) on an integral, it may be necessary to first perform an algebraic manipulation. Example 3 will illustrate.

EXAMPLE 3   *Find the following integrals.*

a. $\int \dfrac{1}{4 + t^2}\ dt$.

If the 4 were a 1, we could use Formula (2). Thus we remove a factor of 4 from the denominator.

$$\int \frac{1}{4 + t^2}\ dt = \int \frac{1}{4\left[1 + \dfrac{t^2}{4}\right]}\ dt = \frac{1}{4}\int \frac{1}{1 + \left(\dfrac{t}{2}\right)^2}\ dt.$$

If $u = \dfrac{t}{2}$, then $du = \dfrac{1}{2}\ dt$, and by Formula (2) we have

$$\frac{1}{4}\int \frac{1}{1 + \left(\dfrac{t}{2}\right)^2}\ dt = 2 \cdot \frac{1}{4}\int \frac{1}{1 + \left(\dfrac{t}{2}\right)^2}\left[\frac{1}{2}\ dt\right]$$

$$= \frac{1}{2}\tan^{-1}\frac{t}{2} + C.$$

b. $\displaystyle\int \frac{dt}{\sqrt{9 - 4t^2}}.$

We shall rewrite this integral so that we can use Formula (1).

$$\int \frac{dt}{\sqrt{9\left[1 - \dfrac{4t^2}{9}\right]}} = \frac{1}{3} \int \frac{dt}{\sqrt{1 - \left(\dfrac{2t}{3}\right)^2}}$$

$$= \frac{3}{2} \cdot \frac{1}{3} \int \frac{dt}{\sqrt{1 - \left(\dfrac{2t}{3}\right)^2}} \left[\frac{2}{3} dt\right]$$

$$= \frac{1}{2} \sin^{-1} \frac{2t}{3} + C.$$

By a technique called *completing the square,* we may be able to match an integral with one of those in Formulas (1)–(3).

EXAMPLE 4   *Find* $\displaystyle\int \frac{1}{t^2 + 6t + 13} dt.$

To complete the square, we express $t^2 + 6t + 13$ as $(t^2 + 6t + a) + b$, where a and b are constants and $t^2 + 6t + a$ is the square of a binomial. The constant a is the square of half the coefficient of the t-term: $a = \left(\dfrac{6}{2}\right)^2 = 3^2 = 9$. Because $a + b$ must equal 13, we have $b = 4$. Thus

$$\int \frac{1}{t^2 + 6t + 13} dt = \int \frac{1}{(t^2 + 6t + 9) + 4} dt$$

$$= \int \frac{1}{(t + 3)^2 + 4} dt.$$

Now we remove a factor of 4 from the denominator.

$$\int \frac{1}{(t + 3)^2 + 4} dt = \int \frac{1}{4\left[\dfrac{(t + 3)^2}{4} + 1\right]} dt$$

$$= \frac{1}{4} \int \frac{1}{\left(\dfrac{t + 3}{2}\right)^2 + 1} dt.$$

If $u = \dfrac{t + 3}{2}$, then $du = \dfrac{1}{2} dt$. By Formula (2),

$$\frac{1}{4} \int \frac{1}{\left(\dfrac{t+3}{2}\right)^2 + 1} \, dt = 2 \cdot \frac{1}{4} \int \frac{1}{\left(\dfrac{t+3}{2}\right)^2 + 1} \left[\frac{1}{2} dt\right]$$

$$= \frac{1}{2} \tan^{-1} \frac{t + 3}{2} + C.$$

EXERCISE 9.7

In Problems 1–24, find the integrals.

1. $\displaystyle\int \frac{1}{1 + t^2} \, dt.$

2. $\displaystyle\int \frac{1}{t\sqrt{t^2 - 1}} \, dt.$

3. $\displaystyle\int \frac{dt}{1 + 25t^2}.$

4. $\displaystyle\int \frac{dt}{\sqrt{1 - 9t^2}}.$

5. $\displaystyle\int \frac{dx}{\sqrt{1 - 4x^2}}.$

6. $\displaystyle\int \frac{dx}{1 + 0.01x^2}.$

7. $\displaystyle\int \frac{dt}{4t\sqrt{4t^2 - 1}}.$

8. $\displaystyle\int \frac{e^t}{e^{2t} + 1} \, dt.$

9. $\displaystyle\int_0^{1/8} \frac{1}{\sqrt{1 - 16t^2}} \, dt.$

10. $\displaystyle\int_{2/3}^{2/\sqrt{3}} \frac{dt}{t\sqrt{3t^2 - 1}}.$

11. $\displaystyle\int \frac{dx}{\sqrt{16 - x^2}}.$

12. $\displaystyle\int \frac{x^2}{\sqrt{1 - 4x^6}} \, dx.$

13. $\displaystyle\int \frac{dt}{25t^2 + 9}.$

14. $\displaystyle\int \frac{dt}{t\sqrt{t^2 - 4}}.$

15. $\displaystyle\int_{-3/2}^{3/2} \frac{1}{4t^2 + 9} \, dt.$

16. $\displaystyle\int \frac{dt}{\sqrt{12 - 3t^2}}.$

17. $\displaystyle\int \frac{x}{\sqrt{3 - 4x^2}} \, dx.$

18. $\displaystyle\int_1^2 \frac{dx}{\sqrt{2 - (x - 1)^2}}.$

19. $\displaystyle\int \frac{dt}{t\sqrt{t^4 - 1}}.$

20. $\displaystyle\int \frac{t}{10t^2 + 9} \, dt.$

21. $\displaystyle\int \frac{dt}{t^2 + 4t + 5}.$

22. $\displaystyle\int \frac{dt}{t^2 - 8t + 25}.$

23. $\displaystyle\int \frac{dt}{\sqrt{8 - 2t - t^2}}.$

24. $\displaystyle\int \frac{dt}{(t - 5)\sqrt{t^2 - 10t + 21}}.$

25. Find the area bounded above by $y = \dfrac{1}{\sqrt{1 - x^2}}$ and below by $y = 0$ from $x = -\frac{1}{2}$ to $x = \frac{1}{2}$.

26. Find the area bounded above by $y = \dfrac{1}{x\sqrt{x^2 - 9}}$ and below by $y = 0$ from $x = 3\sqrt{2}$ to $x = 6$.

27. Determine the average value of $f(x) = \dfrac{1}{x^2 + 4}$ over the interval $[0, 2]$.

28. Determine the average value of $f(x) = \dfrac{1}{\sqrt{9 - 16x^2}}$ over the interval $[-\frac{3}{8}, \frac{3}{8}]$.

29. Suppose the demand equation for a manufacturer's product is

$$p = \frac{1000}{q^2 - 2q + 2},$$

where p is price per unit at which q units are demanded. If the equilibrium quantity is 8, determine the consumers' surplus.

9.8 REVIEW

Important Terms and Symbols

Section 9.1 initial side   terminal side   vertex   positive angle   negative angle
standard position   quadrant of an angle   quadrantal angle   degree   radian
acute angle   obtuse angle   right angle   coterminal angles

Section 9.2 $\sin t$ $\cos t$ opposite side   adjacent side   special angles   angle of elevation
trigonometric identity   reference angle

Section 9.3 periodic function   period   cycle   amplitude $\tan t$ $\cot t$ $\sec t$ $\csc t$

Section 9.6 $\sin^{-1} x$ $\cos^{-1} x$ $\tan^{-1} x$ $\cot^{-1} x$ $\sec^{-1} x$ $\csc^{-1} x$

Summary

Two units for measuring angles are the degree and radian. One radian is the measure of the central angle that subtends an arc of length r on a circle of radius r. The conversion formula between radians and degrees is

$$\pi \text{ rad} = 180°.$$

The six trigonometric functions are defined by using a point (x, y) on the terminal side of an angle t in standard position (t measured in radians). If r is the distance of (x, y) from the origin, then $r = \sqrt{x^2 + y^2}$ and

$$\sin t = \frac{y}{r}, \qquad \cos t = \frac{x}{r},$$

$$\tan t = \frac{y}{x}, \qquad \cot t = \frac{x}{y},$$

$$\sec t = \frac{r}{x}, \qquad \csc t = \frac{r}{y}.$$

Some relationships between these functions are

$$\tan t = \frac{\sin t}{\cos t}, \quad \cot t = \frac{\cos t}{\sin t}, \quad \sec t = \frac{1}{\cos t},$$

$$\csc t = \frac{1}{\sin t}, \quad \cot t = \frac{1}{\tan t}.$$

The sine and cosine functions can also be defined in another way. If an angle t is in standard position, then its terminal side intersects the unit circle at the point $(\cos t, \sin t)$. With this interpretation of the sine and cosine, it is easy to determine the trigonometric values of quadrantal angles.

If t is a first-quadrant angle, then all six trigonometric functions are positive. For the second quadrant, the only positive functions are the sine and cosecant; for the third-quadrant, they are the tangent and cotangent; for the fourth-quadrant, they are the cosine and secant.

If t is an acute angle, then the trigonometric functions of t can be defined as ratios of the sides of a right triangle containing angle t.

$$\sin t = \frac{\text{opp}}{\text{hyp}}, \qquad \cos t = \frac{\text{adj}}{\text{hyp}},$$

$$\tan t = \frac{\text{opp}}{\text{adj}}, \qquad \cot t = \frac{\text{adj}}{\text{opp}},$$

$$\sec t = \frac{\text{hyp}}{\text{adj}}, \qquad \csc t = \frac{\text{hyp}}{\text{opp}},$$

By using special triangles, it is quite simple to determine the trigonometric values of the special angles $\pi/6$, $\pi/4$, and $\pi/3$ (see Fig. 9.14).

The trigonometric values for a nonquadrantal angle t can be determined from those for its reference angle t_R, which is acute. That is, a trigonometric function of t and of t_R can differ only in sign. The correct sign depends on the quadrant of t. For example, if $t = 7\pi/6$, then $t_R = \pi/6$ and

$$\sin \frac{7\pi}{6} = -\sin \frac{\pi}{6} = -\frac{1}{2},$$

because $7\pi/6$ is a third-quadrant angle and the sine is negative there.

Three of the most basic trigonometric identities are

$$\sin^2 t + \cos^2 t = 1,$$

$$1 + \tan^2 t = \sec^2 t,$$

$$\text{and} \qquad 1 + \cot^2 t = \csc^2 t.$$

Other identities are found in Tables 9.3 and 9.5.

The trigonometric functions are periodic; that is, their graphs are repeating. The sine, cosine, secant, and cosecant functions have period 2π, while the tangent and cotangent functions have period π. This means, for example, that the graph of $y = \sin t$ repeats itself every 2π radians.

The graph of $y = a \sin bt$ (for $b > 0$) is that of a sine curve if a is positive, but is inverted if a is negative. The amplitude is $|a|$ and the period is $2\pi/b$. Similar results are true for the cosine function.

The derivatives of the trigonometric functions are

$$\frac{d}{dt}(\sin u) = \cos t \frac{du}{dt},$$

$$\frac{d}{dt}(\cos t) = -\sin u \frac{du}{dt},$$

$$\frac{d}{dt}(\tan u) = \sec^2 u \frac{du}{dt},$$

$$\frac{d}{dt}(\cot u) = -\csc^2 u \frac{du}{dt},$$

$$\frac{d}{dt}(\sec u) = \sec u \tan u \frac{du}{dt},$$

$$\frac{d}{dt}(\csc u) = -\csc u \cot u \frac{du}{dt}.$$

The integral formulas are

$$\int \sin u \, du = -\cos u + C,$$

$$\int \cos u \, du = \sin u + C,$$

$$\int \sec^2 u \, du = \tan u + C,$$

$$\int \csc^2 u \, du = -\cot u + C,$$

$$\int \sec u \tan u \, du = \sec u + C,$$

$$\int \csc u \cot u \, du = -\csc u + C.$$

By restricting the domains of the trigonometric functions, we are able to define the inverse trigonometric functions. Formulas for their derivatives are

$$\frac{d}{dx}(\sin^{-1} u) = \frac{1}{\sqrt{1 - u^2}} \frac{du}{dx},$$

$$\frac{d}{dx}(\cos^{-1} u) = -\frac{1}{\sqrt{1 - u^2}} \frac{du}{dx},$$

$$\frac{d}{dx}(\tan^{-1} u) = \frac{1}{1 + u^2} \frac{du}{dx},$$

$$\frac{d}{dx}(\cot^{-1} u) = -\frac{1}{1 + u^2} \frac{du}{dx},$$

$$\frac{d}{dx}(\sec^{-1} u) = \frac{1}{u\sqrt{u^2 - 1}} \frac{du}{dx},$$

$$\frac{d}{dx}(\csc^{-1} u) = -\frac{1}{u\sqrt{u^2 - 1}} \frac{du}{dx}.$$

Integrals involving inverse trigonometric functions are

$$\int \frac{1}{\sqrt{1 - u^2}} \, du = \sin^{-1} u + C,$$

$$\int \frac{1}{1 + u^2} \, du = \tan^{-1} u + C,$$

$$\int \frac{1}{u\sqrt{u^2 - 1}} \, du = \sec^{-1} u + C.$$

Review Problems

In Problems 1–3, convert each angle to degrees.

1. $\dfrac{5\pi}{6}$.

2. $\dfrac{9\pi}{4}$.

3. $-\dfrac{4\pi}{3}$.

In Problems 4–6, convert each angle to radians.

4. $300°$.

5. $270°$.

6. $-135°$.

In Problems 7–10, find the quadrant of the angle.

7. $\dfrac{7\pi}{3}$.

8. $\dfrac{5\pi}{6}$.

9. $-\dfrac{5\pi}{4}$.

10. $\dfrac{4\pi}{3}$.

In Problems 11 and 12, find an angle t that is coterminal with the given angle and such that $0 \le t < 2\pi$.

11. $-\dfrac{\pi}{3}$.

12. $\dfrac{13\pi}{2}$.

In Problems **13–16**, *the given point lies on the terminal side of an angle t in standard position. Find the values of* sin *t*, cos *t*, tan *t*, cot *t*, sec *t*, *and* csc *t*.

13. $(-4, 3)$.

14. $(-2, 0)$.

15. $(-\sqrt{11}, -5)$.

16. $(1, -2)$.

In Problems **17–34**, *find the given value. Do not use Appendix G or a calculator.*

17. $\sin \dfrac{\pi}{3}$.

18. $\cos \dfrac{\pi}{6}$.

19. $\tan 0$.

20. $\sec \dfrac{\pi}{4}$.

21. $\cot \dfrac{\pi}{6}$.

22. $\csc \dfrac{\pi}{2}$.

23. $\cos \dfrac{2\pi}{3}$.

24. $\tan \dfrac{5\pi}{6}$.

25. $\sec \pi$.

26. $\sin\left(-\dfrac{2\pi}{3}\right)$.

27. $\csc \dfrac{3\pi}{4}$.

28. $\cot \dfrac{3\pi}{2}$.

29. $\tan \dfrac{7\pi}{6}$.

30. $\sec \dfrac{9\pi}{4}$.

31. $\sin\left(-\dfrac{\pi}{2}\right)$.

32. $\cos \dfrac{11\pi}{6}$.

33. $\csc \dfrac{4\pi}{3}$.

34. $\cot\left(-\dfrac{3\pi}{2}\right)$.

In Problems **35–38**, *ABC is a right triangle with right angle C. Determine the remaining sides to one decimal place.*

35. $b = 3$, $A = 0.75$.

36. $b = 2$, $B = 0.40$.

37. $c = 10$, $B = 1.32$.

38. $c = 4$, $A = 0.85$.

In Problems **39** *and* **40**, *use trigonometric identities.*

39. If $\cos t = 0.6$ and t is a fourth-quadrant angle, find $\sin t$.

40. If $\sin \dfrac{9\pi}{8} = -0.3827$, find $\cos \dfrac{9\pi}{8}$.

In Problems **41–46**, *draw one cycle of each curve. Give the amplitude A and the period P.*

41. $y = 4 \cos t$.

42. $y = 3 \cos 6t$.

43. $y = -\sin 3t$.

44. $y = -\cos \dfrac{t}{2}$.

45. $y = -2 \sin \dfrac{t}{2}$.

46. $y = \sin \pi t$.

In Problems **47–62**, *differentiate.*

47. $y = \cos(2\pi t + 3)$.

48. $y = \sin(t^2 + 7)$.

49. $y = \sin \dfrac{t - 3}{2}$.

50. $y = \cos \dfrac{1}{t}$.

51. $y = \tan \sqrt{t}$.

52. $y = \tan(t^2 + 3)^3$.

53. $y = \csc\left(\dfrac{t + 1}{t + 2}\right)$.

54. $y = \cot \dfrac{1}{(2t + 1)^2}$.

55. $y = \cot e^{3t}$.

56. $y = \sec(\ln t)$.

57. $y = \csc \sqrt{t^2 + 4}$.

58. $y = \cos^3(4t^2 + 5)$.

59. $y = t^2 \tan t^2$.

60. $y = \dfrac{\cot \pi t}{\pi t}$.

61. $y = \sqrt{2t + \sec 3t}$.

62. $y = \dfrac{\sin t}{1 - \cos t}$.

In Problems **63–72**, *find the integrals.*

63. $\displaystyle\int 2 \sin 7t \, dt$.

64. $\displaystyle\int \cos \dfrac{t}{4} \, dt$.

65. $\displaystyle\int e^t \csc^2 e^t \, dt$.

66. $\displaystyle\int x \csc \dfrac{2x^2}{3} \cot \dfrac{2x^2}{3} \, dx$.

67. $\displaystyle\int \sin^6 y \cos y \, dy$.

68. $\displaystyle\int x \sec x^2 \tan x^2 \, dx$.

69. $\displaystyle\int (8 - \cos 5t)^4 \sin 5t \, dt$.

70. $\displaystyle\int \dfrac{\sec^2 \sqrt{x}}{\sqrt{x}} \, dx$.

71. $\displaystyle\int_0^{\pi/4} (\sin t + \cos t) \, dt$.

72. $\displaystyle\int_0^{\pi/4} \sin^2 2t \cos 2t \, dt$.

In Problems **73–82**, *find the value of the expression.*

73. $\sin^{-1}(-1)$.

74. $\cos^{-1} \dfrac{1}{2}$.

75. $\tan^{-1} \dfrac{\sqrt{3}}{3}$.

76. $\sin^{-1}\left(-\dfrac{\sqrt{3}}{2}\right)$.

77. $\cos^{-1}\left(-\dfrac{\sqrt{3}}{2}\right)$.

78. $\tan^{-1}(-1)$.

79. $\tan(\cos^{-1} 1)$.

80. $\cos\left(\sin^{-1} \dfrac{\sqrt{2}}{2}\right)$.

81. $\sin(\sin^{-1} 0.8)$.

82. $\tan^{-1}(\tan 2\pi)$.

In Problems **83–94**, *differentiate the given function.*

83. $y = \cos^{-1} 7x$.

84. $y = \sin^{-1} x^2$.

85. $y = \tan^{-1} e^x$.

86. $y = \cot^{-1} 3x$.

87. $y = \sin^{-1} \dfrac{x}{4}$.

88. $y = \tan^{-1} \dfrac{x}{x+1}$.

89. $y = \sec^{-1} 2x + \csc^{-1} 2x$.

90. $y = \sin^{-1}(\cos x)$.

91. $y = \ln(\sin^{-1} x)$.

92. $y = \sin x \tan^{-1} x$.

93. $y = e^{\cos^{-1} x}$.

94. $y = \dfrac{1}{\tan^{-1} e^x}$.

In Problems **95–104**, *find the integrals.*

95. $\displaystyle \int \frac{dt}{\sqrt{1 - 25t^2}}$.

96. $\displaystyle \int \frac{dt}{t\sqrt{4t^2 - 1}}$.

97. $\displaystyle \int_0^{1/\sqrt{3}} \frac{dt}{1 + 9t^2}$.

98. $\displaystyle \int_0^{1/3} \frac{dt}{\sqrt{4 - 9t^2}}$.

99. $\displaystyle \int \frac{dt}{t\sqrt{t^2 - 16}}$.

100. $\displaystyle \int \frac{dt}{2t^2 + 25}$.

101. $\displaystyle \int \frac{t}{\sqrt{4t^2 - 15}}\, dt$.

102. $\displaystyle \int \frac{dt}{\sqrt{49 - 4t^2}}$.

103. $\displaystyle \int \frac{dt}{t^2 - 4t + 20}$.

104. $\displaystyle \int \frac{dt}{\sqrt{6t - t^2}}$.

105. A wrecking company plans to make a bid for demolishing a smokestack and needs to know its height. At a point 100 ft from the base of the stack, the angle of elevation of its top is found to be 0.56. To the nearest foot, what is the height of the smokestack?

106. The inclination of an agricultural grain conveyor can be set between the angles of 0.25 and 0.85. If the length of the conveyor is 7 m, what are the minimum and maximum heights the conveyor can reach?

107. An antenna is situated on the edge of a cliff. From a point on level ground 480 m from the base of the cliff, the angle of elevation of the top of the antenna is 0.26; the angle of elevation of the bottom of the antenna is 0.22. How tall is the antenna?

108. To determine the height of a building, a surveyor determines that the angle of elevation to the top is 0.93. If the sighting instrument is 30 m from the building and 1.5 m above the ground, find the height of the building.

109. Find an equation of the tangent line to the graph of $y = \cos 2t$ at the point where $t = \pi/4$.

110. Find the rate of change of

$$y = \cos(4t + \pi) + 2 \tan 2t$$

when $t = 0$.

111. Determine (a) the maximum and (b) the minimum values of $y = \dfrac{t}{2} + \sin t$ on the interval $[0, \pi]$.

112. Suppose that the blood pressure P of a person after t seconds is given by

$$P = 90 + 20 \sin \frac{9\pi t}{4},$$

where P is in millimeters of mercury. At what rate is the pressure changing after 60 seconds?

113. A manufacturer sells q units of a product per month after t months from the beginning of the year. If

$$q = 100 + 500 \sin \frac{\pi t}{6}, \qquad 0 \le t \le 12,$$

determine the maximum number of units sold per month and the month in which these sales occur.

114. If $y = \sin 2t$, find $\dfrac{d^3 y}{dt^3}$.

115. Find the area of the region bounded by the graphs of $y = \sin t$ and $y = \cos t$ between $t = \pi/4$ and $t = 5\pi/4$.

References

[1] Babkoff, H. "Magnitude estimation of short electrocutaneous pulses," *Psychological Research*, vol. 39, no. 1 (1976), 39–49.

[2] Laming, D. *Mathematical Psychology*. New York: Academic Press, 1973.

[3] Imrie, F. K. E., and A. J. Vlitos. "Production of fungal protein from carob," in S. R. Tannenbaum and D. I. C. Wang (eds.), *Single-Cell Protein II*. Cambridge, Mass.: MIT Press, 1975.

[4] Leik, R. K., and B. F. Meeker. *Mathematical Sociology*. Englewood Cliffs, N.J.: Prentice-Hall, 1975.

[5] Loftus, G. R., and E. F. Loftus. *Human Memory: the Processing of Information*. Hillsdale, N.J.: Lawrence Erlbaum, 1976.

[6] Rha, C. "Utilization of single-cell protein for human food," in S. R. Tannenbaum and D. I. C. Wang (eds.), *Single-Cell Protein II*. Cambridge, Mass.: MIT Press, 1975.

[7] Folk, G. E., Jr. *Textbook of Environmental Physiology*, 2nd ed. Philadelphia: Lea & Febiger, 1974.

[8] Hintzman, D. L. "Repetition and learning," in G. H. Bower (ed.), *The Psychology of Learning and Motivation*, Vol. 10. New York: Academic Press, 1976. p. 77.

[9] Bressani, R. "The use of yeast in human foods," in R. J. Mateles and S. R. Tannenbaum (eds.), *Single-Cell Protein*. Cambridge, Mass.: MIT Press, 1968.

[10] Shonle, J. I. *Environmental Applications of General Physics*. Reading, Mass.: Addison-Wesley, 1975.

[11] Christofi, A., and A. Agapos. "Interest rate deregulation: an empirical justification," *Review of Business and Economic Research*, vol. 20, no. 1 (1984), 39–49.

[12] Eswaran, M., and A. Kotwal. "A theory of two-tier labor markets in agrarian economies," *The American Economic Review*, vol. 75, no. 1 (1985), 162–177.

[13] Stacy, R. W., D. T. Williams, R. E. Worden, and R. O. McMorris. *Essentials of Biological and Medical Physics*. New York: McGraw-Hill, 1955.

[14] Dean, J. "Statistical cost functions of a hosiery mill," *Studies in Business Administration*, vol. 11, no. 4 (1941).

[15] Nordin, J. A. "Note on a light plant's cost curves," *Econometrica*, vol. 15 (1947), 231–235.

[16] Lotka, A. J. *Elements of Mathematical Biology*. New York: Dover, 1956.

[17] Embree, D. G. "The population dynamics of the winter moth in Nova Scotia, 1954–1962," *Memoirs of the Entomological Society of Canada*, no. 46 (1965).

[18] Odum, H. T. "Biological circuits and the marine systems of Texas," in T. A. Olsen and F. J. Burgess (eds.), *Pollution and Marine Biology*. New York: Interscience, 1967.

[19] Haavelmo, T. "Methods of measuring the marginal propensity to consume," *Journal of the American Statistical Association*, vol. 42 (1947), 105–122.

[20] Doelle, L. L. *Environmental Acoustics*. New York: McGraw-Hill, 1972.

[21] Holling, C. S. "Some characteristics of simple types of predation and parasitism," *The Canadian Entomologist*, vol. 91, no. 7 (1959), 385–398.

[22] Feldstein, M. "The optimal level of social security benefits," *The Quarterly Journal of Economics*, vol. C, issue 2 (1985), 303–320.

[23] Mantell, L. H., and F. P. Sing. *Economics for Business Decisions*. New York: McGraw-Hill, 1972.

[24] Renshaw, E. "A note of equity and efficiency in the pricing of local telephone services," *The American Economic Review*, vol. 75, no. 3 (1985), 515–518.

[25] Poole, R. W. *An Introduction to Quantitative Ecology*. New York: McGraw-Hill, 1974.

[26] Graesser, A., and G. Mandler. "Limited processing capacity constrains

the storage of unrelated sets of words and retrieval from natural categories," *Human Learning and Memory*, vol. 4, no. 1 (1978), 86–100.

[27] Lancaster, P. *Mathematics: Models of the Real World*. Englewood Cliffs, N.J.: Prentice-Hall, 1976.

[28] Thrall, R. M., J. A. Mortimer, K. R. Rebman, and R. F. Baum, eds. *Some Mathematical Models in Biology*, rev. ed. Report 40241-R-7. Prepared at University of Michigan, 1967.

[29] Smith, J. M. *Mathematical Ideas in Biology*. London: Cambridge University Press, 1968.

[30] Ewens, W. J. *Population Genetics*. London: Methuen, 1969.

[31] Taagepera, R., and J. P. Hayes. "How trade/GNP ratio decreases with country size," *Social Science Research*, vol. 6 (1977), 108–132.

[32] Tintner, G. *Methodology of Economics and Econometrics*. Chicago: University of Chicago Press, 1967.

[33] DeCanio, S. J. "Delivered pricing and multiple basing point equilibria: a reevaluation," *The Quarterly Journal of Economics*, vol. 99, no. 2 (1984), 329–349.

[34] Stigler, G. *The Theory of Price*, 3rd ed. New York: Macmillan, 1966.

[35] Roberts, A. M. "The origins of fluctuations in the human secondary sex ratio," *Journal of Biosocial Science*, vol. 10, no. 2 (1978), 169–182.

[36] Sokal, R. R., and F. J. Rohlf. *Introduction to Biostatistics*. New York: W. H. Freeman, 1973.

[37] Holling, C. S. "The functional response of invertebrate predators to prey density," *Memoirs of the Entomological Society of Canada*, no. 48 (1966).

[38] Lewis, A. E. *Biostatistics*. New York: Reinhold, 1966.

[39] Bullen, K. E. *An Introduction to the Theory of Seismology*. Cambridge at the University Press, 1963.

[40] Hurter, A. P., Jr., A. H. Rubenstein, et al. "Market penetration by new innovations: the technological literature," *Technological Forecasting and Social Change*, vol. 11 (1978), 197–221.

[41] Richter, C. F. *Elementary Seismology*. New York: W. H. Freeman, 1958.

[42] Persky, A. L. "An inferior good and a novel indifference map," *The American Economist*, vol. 29, no. 1 (1985), 67–69.

[43] Brown, C. "Military enlistments: what can we learn from geographic variation?" *The American Economic Review*, vol. 75, no. 1 (1985), 228–234.

[44] Peterson, L. R., and M. J. Peterson. "Short-term retention of individual verbal items," *Journal of Experimental Psychology*, vol. 58 (1959), 193–198.

[45] Hovland, C. I. "The generalization of conditioned responses: I. The sensory generalization of conditioned responses with varying frequencies of tone," *Journal of General Psychology*, vol. 17 (1937), 125–148.

[46] Eaton, W. W., and G. A. Whitmore. "Length of stay as a stochastic process: a general approach and application to hospitalization for schizophrenia," *Journal of Mathematical Sociology*, vol. 5 (1977), 273–292.

[47] Simon, W. *Mathematical Techniques for Physiology and Medicine*. New York: Academic Press, 1972.

[48] Mather, W. B. *Principles of Quantitative Genetics*. Minneapolis: Burgess, 1964.

[49] Taagepera, R. "Why the trade/GNP ratio decreases with country size," *Social Science Research*, vol. 5 (1976), 385–404.

[50] Barbosa, L. C., and M. Friedman. "Deterministic inventory lot size models—a general root law," *Management Science*, vol. 24, no. 8 (1978), 819–826.

[51] Wilson, E. O., and W. H. Bossert. *A Primer of Population Biology*. Sunderland, Conn.: Sinauer Associates, 1971.

[52] Keyfitz, N., and W. Flieger. *Population Facts and Methods of Demography*. New York: W. H. Freeman, 1971.

[53] Keyfitz, N. *Introduction to the Mathematics of Population*. Reading, Mass.: Addison-Wesley, 1968.

[54] Gause, G. F. *The Struggle for Existence*. New York: Hafner, 1964.

[55] Bailey, N. T. J. *The Mathematical Approach to Biology and Medicine*. New York: Wiley, 1967.

[56] Odum, E. P. *Ecology*. New York: Holt, Rinehart and Winston, 1966.

[57] Swanson, P. E. "Integer constraints on the inventory theory of money demand," *Quarterly Journal of Business and Economics*, vol. 23, no. 1 (1984), 32–37.

[58] Palmon, D., and U. Yaari. "Taxation of capital gains and the behavior of stock prices over the dividend cycle," *The American Economist*, vol. 27, no. 1 (1983), 13–22.

[59] Swales, J. K. "Advertising as an intangible asset: profitability and entry barriers. A comment on Reekie and Bhoyrub," *Applied Economics*, vol. 17, no. 4 (1985), 603–617.

[60] Daly, P., and P. Douglas. "The production function for Canadian man-

ufactures," *Journal of the American Statistical Association,* vol. 38 (1943), 178–186.

[61] Tintner, G., and O. H. Brownlee. "Production functions derived from farm records," *American Journal of Agricultural Economics*, vol. 26 (1944), 566–571.

[62] Weinstein, A. G., and V. Srinivasen. "Predicting managerial success of master of business administration (MBA) graduates," *Journal of Applied Psychology*, vol. 59, no. 2 (1974), 207–212.

[63] Flesch, R. *The Art of Readable Writing.* New York: Harper & Row, 1949.

[64] Hurst, P. M., K. Perchonok, and E. L. Seguin. "Vehicle kinematics and gap acceptance," *Journal of Applied Psychology*, vol. 52, no. 4 (1968), 321–324.

[65] Perchonok, K., and P. M. Hurst. "Effect of lane-closure signals upon driver decision making and traffic flow," *Journal of Applied Psychology*, vol. 52, no. 5 (1968), 410–413.

[66] Silberman, J., and G. Yochum. "The role of money in determining election outcomes," *Social Science Quarterly*, vol. 58, no. 4 (1978), 671–682.

[67] Covert, J. B., and W. W. Reynolds. "Survival value of fever in fish," *Nature*, vol. 267, no. 5606 (1977), 43–45.

[68] Batschelet, E. *Introduction to Mathematics for Life Scientists*, pp. 235–239. New York-Heidelberg-Berlin: Springer-Verlag, 1971.

Algebra Review

A.1 REAL NUMBERS

The numbers 1, 2, 3, . . . form the set of **positive integers**:

$$\text{set of positive integers} = \{1, 2, 3, \ldots\}.$$

The positive integers together with the number 0 and the **negative integers** $-1, -2, -3, \ldots$ form the set of **integers**:

$$\text{set of integers} = \{\ldots, -3, -2, -1, 0, 1, 2, 3, \ldots\}.$$

The set of **rational numbers** consists of numbers that can be written as a quotient of two integers. That is, a rational number is a number that can be written as p/q, where p and q are integers and $q \neq 0$. (The symbol "$\neq$" is read "is not equal to.") **We cannot divide by zero.** Some rational numbers are $\frac{19}{20}$, $\frac{-2}{7}$, and $\frac{-6}{-2}$. The integer 2 is rational because $2 = \frac{2}{1}$. In fact, every integer is rational.

Rational numbers can be represented by decimal numbers that *terminate*, such as $\frac{3}{4} = 0.75$ and $\frac{3}{2} = 1.5$, or by *nonterminating repeating* decimals (a group of digits repeats without end), such as $\frac{2}{3} = 0.666\ldots$ and $\frac{-4}{11} = -0.3636\ldots$. Numbers that are represented by *nonterminating nonrepeating* decimals are called **irrational numbers**. An irrational number cannot be written as an integer divided by an integer. The numbers π (pi) and $\sqrt{2}$ are irrational.

Together, the rational numbers and irrational numbers form the set of **real numbers**. Real numbers can be represented by points on a line. In Fig. A.1 some points and their associated real numbers are identified. To each point on the line there corresponds a unique real number, and to each real

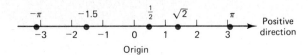

The Real Number Line

FIGURE A.1

number there corresponds a unique point on the line. We call this line the **real number line**.

If a, b, and c are real numbers, here are three important properties of real numbers.

Commutative Laws

$$a + b = b + a \quad \text{and} \quad ab = ba.$$

This means that we can add or multiply two real numbers in any order. For example, $3 + 4 = 4 + 3$ and $7(-4) = (-4)(7)$.

Associative Laws

$$a + (b + c) = (a + b) + c \quad \text{and} \quad a(bc) = (ab)c.$$

This means that in addition or multiplication, numbers can be grouped in any order. For example, $2 + (3 + 4) = (2 + 3) + 4$ and $6(\frac{1}{3} \cdot 5) = (6 \cdot \frac{1}{3}) \cdot 5$. Also, $2x + (x + y) = (2x + x) + y$.

Distributive Laws

$$a(b + c) = ab + ac \quad \text{and} \quad (b + c)a = ba + ca.$$

For example,

$$x(z + 4) = x(z) + x(4) = xz + 4x.$$

The distributive law can be extended to the form

$$a(b + c + d) = ab + ac + ad.$$

In fact, it can be extended to sums involving any number of terms.

The list below includes other properties of real numbers along with numerical examples. All denominators are different from zero.

PROPERTY	EXAMPLE
1. $a - b = a + (-b)$.	$2 - 7 = 2 + (-7) = -5$.
2. $a - (-b) = a + b$.	$2 - (-7) = 2 + 7 = 9$.
3. $-a = (-1)(a)$.	$-7 = (-1)(7)$.
4. $a(b + c) = ab + ac$.	$6(7 + 2) = 6 \cdot 7 + 6 \cdot 2 = 54$.
5. $a(b - c) = ab - ac$.	$6(7 - 2) = 6 \cdot 7 - 6 \cdot 2 = 30$.

6. $-(a + b) = -a - b.$ $\qquad$ $-(7 + 2) = -7 - 2 = -9.$

7. $-(a - b) = -a + b.$ $\qquad$ $-(2 - 7) = -2 + 7 = 5.$

8. $-(-a) = a.$ $\qquad$ $-(-2) = 2.$

9. $a(0) = (-a)(0) = 0.$ $\qquad$ $2(0) = (-2)(0) = 0.$

10. $(-a)(b) = -(ab) = a(-b).$ $\qquad$ $(-2)(7) = -(2 \cdot 7) = 2(-7)$
$$= -14.$$

11. $(-a)(-b) = ab.$ $\qquad$ $(-2)(-7) = 2 \cdot 7 = 14.$

12. $\dfrac{a}{1} = a.$ $\qquad$ $\dfrac{7}{1} = 7; \quad \dfrac{-2}{1} = -2.$

13. $\dfrac{a}{b} = a\left(\dfrac{1}{b}\right).$ $\qquad$ $\dfrac{2}{7} = 2\left(\dfrac{1}{7}\right).$

14. $\dfrac{a}{-b} = -\dfrac{a}{b} = \dfrac{-a}{b}.$ $\qquad$ $\dfrac{2}{-7} = -\dfrac{2}{7} = \dfrac{-2}{7}.$

15. $\dfrac{-a}{-b} = \dfrac{a}{b}.$ $\qquad$ $\dfrac{-2}{-7} = \dfrac{2}{7}.$

16. $\dfrac{0}{a} = 0$ when $a \neq 0.$ $\qquad$ $\dfrac{0}{7} = 0.$

17. $\dfrac{a}{a} = 1$ when $a \neq 0.$ $\qquad$ $\dfrac{2}{2} = 1; \quad \dfrac{-5}{-5} = 1.$

18. $a\left(\dfrac{b}{a}\right) = b.$ $\qquad$ $2\left(\dfrac{7}{2}\right) = 7.$

19. $a \cdot \dfrac{1}{a} = 1$ when $a \neq 0.$ $\qquad$ $2 \cdot \dfrac{1}{2} = 1.$

20. $\dfrac{a}{b} \cdot \dfrac{c}{d} = \dfrac{ac}{bd}.$ $\qquad$ $\dfrac{2}{3} \cdot \dfrac{4}{5} = \dfrac{2 \cdot 4}{3 \cdot 5} = \dfrac{8}{15}.$

21. $\dfrac{ab}{c} = \left(\dfrac{a}{c}\right)b = a\left(\dfrac{b}{c}\right).$ $\qquad$ $\dfrac{2 \cdot 7}{3} = \dfrac{2}{3} \cdot 7 = 2 \cdot \dfrac{7}{3}.$

22. $\dfrac{a}{bc} = \left(\dfrac{a}{b}\right)\left(\dfrac{1}{c}\right) = \left(\dfrac{1}{b}\right)\left(\dfrac{a}{c}\right).$ $\qquad$ $\dfrac{2}{3 \cdot 7} = \dfrac{2}{3} \cdot \dfrac{1}{7} = \dfrac{1}{3} \cdot \dfrac{2}{7}.$

23. $\dfrac{a}{b} = \left(\dfrac{a}{b}\right)\left(\dfrac{c}{c}\right) = \dfrac{ac}{bc}$ when $c \neq 0.$ $\qquad$ $\dfrac{2}{7} = \left(\dfrac{2}{7}\right)\left(\dfrac{5}{5}\right) = \dfrac{2 \cdot 5}{7 \cdot 5}.$

24. $\dfrac{a}{b(-c)} = \dfrac{a}{(-b)(c)} = \dfrac{-a}{bc}$ $\qquad$ $\dfrac{2}{3(-5)} = \dfrac{2}{(-3)(5)} = \dfrac{-2}{3(5)}$

$\qquad = \dfrac{-a}{(-b)(-c)} = -\dfrac{a}{bc}.$ $\qquad$ $= \dfrac{-2}{(-3)(-5)} = -\dfrac{2}{3(5)} = -\dfrac{2}{15}.$

25. $\dfrac{a}{c} + \dfrac{b}{c} = \dfrac{a+b}{c}.$ $\qquad\qquad$ $\dfrac{2}{9} + \dfrac{3}{9} = \dfrac{2+3}{9} = \dfrac{5}{9}.$

26. $\dfrac{a}{c} - \dfrac{b}{c} = \dfrac{a-b}{c}.$ $\qquad\qquad$ $\dfrac{2}{9} - \dfrac{3}{9} = \dfrac{2-3}{9} = \dfrac{-1}{9}.$

27. $\dfrac{a}{b} + \dfrac{c}{d} = \dfrac{ad+bc}{bd}.$ $\qquad\qquad$ $\dfrac{4}{5} + \dfrac{2}{3} = \dfrac{4\cdot 3 + 5\cdot 2}{5\cdot 3} = \dfrac{22}{15}.$

28. $\dfrac{a}{b} - \dfrac{c}{d} = \dfrac{ad-bc}{bd}.$ $\qquad\qquad$ $\dfrac{4}{5} - \dfrac{2}{3} = \dfrac{4\cdot 3 - 5\cdot 2}{5\cdot 3} = \dfrac{2}{15}.$

29. $\dfrac{\dfrac{a}{b}}{\dfrac{c}{d}} = \dfrac{a}{b} \div \dfrac{c}{d} = \dfrac{a}{b}\cdot\dfrac{d}{c} = \dfrac{ad}{bc}.$ $\qquad$ $\dfrac{\dfrac{2}{3}}{\dfrac{7}{5}} = \dfrac{2}{3} \div \dfrac{7}{5} = \dfrac{2}{3}\cdot\dfrac{5}{7} = \dfrac{10}{21}.$

30. $\dfrac{a}{\dfrac{b}{c}} = a \div \dfrac{b}{c} = a\cdot\dfrac{c}{b} = \dfrac{ac}{b}.$ $\qquad$ $\dfrac{2}{\dfrac{3}{5}} = 2 \div \dfrac{3}{5} = 2\cdot\dfrac{5}{3} = \dfrac{2\cdot 5}{3} = \dfrac{10}{3}.$

31. $\dfrac{\dfrac{a}{b}}{c} = \dfrac{a}{b} \div c = \dfrac{a}{b}\cdot\dfrac{1}{c} = \dfrac{a}{bc}.$ $\qquad$ $\dfrac{\dfrac{2}{3}}{5} = \dfrac{2}{3} \div 5 = \dfrac{2}{3}\cdot\dfrac{1}{5} = \dfrac{2}{3\cdot 5} = \dfrac{2}{15}.$

Property 23 is essentially the **fundamental principle of fractions,** which states that *multiplying or dividing both the numerator and denominator of a fraction by the same number, except* 0, *results in a fraction which is equivalent to (that is, it has the same value as) the original fraction.* Thus

$$\frac{7}{\frac{1}{8}} = \frac{7\cdot 8}{\frac{1}{8}\cdot 8} = \frac{56}{1} = 56.$$

By Properties 27 and 23 we have

$$\frac{2}{5} + \frac{4}{15} = \frac{2\cdot 15 + 5\cdot 4}{5\cdot 15} = \frac{50}{75} = \frac{2\cdot 25}{3\cdot 25} = \frac{2}{3}.$$

We can also do this problem by first converting $\frac{2}{5}$ and $\frac{4}{15}$ into equivalent fractions that have the same denominators and then using Property 25. The fractions $\frac{2}{5}$ and $\frac{4}{15}$ can be written with a common denominator of $5 \cdot 15$: $\frac{2}{5} = \frac{2\cdot 15}{5\cdot 15}$ and $\frac{4}{15} = \frac{4\cdot 5}{15\cdot 5}$. However, 15 is the *least common denominator* (L.C.D.) of $\frac{2}{5}$ and $\frac{4}{15}$. Thus

$$\frac{2}{5} + \frac{4}{15} = \frac{2\cdot 3}{5\cdot 3} + \frac{4}{15} = \frac{6}{15} + \frac{4}{15} = \frac{10}{15} = \frac{2}{3}.$$

Similarly,

$$\frac{3}{8} - \frac{5}{12} = \frac{3 \cdot 3}{8 \cdot 3} - \frac{5 \cdot 2}{12 \cdot 2}$$

$$= \frac{9}{24} - \frac{10}{24} = \frac{9 - 10}{24} = -\frac{1}{24}.$$

EXERCISE A.1

Simplify each of the following if possible.

1. $-2 + (-4)$.

2. $-6 + 2$.

3. $6 + (-4)$.

4. $7 - 2$.

5. $7 - (-4)$.

6. $-7 - (-4)$.

7. $-8 - (-6)$.

8. $(-2)(9)$.

9. $7(-9)$.

10. $(-2)(-12)$.

11. $(-1)6$.

12. $-(-9)$.

13. $-(-6 + x)$.

14. $-7(x)$.

15. $-12(x - y)$.

16. $-[-6 + (-y)]$.

17. $-2 \div 6$.

18. $-2 \div (-4)$.

19. $4 \div (-2)$.

20. $2(-6 + 2)$.

21. $3[-2(3) + 6(2)]$.

22. $(-2)(-4)(-1)$.

23. $(-5)(-5)$.

24. $x(0)$.

25. $3(x - 4)$.

26. $4(5 + x)$.

27. $-(x - 2)$.

28. $0(-x)$.

29. $8\left(\dfrac{1}{11}\right)$.

30. $\dfrac{7}{1}$.

31. $\dfrac{-5x}{7y}$.

32. $\dfrac{3}{-2x}$.

33. $\dfrac{2}{3} \cdot \dfrac{1}{x}$.

34. $\dfrac{x}{y}(2z)$.

35. $(2x)\left(\dfrac{3}{2x}\right)$.

36. $\dfrac{-15x}{-3y}$.

37. $\dfrac{7}{y} \cdot \dfrac{1}{x}$.

38. $\dfrac{2}{x} \cdot \dfrac{5}{y}$.

39. $\dfrac{1}{2} + \dfrac{1}{3}$.

40. $\dfrac{5}{12} + \dfrac{3}{4}$.

41. $\dfrac{3}{10} - \dfrac{7}{15}$.

42. $\dfrac{2}{3} + \dfrac{7}{3}$.

43. $\dfrac{x}{9} - \dfrac{y}{9}$.

44. $\dfrac{3}{2} - \dfrac{1}{4} + \dfrac{1}{6}$.

45. $\dfrac{2}{3} - \dfrac{5}{8}$.

46. $\dfrac{\dfrac{6}{x}}{y}$.

47. $\dfrac{\dfrac{x}{6}}{y}$.

48. $\dfrac{\dfrac{-7}{2}}{\dfrac{5}{8}}$.

49. $\dfrac{7}{0}$.

50. $\dfrac{0}{7}$.

51. $\dfrac{0}{0}$.

52. $0 \cdot 0$.

A.2 EXPONENTS AND RADICALS

The product $x \cdot x \cdot x$ is abbreviated x^3. In general, for n a positive integer, x^n is the abbreviation for the product of n x's. The letter n in x^n is called the *exponent* and x is called the *base*. More specifically, if n is a positive integer we have:

1. $x^n = \underbrace{x \cdot x \cdot x \cdot \ldots \cdot x}_{n \text{ factors}}$

2. $x^{-n} = \dfrac{1}{x^n} = \dfrac{1}{\underbrace{x \cdot x \cdot x \cdot \ldots \cdot x}_{n \text{ factors}}}.$

3. $\dfrac{1}{x^{-n}} = x^n.$

4. $x^0 = 1$ if $x \neq 0$. 0^0 is not defined.

EXAMPLE 1

a. $\left(\dfrac{1}{2}\right)^4 = \left(\dfrac{1}{2}\right)\left(\dfrac{1}{2}\right)\left(\dfrac{1}{2}\right)\left(\dfrac{1}{2}\right) = \dfrac{1}{16}.$

b. $3^{-5} = \dfrac{1}{3^5} = \dfrac{1}{3 \cdot 3 \cdot 3 \cdot 3 \cdot 3} = \dfrac{1}{243}.$

c. $\dfrac{1}{3^{-5}} = 3^5 = 243.$

d. $2^0 = 1, \quad \pi^0 = 1, \quad (-5)^0 = 1.$

e. $x^1 = x.$

If $r^n = x$ where n is a positive integer, then r is an nth *root* of x. For example, $3^2 = 9$, so 3 is a second root (usually called a *square root*) of 9. Since $(-3)^2 = 9$, -3 is also a square root of 9. Similarly, -2 is a *cube root* of -8 because $(-2)^3 = -8$.

Some numbers do not have an nth root that is a real number. For example, since the square of any real number is nonnegative, there is no real number that is a square root of -4.

The **principal nth root** of x is that nth root of x which is positive if x is positive, and is negative if x is negative and n is odd. We denote it by $\sqrt[n]{x}$. Thus

$$\sqrt[n]{x} \text{ is } \begin{cases} \text{positive if } x \text{ is positive,} \\ \text{negative if } x \text{ is negative and } n \text{ is odd.} \end{cases}$$

For example, $\sqrt[2]{9} = 3$, $\sqrt[3]{-8} = -2$, and $\sqrt[3]{\frac{1}{27}} = \frac{1}{3}$. We define $\sqrt[n]{0} = 0$.

The symbol $\sqrt[n]{x}$ is called a **radical**. Here n is the *index*, x is the *radicand*, and $\sqrt{}$ is the *radical sign*. With principal square roots we usually omit the index and write $\sqrt{x}$ instead of $\sqrt[2]{x}$. Thus $\sqrt{9} = 3$.

If x is positive, the expression $x^{p/q}$ where p and q are integers and q is

Pitfall
Although 2 and -2 are square roots of 4, the **principal** square root of 4 is 2, not -2. Hence $\sqrt{4} = 2$.

positive is defined to be $\sqrt[q]{x^p}$. Thus

$$x^{3/4} = \sqrt[4]{x^3}; \qquad 8^{2/3} = \sqrt[3]{8^2} = \sqrt[3]{64} = 4;$$

$$4^{-1/2} = \sqrt{4^{-1}} = \sqrt{\tfrac{1}{4}} = \tfrac{1}{2}.$$

Here are the basic laws of exponents and radicals:*

LAW	EXAMPLE
1. $x^m \cdot x^n = x^{m+n}$.	$2^3 \cdot 2^5 = 2^8 = 256$; $\ x^2 \cdot x^3 = x^5$.
2. $x^0 = 1$ if $x \neq 0$.	$2^0 = 1$.
3. $x^{-n} = \dfrac{1}{x^n}$.	$2^{-3} = \dfrac{1}{2^3} = \dfrac{1}{8}$.
4. $\dfrac{1}{x^{-n}} = x^n$.	$\dfrac{1}{2^{-3}} = 2^3 = 8$; $\ \dfrac{1}{x^{-5}} = x^5$.
5. $\dfrac{x^m}{x^n} = x^{m-n} = \dfrac{1}{x^{n-m}}$.	$\dfrac{2^{12}}{2^8} = 2^4 = 16$; $\ \dfrac{x^8}{x^{12}} = \dfrac{1}{x^4}$.
6. $\dfrac{x^m}{x^m} = 1$ if $x \neq 0$.	$\dfrac{2^4}{2^4} = 1$.
7. $(x^m)^n = x^{mn}$.	$(2^3)^5 = 2^{15}$; $\ (x^2)^3 = x^6$.
8. $(xy)^n = x^n y^n$.	$(2 \cdot 4)^3 = 2^3 \cdot 4^3 = 8 \cdot 64$.
9. $\left(\dfrac{x}{y}\right)^n = \dfrac{x^n}{y^n}$.	$\left(\dfrac{2}{3}\right)^3 = \dfrac{2^3}{3^3}$; $\ \left(\dfrac{1}{3}\right)^5 = \dfrac{1^5}{3^5} = \dfrac{1}{3^5}$.
10. $\left(\dfrac{x}{y}\right)^{-n} = \left(\dfrac{y}{x}\right)^n$.	$\left(\dfrac{3}{4}\right)^{-2} = \left(\dfrac{4}{3}\right)^2 = \dfrac{16}{9}$.
11. $x^{1/n} = \sqrt[n]{x}$.	$3^{1/5} = \sqrt[5]{3}$.
12. $x^{-1/n} = \dfrac{1}{x^{1/n}} = \dfrac{1}{\sqrt[n]{x}}$.	$4^{-1/2} = \dfrac{1}{4^{1/2}} = \dfrac{1}{\sqrt{4}} = \dfrac{1}{2}$.
13. $\sqrt[n]{x}\,\sqrt[n]{y} = \sqrt[n]{xy}$.	$\sqrt[3]{9}\,\sqrt[3]{2} = \sqrt[3]{18}$.
14. $\dfrac{\sqrt[n]{x}}{\sqrt[n]{y}} = \sqrt[n]{\dfrac{x}{y}}$.	$\dfrac{\sqrt[3]{90}}{\sqrt[3]{10}} = \sqrt[3]{\dfrac{90}{10}} = \sqrt[3]{9}$.
15. $\sqrt[m]{\sqrt[n]{x}} = \sqrt[mn]{x}$.	$\sqrt[3]{\sqrt[4]{2}} = \sqrt[12]{2}$.
16. $x^{m/n} = \sqrt[n]{x^m} = (\sqrt[n]{x})^m$.	$8^{2/3} = \sqrt[3]{8^2} = (\sqrt[3]{8})^2 = 2^2 = 4$.
17. $(\sqrt[n]{x})^n = \sqrt[n]{x^n} = x$.	$(\sqrt[8]{7})^8 = \sqrt[8]{7^8} = 7$.

* Although some laws involve restrictions, they are not vital to our discussion.

EXAMPLE 2

a. By Law 1,

$$x^6x^8 = x^{6+8} = x^{14},$$

$$a^3b^2a^5b = a^3a^5b^2b = a^8b^3,$$

$$x^{11}x^{-5} = x^{11-5} = x^6,$$

$$z^{2/5}z^{3/5} = z^1 = z,$$

$$xx^{1/2} = x^1x^{1/2} = x^{3/2}.$$

b. By Law 16,

$$\left(\frac{1}{4}\right)^{3/2} = \left(\sqrt{\frac{1}{4}}\right)^3 = \left(\frac{1}{2}\right)^3 = \frac{1}{8}.$$

c. $\left(-\dfrac{8}{27}\right)^{4/3} = \left(\sqrt[3]{\dfrac{-8}{27}}\right)^4 = \left(\dfrac{\sqrt[3]{-8}}{\sqrt[3]{27}}\right)^4$ (Laws 16 and 14)

$$= \left(\frac{-2}{3}\right)^4$$

$$= \frac{(-2)^4}{3^4} = \frac{16}{81}$$ (Law 9).

d. $(64a^3)^{2/3} = 64^{2/3}(a^3)^{2/3}$ (Law 8)

$$= (\sqrt[3]{64})^2a^2$$ (Laws 16 and 7)

$$= (4)^2a^2 = 16a^2.$$

Rationalizing the denominator of a fraction is a procedure in which a fraction having a radical in its denominator is expressed as an equivalent fraction without a radical in its denominator. We use the fundamental principle of fractions.

EXAMPLE 3   *Rationalize the denominators.*

a. $\dfrac{2}{\sqrt{5}} = \dfrac{2}{5^{1/2}} = \dfrac{2 \cdot 5^{1/2}}{5^{1/2} \cdot 5^{1/2}} = \dfrac{2 \cdot 5^{1/2}}{5^1} = \dfrac{2\sqrt{5}}{5}.$

b. $\dfrac{2}{\sqrt[6]{3x^5}} = \dfrac{2}{\sqrt[6]{3} \cdot \sqrt[6]{x^5}} = \dfrac{2}{3^{1/6}x^{5/6}} = \dfrac{2 \cdot 3^{5/6}x^{1/6}}{3^{1/6}x^{5/6} \cdot 3^{5/6}x^{1/6}}$

$$= \dfrac{2(3^5x)^{1/6}}{3x} = \dfrac{2\sqrt[6]{3^5x}}{3x}.$$

The following examples illustrate various applications of the laws of exponents and radicals.

EXAMPLE 4

a. *Eliminate negative exponents in $\dfrac{x^{-2}y^3}{z^{-2}}$.*

$$\frac{x^{-2}y^3}{z^{-2}} = x^{-2} \cdot y^3 \cdot \frac{1}{z^{-2}} = \frac{1}{x^2} \cdot y^3 \cdot z^2 = \frac{y^3 z^2}{x^2}.$$

Thus we can bring a factor of the numerator down to the denominator, and vice versa, by changing the sign of the exponent.

b. *Simplify $\dfrac{x^2 y^7}{x^3 y^5}$.*

$$\frac{x^2 y^7}{x^3 y^5} = \frac{y^{7-5}}{x^{3-2}} = \frac{y^2}{x}.$$

c. *Simplify $x^{3/2} - x^{1/2}$ by using the distributive law.*

$$x^{3/2} - x^{1/2} = x^{1/2}(x - 1).$$

d. *Simplify $(x^5 y^8)^5$.*

$$(x^5 y^8)^5 = (x^5)^5 (y^8)^5 = x^{25} y^{40}.$$

e. *Simplify $(x^{5/9} y^{4/3})^{18}$.*

$$(x^{5/9} y^{4/3})^{18} = (x^{5/9})^{18} (y^{4/3})^{18} = x^{10} y^{24}.$$

f. *Simplify $\left(\dfrac{x^{1/5} y^{6/5}}{z^{2/5}}\right)^5$.*

$$\left(\frac{x^{1/5} y^{6/5}}{z^{2/5}}\right)^5 = \frac{(x^{1/5} y^{6/5})^5}{(z^{2/5})^5} = \frac{xy^6}{z^2}.$$

g. *Eliminate negative exponents in $7x^{-2} + (7x)^{-2}$.*

$$7x^{-2} + (7x)^{-2} = \frac{7}{x^2} + \frac{1}{(7x)^2} = \frac{7}{x^2} + \frac{1}{49x^2}.$$

h. *Apply the distributive law to $x^{2/5}(y^{1/2} + 2z^{6/5})$.*

$$x^{2/5}(y^{1/2} + 2z^{6/5}) = x^{2/5} y^{1/2} + 2x^{2/5} z^{6/5}.$$

i. *Simplify $\dfrac{x^3}{y^2} \div \dfrac{x^6}{y^5}$.*

$$\frac{x^3}{y^2} \div \frac{x^6}{y^5} = \frac{x^3}{y^2} \cdot \frac{y^5}{x^6} = \frac{y^3}{x^3}.$$

EXAMPLE 5

a. *Simplify* $\sqrt[4]{48}$.

$$\sqrt[4]{48} = \sqrt[4]{16 \cdot 3} = \sqrt[4]{16}\,\sqrt[4]{3} = 2\sqrt[4]{3}.$$

b. *Rewrite* $\sqrt{2 + 5x}$ *without using a radical sign.*

$$\sqrt{2 + 5x} = (2 + 5x)^{1/2}.$$

c. *Rationalize the denominator of* $\dfrac{\sqrt[5]{2}}{\sqrt[3]{6}}$ *and simplify.*

$$\frac{\sqrt[5]{2}}{\sqrt[3]{6}} = \frac{2^{1/5} \cdot 6^{2/3}}{6^{1/3} \cdot 6^{2/3}} = \frac{2^{3/15}6^{10/15}}{6} = \frac{(2^3 6^{10})^{1/15}}{6} = \frac{\sqrt[15]{2^3 6^{10}}}{6}.$$

d. *Simplify* $\dfrac{\sqrt{20}}{\sqrt{5}}$.

$$\frac{\sqrt{20}}{\sqrt{5}} = \sqrt{\frac{20}{5}} = \sqrt{4} = 2.$$

e. *Simplify* $\sqrt[3]{x^6 y^4}$.

$$\sqrt[3]{x^6 y^4} = \sqrt[3]{(x^2)^3 y^3 y} = \sqrt[3]{(x^2)^3} \cdot \sqrt[3]{y^3} \cdot \sqrt[3]{y}$$
$$= x^2 y \sqrt[3]{y}.$$

f. *Simplify* $\sqrt{\dfrac{2}{7}}$.

$$\sqrt{\frac{2}{7}} = \sqrt{\frac{2}{7} \cdot \frac{7}{7}} = \sqrt{\frac{14}{7^2}} = \frac{\sqrt{14}}{\sqrt{7^2}} = \frac{\sqrt{14}}{7}.$$

g. *Simplify* $\sqrt{250} - \sqrt{50} + 15\sqrt{2}$.

$$\sqrt{250} - \sqrt{50} + 15\sqrt{2} = \sqrt{25 \cdot 10} - \sqrt{25 \cdot 2} + 15\sqrt{2}$$
$$= 5\sqrt{10} - 5\sqrt{2} + 15\sqrt{2}$$
$$= 5\sqrt{10} + 10\sqrt{2}.$$

h. *If x is any real number, simplify* $\sqrt{x^2}$.

$$\sqrt{x^2} = \begin{cases} x, & \text{if } x \text{ is positive,} \\ -x, & \text{if } x \text{ is negative,} \\ 0, & \text{if } x = 0. \end{cases}$$

Thus $\sqrt{2^2} = 2$ and $\sqrt{(-3)^2} = -(-3) = 3$.

EXERCISE A.2

*In Problems **1–14**, simplify and express all answers in terms of positive exponents.*

1. $(2^3)(2^2)$.

2. x^6x^9.

3. w^4w^8.

4. $x^6x^4x^3$.

5. $\dfrac{x^2x^6}{y^7y^{10}}$.

6. $(x^{12})^4$.

7. $\dfrac{(x^2)^5}{(y^5)^{10}}$.

8. $\left(\dfrac{x^2}{y^3}\right)^5$.

9. $(2x^2y^3)^3$.

10. $\left(\dfrac{w^2s^3}{y^2}\right)^2$.

11. $\dfrac{x^8}{x^2}$.

12. $\left(\dfrac{2x^2}{4x^4}\right)^3$.

13. $\dfrac{(x^3)^6}{x(x^3)}$.

14. $\dfrac{(x^2)^3(x^3)^2}{(x^3)^4}$.

*In Problems **15–28**, evaluate the expressions.*

15. $\sqrt{25}$.

16. $\sqrt[3]{64}$.

17. $\sqrt[5]{-32}$.

18. $\sqrt{0.04}$.

19. $\sqrt[4]{\frac{1}{16}}$.

20. $\sqrt[3]{-\frac{8}{27}}$.

21. $(100)^{1/2}$.

22. $(64)^{1/3}$.

23. $4^{3/2}$.

24. $(25)^{-3/2}$.

25. $(32)^{-2/5}$.

26. $(0.09)^{-1/2}$.

27. $\left(\dfrac{1}{16}\right)^{5/4}$.

28. $\left(-\dfrac{27}{64}\right)^{2/3}$.

*In Problems **29–38**, simplify the expressions.*

29. $\sqrt{32}$.

30. $\sqrt[3]{24}$.

31. $\sqrt[3]{2x^3}$.

32. $\sqrt{4x}$.

33. $\sqrt{16x^4}$.

34. $\sqrt[4]{x/16}$.

35. $(9z^4)^{1/2}$.

36. $(16y^8)^{3/4}$.

37. $\left(\dfrac{27t^3}{8}\right)^{2/3}$.

38. $\left(\dfrac{1000}{a^9}\right)^{-2/3}$.

*In Problems **39–50**, write the expressions in terms of positive exponents only. Avoid all radicals in the final form. For example, $y^{-1}\sqrt{x} = \dfrac{x^{1/2}}{y}$.*

39. $\dfrac{x^3y^{-2}}{z^2}$.

40. $\sqrt[5]{x^2y^3z^{-10}}$.

41. $2x^{-1}x^{-3}$.

42. $x + y^{-1}$.

43. $(3t)^{-2}$.

44. $(3-z)^{-4}$.

45. $\sqrt[3]{7s^2}$.

46. $(x^{-2}y^2)^{-2}$.

47. $\sqrt{x} - \sqrt{y}$.

48. $\dfrac{x^{-2}y^{-6}z^2}{xy^{-1}}$.

49. $x^2\sqrt[4]{xy^{-2}z^3}$.

50. $(\sqrt[5]{xy^{-3}})x^{-1}y^{-2}$.

*In Problems **51–56**, write the exponential forms in equivalent forms involving radicals.*

51. $(8x - y)^{4/5}$.

52. $(ab^2c^3)^{3/4}$.

53. $x^{-4/5}$.

54. $2x^{1/2} - (2y)^{1/2}$.

55. $2x^{-2/5} - (2x)^{-2/5}$.

56. $[(x^{-4})^{1/5}]^{1/6}$.

*In Problems **57–78**, simplify. Express all answers in terms of positive exponents. Rationalize the denominator where necessary to avoid fractional exponents in the denominator.*

57. $2x^2y^{-3}x^4$.

58. $\dfrac{2}{x^{3/2}y^{1/3}}$.

59. $\sqrt{\sqrt[3]{t^4}}$.

60. $\{[(2x^2)^3]^{-4}\}^{-1}$.

61. $\dfrac{2^0}{(2^{-2}x^{1/2}y^{-2})^3}$.

62. $\dfrac{\sqrt{s^5}}{\sqrt[3]{s^2}}$.

63. $\sqrt[3]{x^2yz^3}\sqrt[3]{xy^2}$.

64. $(\sqrt[5]{2})^{10}$.

65. $3^2(27)^{-4/3}$.

66. $(\sqrt[5]{x^2y})^{2/5}$.

67. $(2x^{-1}y^2)^2$.

68. $\dfrac{3}{\sqrt[3]{y}\,\sqrt[4]{x}}$.

69. $\sqrt{x}\,\sqrt{x^2y^3}\,\sqrt{xy^2}$.

70. $\sqrt{75k^4}$.

71. $\dfrac{(x^2y^{-1}z)^{-2}}{(xy^2)^{-4}}$.

72. $\sqrt{6(6)}$.

73. $\dfrac{(x^2)^3}{x^4} \div \left[\dfrac{x^3}{(x^3)^2}\right]^{-2}$.

74. $\sqrt{(-6)(-6)}$.

75. $-\dfrac{8s^{-2}}{2s^3}$.

76. $(x^{-1}y^{-2}\sqrt{z})^4$.

77. $(2x^2y \div 3y^3z^{-2})^2$.

78. $\dfrac{1}{\left(\dfrac{\sqrt{2}x^{-2}}{\sqrt{16x^3}}\right)^2}$.

A.3 OPERATIONS WITH ALGEBRAIC EXPRESSIONS

The algebraic expression $5ax^3 - 2bx + 3$ consists of three *terms:* $+5ax^3$, $-2bx$, and $+3$. Some of the *factors* of the first term $5ax^3$ are 5, a, x, x^2, x^3, $5ax$, and ax^2. Also, $5a$ is the *coefficient of* x^3 and 5 is the *numerical coefficient* of ax^3. If a and b represent fixed numbers throughout a discussion, then a and b are called *constants*.

Algebraic expressions with exactly one term are called *monomials*. Those having exactly two terms are *binomials,* and those with exactly three terms are *trinomials*. Algebraic expressions with more than one term are called *multinomials*. Thus the multinomial $2x - 5$ is a binomial, and the multinomial $3\sqrt{y} + 2y - 4y^2$ is a trinomial.

A *polynomial in* x is an algebraic expression of the form*

$$c_n x^n + c_{n-1} x^{n-1} + \cdots + c_1 x + c_0$$

where n is a nonnegative integer and $c_0, c_1, \ldots, c_n$ are constants with $c_n \neq 0$. We call n the *degree* of the polynomial. Hence $4x^3 - 5x^2 + x - 2$ is a polynomial in x of degree 3, and $y^5 - 2$ is a polynomial in y of degree 5. A nonzero constant is a polynomial of degree zero; thus 5 is a polynomial of degree zero. The constant 0 is considered to be a polynomial; however, no degree is assigned to it.

Below is a list of special products which may be obtained from the distributive law and are useful in multiplying algebraic expressions.

SPECIAL PRODUCTS

1. $x(y + z) = xy + xz$ (distributive law).

2. $(x + a)(x + b) = x^2 + (a + b)x + ab$.

3. $(ax + c)(bx + d) = abx^2 + (ad + cb)x + cd$.

4. $(x + a)^2 = x^2 + 2ax + a^2$ (square of a binomial).

5. $(x - a)^2 = x^2 - 2ax + a^2$ (square of a binomial).

6. $(x + a)(x - a) = x^2 - a^2$ (product of sum and difference).

7. $(x + a)^3 = x^3 + 3ax^2 + 3a^2x + a^3$ (cube of a binomial).

8. $(x - a)^3 = x^3 - 3ax^2 + 3a^2x - a^3$ (cube of a binomial).

EXAMPLE 1

a. By Rule 2, $(x + 2)(x - 5) = [x + 2][x + (-5)]$

$$= x^2 + (2 - 5)x + 2(-5)$$
$$= x^2 - 3x - 10.$$

---

* The three dots indicate the terms that are understood to be included in the sum.

b. By Rule 3, $(3z + 5)(7z + 4) = 3 \cdot 7z^2 + (3 \cdot 4 + 5 \cdot 7)z + 5 \cdot 4$
$$= 21z^2 + 47z + 20.$$

c. By Rule 5, $(x - 4)^2 = x^2 - 2(4)x + 4^2$
$$= x^2 - 8x + 16.$$

d. By Rule 6,

$$(\sqrt{y^2 + 1} + 3)(\sqrt{y^2 + 1} - 3) = (\sqrt{y^2 + 1})^2 - 3^2$$
$$= (y^2 + 1) - 9$$
$$= y^2 - 8.$$

EXAMPLE 2

Multiply: $(2t - 3)(5t^2 + 3t - 1)$.

We treat $2t - 3$ as a single number and apply the distributive law.

$$(2t - 3)(5t^2 + 3t - 1) = (2t - 3)5t^2 + (2t - 3)3t - (2t - 3)1$$
$$= 10t^3 - 15t^2 + 6t^2 - 9t - 2t + 3$$
$$= 10t^3 - 9t^2 - 11t + 3.$$

To divide a multinomial by a monomial, divide each term in the multinomial by the monomial.

EXAMPLE 3

a. $\dfrac{x^3 + 3x}{x} = \dfrac{x^3}{x} + \dfrac{3x}{x} = x^2 + 3.$

b. $\dfrac{4z^3 - 8z^2 + 3z - 6}{2z} = \dfrac{4z^3}{2z} - \dfrac{8z^2}{2z} + \dfrac{3z}{2z} - \dfrac{6}{2z}$

$$= 2z^2 - 4z + \frac{3}{2} - \frac{3}{z}.$$

To divide a polynomial by a polynomial we use so-called "long division."

EXAMPLE 4   *Divide $2x^3 - 14x - 5$ by $x - 3$.*

Here $2x^3 - 14x - 5$ is the *dividend* and $x - 3$ is the *divisor*. To avoid errors it is best to write the dividend as $2x^3 + 0x^2 - 14x - 5$. Note that the powers of x are in decreasing order.

$$2x^2 + 6x + 4 \leftarrow \text{quotient}$$
$$x - 3 \overline{)2x^3 + 0x^2 - 14x - 5}$$
$$\underline{2x^3 - 6x^2}$$
$$6x^2 - 14x$$
$$\underline{6x^2 - 18x}$$
$$4x - 5$$
$$\underline{4x - 12}$$
$$7 \leftarrow \text{remainder.}$$

Here we divided x (the first term of the divisor) into $2x^3$ and got $2x^2$. Then we multiplied $2x^2$ by $x - 3$, getting $2x^3 - 6x^2$. After subtracting $2x^3 - 6x^2$ from $2x^3 + 0x^2$, we obtained $6x^2$ and then "brought down" the term $-14x$. This process is continued until we arrive at 7, the *remainder*. We always stop when the remainder is 0 or is a polynomial whose degree is less than the degree of the divisor. Our answer may be written as

$$2x^2 + 6x + 4 + \frac{7}{x - 3}.$$

A way of checking a division is to verify that

$$(\text{quotient})(\text{divisor}) + \text{remainder} = \text{dividend}.$$

By using this equation you should verify the result of the example.

EXERCISE A.3

Simplify the following expressions.

1. $(x + 4)(x + 5)$.

2. $(x + 3)(x + 2)$.

3. $(x + 3)(x - 2)$.

4. $(z - 7)(z - 3)$.

5. $(2x + 3)(5x + 2)$.

6. $(y - 4)(2y + 3)$.

7. $(x + 3)^2$.

8. $(2x - 1)^2$.

9. $(x - 5)^2$.

10. $(\sqrt{x} - 1)(2\sqrt{x} + 5)$.

11. $(\sqrt{2y} + 3)^2$.

12. $(y - 3)(y + 3)$.

13. $(2s - 1)(2s + 1)$.

14. $(z^2 - 3w)(z^2 + 3w)$.

15. $(x^2 - 3)(x + 4)$.

16. $(x + 1)(x^2 + x + 3)$.

17. $(x^2 - 1)(2x^2 + 2x - 3)$.

18. $(2x - 1)(3x^3 + 7x^2 - 5)$.

19. $x\{3(x - 1)(x - 2) + 2[x(x + 7)]\}$.

20. $[(2z + 1)(2z - 1)](4z^2 + 1)$.

21. $(x + y + 2)(3x + 2y - 4)$.

22. $(x^2 + x + 1)^2$.

23. $\dfrac{z^2 - 4z}{z}$.

24. $\dfrac{2x^3 - 7x + 4}{x}$.

25. $\dfrac{6x^5 + 4x^3 - 1}{2x^2}$.

26. $\dfrac{(3x - 4) - (x + 8)}{4x}$.

27. $(x^2 + 3x - 1) \div (x + 3)$.

28. $(x^2 - 5x + 4) \div (x - 4)$.

29. $(3x^3 - 2x^2 + x - 3) \div (x + 2)$.

30. $(x^4 + 2x^2 + 1) \div (x - 1)$.

31. $t^2 \div (t - 8)$.

32. $(4x^2 + 6x + 1) \div (2x - 1)$.

33. $(3x^2 - 4x + 3) \div (3x + 2)$.

34. $(z^3 + z^2 + z) \div (z^2 - z + 1)$.

A.4 FACTORING _____

If two or more expressions are multiplied together, the expressions are called *factors* of the product. Thus if $c = ab$, then a and b are both factors of the product c. The process by which an expression is written as a product of its factors is called *factoring*.

Listed below are special products discussed in Sec. A.3. The right side of each identity is the factored form of the left side.

1. $xy + xz = x(y + z)$ (common factor).

2. $x^2 + (a + b)x + ab = (x + a)(x + b)$.

3. $abx^2 + (ad + cb)x + cd = (ax + c)(bx + d)$.

4. $x^2 + 2ax + a^2 = (x + a)^2$ (perfect-square trinomial).

5. $x^2 - 2ax + a^2 = (x - a)^2$ (perfect-square trinomial).

6. $x^2 - a^2 = (x + a)(x - a)$ (difference of two squares).

When factoring a polynomial we usually choose factors which themselves are polynomials. For example, $x^2 - 4 = (x + 2)(x - 2)$. We shall not write $x - 4$ as $(\sqrt{x} + 2)(\sqrt{x} - 2)$.

Always factor completely. For example,

$$2x^2 - 8 = 2(x^2 - 4) = 2(x + 2)(x - 2).$$

EXAMPLE 1

a. *Completely factor* $3k^2x^2 + 9k^3x$.

Since $3k^2x^2 = (3k^2x)(x)$ and $9k^3x = (3k^2x)(3k)$, each term of the original expression contains the common factor $3k^2x$. Thus by Rule 1,

$$3k^2x^2 + 9k^3x = 3k^2x(x + 3k).$$

Note that although $3k^2x^2 + 9k^3x = 3(k^2x^2 + 3k^3x)$, we do not say that the expression is completely factored since $k^2x^2 + 3k^3x$ can yet be factored.

b. *Completely factor* $8a^5x^2y^3 - 6a^2b^3yz - 2a^4b^4xy^2z^2$.

$$8a^5x^2y^3 - 6a^2b^3yz - 2a^4b^4xy^2z^2 = 2a^2y(4a^3x^2y^2 - 3b^3z - a^2b^4xyz^2).$$

c. *Completely factor* $3x^2 + 6x + 3$.

$$3x^2 + 6x + 3 = 3(x^2 + 2x + 1)$$
$$= 3(x + 1)^2 \qquad \text{(Rule 4)}.$$

EXAMPLE 2

a. *Completely factor* $x^2 - x - 6$.

If this trinomial factors into the form $x^2 - x - 6 = (x + a)(x + b)$, which is a product of two binomials, then we must determine a and b. Since $(x + a)(x + b) = x^2 + (a + b)x + ab$, then

$$x^2 + (-1)x + (-6) = x^2 + (a + b)x + ab.$$

By equating corresponding coefficients, we want

$$a + b = -1 \quad \text{and} \quad ab = -6.$$

If $a = -3$ and $b = 2$, then both conditions are met and hence

$$x^2 - x - 6 = (x - 3)(x + 2).$$

b. *Completely factor* $x^2 - 7x + 12$.

$$x^2 - 7x + 12 = (x - 3)(x - 4).$$

EXAMPLE 3

Listed below are expressions that are completely factored. The numbers in parentheses refer to the rules used.

a. $x^2 + 8x + 16 = (x + 4)^2$ (4).

b. $9x^2 + 9x + 2 = (3x + 1)(3x + 2)$ (3).

c. $6y^3 + 3y^2 - 18y = 3y(2y^2 + y - 6)$ (1)

$$= 3y(2y - 3)(y + 2)$$ (3).

d. $x^2 - 6x + 9 = (x - 3)^2$ (5).

e. $z^{1/4} + z^{5/4} = z^{1/4}(1 + z)$ (1).

f. $x^4 - 1 = (x^2 + 1)(x^2 - 1)$ (6)

$$= (x^2 + 1)(x + 1)(x - 1)$$ (6).

g. $x^{2/3} - 5x^{1/3} + 4 = (x^{1/3} - 1)(x^{1/3} - 4)$ (2).

h. $ax^2 - ay^2 + bx^2 - by^2 = (ax^2 - ay^2) + (bx^2 - by^2)$

$$= a(x^2 - y^2) + b(x^2 - y^2)$$ (1)

$$= (x^2 - y^2)(a + b)$$ (1)

$$= (x + y)(x - y)(a + b)$$ (6).

EXERCISE A.4

Completely factor the expressions.

1. $6x + 4$.

2. $6y^2 - 4y$.

3. $10xy + 5xz$.

4. $3x^2y - 9x^3y^3$.

5. $8a^3bc - 12ab^3cd + 4b^4c^2d^2$.

6. $6z^2t^3 + 3zst^4 - 12z^2t^3$.

7. $x^2 - 25$.

8. $x^2 + 3x - 4$.

9. $p^2 + 4p + 3$.

10. $s^2 - 6s + 8$.

11. $16x^2 - 9$.

12. $x^2 + 5x - 24$.

13. $z^2 + 6z + 8$.

14. $4t^2 - 9s^2$.

15. $x^2 + 6x + 9$.

16. $y^2 - 15y + 50$.

17. $2x^2 + 12x + 16$.

18. $2x^2 + 7x - 15$.

19. $3x^2 - 3$.

20. $4y^2 - 8y + 3$.

21. $6y^2 + 13y + 2$.

22. $4x^2 - x - 3$.

23. $12s^3 + 10s^2 - 8s$.

24. $9z^2 + 24z + 16$.

25. $x^{2/3}y - 4x^{8/3}y^3$.

26. $9x^{4/7} - 1$.

27. $2x^3 + 2x^2 - 12x$.

28. $x^2y^2 - 4xy + 4$.

29. $(4x + 2)^2$.

30. $3s^2(3s - 9s^2)^2$.

31. $(x + 3)^3(x - 1) + (x + 3)^2(x - 1)^2$.

32. $(x + 5)^2(x + 1)^3 + (x + 5)^3(x + 1)^2$.

33. $(x + 4)(2x + 1) + (x + 4)$.

34. $(x - 3)(2x + 3) - (2x + 3)(x + 5)$.

35. $x^4 - 16$.

36. $81x^4 - y^4$.

37. $y^8 - 1$.

38. $t^4 - 4$.

39. $x^4 + x^2 - 2$.

40. $x^4 - 5x^2 + 4$.

41. $x^5 - 2x^3 + x$.

42. $4x^3 - 6x^2 - 4x$.

43. $x^3y^2 - 10x^2y + 25x$.

44. $(3x^2 + x) + (6x + 2)$.

45. $(x^3 - 4x) + (8 - 2x^2)$.

46. $(x^2 - 1) + (x^2 - x - 2)$.

47. $(y^{10} + 8y^6 + 16y^2) - (y^8 + 8y^4 + 16)$.

48. $x^3y - xy + z^2x^2 - z^2$.

A.5 FRACTIONS

By using the fundamental principle of fractions (Sec. A.1), we may be able to simplify fractions. That principle allows us to multiply or divide both numerator and denominator of a fraction by the same nonzero quantity. The resulting fraction will be equivalent to the original one. The fractions that we shall consider are assumed to have nonzero denominators.

EXAMPLE 1

a. *Simplify* $\dfrac{x^2 - x - 6}{x^2 - 7x + 12}$.

First, we completely factor the numerator and denominator.

$$\frac{x^2 - x - 6}{x^2 - 7x + 12} = \frac{(x - 3)(x + 2)}{(x - 3)(x - 4)}.$$

Dividing both numerator and denominator by the common factor $x - 3$, we have

$$\frac{(x - 3)(x + 2)}{(x - 3)(x - 4)} = \frac{1(x + 2)}{1(x - 4)} = \frac{x + 2}{x - 4}.$$

Usually, we just write

$$\frac{x^2 - x - 6}{x^2 - 7x + 12} = \frac{(x - 3)(x + 2)}{(x - 3)(x - 4)} = \frac{x + 2}{x - 4}.$$

b. *Simplify* $\dfrac{2x^2 + 6x - 8}{8 - 4x - 4x^2}$.

$$\frac{2x^2 + 6x - 8}{8 - 4x - 4x^2} = \frac{2(x^2 + 3x - 4)}{4(2 - x - x^2)} = \frac{2(x - 1)(x + 4)}{4(1 - x)(2 + x)}$$

$$= \frac{2(x - 1)(x + 4)}{2(2)[(-1)(x - 1)](2 + x)}$$

$$= \frac{x + 4}{-2(x + 2)} = -\frac{x + 4}{2(x + 2)}.$$

By Property 20 of Section A.1, if we wish to multiply $\dfrac{a}{b}$ and $\dfrac{c}{d}$, then

$$\frac{a}{b} \cdot \frac{c}{d} = \frac{ac}{bd}.$$

By Property 29, to divide $\dfrac{a}{b}$ by $\dfrac{c}{d}$, where $c \neq 0$, we have

$$\frac{a}{b} \div \frac{c}{d} = \frac{\dfrac{a}{b}}{\dfrac{c}{d}} = \frac{a}{b} \cdot \frac{d}{c}.$$

EXAMPLE 2

a. $\dfrac{x}{x + 2} \cdot \dfrac{x + 3}{x - 5} = \dfrac{x(x + 3)}{(x + 2)(x - 5)}.$

b. $\dfrac{x^2 - 4x + 4}{x^2 + 2x - 3} \cdot \dfrac{6x^2 - 6}{x^2 + 2x - 8} = \dfrac{[(x - 2)^2][6(x + 1)(x - 1)]}{[(x + 3)(x - 1)][(x + 4)(x - 2)]}$

$$= \frac{6(x - 2)(x + 1)}{(x + 3)(x + 4)}.$$

c. $\dfrac{\dfrac{x - 5}{x - 3}}{2x} = \dfrac{\dfrac{x - 5}{x - 3}}{\dfrac{2x}{1}} = \dfrac{x - 5}{x - 3} \cdot \dfrac{1}{2x} = \dfrac{x - 5}{2x(x - 3)}.$

d. $\dfrac{\dfrac{4x}{x^2 - 1}}{\dfrac{2x^2 + 8x}{x - 1}} = \dfrac{4x}{x^2 - 1} \cdot \dfrac{x - 1}{2x^2 + 8x} = \dfrac{4x(x - 1)}{[(x + 1)(x - 1)][2x(x + 4)]}$

$$= \frac{2}{(x + 1)(x + 4)}.$$

e. *Rationalize the denominator of* $\dfrac{x}{\sqrt{2} - 6}$.

$$\frac{x}{\sqrt{2} - 6} = \frac{x}{\sqrt{2} - 6} \cdot \frac{\sqrt{2} + 6}{\sqrt{2} + 6}$$

$$= \frac{x(\sqrt{2} + 6)}{(\sqrt{2})^2 - 6^2} = \frac{x(\sqrt{2} + 6)}{2 - 36}$$

$$= -\frac{x(\sqrt{2} + 6)}{34}.$$

By Property 25 of Section A.1, to add two fractions having a common denominator, we have $\dfrac{a}{c} + \dfrac{b}{c} = \dfrac{a + b}{c}$. Similarly, $\dfrac{a}{c} - \dfrac{b}{c} = \dfrac{a - b}{c}$.

EXAMPLE 3

$$\frac{p^2 - 5}{p - 2} + \frac{3p + 2}{p - 2} = \frac{(p^2 - 5) + (3p + 2)}{p - 2} = \frac{p^2 + 3p - 3}{p - 2}.$$

To add (or subtract) two fractions with different denominators, rewrite the fractions as equivalent fractions that have the same denominator. Then proceed with the addition (or subtraction) by the method described above.

For example, to find

$$\frac{2}{x^3(x - 3)} + \frac{3}{x(x - 3)^2},$$

we can convert the first fraction into an equivalent fraction by multiplying the numerator and denominator by $x - 3$:

$$\frac{2(x - 3)}{x^3(x - 3)^2};$$

we can convert the second fraction by multiplying the numerator and denominator by x^2:

$$\frac{3x^2}{x^3(x - 3)^2}.$$

These fractions have the same denominator. Hence

$$\frac{2}{x^3(x - 3)} + \frac{3}{x(x - 3)^2} = \frac{2(x - 3)}{x^3(x - 3)^2} + \frac{3x^2}{x^3(x - 3)^2}$$

$$= \frac{3x^2 + 2x - 6}{x^3(x - 3)^2}.$$

We could have converted the original fractions into equivalent fractions with many possible common denominators. However, we chose to convert them into fractions with the denominator $x^3(x - 3)^2$. This is the **least common denominator (L.C.D.)** of the fractions $2/[x^3(x - 3)]$ and $3/[x(x - 3)^2]$.

In general, to find the L.C.D. of two or more fractions, first factor each denominator completely. *The L.C.D. is the product of each of the distinct factors appearing in the denominators, each raised to the highest power to which it occurs in any single denominator.*

EXAMPLE 4

a. *Subtract:* $\dfrac{t}{3t + 2} - \dfrac{4}{t - 1}$.

Here the denominators are already factored. The L.C.D. is the product $(3t + 2)(t - 1)$.

$$\frac{t}{3t + 2} - \frac{4}{t - 1} = \frac{t(t - 1)}{(3t + 2)(t - 1)} - \frac{4(3t + 2)}{(3t + 2)(t - 1)}$$

$$= \frac{t(t - 1) - 4(3t + 2)}{(3t + 2)(t - 1)}$$

$$= \frac{t^2 - t - 12t - 8}{(3t + 2)(t - 1)} = \frac{t^2 - 13t - 8}{(3t + 2)(t - 1)}.$$

b. $\dfrac{4}{q - 1} + 3 = \dfrac{4}{q - 1} + \dfrac{3(q - 1)}{q - 1}$

$$= \frac{4 + 3(q - 1)}{q - 1} = \frac{3q + 1}{q - 1}.$$

EXAMPLE 5

$\dfrac{x - 2}{x^2 + 6x + 9} - \dfrac{x + 2}{2(x^2 - 9)}$

$$= \frac{x - 2}{(x + 3)^2} - \frac{x + 2}{2(x + 3)(x - 3)} \qquad [\text{L.C.D.} = 2(x + 3)^2(x - 3)]$$

$$= \frac{(x - 2)(2)(x - 3)}{(x + 3)^2(2)(x - 3)} - \frac{(x + 2)(x + 3)}{2(x + 3)(x - 3)(x + 3)}$$

$$= \frac{(x - 2)(2)(x - 3) - (x + 2)(x + 3)}{2(x + 3)^2(x - 3)}$$

$$= \frac{2(x^2 - 5x + 6) - (x^2 + 5x + 6)}{2(x + 3)^2(x - 3)}$$

$$= \frac{2x^2 - 10x + 12 - x^2 - 5x - 6}{2(x + 3)^2(x - 3)}$$

$$= \frac{x^2 - 15x + 6}{2(x + 3)^2(x - 3)}.$$

EXAMPLE 6

$$\frac{\dfrac{1}{x + h} - \dfrac{1}{x}}{h} = \frac{\dfrac{x}{x(x + h)} - \dfrac{x + h}{x(x + h)}}{h} = \frac{\dfrac{x - (x + h)}{x(x + h)}}{h}$$

$$= \frac{\dfrac{-h}{x(x + h)}}{\dfrac{h}{1}} = \frac{-h}{x(x + h)} \cdot \frac{1}{h} = \frac{-h}{x(x + h)h} = -\frac{1}{x(x + h)}.$$

EXERCISE A.5

*In Problems **1–28**, perform the operations and simplify as much as possible.*

1. $\dfrac{y^2}{y - 3} \cdot \dfrac{-1}{y + 2}.$

2. $\dfrac{z^2 - 4}{z^2 + 2z} \cdot \dfrac{z^2}{z - 2}.$

3. $\dfrac{2x - 3}{x - 2} \cdot \dfrac{2 - x}{2x + 3}.$

4. $\dfrac{x^2 - y^2}{x + y} \cdot \dfrac{x^2 + 2xy + y^2}{y - x}.$

5. $\dfrac{\dfrac{x^2}{6}}{\dfrac{x}{3}}.$

6. $\dfrac{\dfrac{4x^3}{9x}}{\dfrac{x}{18}}.$

7. $\dfrac{\dfrac{4x}{3}}{2x}.$

8. $\dfrac{-9x^3}{\dfrac{x}{3}}.$

9. $\dfrac{2x - 2}{x^2 - 2x - 8} \div \dfrac{x^2 - 1}{x^2 + 5x + 4}.$

10. $\dfrac{x^2 + 2x}{3x^2 - 18x + 24} \div \dfrac{x^2 - x - 6}{x^2 - 4x + 4}.$

11. $\dfrac{\dfrac{4x^2 - 9}{x^2 + 3x - 4}}{\dfrac{2x - 3}{1 - x^2}}.$

12. $\dfrac{\dfrac{6x^2y + 7xy - 3y}{xy - x + 5y - 5}}{\dfrac{x^3y + 4x^2y}{xy - x + 4y - 4}}.$

13. $\dfrac{x^2}{x + 3} + \dfrac{5x + 6}{x + 3}.$

14. $\dfrac{2}{x + 2} + \dfrac{x}{x + 2}.$

15. $\dfrac{1}{t} + \dfrac{2}{3t}.$

16. $\dfrac{4}{x^2} - \dfrac{1}{x}.$

17. $1 - \dfrac{p^2}{p^2 - 1}.$

18. $\dfrac{4}{s + 4} + s.$

19. $\dfrac{4}{2x - 1} + \dfrac{x}{x + 3}.$

20. $\dfrac{x + 1}{x - 1} - \dfrac{x - 1}{x + 1}.$

21. $\dfrac{1}{x^2 - x - 2} + \dfrac{1}{x^2 - 1}.$

22. $\dfrac{y}{3y^2 - 5y - 2} - \dfrac{2}{3y^2 - 7y + 2}.$

23. $\dfrac{4}{x - 1} - 3 + \dfrac{-3x^2}{5 - 4x - x^2}.$

24. $\dfrac{2x - 3}{2x^2 + 11x - 6} - \dfrac{3x + 1}{3x^2 + 16x - 12} + \dfrac{1}{3x - 2}.$

25. $\dfrac{1 + \dfrac{1}{x}}{3}.$

26. $\dfrac{\dfrac{x + 3}{x}}{x - \dfrac{9}{x}}.$

27. $\dfrac{3 - \dfrac{1}{2x}}{x + \dfrac{x}{x + 2}}.$

28. $\dfrac{\dfrac{x-1}{x^2+5x+6}-\dfrac{1}{x+2}}{3+\dfrac{x-7}{3}}.$

In Problems **29** *and* **30**, *simplify and express your answers in a form that is free of radicals in the denominator.*

29. $\dfrac{1}{x+\sqrt{5}}.$

30. $\dfrac{4}{\sqrt{x}+2}\cdot\dfrac{x^2}{3}.$

A.6 EQUATIONS

The basic operations that are used to solve equations are:

1. Adding (or subtracting) the same number to (or from) both sides of an equation.

2. Multiplying (or dividing) both sides of an equation by the same *constant*, except zero.

When any one of these operations is performed on a given equation, the resulting equation is equivalent to (that is, has the same solution as) the given equation.

EXAMPLE 1   *Solve the following equations for p or x.*

a. $2(p+4)=7p+2.$

$$2(p+4)=7p+2,$$
$$2p+8=7p+2 \quad \text{(distributive law)},$$
$$2p=7p-6 \quad \text{(subtracting 8 from both sides)},$$
$$-5p=-6 \quad \text{(subtracting } 7p \text{ from both sides)},$$
$$p=\frac{-6}{-5} \quad \text{(dividing both sides by } -5),$$
$$p=\frac{6}{5}.$$

b. $\dfrac{7x+3}{2}-\dfrac{9x-8}{4}=6.$

We first clear the equation of fractions by multiplying *both* sides by the least common denominator (L.C.D.), which is 4.

$$4\left(\frac{7x+3}{2}-\frac{9x-8}{4}\right)=4(6),$$
$$2(7x+3)-(9x-8)=24,$$
$$14x+6-9x+8=24,$$
$$5x+14=24,$$
$$5x=10,$$
$$x=2.$$

c. $(a + c)x + x^2 = (x + a)^2$.

$$(a + c)x + x^2 = (x + a)^2,$$
$$ax + cx + x^2 = x^2 + 2ax + a^2,$$
$$cx - ax = a^2,$$
$$x(c - a) = a^2,$$
$$x = \frac{a^2}{c - a}.$$

Actually, we assumed $c - a \neq 0$ to avoid division by 0.

EXAMPLE 2   *Solve* $\dfrac{5}{x - 4} = \dfrac{6}{x - 3}$.

To solve this equation we first clear it of fractions. Multiplying both sides by the L.C.D., $(x - 4)(x - 3)$, we have

$$(x - 4)(x - 3)\left(\frac{5}{x - 4}\right) = (x - 4)(x - 3)\left(\frac{6}{x - 3}\right),$$
$$5(x - 3) = 6(x - 4),$$
$$5x - 15 = 6x - 24,$$
$$9 = x.$$

In the first step we multiplied each side by an expression involving a *variable*, x. This means that we are not guaranteed that the last equation, $9 = x$, is equivalent to the *original* equation. Thus we must check whether or not 9 satisfies the original equation. If 9 is substituted for x there, the left side is

$$\frac{5}{9 - 4} = \frac{5}{5} = 1,$$

and the right side is

$$\frac{6}{9 - 3} = \frac{6}{6} = 1.$$

Since both sides are equal, 9 is a root (solution).

Quadratic equations have the form

$$ax^2 + bx + c = 0, \qquad (1)$$

where a, b, and c are constants and $a \neq 0$. Equation (1) may be solved by first factoring the left side and then setting each factor equal to 0. This method also applies to many equations that are not quadratic. The roots of (1) are also given by the **quadratic formula**:

$$x = \frac{-b \pm \sqrt{b^2 - 4ac}}{2a}.$$

EXAMPLE 3   *Solve each equation by the method of factoring.*

a. $3x^2 - x - 4 = -2$.

$$3x^2 - x - 4 = -2,$$
$$3x^2 - x - 2 = 0,$$
$$(3x + 2)(x - 1) = 0.$$

Setting each factor equal to 0, we have

$$
\begin{array}{c|c}
3x + 2 = 0, & x - 1 = 0, \\
3x = -2, & x = 1. \\
x = -\dfrac{2}{3}. &
\end{array}
$$

Thus $x = -\frac{2}{3}, 1$.

b. $4x - 4x^3 = 0$.

Although the equation is not quadratic, the method of factoring applies.

$$4x - 4x^3 = 0,$$
$$4x(1 - x^2) = 0,$$
$$4x(1 + x)(1 - x) = 0.$$

Setting each factor equal to 0, we get

$$x = 0, -1, 1,$$

which we can write as $x = 0, \pm 1$.

c. $x^2 = 3$.

$$x^2 - 3 = 0,$$
$$(x - \sqrt{3})(x + \sqrt{3}) = 0.$$

Thus $x - \sqrt{3} = 0$ or $x + \sqrt{3} = 0$. The roots are $\pm\sqrt{3}$. More generally,

$$\text{if } u^2 = k, \quad \text{then} \quad u = \pm\sqrt{k}.$$

For example, the solution of $y^2 = 4$ is $y = \pm\sqrt{4} = \pm 2$.

EXAMPLE 4   *Solve each equation by the quadratic formula.*

a. $4x^2 - 17x + 15 = 0$.

Here $a = 4$, $b = -17$, and $c = 15$.

$$x = \frac{-b \pm \sqrt{b^2 - 4ac}}{2a} = \frac{-(-17) \pm \sqrt{(-17)^2 - 4(4)(15)}}{2(4)}$$

$$= \frac{17 \pm \sqrt{49}}{8} = \frac{17 \pm 7}{8}.$$

The roots are $\dfrac{17 + 7}{8} = \dfrac{24}{8} = 3$ and $\dfrac{17 - 7}{8} = \dfrac{10}{8} = \dfrac{5}{4}$.

b. $2 + 6\sqrt{2}y + 9y^2 = 0$.

Look at the arrangement of the terms. Here $a = 9$, $b = 6\sqrt{2}$, and $c = 2$.

$$y = \frac{-b \pm \sqrt{b^2 - 4ac}}{2a} = \frac{-6\sqrt{2} \pm \sqrt{0}}{2(9)}.$$

Thus $y = \dfrac{-6\sqrt{2} + 0}{18} = -\dfrac{\sqrt{2}}{3}$ or $y = \dfrac{-6\sqrt{2} - 0}{18} = -\dfrac{\sqrt{2}}{3}$. The only root is $-\dfrac{\sqrt{2}}{3}$.

c. $z^2 + z + 1 = 0$.

Here $a = 1$, $b = 1$, and $c = 1$. The roots are

$$\frac{-b \pm \sqrt{b^2 - 4ac}}{2a} = \frac{-1 \pm \sqrt{-3}}{2}.$$

Since $\sqrt{-3}$ is not a real number, there are no real roots.*

As Example 4 illustrates, a quadratic equation may have two real distinct roots, one real root, or no real root.

EXERCISE A.6

In Problems 1–22, solve the equations.

1. $6x = 45$.

2. $3y = 0$.

3. $5x - 3 = 9$.

4. $7x + 7 = 2(x + 1)$.

5. $2(p - 1) - 3(p - 4) = 4p$.

6. $\dfrac{x}{5} = 2x - 6$.

7. $5 + \dfrac{4x}{9} = \dfrac{x}{2}$.

8. $q = \dfrac{3}{2}q - 4$.

9. $3x + \dfrac{x}{5} - 5 = \dfrac{1}{5} + 5x$.

10. $w + \dfrac{w}{2} - \dfrac{w}{3} + \dfrac{w}{4} = 5$.

11. $\dfrac{2y - 3}{4} = \dfrac{6y + 7}{3}$.

12. $\dfrac{x + 2}{3} - \dfrac{2 - x}{6} = x - 2$.

13. $\frac{9}{5}(3 - x) = \frac{3}{4}(x - 3)$.

14. $\frac{3}{2}(4x - 3) = 2[x - (4x - 3)]$.

15. $\dfrac{4}{x} = 16$.

16. $\dfrac{4}{8 - x} = \dfrac{3}{4}$.

17. $\dfrac{x}{3x - 4} = 3$.

18. $\dfrac{1}{p - 1} = \dfrac{2}{p - 2}$.

19. $\dfrac{1}{x} + \dfrac{1}{5} = \dfrac{4}{5}$.

20. $\dfrac{3x - 2}{2x + 3} = \dfrac{3x - 1}{2x + 1}$.

* $\dfrac{-1 \pm \sqrt{-3}}{2}$ can be expressed as $\dfrac{-1 \pm i\sqrt{3}}{2}$, where $i \ (=\sqrt{-1})$ is called the *imaginary unit*.

21. $\dfrac{y-6}{y} - \dfrac{6}{y} = \dfrac{y+6}{y-6}$.   **22.** $\dfrac{9}{x-3} = \dfrac{3x}{x-3}$.

In Problems 23–28, express the indicated symbol in terms of the remaining symbols.

23. $I = Prt$; P.

24. $ax + b = 0$; x.

25. $p = 6x - 1$; x.

26. $p = -3x + 6$; x.

27. $S = P(1 + rt)$; r.

28. $S = \dfrac{R[(1+i)^n - 1]}{i}$; R.

In Problems 29–48, solve by factoring.

29. $x^2 + 3x + 2 = 0$.

30. $x^2 - 2x - 3 = 0$.

31. $y^2 - 7y + 12 = 0$.

32. $x^2 - 12x = -36$.

33. $x^2 - 4 = 0$.

34. $y(2y + 3) = 5$.

35. $z^2 - 8z = 0$.

36. $-x^2 + 3x + 10 = 0$.

37. $4x^2 + 1 = 4x$.

38. $2p^2 = 3p$.

39. $x^3 - 64x = 0$.

40. $6x^3 + 5x^2 - 4x = 0$.

41. $x(x - 1)(x + 2) = 0$.

42. $x^2(x - 4) = 0$.

43. $(x - 2)^2(x + 1)^2 = 0$.

44. $x(x - 1)(x + 1) = 0$.

45. $7x^2(x - 2)^2(x + 3)(x - 4) = 0$.

46. $x(x^2 - 1)(x^2 - 4) = 0$.

47. $x(x^2 - 1)(x^2 - 1) = 0$.

48. $x^4 - x^2 = 0$.

In Problems 49–62, find all real roots by using the quadratic formula.

49. $x^2 + 2x - 15 = 0$.   **50.** $x^2 - 2x - 24 = 0$.

51. $4x^2 - 12x + 9 = 0$.   **52.** $p^2 + 2p = 0$.

53. $p^2 - 5p + 3 = 0$.   **54.** $2 - 2x + x^2 = 0$.

55. $4 - 2n + n^2 = 0$.   **56.** $2x^2 + x = 5$.

57. $6x^2 + 7x - 5 = 0$.

58. $w^2 - 2\sqrt{2}w + 2 = 0$.

59. $2x^2 - 3x = 20$.

60. $0.01x^2 + 0.2x - 0.6 = 0$.

61. $2x^2 + 4x = 5$.

62. $-2x^2 - 6x + 5 = 0$.

INEQUALITIES

The basic rules that are used to solve inequalities are

1. If $a < b$, then $a + c < b + c$ and $a - c < b - c$.

2. If $a < b$ and c is a positive number, then $ac < bc$ and $\dfrac{a}{c} < \dfrac{b}{c}$.

3. If $a < b$ and $-c$ is a negative number, then $a(-c) > b(-c)$ and $\dfrac{a}{-c} > \dfrac{b}{-c}$.

These rules are also true if you replace $<$ by $\leq$, $>$, or $\geq$. In Rule 3, however, if you started with the symbol $\leq$, it would change to $\geq$, etc.

In the following examples, the rule used is indicated to the right.

EXAMPLE 1

a. *Solve* $2(x - 3) < 4$.

$$2(x - 3) < 4,$$
$$2x - 6 < 4 \qquad \text{(distributive law)},$$
$$2x - 6 + 6 < 4 + 6 \qquad (1),$$
$$2x < 10,$$
$$\frac{2x}{2} < \frac{10}{2} \qquad (2),$$
$$x < 5.$$

Thus the original inequality is true for *all* real numbers x such that $x < 5$.

b. *Solve* $3 - 2x \le 6$.

$$3 - 2x \le 6,$$
$$-2x \le 3 \qquad (1),$$
$$x \ge -\frac{3}{2} \qquad (3).$$

The solution is $x \ge -3/2$.

c. *Solve* $\frac{3}{2}(s - 2) + 1 > -2(s - 4)$.

$$\frac{3}{2}(s - 2) + 1 > -2(s - 4),$$
$$3(s - 2) + 2 > -4(s - 4) \qquad (2),$$
$$3s - 4 > -4s + 16,$$
$$7s > 20 \qquad (1),$$
$$s > \frac{20}{7} \qquad (2).$$

EXAMPLE 2

a. *Solve* $2(x - 4) - 3 > 2x - 1$.

$$2(x - 4) - 3 > 2x - 1,$$
$$2x - 8 - 3 > 2x - 1,$$
$$-11 > -1 \qquad (1).$$

Since $-11 > -1$ is never true, there is no solution.

b. *Solve* $2(x - 4) - 3 < 2x - 1$.

Proceeding in the same manner as in part (a), we obtain $-11 < -1$. This is always true, so the inequality is true for all real numbers x. We write our solution as $-\infty < x < \infty$. The symbols $-\infty$ and ∞ are not numbers but are merely used as a convenience for indicating that the solution is all real numbers.

Suppose that $a < b$ and x is between a and b. Then not only is $a < x$, but $x < b$. We indicate this by writing $a < x < b$. Thus $0 < 7 < 9$. Similarly, $-1 \leq x < 3$ represents all real numbers x such that $x \geq -1$ and $x < 3$ simultaneously.

EXERCISE A.7

Solve the following inequalities.

1. $4x > 8$.

2. $8x < -2$.

3. $3x - 4 \leq 2$.

4. $5x \geq 0$.

5. $3 < 2y + 3$.

6. $2y + 1 > 0$.

7. $-4x \geq 2$.

8. $6 \leq 5 - 3y$.

9. $3 - 5s > 5$.

10. $4s - 1 < -5$.

11. $2x - 3 \leq 4 + 7x$.

12. $-3 \geq 8(2 - x)$.

13. $3(2 - 3x) > 4(1 - 4x)$.

14. $8(x + 1) + 1 < 3(2x) + 1$.

15. $2(3x - 2) > 3(2x - 1)$.

16. $3 - 2(x - 1) \leq 2(4 + x)$.

17. $x + 2 < \sqrt{3} - x$.

18. $\sqrt{2}(x + 2) > \sqrt{8}(3 - x)$.

19. $\dfrac{9y + 1}{4} \leq 2y - 1$.

20. $\dfrac{4y - 3}{2} \geq \dfrac{1}{3}$.

21. $4x - 1 \geq 4(x - 2) + 7$.

22. $0x \leq 0$.

23. $\dfrac{1 - t}{2} < \dfrac{3t - 7}{3}$.

24. $\dfrac{3(2t - 2)}{2} > \dfrac{6t - 3}{5} + \dfrac{t}{10}$.

25. $2x + 3 \geq \frac{1}{2}x - 4$.

26. $4x - \frac{1}{2} \leq \frac{3}{2}x$.

27. $\frac{2}{3}r < \frac{5}{6}r$.

28. $\frac{7}{4}t > -\frac{2}{3}t$.

29. $0.1(0.03x + 4) \geq 0.02x + 0.434$.

30. $9 - 0.1x \leq \dfrac{2 - 0.01x}{0.2}$.

31. $\dfrac{y}{2} + \dfrac{y}{3} > y + \dfrac{y}{5}$.

32. $\dfrac{5y - 1}{-3} < \dfrac{7(y + 1)}{-2}$.

Proofs of Differentiation Rules

The following are proofs of some of the differentiation rules in Chapters 2 and 9. We begin with Rules 4, 5, and 6 in Chapter 2.

Rule 4

If f and g are differentiable functions, then

$$\frac{d}{dx}[f(x) + g(x)] = f'(x) + g'(x).$$

Proof. If $F(x) = f(x) + g(x)$, applying the definition of the derivative gives

$$F'(x) = \lim_{h \to 0} \frac{F(x + h) - F(x)}{h}$$

$$= \lim_{h \to 0} \frac{[f(x + h) + g(x + h)] - [f(x) + g(x)]}{h}$$

$$= \lim_{h \to 0} \frac{[f(x + h) - f(x)] + [g(x + h) - g(x)]}{h} \qquad \text{(regrouping)}$$

$$= \lim_{h \to 0} \left[\frac{f(x + h) - f(x)}{h} + \frac{g(x + h) - g(x)}{h} \right]$$

$$= \lim_{h \to 0} \frac{f(x + h) - f(x)}{h} + \lim_{h \to 0} \frac{g(x + h) - g(x)}{h}.$$

These two limits are $f'(x)$ and $g'(x)$. Thus

$$F'(x) = f'(x) + g'(x).$$

The proof of the rule for the derivative of a difference of two functions is similar to the above.

431

432 B PROOFS OF DIFFERENTIATION RULES

Rule 5

If f and g are differentiable functions, then

$$\frac{d}{dx}[f(x)g(x)] = f(x)g'(x) + g(x)f'(x).$$

Proof. If $F(x) = f(x)g(x)$, then by the definition of the derivative,

$$F'(x) = \lim_{h \to 0} \frac{F(x + h) - F(x)}{h}$$

$$= \lim_{h \to 0} \frac{f(x + h)g(x + h) - f(x)g(x)}{h}.$$

Now we use a "trick." Adding and subtracting $f(x + h)g(x)$ in the numerator, we have

$$F'(x) = \lim_{h \to 0} \frac{f(x + h)g(x + h) - f(x)g(x) + [f(x + h)g(x) - f(x + h)g(x)]}{h}.$$

Regrouping gives

$$F'(x) = \lim_{h \to 0} \frac{[f(x + h)g(x + h) - f(x + h)g(x)] + [f(x + h)g(x) - f(x)g(x)]}{h}$$

$$= \lim_{h \to 0} \frac{f(x + h)[g(x + h) - g(x)] + g(x)[f(x + h) - f(x)]}{h}$$

$$= \lim_{h \to 0} \frac{f(x + h)[g(x + h) - g(x)]}{h} + \lim_{h \to 0} \frac{g(x)[f(x + h) - f(x)]}{h}$$

$$= \lim_{h \to 0} f(x + h) \cdot \lim_{h \to 0} \frac{g(x + h) - g(x)}{h} + \lim_{h \to 0} g(x) \cdot \lim_{h \to 0} \frac{f(x + h) - f(x)}{h}.$$

Since we assumed that f and g are differentiable, then

$$\lim_{h \to 0} \frac{g(x + h) - g(x)}{h} = g'(x)$$

and $$\lim_{h \to 0} \frac{f(x + h) - f(x)}{h} = f'(x).$$

The differentiability of f implies that f is continuous, and because of continuity, it can be shown that

$$\lim_{h \to 0} f(x + h) = f(x).$$

Thus

$$F'(x) = f(x)g'(x) + g(x)f'(x).$$

Rule 6

If f and g are differentiable functions and $g(x) \neq 0$, then

$$\frac{d}{dx}\left[\frac{f(x)}{g(x)}\right] = \frac{g(x)f'(x) - f(x)g'(x)}{[g(x)]^2}.$$

Proof. If $F(x) = \dfrac{f(x)}{g(x)}$, then

$$F(x)g(x) = f(x).$$

By the product rule (Rule 5),

$$F(x)g'(x) + g(x)F'(x) = f'(x).$$

Solving for $F'(x)$, we have

$$F'(x) = \frac{f'(x) - F(x)g'(x)}{g(x)}.$$

But $F(x) = f(x)/g(x)$. Thus

$$F'(x) = \frac{f'(x) - \dfrac{f(x)g'(x)}{g(x)}}{g(x)}.$$

Simplifying, we have

$$F'(x) = \frac{g(x)f'(x) - f(x)g'(x)}{[g(x)]^2}. \text{ *}$$

To prove the differentiation rule in Chapter 9 for $\sin t$, we shall first consider some limits.

Our first limit is

$$\lim_{t \to 0} \frac{\sin t}{t},$$

where t is in radians. As $t \to 0$, both $\sin t$ and t approach 0. Thus this limit has the meaningless form 0/0. In Table B.1 are values of t, $\sin t$, and the

TABLE B.1

t	$\sin t$	$(\sin t)/t$
0.30	0.2955	0.9850
0.20	0.1987	0.9935
0.10	0.0998	0.9980
0.05	0.0500	1.0000
0.01	0.0100	1.0000

* You may have observed that this proof assumes the existence of $F'(x)$. However, this rule can be proven without this assumption.

ratio $(\sin t)/t$. As $t \to 0$, it appears that $(\sin t)/t \to 1$, and in fact, it can be shown that this is indeed true. Thus, if t is in radians, then

$$\lim_{t \to 0} \frac{\sin t}{t} = 1. \tag{1}$$

Another limit of interest to us is

$$\lim_{t \to 0} \frac{\cos t - 1}{t},$$

where t is in radians. As $t \to 0$, both the numerator and denominator approach 0, so we have the meaningless form $0/0$. To determine the limit, we shall express the quotient in a different form.

$$\begin{aligned}
\lim_{t \to 0} \frac{\cos t - 1}{t} &= \lim_{t \to 0} \frac{(\cos t - 1)(\cos t + 1)}{t(\cos t + 1)} \\
&= \lim_{t \to 0} \frac{\cos^2 t - 1}{t(\cos t + 1)} \\
&= \lim_{t \to 0} \frac{-\sin^2 t}{t(\cos t + 1)} \qquad (\text{since } \sin^2 t + \cos^2 t = 1) \\
&= -\lim_{t \to 0} \frac{\sin t}{t} \cdot \lim_{t \to 0} \frac{\sin t}{\cos t + 1} \\
&= -1 \cdot \frac{0}{2} \qquad [\text{by Eq. (1)}] \\
&= 0.
\end{aligned}$$

Thus

$$\lim_{t \to 0} \frac{\cos t - 1}{t} = 0. \tag{2}$$

Rule

If t is in radians, then $\dfrac{d}{dt}(\sin t) = \cos t$.

    Proof. If $f(t) = \sin t$, applying the definition of the derivative gives

$$\begin{aligned}
f'(t) &= \lim_{h \to 0} \frac{f(t + h) - f(t)}{h} \\
&= \lim_{h \to 0} \frac{\sin(t + h) - \sin t}{h} \\
&= \lim_{h \to 0} \frac{(\sin t \cos h + \cos t \sin h) - \sin t}{h} \qquad (\text{Identity 6, Table 9.3}) \\
&= \lim_{h \to 0} \frac{(\sin t \cos h - \sin t) + \cos t \sin h}{h} \qquad (\text{regrouping})
\end{aligned}$$

$$= \lim_{h \to 0} \frac{\sin t(\cos h - 1) + \cos t \sin h}{h}$$

$$= \lim_{h \to 0} \left[\sin t \cdot \frac{\cos h - 1}{h} + \cos t \cdot \frac{\sin h}{h} \right]$$

$$= \sin t \cdot \lim_{h \to 0} \frac{\cos h - 1}{h} + \cos t \cdot \lim_{h \to 0} \frac{\sin h}{h}$$

$$= (\sin t) \cdot 0 + (\cos t) \cdot 1 \qquad \text{[by Eqs. (1) and (2)]}$$

$$= \cos t.$$

To prove the differentiation rule in Chapter 9 for cos t, we shall first express the cosine in terms of the sine. By Identity 6 in Table 9.3,

$$\sin\left(t + \frac{\pi}{2}\right) = \sin t \cos \frac{\pi}{2} + \cos t \sin \frac{\pi}{2}$$

$$= (\sin t) \cdot 0 + (\cos t) \cdot 1 = \cos t.$$

Thus

$$\cos t = \sin\left(t + \frac{\pi}{2}\right). \tag{3}$$

Rule

If t is in radians, then $\dfrac{d}{dt}(\cos t) = -\sin t$.

Proof.

$$\frac{d}{dt}(\cos t) = \frac{d}{dt}\left[\sin\left(t + \frac{\pi}{2}\right) \right] \qquad \text{[Eq. (3)]}$$

$$= \cos\left(t + \frac{\pi}{2}\right) D_t\left(t + \frac{\pi}{2}\right) \qquad \text{(Rule 1, Table 9.6)}$$

$$= \cos\left(t + \frac{\pi}{2}\right) \cdot 1$$

$$= \cos t \cos \frac{\pi}{2} - \sin t \sin \frac{\pi}{2} \qquad \text{(Identity 7, Table 9.3)}$$

$$= (\cos t) \cdot 0 - (\sin t) \cdot 1$$

$$= -\sin t.$$

Table of Powers— Roots—Reciprocals

n	n^2	$\sqrt{n}$	$\sqrt{10n}$	n^3	$\sqrt[3]{n}$	$\sqrt[3]{10n}$	$\sqrt[3]{100n}$	$1/n$
1.0	1.0000	1.0000	3.1623	1.0000	1.0000	2.1544	4.6416	1.0000
1.1	1.2100	1.0488	3.3166	1.3310	1.0323	2.2240	4.7914	0.9091
1.2	1.4400	1.0954	3.4641	1.7280	1.0627	2.2894	4.9324	0.8333
1.3	1.6900	1.1402	3.6056	2.1970	1.0914	2.3513	5.0658	0.7692
1.4	1.9600	1.1832	3.7417	2.7440	1.1187	2.4101	5.1925	0.7143
1.5	2.2500	1.2247	3.8730	3.3750	1.1447	2.4662	5.3133	0.6667
1.6	2.5600	1.2649	4.0000	4.0960	1.1696	2.5198	5.4288	0.6250
1.7	2.8900	1.3038	4.1231	4.9130	1.1935	2.5713	5.5397	0.5882
1.8	3.2400	1.3416	4.2426	5.8320	1.2164	2.6207	5.6462	0.5556
1.9	3.6100	1.3784	4.3589	6.8590	1.2386	2.6684	5.7489	0.5263
2.0	4.0000	1.4142	4.4721	8.0000	1.2599	2.7144	5.8480	0.5000
2.1	4.4100	1.4491	4.5826	9.2610	1.2806	2.7589	5.9439	0.4762
2.2	4.8400	1.4832	4.6904	10.6480	1.3006	2.8020	6.0368	0.4545
2.3	5.2900	1.5166	4.7958	12.1670	1.3200	2.8439	6.1269	0.4348
2.4	5.7600	1.5492	4.8990	13.8240	1.3389	2.8845	6.2145	0.4167
2.5	6.2500	1.5811	5.0000	15.6250	1.3572	2.9240	6.2996	0.4000
2.6	6.7600	1.6125	5.0990	17.5760	1.3751	2.9625	6.3825	0.3846
2.7	7.2900	1.6432	5.1962	19.6830	1.3925	3.0000	6.4633	0.3704
2.8	7.8400	1.6733	5.2915	21.9520	1.4095	3.0366	6.5421	0.3571
2.9	8.4100	1.7029	5.3852	24.3890	1.4260	3.0723	6.6191	0.3448
3.0	9.0000	1.7321	5.4772	27.0000	1.4422	3.1072	6.6943	0.3333
3.1	9.6100	1.7607	5.5678	29.7910	1.4581	3.1414	6.7679	0.3226
3.2	10.2400	1.7889	5.6569	32.7680	1.4736	3.1748	6.8399	0.3125
3.3	10.8900	1.8166	5.7446	35.9370	1.4888	3.2075	6.9104	0.3030
3.4	11.5600	1.8439	5.8310	39.3040	1.5037	3.2396	6.9795	0.2941
3.5	12.2500	1.8708	5.9161	42.8750	1.5183	3.2711	7.0473	0.2857
3.6	12.9600	1.8974	6.0000	46.6560	1.5326	3.3019	7.1138	0.2778
3.7	13.6900	1.9235	6.0828	50.6530	1.5467	3.3322	7.1791	0.2703
3.8	14.4400	1.9494	6.1644	54.8720	1.5605	3.3620	7.2432	0.2632
3.9	15.2100	1.9748	6.2450	59.3190	1.5741	3.3912	7.3061	0.2564

n	n^2	$\sqrt{n}$	$\sqrt{10n}$	n^3	$\sqrt[3]{n}$	$\sqrt[3]{10n}$	$\sqrt[3]{100n}$	$1/n$
4.0	16.0000	2.0000	6.3246	64.0000	1.5874	3.4200	7.3681	0.2500
4.1	16.8100	2.0248	6.4031	68.9210	1.6005	3.4482	7.4290	0.2439
4.2	17.6400	2.0494	6.4807	74.0880	1.6134	3.4760	7.4889	0.2381
4.3	18.4900	2.0736	6.5574	79.5070	1.6261	3.5034	7.5478	0.2326
4.4	19.3600	2.0976	6.6333	85.1840	1.6386	3.5303	7.6059	0.2273
4.5	20.2500	2.1213	6.7082	91.1250	1.6510	3.5569	7.6631	0.2222
4.6	21.1600	2.1448	6.7823	97.3360	1.6631	3.5830	7.7194	0.2174
4.7	22.0900	2.1679	6.8557	103.823	1.6751	3.6088	7.7750	0.2128
4.8	23.0400	2.1909	6.9282	110.592	1.6869	3.6342	7.8297	0.2083
4.9	24.0100	2.2136	7.0000	117.649	1.6985	3.6593	7.8837	0.2041
5.0	25.0000	2.2361	7.0711	125.000	1.7100	3.6840	7.9370	0.2000
5.1	26.0100	2.2583	7.1414	132.651	1.7213	3.7084	7.9896	0.1961
5.2	27.0400	2.2804	7.2111	140.608	1.7325	3.7325	8.0415	0.1923
5.3	28.0900	2.3022	7.2801	148.877	1.7435	3.7563	8.0927	0.1887
5.4	29.1600	2.3238	7.3485	157.464	1.7544	3.7798	8.1433	0.1852
5.5	30.2500	2.3452	7.4162	166.375	1.7652	3.8030	8.1932	0.1818
5.6	31.3600	2.3664	7.4833	175.616	1.7758	3.8259	8.2426	0.1786
5.7	32.4900	2.3875	7.5498	185.193	1.7863	3.8485	8.2913	0.1754
5.8	33.6400	2.4083	7.6158	195.112	1.7967	3.8709	8.3396	0.1724
5.9	34.8100	2.4290	7.6811	205.379	1.8070	3.8930	8.3872	0.1695
6.0	36.0000	2.4495	7.7460	216.000	1.8171	3.9149	8.4343	0.1667
6.1	37.2100	2.4698	7.8102	226.981	1.8272	3.9365	8.4809	0.1639
6.2	38.4400	2.4900	7.8740	238.328	1.8371	3.9579	8.5270	0.1613
6.3	39.6900	2.5100	7.9372	250.047	1.8469	3.9791	8.5726	0.1587
6.4	40.9600	2.5298	8.0000	262.144	1.8566	4.0000	8.6177	0.1563
6.5	42.2500	2.5495	8.0623	274.625	1.8663	4.0207	8.6624	0.1538
6.6	43.5600	2.5690	8.1240	287.496	1.8758	4.0412	8.7066	0.1515
6.7	44.8900	2.5884	8.1854	300.763	1.8852	4.0615	8.7503	0.1493
6.8	46.2400	2.6077	8.2462	314.432	1.8945	4.0817	8.7937	0.1471
6.9	47.6100	2.6268	8.3066	328.509	1.9038	4.1016	8.8366	0.1449
7.0	49.0000	2.6458	8.3666	343.000	1.9129	4.1213	8.8790	0.1429
7.1	50.4100	2.6646	8.4261	357.911	1.9220	4.1408	8.9211	0.1408
7.2	51.8400	2.6833	8.4853	373.248	1.9310	4.1602	8.9628	0.1389
7.3	53.2900	2.7019	8.5440	389.017	1.9399	4.1793	9.0041	0.1370
7.4	54.7600	2.7203	8.6023	405.224	1.9487	4.1983	9.0450	0.1351
7.5	56.2500	2.7386	8.6603	421.875	1.9574	4.2172	9.0856	0.1333
7.6	57.7600	2.7568	8.7178	438.976	1.9661	4.2358	9.1258	0.1316
7.7	59.2900	2.7749	8.7750	456.533	1.9747	4.2543	9.1657	0.1299
7.8	60.8400	2.7928	8.8318	474.552	1.9832	4.2727	9.2052	0.1282
7.9	62.4100	2.8107	8.8882	493.039	1.9916	4.2908	9.2443	0.1266
8.0	64.0000	2.8284	8.9443	512.000	2.0000	4.3089	9.2832	0.1250
8.1	65.6100	2.8460	9.0000	531.441	2.0083	4.3267	9.3217	0.1235
8.2	67.2400	2.8636	9.0554	551.368	2.0165	4.3445	9.3599	0.1220
8.3	68.8900	2.8810	9.1104	571.787	2.0247	4.3621	9.3978	0.1205
8.4	70.5600	2.8983	9.1652	592.704	2.0328	4.3795	9.4354	0.1190
8.5	72.2500	2.9155	9.2195	614.125	2.0408	4.3968	9.4727	0.1176

n	n^2	$\sqrt{n}$	$\sqrt{10n}$	n^3	$\sqrt[3]{n}$	$\sqrt[3]{10n}$	$\sqrt[3]{100n}$	$1/n$
8.0	64.0000	2.8284	8.9443	512.000	2.0000	4.3089	9.2832	0.1250
8.6	73.9600	2.9326	9.2736	636.056	2.0488	4.4140	9.5097	0.1163
8.7	75.6900	2.9496	9.3274	658.503	2.0567	4.4310	9.5464	0.1149
8.8	77.4400	2.9665	9.3808	681.472	2.0646	4.4480	9.5828	0.1136
8.9	79.2100	2.9833	9.4340	704.969	2.0723	4.4647	9.6190	0.1124
9.0	81.000	3.0000	9.4868	729.000	2.0801	4.4814	9.6549	0.1111
9.1	82.8100	3.0166	9.5394	573.571	2.0878	4.4979	9.6905	0.1099
9.2	84.6400	3.0332	9.5917	778.688	2.0954	4.5144	9.7259	0.1087
9.3	86.4900	3.0496	9.6436	804.357	2.1029	4.5307	9.7610	0.1075
9.4	88.3600	3.0659	9.6954	830.584	2.1105	4.5468	9.7959	0.1064
9.5	90.2500	3.0822	9.7468	857.375	2.1179	4.5629	9.8305	0.1053
9.6	92.1600	3.0984	9.7980	884.736	2.1253	4.5789	9.8648	0.1042
9.7	94.0900	3.1145	9.8489	912.673	2.1327	4.5947	9.8990	0.1031
9.8	96.0400	3.1305	9.8995	941.192	2.1400	4.6014	9.9329	0.1020
9.9	98.0100	3.1464	9.9499	970.299	2.1472	4.6261	9.9666	0.1010
10.0	100.000	3.1623	10.000	1000.00	2.1544	4.6416	10.0000	0.1000

Table of e^x and e^{-x}

x	e^x	e^{-x}	x	e^x	e^{-x}
0.00	1.0000	1.00000	0.35	1.4191	.70469
0.01	1.0101	0.99005	0.36	1.4333	.69768
0.02	1.0202	.98020	0.37	1.4477	.69073
0.03	1.0305	.97045	0.38	1.4623	.68386
0.04	1.0408	.96079	0.39	1.4770	.67706
0.05	1.0513	.95123	0.40	1.4918	.67032
0.06	1.0618	.94176	0.41	1.5068	.66365
0.07	1.0725	.93239	0.42	1.5220	.65705
0.08	1.0833	.92312	0.43	1.5373	.65051
0.09	1.0942	.91393	0.44	1.5527	.64404
0.10	1.1052	.90484	0.45	1.5683	.63763
0.11	1.1163	.89583	0.46	1.5841	.63128
0.12	1.1275	.88692	0.47	1.6000	.62500
0.13	1.1388	.87809	0.48	1.6161	.61878
0.14	1.1503	.86936	0.49	1.6323	.61263
0.15	1.1618	.86071	0.50	1.6487	.60653
0.16	1.1735	.85214	0.51	1.6653	.60050
0.17	1.1853	.84366	0.52	1.6820	.59452
0.18	1.1972	.83527	0.53	1.6989	.58860
0.19	1.2092	.82696	0.54	1.7160	.58275
0.20	1.2214	.81873	0.55	1.7333	.57695
0.21	1.2337	.81058	0.56	1.7507	.57121
0.22	1.2461	.80252	0.57	1.7683	.56553
0.23	1.2586	.79453	0.58	1.7860	.55990
0.24	1.2712	.78663	0.59	1.8040	.55433
0.25	1.2840	.77880	0.60	1.8221	.54881
0.26	1.2969	.77105	0.61	1.8404	.54335
0.27	1.3100	.76338	0.62	1.8589	.53794
0.28	1.3231	.75578	0.63	1.8776	.53259
0.29	1.3364	.74826	0.64	1.8965	.52729
0.30	1.3499	.74082	0.65	1.9155	.52205
0.31	1.3634	.73345	0.66	1.9348	.51685
0.32	1.3771	.72615	0.67	1.9542	.51171
0.33	1.3910	.71892	0.68	1.9739	.50662
0.34	1.4049	.71177	0.69	1.9937	.50158

x	e^x	e^{-x}	x	e^x	e^{-x}
0.70	2.0138	.49659	2.50	12.182	.08208
0.71	2.0340	.49164	2.60	13.464	.07427
0.72	2.0544	.48675	2.70	14.880	.06721
0.73	2.0751	.48191	2.80	16.445	.06081
0.74	2.0959	.47711	2.90	18.174	.05502
0.75	2.1170	.47237	3.00	20.086	.04979
0.76	2.1383	.46767	3.10	22.198	.04505
0.77	2.1598	.46301	3.20	24.533	.04076
0.78	2.1815	.45841	3.30	27.113	.03688
0.79	2.2034	.45384	3.40	29.964	.03337
0.80	2.2255	.44933	3.50	33.115	.03020
0.81	2.2479	.44486	3.60	36.598	.02732
0.82	2.2705	.44043	3.70	40.447	.02472
0.83	2.2933	.43605	3.80	44.701	.02237
0.84	2.3164	.43171	3.90	49.402	.02024
0.85	2.3396	.42741	4.00	54.598	.01832
0.86	2.3632	.42316	4.10	60.340	.01657
0.87	2.3869	.41895	4.20	66686	.01500
0.88	2.4109	.41478	.430	73.700	.01357
0.89	2.4351	.41066	4.40	81.451	.01227
0.90	2.4596	.40657	4.50	90.107	.01111
0.91	2.4843	.40252	4.60	99.484	.01005
0.92	2.5093	.39852	4.70	109.95	.00910
0.93	2.5345	.39455	4.80	121.51	.00823
0.94	2.5600	.39063	4.90	134.29	.00745
0.95	2.5857	.38674	5.00	148.41	.00674
0.96	2.6117	.38289	5.10	164.02	.00610
0.97	2.6379	.37908	5.20	181.27	.00552
0.98	2.6645	.37531	5.30	200.34	.00499
0.99	2.6912	.37158	5.40	221.41	.00452
1.00	2.7183	.36788	5.50	244.69	.00409
1.10	3.0042	.33287	5.60	270.43	.00370
1.20	3.3201	.30119	5.70	298.87	.00335
1.30	3.6693	.27253	5.80	330.30	.00303
1.40	4.0552	.24660	5.90	365.04	.00274
1.50	4.4817	.22313	6.00	403.43	.00248
1.60	4.9530	.20190	6.25	518.01	.00193
1.70	5.4739	.18268	6.50	665.14	.00150
1.80	6.0496	.16530	6.75	854.06	.00117
1.90	6.6859	.14957	7.00	1096.6	.00091
2.00	7.3891	.13534	7.50	1808.0	.00055
2.10	8.1662	.12246	8.00	2981.0	.00034
2.20	9.0250	.11080	8.50	4914.8	.00020
2.30	9.9742	.10026	9.00	8103.1	.00012
2.40	11.023	.09072	9.50	13360.	.00007
			10.00	22026.	.00005

Table of Natural Logarithms

In the body of the table the first two digits (and decimal point) of most entries are carried over from a preceding entry in the first column. For example, ln 3.32 ≈ 1.19996. However, an asterisk (*) indicates that the first two digits are those of a following entry in the first column. For example, ln 3.33 ≈ 1.20297.

To extend this table for a number less than 1.0 or greater than 10.09, write the number in the form $x = y \cdot 10^n$ where $1.0 \leq y < 10$ and use the fact that $\ln x = \ln y + n \ln 10$. Some values of $n \ln 10$ are

$$1 \ln 10 \approx 2.30259, \qquad 6 \ln 10 \approx 13.81551,$$
$$2 \ln 10 \approx 4.60517, \qquad 7 \ln 10 \approx 16.11810,$$
$$3 \ln 10 \approx 6.90776, \qquad 8 \ln 10 \approx 18.42068,$$
$$4 \ln 10 \approx 9.21034, \qquad 9 \ln 10 \approx 20.72327,$$
$$5 \ln 10 \approx 11.51293, \qquad 10 \ln 10 \approx 23.02585.$$

For example,

$$\ln 332 = \ln[(3.32)(10^2)] = \ln 3.32 + 2 \ln 10$$
$$\approx 1.19996 + 4.60517 = 5.80513$$

and
$$\ln 0.0332 = \ln[(3.32)(10^{-2})] = \ln 3.32 - 2 \ln 10$$
$$\approx 1.19996 - 4.60517 = -3.40521.$$

Properties of logarithms may be used to find the logarithm of a number such as $\frac{3}{8}$:

$$\ln \tfrac{3}{8} = \ln 3 - \ln 8 \approx 1.09861 - 2.07944$$
$$= -0.98083.$$

N	0	1	2	3	4	5	6	7	8	9
1.0	0.0 0000	0995	1980	2956	3922	4879	5827	6766	7696	8618
1.1	9531	*0436	*1333	*2222	*3103	*3976	*4842	*5700	*6551	*7395
1.2	0.1 8232	9062	9885	*0701	*1511	*2314	*3111	*3902	*4686	*5464
1.3	0.2 6236	7003	7763	8518	9267	*0010	*0748	*1481	*2208	*2930
1.4	0.3 3647	4359	5066	5767	6464	7156	7844	8526	9204	9878

N	0	1	2	3	4	5	6	7	8	9
1.5	0.4 0547	1211	1871	2527	3178	3825	4469	5108	5742	6373
1.6	7000	7623	8243	8858	9470	*0078	*0672	*1282	*1879	*2473
1.7	0.5 3063	3649	4232	4812	5389	5962	6531	7098	7661	8222
1.8	8779	9333	9884	*0432	*0977	*1519	*2058	*2594	*3127	*3658
1.9	0.6 4185	4710	5233	5752	6269	6783	7294	7803	8310	8813
2.0	9315	9813	*0310	*0804	*1295	*1784	*2271	*2755	*3237	*3716
2.1	0.7 4194	4669	5142	5612	6081	6547	7011	7473	7932	8390
2.2	8846	9299	9751	*0200	*0648	*1093	*1536	*1978	*2418	*2855
2.3	0.8 3291	3725	4157	4587	5015	5442	5866	6289	6710	7129
2.4	7547	7963	8377	8789	9200	9609	*0016	*0422	*0826	*1228
2.5	0.9 1629	2028	2426	2822	3216	3609	4001	4391	4779	5166
2.6	5551	5935	6317	6698	7078	7456	7833	8208	8582	8954
2.7	9325	9695	*0063	*0430	*0796	*1160	*1523	*1885	*2245	*2604
2.8	1.0 2962	3318	3674	4028	4380	4732	5082	5431	5779	6126
2.9	6471	6815	7158	7500	7841	8181	8519	8856	9192	9527
3.0	9861	*0194	*0526	*0856	*1186	*1514	*1841	*2168	*2493	*2817
3.1	1.1 3140	3462	3783	4103	4422	4740	5057	5373	5688	6002
3.2	6315	6627	6938	7248	7557	7865	8173	8479	8784	9089
3.3	9392	9695	9996	*0297	*0597	*0896	*1194	*1491	*1788	*2083
3.4	1.2 2378	2671	2964	3256	3547	3837	4127	4415	4703	4990
3.5	5276	5562	5846	6130	6413	6695	6976	7257	7536	7815
3.6	8093	8371	8647	8923	9198	9473	9746	*0019	*0291	*0563
3.7	1.3 0833	1103	1372	1641	1909	2176	2442	2708	2972	3237
3.8	3500	3763	4025	4286	4547	4807	5067	5325	5584	5841
3.9	6098	6354	6609	6864	7118	7372	7624	7877	8128	8379
4.0	8629	8879	9128	9377	9624	9872	*0118	*0364	*0610	*0854
4.1	1.4 1099	1342	1585	1828	2070	2311	2552	2792	3031	3270
4.2	3508	3746	3984	4220	4456	4692	4927	5161	5395	5629
4.3	5862	6094	6326	6557	6787	7018	7247	7476	7705	7933
4.4	8160	8387	8614	8840	9065	9290	9515	9739	9962	*0185
4.5	1.5 0408	0630	0851	1072	1293	1513	1732	1951	2170	2388
4.6	2606	2823	3039	3256	3471	3687	3902	4116	4330	4543
4.7	4756	4969	5181	5393	5604	5814	6025	6235	6444	6653
4.8	6862	7070	7277	7485	7691	7898	8104	8309	8515	8719
4.9	8924	9127	9331	9534	9737	9939	*0141	*0342	*0543	*0744
5.0	1.6 0944	1144	1343	1542	1741	1939	2137	2334	2531	2728
5.1	2924	3120	3315	3511	3705	3900	4094	4287	4481	4673
5.2	4866	5058	5250	5441	5632	5823	6013	6203	6393	6582
5.3	6771	6959	7147	7335	7523	7710	7896	8083	8269	8455
5.4	8640	8825	9010	9194	9378	9562	9745	9928	*0111	*0293
5.5	1.7 0475	0656	0838	1019	1199	1380	1560	1740	1919	2098
5.6	2277	2455	2633	2811	2988	3166	3342	3519	3695	3871
5.7	4047	4222	4397	4572	4746	4920	5094	5267	5440	5613
N	0	1	2	3	4	5	6	7	8	9

N	0	1	2	3	4	5	6	7	8	9
5.8	5786	5958	6130	6302	6473	6644	6815	6985	7156	7326
5.9	7495	7665	7834	8002	8171	8339	8507	8675	8842	9009
6.0	1.7 9176	9342	9509	9675	9840	*0006	*0171	*0336	*0500	*0665
6.1	1.8 0829	0993	1156	1319	1482	1645	1808	1970	2132	2294
6.2	2455	2616	2777	2938	3098	3258	3418	3578	3737	3896
6.3	4055	4214	4372	4530	4688	4845	5003	5160	5317	5473
6.4	5630	5786	5942	6097	6253	6408	6563	6718	6872	7026
6.5	7180	7334	7487	7641	7794	7947	8099	8251	8403	8555
6.6	8707	8858	9010	9160	9311	9462	9612	9762	9912	*0061
6.7	1.9 0211	0360	0509	0658	0806	0954	1102	1250	1398	1545
6.8	1692	1839	1986	2132	2279	2425	2571	2716	2862	3007
6.9	3152	3297	3442	3586	3730	3874	4018	4162	4305	4448
7.0	4591	4734	4876	5019	5161	5303	5445	5586	5727	5869
7.1	6009	6150	6291	6431	6571	6711	6851	6991	7130	7269
7.2	7408	7547	7685	7824	7962	8100	8238	8376	8513	8650
7.3	8787	8924	9061	9198	9334	9470	9606	9742	9877	*0013
7.4	2.0 0148	0283	0418	0553	0687	0821	0956	1089	1223	1357
7.5	1490	1624	1757	1890	2022	2155	2287	2419	2551	2683
7.6	2815	2946	3078	3209	3340	3471	3601	3732	3862	3992
7.7	4122	4252	4381	4511	4640	4769	4898	5027	5156	5284
7.8	5412	5540	5668	5796	5924	6051	6179	6306	6433	6560
7.9	6686	6813	6939	7065	7191	7317	7443	7568	7694	7819
8.0	7944	8069	8194	8318	8443	8567	8691	8815	8939	9063
8.1	9186	9310	9433	9556	9679	9802	9924	*0047	*0169	*0291
8.2	2.1 0413	0535	0657	0779	0900	1021	1142	1263	1384	1505
8.3	1626	1746	1866	1986	2106	2226	2346	2465	2585	2704
8.4	2823	2942	3061	3180	3298	3417	3535	3653	3771	3889
8.5	4007	4124	4242	4359	4476	4593	4710	4827	4943	5060
8.6	5176	5292	5409	5524	5640	5756	5871	5987	6102	6217
8.7	6332	6447	6562	6677	6791	6905	7020	7134	7248	7361
8.8	7475	7589	7702	7816	7929	8042	8155	8267	8380	8493
8.9	8605	8717	8830	8942	9054	9165	9277	9389	9500	9611
9.0	9722	9834	9944	*0055	*0166	*0276	*0387	*0497	*0607	*0717
9.1	2.2 0827	0937	1047	1157	1266	1375	1485	1594	1703	1812
9.2	1920	2029	2138	2246	2354	2462	2570	2678	2786	2894
9.3	3001	3109	3216	3324	3431	3538	3645	3751	3858	3965
9.4	4071	4177	4284	4390	4496	4601	4707	4813	4918	5024
9.5	5129	5234	5339	5444	5549	5654	5759	5863	5968	6072
9.6	6176	6280	6384	6488	6592	6696	6799	6903	7006	7109
9.7	7213	7316	7419	7521	7624	7727	7829	7932	8034	8136
9.8	8238	8340	8442	8544	8646	8747	8849	8950	9051	9152
9.9	9253	9354	9455	9556	9657	9757	9858	9958	*0058	*0158
10.0	2.3 0259	0358	0458	0558	0658	0757	0857	0956	1055	1154
N	0	1	2	3	4	5	6	7	8	9

Table of Selected Integrals

Rational Forms Containing $(a + bu)$

1. $\displaystyle\int u^n\,du = \frac{u^{n+1}}{n+1} + C, \quad n \neq -1.$

2. $\displaystyle\int \frac{du}{a+bu} = \frac{1}{b}\ln|a+bu| + C.$

3. $\displaystyle\int \frac{u\,du}{a+bu} = \frac{u}{b} - \frac{a}{b^2}\ln|a+bu| + C$

4. $\displaystyle\int \frac{u^2\,du}{a+bu} = \frac{u^2}{2b} - \frac{au}{b^2} + \frac{a^2}{b^3}\ln|a+bu| + C.$

5. $\displaystyle\int \frac{du}{u(a+bu)} = \frac{1}{a}\ln\left|\frac{u}{a+bu}\right| + C.$

6. $\displaystyle\int \frac{du}{u^2(a+bu)} = -\frac{1}{au} + \frac{b}{a^2}\ln\left|\frac{a+bu}{u}\right| + C.$

7. $\displaystyle\int \frac{u\,du}{(a+bu)^2} = \frac{1}{b^2}\left(\ln|a+bu| + \frac{a}{a+bu}\right) + C.$

8. $\displaystyle\int \frac{u^2\,du}{(a+bu)^2} = \frac{u}{b^2} - \frac{a^2}{b^3(a+bu)} - \frac{2a}{b^3}\ln|a+bu| + C.$

9. $\displaystyle\int \frac{du}{u(a+bu)^2} = \frac{1}{a(a+bu)} + \frac{1}{a^2}\ln\left|\frac{u}{a+bu}\right| + C.$

10. $\displaystyle\int \frac{du}{u^2(a+bu)^2} = -\frac{a+2bu}{a^2u(a+bu)} + \frac{2b}{a^3}\ln\left|\frac{a+bu}{u}\right| + C.$

11. $\displaystyle\int \frac{du}{(a+bu)(c+ku)} = \frac{1}{bc-ak}\ln\left|\frac{a+bu}{c+ku}\right| + C.$

12. $\int \dfrac{u \, du}{(a + bu)(c + ku)} = \dfrac{1}{bc - ak} \left[\dfrac{c}{k} \ln |c + ku| - \dfrac{a}{b} \ln |a + bu| \right] + C.$

Forms Containing $\sqrt{a + bu}$

13. $\int u\sqrt{a + bu} \, du = \dfrac{2(3bu - 2a)(a + bu)^{3/2}}{15b^2} + C.$

14. $\int u^2\sqrt{a + bu} \, du = \dfrac{2(8a^2 - 12abu + 15b^2u^2)(a + bu)^{3/2}}{105b^3} + C.$

15. $\int \dfrac{u \, du}{\sqrt{a + bu}} = \dfrac{2(bu - 2a)\sqrt{a + bu}}{3b^2} + C.$

16. $\int \dfrac{u^2 \, du}{\sqrt{a + bu}} = \dfrac{2(3b^2u^2 - 4abu + 8a^2)\sqrt{a + bu}}{15b^3} + C.$

17. $\int \dfrac{du}{u\sqrt{a + bu}} = \dfrac{1}{\sqrt{a}} \ln \left| \dfrac{\sqrt{a + bu} - \sqrt{a}}{\sqrt{a + bu} + \sqrt{a}} \right| + C, \ a > 0.$

18. $\int \dfrac{\sqrt{a + bu} \, du}{u} = 2\sqrt{a + bu} + a \int \dfrac{du}{u\sqrt{a + bu}}.$

Forms Containing $\sqrt{a^2 - u^2}$

19. $\int \dfrac{du}{(a^2 - u^2)^{3/2}} = \dfrac{u}{a^2\sqrt{a^2 - u^2}} + C.$

20. $\int \dfrac{du}{u\sqrt{a^2 - u^2}} = -\dfrac{1}{a} \ln \left| \dfrac{a + \sqrt{a^2 - u^2}}{u} \right| + C.$

21. $\int \dfrac{du}{u^2\sqrt{a^2 - u^2}} = -\dfrac{\sqrt{a^2 - u^2}}{a^2u} + C.$

22. $\int \dfrac{\sqrt{a^2 - u^2} \, du}{u} = \sqrt{a^2 - u^2} - a \ln \left| \dfrac{a + \sqrt{a^2 - u^2}}{u} \right| + C, \ a > 0.$

Forms Containing $\sqrt{u^2 \pm a^2}$

23. $\int \sqrt{u^2 \pm a^2} \, du = \dfrac{1}{2}(u\sqrt{u^2 \pm a^2} \pm a^2 \ln |u + \sqrt{u^2 \pm a^2}|) + C.$

24. $\int u^2\sqrt{u^2 \pm a^2}\, du = \dfrac{u}{8}(2u^2 \pm a^2)\sqrt{u^2 \pm a^2} -$

$$\dfrac{a^4}{8}\ln\left|u + \sqrt{u^2 \pm a^2}\right| + C.$$

25. $\int \dfrac{\sqrt{u^2 + a^2}\, du}{u} = \sqrt{u^2 + a^2} - a\ln\left|\dfrac{a + \sqrt{u^2 + a^2}}{u}\right| + C.$

26. $\int \dfrac{\sqrt{u^2 \pm a^2}\, du}{u^2} = -\dfrac{\sqrt{u^2 \pm a^2}}{u} + \ln\left|u + \sqrt{u^2 \pm a^2}\right| + C.$

27. $\int \dfrac{du}{\sqrt{u^2 \pm a^2}} = \ln\left|u + \sqrt{u^2 \pm a^2}\right| + C.$

28. $\int \dfrac{du}{u\sqrt{u^2 + a^2}} = \dfrac{1}{a}\ln\left|\dfrac{\sqrt{u^2 + a^2} - a}{u}\right| + C.$

29. $\int \dfrac{u^2\, du}{\sqrt{u^2 \pm a^2}} = \dfrac{1}{2}\left(u\sqrt{u^2 \pm a^2} \mp a^2\ln\left|u + \sqrt{u^2 \pm a^2}\right|\right) + C.$

30. $\int \dfrac{du}{u^2\sqrt{u^2 \pm a^2}} = -\dfrac{\pm\sqrt{u^2 \pm a^2}}{a^2 u} + C.$

31. $\int (u^2 \pm a^2)^{3/2}\, du = \dfrac{u}{8}(2u^2 \pm 5a^2)\sqrt{u^2 \pm a^2} +$

$$\dfrac{3a^4}{8}\ln\left|u + \sqrt{u^2 \pm a^2}\right| + C.$$

32. $\int \dfrac{du}{(u^2 \pm a^2)^{3/2}} = \dfrac{\pm u}{a^2\sqrt{u^2 \pm a^2}} + C.$

33. $\int \dfrac{u^2\, du}{(u^2 \pm a^2)^{3/2}} = \dfrac{-u}{\sqrt{u^2 \pm a^2}} + \ln\left|u + \sqrt{u^2 \pm a^2}\right| + C.$

Rational Forms Containing $a^2 - u^2$ and $u^2 - a^2$

34. $\int \dfrac{du}{a^2 - u^2} = \dfrac{1}{2a}\ln\left|\dfrac{a + u}{a - u}\right| + C.$

35. $\int \dfrac{du}{u^2 - a^2} = \dfrac{1}{2a}\ln\left|\dfrac{u - a}{u + a}\right| + C.$

Exponential and Logarithmic Forms

36. $\int e^u\, du = e^u + C.$

37. $\int a^u\, du = \dfrac{a^u}{\ln a} + C, \quad a > 0, \quad a \neq 1.$

38. $\int ue^{au} du = \dfrac{e^{au}}{a^2}(au - 1) + C.$

39. $\int u^n e^{au} du = \dfrac{u^n e^{au}}{a} - \dfrac{n}{a} \int u^{n-1} e^{au} du.$

40. $\int \dfrac{e^{au} du}{u^n} = -\dfrac{e^{au}}{(n-1)u^{n-1}} + \dfrac{a}{n-1} \int \dfrac{e^{au} du}{u^{n-1}}.$

41. $\int \ln u \, du = u \ln u - u + C.$

42. $\int u^n \ln u \, du = \dfrac{u^{n+1} \ln u}{n+1} - \dfrac{u^{n+1}}{(n+1)^2} + C, \quad n \neq -1.$

43. $\int u^n \ln^m u \, du = \dfrac{u^{n+1}}{n+1} \ln^m u - \dfrac{m}{n+1} \int u^n \ln^{m-1} u \, du, \quad m, n \neq -1.$

44. $\int \dfrac{du}{u \ln u} = \ln |\ln u| + C.$

45. $\int \dfrac{du}{a + be^{cu}} = \dfrac{1}{ac}(cu - \ln |a + be^{cu}|) + C.$

Miscellaneous Forms

46. $\int \sqrt{\dfrac{a+u}{b+u}} \, du = \sqrt{(a+u)(b+u)} +$

$$(a-b) \ln (\sqrt{a+u} + \sqrt{b+u}) + C.$$

47. $\int \dfrac{du}{\sqrt{(a+u)(b+u)}} = \ln \left| \dfrac{a+b}{2} + u + \sqrt{(a+u)(b+u)} \right| + C.$

48. $\int \sqrt{a + bu + cu^2} \, du = \dfrac{2cu + b}{4c} \sqrt{a + bu + cu^2} -$

$$\dfrac{b^2 - 4ac}{8c^{3/2}} \ln |2cu + b + 2\sqrt{c} \sqrt{a + bu + cu^2}| + C, \quad c > 0.$$

Table of Trigonometric Functions (Radians)

t	sin	cos	tan	cot	t	sin	cos	tan	cot
0.00	0.0000	1.0000	0.0000	—	0.25	0.2474	0.9689	0.2553	3.916
0.01	0.0100	1.0000	0.0100	99.997	0.26	0.2571	0.9664	0.2660	3.759
0.02	0.0200	0.9998	0.0200	49.993	0.27	0.2667	0.9638	0.2768	3.613
0.03	0.0300	0.9996	0.0300	33.323	0.28	0.2764	0.9611	0.2876	3.478
0.04	0.0400	0.9992	0.0400	24.987	0.29	0.2860	0.9582	0.2984	3.351
0.05	0.0500	0.9988	0.0500	19.983	0.30	0.2955	0.9553	0.3093	3.233
0.06	0.0600	0.9982	0.0601	16.647	0.31	0.3051	0.9523	0.3203	3.122
0.07	0.0699	0.9976	0.0701	14.262	0.32	0.3146	0.9492	0.3314	3.018
0.08	0.0799	0.9968	0.0802	12.473	0.33	0.3240	0.9460	0.3425	2.920
0.09	0.0899	0.9960	0.0902	11.081	0.34	0.3335	0.9428	0.3537	2.827
0.10	0.0998	0.9950	0.1003	9.967	0.35	0.3429	0.9394	0.3650	2.740
0.11	0.1098	0.9940	0.1104	9.054	0.36	0.3523	0.9359	0.3764	2.657
0.12	0.1197	0.9928	0.1206	8.293	0.37	0.3616	0.9323	0.3879	2.578
0.13	0.1296	0.9916	0.1307	7.649	0.38	0.3709	0.9287	0.3994	2.504
0.14	0.1395	0.9902	0.1409	7.096	0.39	0.3802	0.9249	0.4111	2.433
0.15	0.1494	0.9888	0.1511	6.617	0.40	0.3894	0.9211	0.4228	2.365
0.16	0.1593	0.9872	0.1614	6.197	0.41	0.3986	0.9171	0.4346	2.301
0.17	0.1692	0.9856	0.1717	5.826	0.42	0.4078	0.9131	0.4466	2.239
0.18	0.1790	0.9838	0.1820	5.495	0.43	0.4169	0.9090	0.4586	2.180
0.19	0.1889	0.9820	0.1923	5.200	0.44	0.4259	0.9048	0.4708	2.124
0.20	0.1987	0.9801	0.2027	4.933	0.45	0.4350	0.9004	0.4831	2.070
0.21	0.2085	0.9780	0.2131	4.692	0.46	0.4439	0.8961	0.4954	2.018
0.22	0.2182	0.9759	0.2236	4.472	0.47	0.4529	0.8916	0.5080	1.969
0.23	0.2280	0.9737	0.2341	4.271	0.48	0.4618	0.8870	0.5206	1.921
0.24	0.2377	0.9713	0.2447	4.086	0.49	0.4706	0.8823	0.5334	1.875

t	sin	cos	tan	cot	t	sin	cos	tan	cot
0.50	0.4794	0.8776	0.5463	1.830	0.90	0.7833	0.6216	1.260	0.7936
0.51	0.4882	0.8727	0.5594	1.788	0.91	0.7895	0.6137	1.286	0.7774
0.52	0.4969	0.8678	0.5726	1.747	0.92	0.7956	0.6058	1.313	0.7615
0.53	0.5055	0.8628	0.5859	1.707	0.93	0.8016	0.5978	1.341	0.7458
0.54	0.5141	0.8577	0.5994	1.668	0.94	0.8076	0.5898	1.369	0.7303
0.55	0.5227	0.8525	0.6131	1.631	0.95	0.8134	0.5817	1.398	0.7151
0.56	0.5312	0.8473	0.6269	1.595	0.96	0.8192	0.5735	1.428	0.7001
0.57	0.5396	0.8419	0.6410	1.560	0.97	0.8249	0.5653	1.459	0.6853
0.58	0.5480	0.8365	0.6552	1.526	0.98	0.8305	0.5570	1.491	0.6707
0.59	0.5564	0.8309	0.6696	1.494	0.99	0.8360	0.5487	1.524	0.6563
0.60	0.5646	0.8253	0.6841	1.462	1.00	0.8415	0.5403	1.557	0.6421
0.61	0.5729	0.8196	0.6989	1.431	1.01	0.8468	0.5319	1.592	0.6281
0.62	0.5810	0.8139	0.7139	1.401	1.02	0.8521	0.5234	1.628	0.6142
0.63	0.5891	0.8080	0.7291	1.372	1.03	0.8573	0.5148	1.665	0.6005
0.64	0.5972	0.8021	0.7445	1.343	1.04	0.8624	0.5062	1.704	0.5870
0.65	0.6052	0.7961	0.7602	1.315	1.05	0.8674	0.4976	1.743	0.5736
0.66	0.6131	0.7900	0.7761	1.288	1.06	0.8724	0.4889	1.784	0.5604
0.67	0.6210	0.7838	0.7923	1.262	1.07	0.8772	0.4801	1.827	0.5473
0.68	0.6288	0.7776	0.8087	1.237	1.08	0.8820	0.4713	1.871	0.5344
0.69	0.6365	0.7712	0.8253	1.212	1.09	0.8866	0.4625	1.917	0.5216
0.70	0.6442	0.7648	0.8423	1.187	1.10	0.8912	0.4536	1.965	0.5090
0.71	0.6518	0.7584	0.8595	1.163	1.11	0.8957	0.4447	2.014	0.4964
0.72	0.6594	0.7518	0.8771	1.140	1.12	0.9001	0.4357	2.066	0.4840
0.73	0.6669	0.7452	0.8949	1.117	1.13	0.9044	0.4267	2.120	0.4718
0.74	0.6743	0.7385	0.9131	1.095	1.14	0.9086	0.4176	2.176	0.4596
0.75	0.6816	0.7317	0.9316	1.073	1.15	0.9128	0.4085	2.234	0.4475
0.76	0.6889	0.7248	0.9505	1.052	1.16	0.9168	0.3993	2.296	0.4356
0.77	0.6961	0.7179	0.9697	1.031	1.17	0.9208	0.3902	2.360	0.4237
0.78	0.7033	0.7109	0.9893	1.011	1.18	0.9246	0.3809	2.427	0.4120
0.79	0.7104	0.7038	1.009	0.9908	1.19	0.9284	0.3717	2.498	0.4003
0.80	0.7174	0.6967	1.030	0.9712	1.20	0.9320	0.3624	2.572	0.3888
0.81	0.7243	0.6895	1.050	0.9520	1.21	0.9356	0.3530	2.650	0.3773
0.82	0.7311	0.6822	1.072	0.9331	1.22	0.9391	0.3436	2.733	0.3659
0.83	0.7379	0.6749	1.093	0.9146	1.23	0.9425	0.3342	2.820	0.3546
0.84	0.7446	0.6675	1.116	0.8964	1.24	0.9458	0.3248	2.912	0.3434
0.85	0.7513	0.6600	1.138	0.8785	1.25	0.9490	0.3153	3.010	0.3323
0.86	0.7578	0.6524	1.162	0.8609	1.26	0.9521	0.3058	3.113	0.3212
0.87	0.7643	0.6448	1.185	0.8437	1.27	0.9551	0.2963	3.224	0.3102
0.88	0.7707	0.6372	1.210	0.8267	1.28	0.9580	0.2867	3.341	0.2993
0.89	0.7771	0.6294	1.235	0.8100	1.29	0.9608	0.2771	3.467	0.2884

t	sin	cos	tan	cot	t	sin	cos	tan	cot
1.30	0.9636	0.2675	3.602	0.2776	1.45	0.9927	0.1205	8.238	0.1214
1.31	0.9662	0.2579	3.747	0.2669	1.46	0.9939	0.1106	8.989	0.1113
1.32	0.9687	0.2482	3.903	0.2562	1.47	0.9949	0.1006	9.887	0.1011
1.33	0.9711	0.2385	4.072	0.2456	1.48	0.9959	0.0907	10.983	0.0910
1.34	0.9735	0.2288	4.256	0.2350	1.49	0.9967	0.0807	12.350	0.0810
1.35	0.9757	0.2190	4.455	0.2245	1.50	0.9975	0.0707	14.101	0.0709
1.36	0.9779	0.2092	4.673	0.2140	1.51	0.9982	0.0608	16.428	0.0609
1.37	0.9799	0.1994	4.913	0.2035	1.52	0.9987	0.0508	19.670	0.0508
1.38	0.9819	0.1896	5.177	0.1931	1.53	0.9992	0.0408	24.498	0.0408
1.39	0.9837	0.1798	5.471	0.1828	1.54	0.9995	0.0308	32.461	0.0308
1.40	0.9854	0.1700	5.798	0.1725	1.55	0.9998	0.0208	48.078	0.0208
1.41	0.9871	0.1601	6.165	0.1622	1.56	0.9999	0.0108	92.620	0.0108
1.42	0.9887	0.1502	6.581	0.1519	1.57	1.0000	0.0008	1,255.8	0.0008
1.43	0.9901	0.1403	7.055	0.1417	1.58	1.0000	-0.0092	-108.65	-0.0092
1.44	0.9915	0.1304	7.602	0.1315	1.59	0.9998	-0.0192	-52.067	-0.0192
					1.60	0.9996	-0.0292	-34.233	-0.0292

Answers to Odd-Numbered Problems

Exercise 1.1

1. All real numbers except 0.   **3.** All real numbers ≥ 5.   **5.** All real numbers.

7. All real numbers except $-\frac{5}{2}$.   **9.** All real numbers except 0 and 1.

11. All real numbers except 4 and $-\frac{1}{2}$.   **13.** $0, 15, -20$.   **15.** $-62, 2 - u^2, 2 - u^4$.

17. $2, (2v)^2 + 2v = 4v^2 + 2v, (-x^2)^2 + (-x^2) = x^4 - x^2$.

19. $4, 0, (x + h)^2 + 2(x + h) + 1 = x^2 + 2xh + h^2 + 2x + 2h + 1$.

21. $0, \dfrac{3x - 5}{(3x)^2 + 4} = \dfrac{3x - 5}{9x^2 + 4}, \dfrac{(x + h) - 5}{(x + h)^2 + 4} = \dfrac{x + h - 5}{x^2 + 2xh + h^2 + 4}$.   **23.** $0, 256, \frac{1}{16}$.

25. (a) $3x + 3h - 4$, (b) 3.   **27.** (a) $x^2 + 2hx + h^2 + 2x + 2h$, (b) $2x + h + 2$.

29. y is a function of x; x is a function of y.   **31.** y is a function of x; x is not a

function of y.   **33.** Yes.   **35.** $V = f(t) = 10{,}000 + 400t$.   **37.** Yes; P; q.

39. (a) 4, (b) $8\sqrt[3]{2}$, (c) $2\sqrt[3]{2} f(I_0)$; doubling the intensity increases the response by a factor

of $2\sqrt[3]{2}$.   **41.** (a) 3000, 2900, 2300, 2000; 12, 10; (b) 10, 12, 17, 20; 3000, 2300.

Exercise 1.2

1. All real numbers.   **3.** All real numbers.   **5.** (a) 3, (b) 2.   **7.** (a) 4, (b) -3.

9. 12, 12, 12.   **11.** $1, -1, 0, -1$.   **13.** 8, 3, 3, 1.   **15.** (a) $C = 850 + 3q$,

(b) 250.   **17.** 9/64.   **19.** (a) All T such that $30 \leq T \leq 39$, (b) 4, $\frac{17}{4}$, $\frac{33}{4}$.

Exercise 1.3

1. (a) $2x + 5$, (b) 5, (c) -3, (d) $x^2 + 5x + 4$, (e) -2, (f) $\dfrac{x + 1}{x + 4}$, (g) $x + 5$, (h) 8,

(i) $x + 5$.   **3.** (a) $2x^2 + x$, (b) $-x$, (c) $\frac{1}{2}$, (d) $x^4 + x^3$, (e) $\dfrac{x^2}{x^2 + x} = \dfrac{x}{x + 1}$, (f) -1,

(g) $(x^2 + x)^2 = x^4 + 2x^3 + x^2$, (h) $x^4 + x^2$, (i) 90.   **5.** 53; -32.

7. $\dfrac{4}{(t-1)^2} + \dfrac{6}{t-1} + 1,\ \dfrac{2}{t^2+3t}$.   **9.** $\dfrac{1}{v+3},\ \sqrt{\dfrac{2w^2+3}{w^2+1}}$.   **11.** $f(x) = x^4,\ g(x) = 3x + 1$.

13. $f(x) = \dfrac{1}{x},\ g(x) = x^2 - 2$.   **15.** $f(x) = \sqrt[5]{x},\ g(x) = \dfrac{x+1}{3}$.

17. $400m - 10m^2$; the total revenue received when the total output of m employees is sold.

Exercise 1.4

1. (a) 1, 2, 3, 0; (b) all real numbers; (c) all real numbers.

3. (a) 0, -1, -1; (b) all real numbers; (c) all nonpositive reals.

5. All real numbers; all real numbers.   **7.** All real numbers; all reals ≤ 4.

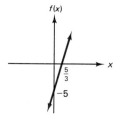   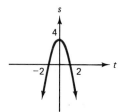

9. All real numbers; 2.   **11.** All real numbers; all reals ≥ -3.

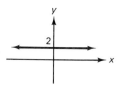   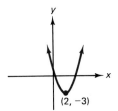

13. All real numbers; all real numbers.   **15.** All real numbers ≥ 5; all nonnegative reals.

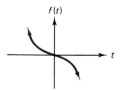   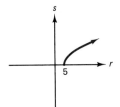

17. All real numbers; all nonnegative reals.   **19.** All nonzero real numbers; all positive real numbers.

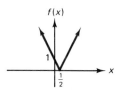

21. All nonnegative reals; all reals x where $0 \le x \le 2$.

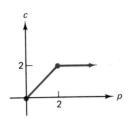

23. All real numbers; all nonnegative reals

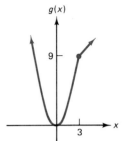

25. a, b, d.

27.

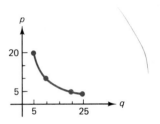

As price decreases, quantity increases; p is a function of q.

29. Domain: all T such that $20 \le T \le 90$; range: all y such that $70 \le y \le 100$.

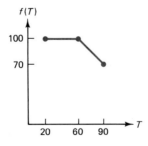

Exercise 1.5

1. 2. **3.** $-\frac{8}{13}$. **5.** Undefined. **7.** 0. **9.** $6x - y - 4 = 0$.

11. $x + 4y - 18 = 0$. **13.** $3x - 7y + 25 = 0$. **15.** $8x - 5y - 29 = 0$.

17. $2x - y + 4 = 0$. **19.** $x + 2y + 6 = 0$. **21.** $y + 2 = 0$. **23.** $x - 2 = 0$.

25. $2; -1$. **27.** $-\frac{1}{2}; \frac{3}{2}$. **29.** Slope undefined; no y-intercept. **31.** $3; 0$.

33. $0; 1$. **35.** $x + 2y - 4 = 0; y = -\frac{1}{2}x + 2$. **37.** $4x + 9y - 5 = 0; y = -\frac{4}{9}x + \frac{5}{9}$. **39.** $3x - 2y + 24 = 0; y = \frac{3}{2}x + 12$. **41.** $(5, -4)$.

Exercise 1.6

1. $-4; 0.$

3. $2; -4.$

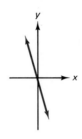

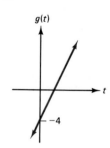

5. $-\frac{1}{2}; \frac{7}{2}.$

7. $f(x) = 5x - 14.$

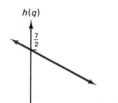

9. $f(x) = -3x + 9.$ **11.** $f(x) = -\frac{1}{2}x + \frac{15}{4}.$ **13.** $f(x) = x - 1.$

15. $p = -\frac{2}{5}q + 28; \$16.$ **17.** $c = 3q + 10; \$115.$

19. $v = -800t + 8000; \text{slope} = -800.$ **21.**   **(a)** $y = \frac{5}{11}x + \frac{600}{11}$, **(b)** 12.

23.   **(a)** $r = -\frac{35}{2}l + 195,$ **(b)** 177.5.   **25.**   **(a)** $p = 0.059t + 0.025,$ **(b)** 0.556.

27.   **(a)** $t = \frac{1}{4}c + 37,$ **(b)** add 37 to the number of chirps in 15 seconds.

Exercise 1.7

1. Not quadratic.   **3.** Quadratic.   **5.** Not quadratic.   **7.** Quadratic.

9. **(a)** (1, 11), **(b)** Highest.   **11.**   **(a)** (0, −8), **(b)** (−4, 0), (2, 0), **(c)** (−1, −9).

13. Intercepts: $(1, 0)$, $(5, 0)$, $(0, 5)$; vertex: $(3, -4)$.

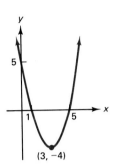

15. Intercepts: $(0, 0)$, $(-3, 0)$; vertex: $(-\frac{3}{2}, \frac{9}{2})$.

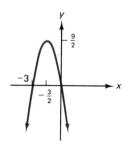

17. Intercepts: $(-1, 0)$, $(0, 1)$; vertex: $(-1, 0)$.

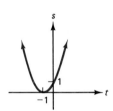

19. Intercept: $(0, -9)$; vertex: $(2, -1)$.

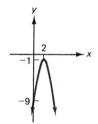

21. Intercepts: $(4 + \sqrt{3}, 0)$, $(4 - \sqrt{3}, 0)$, $(0, 13)$; vertex: $(4, -3)$.

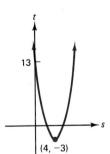

23. Minimum; 24.

25. Maximum; -10.   **27.** $q = 200$; $r = \$120,000$.   **29.** 70 grams.

Exercise 1.8

1. 16.   **3.** 7.   **5.** 20.   **7.** -1.   **9.** $-\frac{5}{2}$.   **11.** 0.   **13.** 5.

15. -2.   **17.** 3.   **19.** 0.   **21.** $\frac{1}{6}$.   **23.** $-\frac{1}{5}$.   **25.** $\frac{11}{9}$.   **27.** 4.

29. $2x$.   **33.** (a) 1, (b) 0.

Exercise 1.9

1. ∞.   **3.** ∞.   **5.** $\frac{1}{4}$.   **7.** 0.   **9.** ∞.   **11.** 0.   **13.** 1.   **15.** 0.

17. $-∞$.   **19.** $-\frac{2}{5}$.   **21.** $\frac{2}{5}$.   **23.** $\frac{1}{2}$.   **25.** 0.   **27.** ∞.   **29.** 20.

31. 6.

Exercise 1.10

1. Continuous at -2 and 0.   **3.** Discontinuous at ± 3.   **5.** f is a polynomial function.

7. f is a rational function and the denominator is never zero.   **9.** None.

11. $x = 4$.   **13.** None.   **15.** $x = -5, 3$.   **17.** $x = 0, \pm 1$.   **19.** None.

21. Discontinuities at $t = 1, 2, 3, 4$.

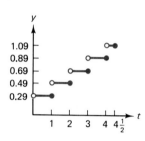

Review Problems—Chapter 1

1. All real numbers except 1 and 2.   **3.** All real numbers.   **5.** All nonnegative reals except 1.   **7.** 7, 46, 62, $3t^2 - 4t + 7$.   **9.** 0, 3, $\sqrt{t}$, $\sqrt{x^2 - 1}$.

11. $\frac{3}{5}$, 0, $\dfrac{\sqrt{x + 4}}{x}$, $\dfrac{\sqrt{u}}{u - 4}$.   **13.** $-8, 4, 4, -92$.   **15.** (a) $3 - 7x - 7h$, (b) -7.

17. (a) $5x + 2$, (b) 22, (c) $x - 4$, (d) $6x^2 + 7x - 3$, (e) 10, (f) $\dfrac{3x - 1}{2x + 3}$,
(g) $3(2x + 3) - 1 = 6x + 8$, (h) 38, (i) $2(3x - 1) + 3 = 6x + 1$.

19. $\dfrac{1}{x - 1}, \dfrac{1}{x} - 1 = \dfrac{1 - x}{x}$.   **21.** $x^3 + 2, (x + 2)^3$.

23. All real numbers; all real numbers; -2; (0, 4).   **25.** All real numbers; all reals ≤ 9; (3, 0), $(-3, 0)$, (0, 9); (0, 9).

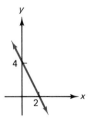

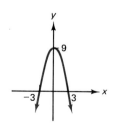

27. All real numbers; all reals ≥ -9;
(5, 0), (−1, 0), (0, −5); (2, −9).

29. All real numbers; all reals ≥ 1.

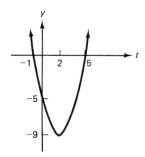

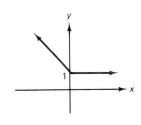

31. All $t \neq 4$; all nonzero real numbers.

33. All real numbers; all reals ≤ -2;
(0, −3); (−1, −2).

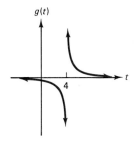

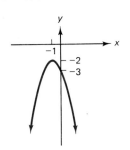

35. All real numbers; all real numbers; 3;
(0, 0).

37. $165,000, yes.

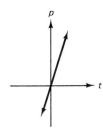

39. $y = -x + 1, x + y - 1 = 0$.   **41.** $y = \frac{1}{2}x - 1, x - 2y - 2 = 0$.

43. $y = 4, y - 4 = 0$.   **45.** $f(x) = -\frac{4}{3}x + \frac{19}{3}$.   **47.** $p = -q + 55$.

49. **(a)** $R = 75L + 1310$, **(b)** 1385 milliseconds, **(c)** the slope is 75. The time necessary to travel from one level to the next level is 75 milliseconds.

51. −5.   **53.** 2.   **55.** x.   **57.** 2.   **59.** −4.   **61.** 0.   **63.** $\frac{3}{5}$.

65. ∞.   **67.** ∞.   **69.** 100.   **71.** Yes; f is a polynomial function.   **73.** $x = -3$.

75. None.   **77.** $x = -4, 1$.

Exercise 2.1

1. 1.   **3.** 2.   **5.** -2.   **7.** 0.   **9.** $2x + 4$.   **11.** $4q + 5$.

13. $-1/x^2$.   **15.** -4.   **17.** 0.   **19.** $y = x + 4$.   **21.** $y = -3x - 7$.

23. $y = -\frac{1}{3}x + \frac{5}{3}$.   **25.** $\dfrac{r}{r_L - r - \dfrac{dC}{dD}}$.

Exercise 2.2

1. 0.   **3.** $8x^7$.   **5.** $45x^4$.   **7.** $-7w^{-8}$.   **9.** 3.   **11.** $\frac{13}{5}$.   **13.** $6x - 5$.

15. $-9q^2 + 9q + 9 = 9(1 + q - q^2)$.   **17.** $1002x^{500} + 0.68x^{2.4}$.   **19.** $-8x^3$.

21. $-\frac{4}{3}x^3$.   **23.** $-4x^{-5} - 3x^{-2/3} - 2x^{-7/5}$.   **25.** $-2(27 - 70x^4) = 2(70x^4 - 27)$.

27. $-4x + \frac{3}{2} + x^3$.   **29.** $-x^{-2} = -1/x^2$.   **31.** $-\frac{5}{4}s^{-6}$.   **33.** $2t^{-1/2} = 2/\sqrt{t}$.

35. $-\frac{3}{5}x^{-8/5}$.   **37.** $9x^2 - 14x + 7$.   **39.** $t + 4t^{-3}$.   **41.** $45x^4$.

43. $\frac{1}{3}x^{-2/3} - \frac{10}{3}x^{-5/3} = \frac{1}{3}x^{-5/3}(x - 10)$.   **45.** $8q + \dfrac{4}{q^2}$.   **47.** $2(x + 2)$.   **49.** 1.

51. $4, 16, -14$.   **53.** $0, 0, 0$.   **55.** $y = 13x - 2$.   **57.** $y = x + 3$.

59. $(0, 0), (2, -\frac{4}{3})$.

Exercise 2.3

1. (a) 4 m, (b) 5 m/s.   **3.** (a) 8 m, (b) 6 m/s.   **5.** (a) 2 m, (b) 9 m/s.

7. $dy/dx = 10x^{3/2}$, 270.   **9.** 0.27.   **11.** $dc/dq = 10, 10$.

13. $dc/dq = 0.6q + 2$, 3.8.   **15.** $dc/dq = 2q + 50$; 80, 82, 84.

17. $dc/dq = 0.02q + 5$; 6, 7.   **19.** $dc/dq = 0.00006q^2 - 0.02q + 6$; 4.6, 11.

21. $dr/dq = 0.7$; 0.7, 0.7, 0.7.   **23.** $dr/dq = 250 + 90q - 3q^2$; 625, 850, 625.

25. $dc/dq = 6.750 - 0.000656q$, 3.47.   **27.** $dP/dR = -4,650,000R^{-1.93}$.

29. (a) -7.5, (b) 4.5.   **31.** (a) 1, (b) $\dfrac{1}{x + 4}$; (c) 1, (d) $\frac{1}{9} \approx 0.111$, (e) 11.1%.

33. (a) $6x$, (b) $\dfrac{2x}{x^2 + 2}$, (c) 12, (d) $\frac{2}{3} \approx 0.667$, (e) 66.7%.

35. (a) $-3x^2$, (b) $-\dfrac{3x^2}{8 - x^3}$, (c) -3, (d) $-\frac{3}{7} \approx -0.429$, (e) -42.9%.   **37.** 3.2, 21.3%.

39. (a) $dr/dq = 30 - 0.6q$, (b) $\frac{4}{45} \approx 0.089$, (c) 9%.   **41.** $\dfrac{0.432}{t}$.

Exercise 2.4

1. $(4x + 1)(6) + (6x + 3)(4) = 48x + 18 = 6(8x + 3)$.

3. $(8 - 7t)(2t) + (t^2 - 2)(-7) = 14 + 16t - 21t^2$.

5. $(3r^2 - 4)(2r - 5) + (r^2 - 5r + 1)(6r) = 12r^3 - 45r^2 - 2r + 20$.

7. $(x^2 + 3x - 2)(4x - 1) + (2x^2 - x - 3)(2x + 3) = 8x^3 + 15x^2 - 20x - 7$.

9. $(8w^2 + 2w - 3)(15w^2) + (5w^3 + 2)(16w + 2) = 200w^4 + 40w^3 - 45w^2 + 32w + 4$.

11. $(x^2 - 1)(9x^2 - 6) + (3x^3 - 6x + 5)(2x) - 4(8x + 2) = 15x^4 - 27x^2 - 22x - 2$.

13. $\frac{3}{2}[(p^{1/2} - 4)(4) + (4p - 5)(\frac{1}{2}p^{-1/2})] = \frac{3}{4}(12p^{1/2} - 5p^{-1/2} - 32)$.   **15.** 0.

17. $\dfrac{(x - 1)(1) - (x + 2)(1)}{(x - 1)^2} = \dfrac{-3}{(x - 1)^2}$.

19. $\dfrac{(z^2 - 4)(-2) - (5 - 2z)(2z)}{(z^2 - 4)^2} = \dfrac{2(z - 4)(z - 1)}{(z^2 - 4)^2}$.

21. $\dfrac{(x^2 - 5x)(16x - 2) - (8x^2 - 2x + 1)(2x - 5)}{(x^2 - 5x)^2} = \dfrac{-38x^2 - 2x + 5}{(x^2 - 5x)^2}$.   **23.** $\dfrac{-100x^{99}}{(x^{100} + 1)^2}$.

25. $\dfrac{4(v^5 + 2)}{v^2}$.   **27.** $\dfrac{15x^2 - 2x + 1}{3x^{4/3}}$.   **29.** $\dfrac{4}{(x - 8)^2} + \dfrac{2}{(3x + 1)^2}$.

31. $\dfrac{(s - 5)(2s - 2) - [(s + 2)(s - 4)](1)}{(s - 5)^2} = \dfrac{s^2 - 10s + 18}{(s - 5)^2}$.   **33.** -6.

35. $y = -\frac{3}{2}x + \frac{15}{2}$.   **37.** 1.5.   **39.** $\dfrac{dr}{dq} = 25 - 0.04q$.   **41.** $\dfrac{dr}{dq} = \dfrac{216}{(q + 2)^2} - 3$.

43. $\dfrac{dC}{dI} = 0.672$.   **45.** $\frac{1}{3}, \frac{2}{3}$.   **47.** $\dfrac{dc}{dq} = \dfrac{5q(q + 6)}{(q + 3)^2}$.   **49.** $\dfrac{0.7355}{(1 + 0.02744x)^2}$.

Exercise 2.5

1. $(2u - 2)(2x - 1) = 4x^3 - 6x^2 - 2x + 2$.   **3.** $\left(-\dfrac{2}{w^3}\right)(-1) = \dfrac{2}{(2 - x)^3}$.

5. $18(3x + 2)^5$.   **7.** $300(3x^2 - 16x + 1)(x^3 - 8x^2 + x)^{99}$.   **9.** $-6x(x^2 - 2)^{-4}$.

11. $\frac{1}{2}(10x - 1)(5x^2 - x)^{-1/2}$.   **13.** $\frac{12}{5}x^2(x^3 + 1)^{-3/5}$.   **15.** $-6(4x - 1)(2x^2 - x + 1)^{-2}$.

17. $-2(2x - 3)(x^2 - 3x)^{-3}$.   **19.** $-8(8x - 1)^{-3/2}$.   **21.** $\frac{7}{3}(7x)^{-2/3} + \sqrt[3]{7}$.

23. $(x^2)[5(x - 4)^4(1)] + (x - 4)^5(2x) = x(x - 4)^4(7x - 8)$.

25. $(2x)[\frac{1}{2}(6x - 1)^{-1/2}(6)] + (\sqrt{6x - 1})(2) = 6x(6x - 1)^{-1/2} + 2\sqrt{6x - 1}$.

27. $(8x - 1)^3[4(2x + 1)^3(2)] + (2x + 1)^4[3(8x - 1)^2(8)] = 16(8x - 1)^2(2x + 1)^3(7x + 1)$.

29. $10\left(\dfrac{x - 7}{x + 4}\right)^9\left[\dfrac{(x + 4)(1) - (x - 7)(1)}{(x + 4)^2}\right] = \dfrac{110(x - 7)^9}{(x + 4)^{11}}$.

31. $\dfrac{1}{2}\left(\dfrac{x - 2}{x + 3}\right)^{-1/2}\left[\dfrac{(x + 3)(1) - (x - 2)(1)}{(x + 3)^2}\right] = \dfrac{5}{2(x + 3)^2}\left(\dfrac{x - 2}{x + 3}\right)^{-1/2}$.

33. $\dfrac{(x^2 + 4)^3(2) - (2x - 5)[3(x^2 + 4)^2(2x)]}{(x^2 + 4)^6} = \dfrac{-2(5x^2 - 15x - 4)}{(x^2 + 4)^4}$.

35. $\dfrac{(x^2 - 7)^4[(2x + 1)(2)(3x - 5)(3) + (3x - 5)^2(2)] - (2x + 1)(3x - 5)^2[4(x^2 - 7)^3(2x)]}{(x^2 - 7)^8}$.

37. 20.   **39.** 13.99.

41. (a) $-\dfrac{q}{\sqrt{q^2 + 20}}$, (b) $-\dfrac{q}{100\sqrt{q^2 + 20} - q^2 - 20}$, (c) $100 - \dfrac{q^2}{\sqrt{q^2 + 20}} - \sqrt{q^2 + 20}$.

43. -325.   **45.** $\dfrac{dc}{dq} = \dfrac{5q(q^2 + 6)}{(q^2 + 3)^{3/2}}$.   **47.** $48\pi(10^{-19})$.   **49.** (a) $-\dfrac{1000}{\sqrt{100 - x}}$, -125;

(b) $-\dfrac{1}{128}$.

Exercise 2.6

1. $-\dfrac{x}{4y}$.   **3.** $\dfrac{5}{12y^3}$.   **5.** $-\dfrac{\sqrt{y}}{\sqrt{x}}$.   **7.** $-\dfrac{y^{1/4}}{x^{1/4}}$.   **9.** $-\dfrac{y}{x}$.   **11.** $\dfrac{4 - y}{x - 1}$.

13. $\dfrac{4y - x^2}{y^2 - 4x}$. **15.** $\dfrac{6y^{2/3}}{3y^{1/6} + 2}$. **17.** $\dfrac{1 - 6xy^3}{1 + 9x^2y^2}$. **19.** $-\frac{3}{5}$. **21.** $y = -\frac{3}{4}x + \frac{5}{4}$.

23. $\dfrac{dq}{dp} = -\dfrac{1}{2q}$. **25.** $\dfrac{dq}{dp} = -\dfrac{(q + 5)^3}{40}$.

Exercise 2.7

1. 12. **3.** 0. **5.** 18. **7.** $12t^2 - t$. **9.** $\frac{105}{16} x^{-9/2}$. **11.** $-\dfrac{10}{p^6}$.

13. $-\dfrac{1}{4(1 - r)^{3/2}}$. **15.** $\dfrac{50}{(5x - 6)^3}$. **17.** $20(x^2 - 4)^8(19x^2 - 4)$. **19.** $\dfrac{4}{(x - 1)^3}$.

21. $12(1 - 2x)(4x - 1)$. **23.** $300(5x - 3)^2$. **25.** 0.6. **27.** ± 1.

Review Problems—Chapter 2

1. $-2x$. **3.** 0. **5.** $28x^3 - 18x^2 + 10x = 2x(14x^2 - 9x + 5)$.

7. $4s^3 + 4s = 4s(s^2 + 1)$. **9.** $\dfrac{2x}{5}$.

11. $(x^2 + 6x)(3x^2 - 12x) + (x^3 - 6x^2 + 4)(2x + 6) = 5x^4 - 108x^2 + 8x + 24$.

13. $100(2x^2 + 4x)^{99}(4x + 4) = 400(x + 1)[(2x)(x + 2)]^{99}$. **15.** $-\dfrac{2}{(2x + 1)^2}$.

17. $(8 + 2x)(4)(x^2 + 1)^3(2x) + (x^2 + 1)^4(2) = 2(x^2 + 1)^3(9x^2 + 32x + 1)$.

19. $\dfrac{(z^2 + 1)(2z) - (z^2 - 1)(2z)}{(z^2 + 1)^2} = \dfrac{4z}{(z^2 + 1)^2}$. **21.** $\frac{4}{3}(4x - 1)^{-2/3}$. **23.** $-\dfrac{y}{x + y}$.

25. $-\frac{1}{2}(1 - x)^{-3/2}(-1) = \frac{1}{2}(1 - x)^{-3/2}$.

27. $(x - 6)^4[3(x + 5)^2] + (x + 5)^3[4(x - 6)^3] = (x - 6)^3(x + 5)^2(7x + 2)$.

29. $\dfrac{(x + 1)(5) - (5x - 4)(1)}{(x + 1)^2} = \dfrac{9}{(x + 1)^2}$.

31. $2(-\frac{3}{8})x^{-11/8} + (-\frac{3}{8})(2x)^{-11/8}(2) = -\frac{3}{4}(1 + 2^{-11/8})x^{-11/8}$.

33. $\dfrac{\sqrt{x^2 + 5}(2x) - (x^2 + 6)(1/2)(x^2 + 5)^{-1/2}(2x)}{x^2 + 5} = \dfrac{x(x^2 + 4)}{(x^2 + 5)^{3/2}}$.

35. $(\frac{2}{3})(x^3 + 6x^2 + 9)^{-2/5}(3x^2 + 12x) = \frac{2}{5}x(x + 4)(x^3 + 6x + 9)^{-2/5}$. **37.** $-\dfrac{y}{x}$.

39. $7(1 - 2z)$. **41.** 12. **43.** $-\frac{1}{128}$. **45.** 192. **47.** $-\frac{1}{4}$.

49. $y = -4x + 3$. **51.** $y = \frac{1}{12}x + \frac{1}{3}$. **53.** $7x - 2\sqrt{10}y - 9 = 0$.

55. $\frac{5}{7} \approx 0.714$; 71.4%. **57.** $dr/dq = 20 - 0.2q$. **59.** 0.569, 0.431.

61. $dr/dq = 450 - q$. **63.** $dc/dq = 0.125 + 0.00878q$; 0.7396. **65.** 84.

67. (a) $\frac{4}{3}$, (b) $\frac{1}{24}$.

Exercise 3.1

1. Sym. about origin. **3.** Sym. about y-axis. **5.** Sym. about x-axis, y-axis, origin.

7. Sym. about x-axis. **9.** No symmetry. **11.** Sym. about origin.

13. Sym. about y-axis.

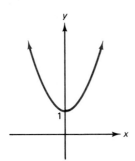

15. Sym. about x-axis.

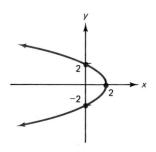

17. Sym. about origin.

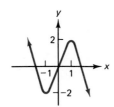

19. Sym. about x-axis, y-axis, origin.

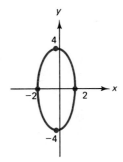

Exercise 3.2

1. $y = \frac{1}{2}, x = -\frac{3}{2}$.   **3.** $y = 0, x = 0$.   **5.** None.   **7.** $y = 2, x = -3, x = 2$.

9. $y = 4, x = 6$.   **11.** $x = 0, x = -1$.

13. $y = 0, x = 1$.

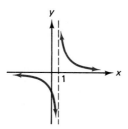

15. Sym. about origin;
$y = 0, x = 0$.

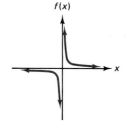

17. Sym. about y-axis;
$y = 0$, $x = 1$, $x = -1$.

19. Sym. about y-axis;
$y = 1$, $x = 2$, $x = -2$.

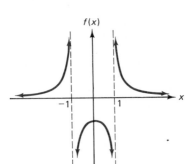

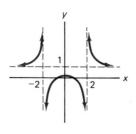

23.

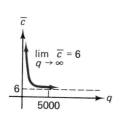

Exercise 3.3

1. Dec. on $(-\infty, 0)$; inc. on $(0, \infty)$; rel. min. when $x = 0$.   **3.** Inc. on $(-\infty, \frac{1}{2})$; dec. on $(\frac{1}{2}, \infty)$; rel. max. when $x = \frac{1}{2}$.   **5.** Dec. on $(-\infty, -5)$ and $(1, \infty)$; inc. on $(-5, 1)$; rel. min. when $x = -5$; rel. max. when $x = 1$.   **7.** Dec. on $(-\infty, -1)$ and $(0, 1)$; inc. on $(-1, 0)$ and $(1, \infty)$; rel. max. when $x = 0$; rel. min. when $x = \pm 1$.   **9.** Inc. on $(-\infty, 1)$ and $(3, \infty)$; dec. on $(1, 3)$; rel. max. when $x = 1$; rel. min. when $x = 3$.   **11.** Inc. on $(-\infty, -1)$ and $(1, \infty)$; dec. on $(-1, 0)$ and $(0, 1)$; rel. max. when $x = -1$; rel. min. when $x = 1$.   **13.** Dec. on $(-\infty, -4)$ and $(0, \infty)$; inc. on $(-4, 0)$; rel. min. when $x = -4$; rel. max. when $x = 0$.   **15.** Dec. on $(-\infty, 1)$ and $(1, \infty)$; no rel. max. or min.
17. Dec. on $(0, \infty)$; no rel. max. or min.   **19.** Dec. on $(-\infty, 0)$ and $(2, \infty)$; inc. on $(0, 1)$ and $(1, 2)$; rel. min. when $x = 0$; rel. max. when $x = 2$.   **21.** Inc. on $(-\infty, -2)$, $(-2, 11/5)$, and $(5, \infty)$; dec. on $(11/5, 5)$; rel. max. when $x = 11/5$; rel. min when $x = 5$.

23. Dec. on $(-\infty, 3)$; inc. on $(3, \infty)$; rel. min. when $x = 3$.

25. Dec. on $(-\infty, -1)$ and $(1, \infty)$; inc. on $(-1, 1)$; rel. min. when $x = -1$; rel. max. when $x = 1$; sym. about origin.

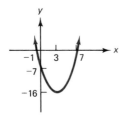

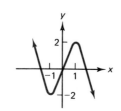

27. Inc. on $(-\infty, 1)$ and $(2, \infty)$; dec. on $(1, 2)$; rel. max. when $x = 1$; rel. min. when $x = 2$.

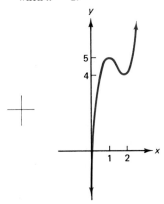

29. Inc. on $(-2, -1)$ and $(0, \infty)$; dec. on $(-\infty, -2)$ and $(-1, 0)$; rel. max. when $x = -1$; rel. min. when $x = -2, 0$.

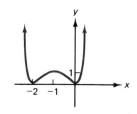

31. Dec. on $(-\infty, 1)$ and $(1, \infty)$; asym. $y = 1, x = 1$.

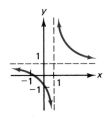

33. Inc. on $(-\infty, -6)$ and $(0, \infty)$; dec. on $(-6, -3)$ and $(-3, 0)$; rel. max. when $x = -6$; rel. min. when $x = 0$; asym. $x = -3$.

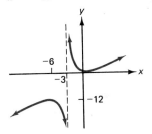

35. Ab. max. when $x = -1$; ab. min. when $x = 1$.

37. Ab. max. when $x = 0$; ab. min. when $x = 2$.

39. Ab. max. when $x = 3$; ab. min. when $x = 1$.   **43.** Never.

47.

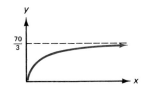

49. (a) $5, \frac{1}{5}$.

Exercise 3.4

1. Conc. down $(-\infty, \infty)$.   **3.** Conc. down $(-\infty, -1)$; conc. up $(-1, \infty)$; inf. pt. when $x = -1$.   **5.** Conc. up $(-\infty, -1)$, $(1, \infty)$; conc. down $(-1, 1)$; inf. pt. when $x = \pm 1$.   **7.** Conc. down $(-\infty, 1)$; conc. up $(1, \infty)$.   **9.** Conc. down $(-\infty, -3)$; conc. up $(-3, \infty)$.

11. Dec. $(-\infty, -2)$; inc. $(-2, \infty)$; rel. min. when $x = -2$; conc. up $(-\infty, \infty)$.

13. Inc. $(-\infty, 2)$; dec. $(2, \infty)$; rel. max. when $x = 2$; conc. down $(-\infty, \infty)$.

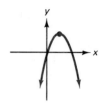

15. Inc. $(-\infty, 2)$, $(4, \infty)$; dec. $(2, 4)$; rel. max. when $x = 2$; rel. min. when $x = 4$; conc. down $(-\infty, 3)$; conc. up $(3, \infty)$; inf. pt. when $x = 3$.

17. Inc. $(-\infty, -2)$, $(2, \infty)$; dec. $(-2, 2)$; rel. max. when $x = -2$; rel. min. when $x = 2$; conc. down $(-\infty, 0)$; conc. up $(0, \infty)$; inf. pt. when $x = 0$; sym. about origin.

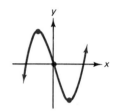

19. Inc. $(-\infty, 1)$, $(1, \infty)$; no rel. max. or min.; conc. down $(-\infty, 1)$; conc. up $(1, \infty)$; inf. pt. when $x = 1$.

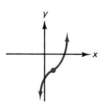

21. Inc. $(-\infty, 0)$, $(0, 1)$; dec. $(1, \infty)$; rel. max. when $x = 1$; conc. up $(0, 2/3)$; conc. down $(-\infty, 0)$, $(2/3, \infty)$; inf. pt. when $x = 0$, $x = 2/3$.

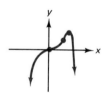

23. Dec. $(-\infty, -1)$, $(1, \infty)$; inc. $(-1, 1)$; rel. min. when $x = -1$; rel. max. when $x = 1$; conc. up $(-\infty, 0)$; conc. down $(0, \infty)$; inf. pt. when $x = 0$; sym. to origin.

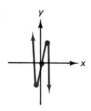

25. Dec. $(-\infty, 0)$, $(0, 1)$; inc. $(1, \infty)$; rel. min. when $x = 1$; conc. up $(-\infty, 0)$, $(2/3, \infty)$; conc. down $(0, 2/3)$; inf. pt. when $x = 0$, $x = 2/3$.

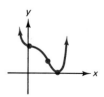

27. Dec. $(-\infty, 0)$, $(0, \infty)$; conc. down $(-\infty, 0)$; conc. up $(0, \infty)$; sym. about origin; asymptotes $x = 0$, $y = 0$.

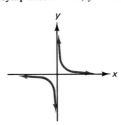

29. Inc. $(-\infty, -1)$, $(-1, \infty)$; conc. up $(-\infty, -1)$; conc. down $(-1, \infty)$; asymptotes $x = -1$, $y = 1$.

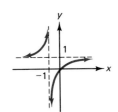

31. Dec. $(-\infty, -1)$, $(0, 1)$; inc. $(-1, 0)$, $(1, \infty)$; rel. min. when $x = \pm 1$; conc. up $(-\infty, 0)$, $(0, \infty)$; sym. to y-axis; asymptote $x = 0$.

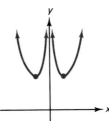

35.

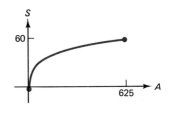

Exercise 3.5

1. Rel. min. when $x = \frac{5}{2}$; abs. min.   3. Rel. max. when $x = \frac{1}{4}$; abs. max.

5. Rel. max. when $x = -3$, rel. min. when $x = 3$.

7. Rel. min. when $x = 0$, rel. max. when $x = 2$.

9. Test fails, when $x = 0$ there is rel. min. by first-deriv. test.

Review Problems—Chapter 3

1. Sym. about origin.   3. $y = 3$, $x = 4$, $x = -4$.   5. $x = 0, 4$.

7. Inc. on $(1, 3)$; dec. on $(-\infty, 1)$ and $(3, \infty)$.

9. Conc. up on $(-\infty, 0)$ and $(\frac{1}{2}, \infty)$; conc. down on $(0, \frac{1}{2})$.   11. Rel. min. at $x = -1$.

13. At $x = 3$.   15. Abs. max. at $x = 2$; abs. min. at $x = 1$.

17. Inc. $(1, \infty)$; dec. $(-\infty, 1)$; rel. min. when $x = 1$; conc. up $(-\infty, \infty)$.

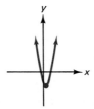

19. Inc. $(-\infty, -2)$, $(2, \infty)$; dec. $(-2, 2)$; rel. max. when $x = -2$; rel. min. when $x = 2$; conc. up $(0, \infty)$; conc. down $(-\infty, 0)$; inf. pt. when $x = 0$.

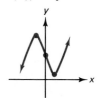

21. Inc. $(-\infty, \infty)$; conc. down $(-\infty, 0)$; conc. up $(0, \infty)$; inf. pt. when $x = 0$; sym. about origin.

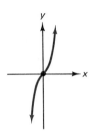

23. Inc. $(-10, 0)$; dec. $(-\infty, -10)$, $(0, \infty)$; rel. min. when $x = -10$; conc. up $(-15, 0)$, $(0, \infty)$; conc. down $(-\infty, -15)$; inf. pt. when $x = -15$; horizon. asym. $y = 0$; vert. asym. $x = 0$.

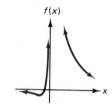

25. $(\frac{1}{3}, 1)$.

Exercise 4.1

1. 100.   **3.** $15.   **5.** 12, 105.   **7.** (a) 110 grams, (b) $51\frac{9}{11}$ grams.

9. 525, $51, $10,525.   **11.** $22.   **13.** 625, $4.   **15.** $3; $9000.   **17.** 20 and 20.

19. 300 ft by 250 ft.   **21.** 4 ft by 4 ft by 2 ft.   **23.** 2 in; 128 in³.

27. 130, $p = 340$, $P = 36,980$; 125, $p = 350$, $P = 34,175$.   **29.** 250 per lot (4 lots).

31. 35.   **33.** 60 mi/h.   **35.** $5 - \sqrt{3}$ tons, $5 - \sqrt{3}$ tons.

Exercise 4.2

1. 0.25410.   **3.** 1.32472.   **5.** -2.38769.   **7.** 0.33767.   **9.** 1.90785.

Exercise 4.3

1. $3\ dx$.   **3.** $\dfrac{2x^3}{\sqrt{x^4 + 2}}\ dx$.   **5.** $-\dfrac{2}{x^3}\ dx$.   **7.** $6x(2x + 1)\ dx$.

9. $\Delta y = -0.14$, $dy = -0.14$.   **11.** $\Delta y = 0.073$, $dy = \frac{3}{40} = 0.075$.   **13.** 10.05.

15. $3\frac{47}{48}$.   **17.** 44; 41.80.   **19.** 2.04.   **21.** 0.7.   **23.** $(1.69 \times 10^{-11})\pi$ cm³.

25. $35\frac{8}{9}$.

Exercise 4.4

1. -3, elastic.   **3.** -1, unit elasticity.   **5.** -1, unit elasticity.

7. -1.02, elastic.

9. $|\eta| = \dfrac{10}{3}$ when $p = 10$, $|\eta| = \dfrac{3}{10}$ when $p = 3$, $|\eta| = 1$ when $p = 6.50$.

Review Problems—Chapter 4

1. 20.   **3.** $2800.   **5.** 2 in.   **7.** 20 in. by 25 in.   **9.** $-12(1 - 3x)^3\ dx$.

11. 5.05.   **13.** Inelastic.   **15.** Elastic.   **17.** 1.7693.

Exercise 5.1

1. $3x + C.$ **3.** $\dfrac{x^9}{9} + C.$ **5.** $-\dfrac{5}{6x^6} + C.$ **7.** $-\dfrac{1}{9x^9} + C.$ **9.** $8u + \dfrac{u^2}{2} + C.$

11. $\dfrac{y^6}{6} - \dfrac{5y^2}{2} + C.$ **13.** $t^3 - 2t^2 + 5t + C.$ **15.** $\dfrac{x^2}{14} - \dfrac{3x^5}{20} + C.$

17. $\dfrac{x^{9.3}}{9.3} - \dfrac{9x^7}{7} - \dfrac{1}{x^3} - \dfrac{1}{2x^2} + C.$ **19.** $-\dfrac{4x^{3/2}}{9} + C.$ **21.** $2\sqrt[8]{x} + C.$

23. $\dfrac{x^4}{12} + \dfrac{3}{2x^2} + C.$ **25.** $\dfrac{w^3}{2} + \dfrac{2}{3w} + C.$ **27.** $\dfrac{4x^{3/2}}{3} - \dfrac{12x^{5/4}}{5} + C.$

29. $-\dfrac{3x^{5/3}}{25} - 7x^{1/2} + 3x^2 + C.$ **31.** $y = \dfrac{3x^2}{2} - 4x + 1.$

33. 18.   **35.** $c = 1.35q + 200.$ **37.** 7715.   **39.** $p = 0.7.$

41. $p = 275 - 0.5q - 0.1q^2.$ **43.** $G = -\dfrac{P^2}{50} + 2P + 20.$

45. $y = -1.5x - 0.5x^2 + 59.3.$

Exercise 5.2

1. $\dfrac{x^4}{4} - x^3 + \dfrac{5x^2}{2} - 15x + C.$ **3.** $\dfrac{2x^{5/2}}{5} + 2x^{3/2} + C.$ **5.** $\dfrac{u^5}{5} + \dfrac{2u^3}{3} + u + C.$

7. $\dfrac{2v^3}{3} + 3v + \dfrac{1}{2v^4} + C.$ **9.** $\dfrac{3x^2}{2} + x + \dfrac{1}{x} + C.$ **11.** $\dfrac{(x+4)^9}{9} + C.$

13. $\dfrac{(x^2+16)^4}{4} + C.$ **15.** $\tfrac{3}{5}(y^3 + 3y^2 + 1)^{5/3} + C.$ **17.** $-\dfrac{(3x-1)^{-2}}{2} + C.$

19. $\dfrac{2(x+10)^{3/2}}{3} + C.$ **21.** $\dfrac{(7x-6)^5}{35} + C.$ **23.** $\dfrac{(x^2+3)^{13}}{26} + C.$

25. $\tfrac{3}{20}(27 + x^5)^{4/3} + C.$ **27.** $-\tfrac{3}{4}(z^2 - 6)^{-4} + C.$

29. $\tfrac{2}{15}(5x)^{3/2} + C = \tfrac{2}{3}x\sqrt{5x} + C.$ **31.** $\sqrt{x^2 - 4} + C.$ **33.** $-\tfrac{1}{24}(3 - 3x^2 - 6x)^4 + C.$

35. $3(y + 1)^{4/3} + C.$ **37.** $-\tfrac{1}{25}(7 - 5x^2)^{5/2} + C.$ **39.** $\sqrt{2x} + C.$

41. $\dfrac{r^8}{8} + 2r^5 + \dfrac{25r^2}{2} + C.$ **43.** $\dfrac{8x}{3} + \dfrac{2}{x} - \dfrac{x^2}{6} + C.$ **45.** $\tfrac{2}{9}(\sqrt{x} + 2)^3 + C.$

47. $-\tfrac{10}{21}(5 - t\sqrt{t})^{1.4} + C.$ **49.** $y = -\tfrac{1}{6}(3 - 2x)^3 + \tfrac{11}{2}.$ **51.** $p = \dfrac{100}{q + 2}.$

53. $C = 2(\sqrt{I} + 1).$ **55.** $C = \tfrac{3}{4}I - \tfrac{1}{3}\sqrt{I} + \tfrac{71}{12}.$

Exercise 5.3

Answers are assumed to be expressed in square units.

1. 12.   **3.** 20.   **5.** $\tfrac{14}{3}.$ **7.** $\tfrac{13}{6}.$ **9.** 16.   **11.** $\tfrac{14}{3}.$ **13.** 8.

Exercise 5.4

1. 12.   **3.** $\tfrac{9}{2}.$ **5.** $-24.$ **7.** $\tfrac{15}{2}.$ **9.** 0.   **11.** $\tfrac{5}{3}.$ **13.** $\tfrac{32}{3}.$

15. $-\tfrac{1}{6}.$ **17.** $\tfrac{65}{8}.$ **19.** $\tfrac{3}{4}.$ **21.** $\tfrac{38}{9}.$ **23.** $\tfrac{15}{28}.$ **25.** $\tfrac{13}{8}.$ **27.** 1.

29. 0.02.   **31.** $\alpha^{5/2}T$.   **33.** 1,973,333.   **35.** \$160.   **37.** \$2000.
39. 696, 492.

Exercise 5.5

*In Problems **1–29**, answers are assumed to be expressed in square units.*

1. $\frac{19}{2}$.   **3.** 8.   **5.** $\frac{19}{3}$.   **7.** 9.   **9.** $\frac{50}{3}$.   **11.** 36.   **13.** 8.   **15.** $\frac{32}{3}$.
17. 18.   **19.** $\frac{3}{2}\sqrt[3]{2}$.   **21.** 36.   **23.** 68.   **25.** 2.   **27.** (a) $\frac{1}{16}$, (b) $\frac{3}{4}$, (c) $\frac{7}{16}$.
29. (a) $\frac{1}{4}$, (b) $\frac{1}{4}$, (c) $\frac{1}{2}$.

Exercise 5.6

*In Problems **1–21**, the answers are assumed to be expressed in square units.*

1. $\frac{4}{3}$.   **3.** $\frac{16}{3}$.   **5.** $8\sqrt{6}$.   **7.** 40.   **9.** $\frac{125}{6}$.   **11.** $\frac{9}{2}$.   **13.** $\frac{125}{12}$.
15. $\frac{32}{81}$.   **17.** $\frac{44}{3}$.   **19.** $\frac{4}{3}(5\sqrt{5} - 2\sqrt{2})$.   **21.** $\frac{1}{2}$.   **23.** $\frac{20}{63}$.

Exercise 5.7

1. $CS = 25.6$, $PS = 38.4$.   **3.** $CS = 90$, $PS = 144$.   **5.** $CS = 800$, $PS = 1000$.

Review Problems—Chapter 5

1. $\dfrac{x^4}{4} + x^2 - 7x + C$.   **3.** $11\frac{7}{2}$.   **5.** $-(x + 5)^{-2} + C$.   **7.** $-\dfrac{7(3x^2 - 8)^{-2}}{12} + C$.

9. $\dfrac{11\sqrt[3]{11}}{4} - 4$.   **11.** $\frac{2}{27}(3x^3 + 2)^{3/2} + C$.   **13.** $\dfrac{y^4}{4} + \dfrac{2y^3}{3} + \dfrac{y^2}{2} + C$.

15. $-\dfrac{1}{2x^2} - \dfrac{2}{x} + C$.   **17.** 11.1.   **19.** $\frac{7}{3}$.   **21.** $4 - 3\sqrt[3]{2}$.   **23.** $\dfrac{3}{t} - \dfrac{2}{\sqrt{t}} + C$.

25. $(0.5x - 0.1)^5 + C$.

*In Problems **27–39**, answers are assumed to be expressed in square units.*

27. $\frac{4}{3}$.   **29.** $\frac{16}{3}$.   **31.** $\frac{125}{6}$.   **33.** $\frac{20}{3}$.   **35.** $\frac{2}{3}$.   **37.** 36.   **39.** $\frac{125}{3}$.
41. $p = 100 - \sqrt{2q}$.   **43.** \$1900.   **45.** $CS = 166\frac{2}{3}$, $PS = 53\frac{1}{3}$.

Exercise 6.1

1.   **3.**   **5.**

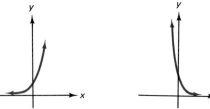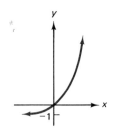

7. ∞.   **9.** 4.   **11.** 3.   **13.** $\frac{1}{4}$.   **15.** 4.4817.   **17.** 0.67032.

19. 140,000.   **21.** 0.2241.   **23.** 50.   **25.** 1.00.   **27.** 0.1466.

29. $1616.10.   **31.** $606,530.   **33.** $26,960.   **35.** 6065.

Exercise 6.2

1. $\log 10{,}000 = 4$.   **3.** $2^6 = 64$.   **5.** $\ln 7.3891 = 2$.   **7.** $e^{1.09861} = 3$.

9.

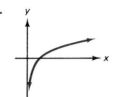

11. 2.   **13.** 3.   **15.** 1.

17. -2.   **19.** 0.   **21.** -3.   **23.** 48.   **25.** -4.   **27.** 3.   **29.** $4x$.

31. $2x$.   **33.** $\frac{1}{10}$.   **35.** e^2.   **37.** 2.   **39.** $\dfrac{\ln 2}{3}$.   **41.** $\log_3 8$.

43. $\dfrac{5 + \ln 3}{2}$.   **45.** $\frac{5}{2}$.   **47.** 1.60944.   **49.** 2.00013.   **51.** $\ln x + 2\ln(x + 1)$.

53. $2\ln x - 3\ln(x + 1)$.   **55.** $3[\ln x - \ln(x + 1)]$.

57. $\ln x - \ln(x + 1) - \ln(x + 2)$.   **59.** $\frac{1}{2}\ln x - 2\ln(x + 1) - 3\ln(x + 2)$.

61. $\frac{2}{3}\ln x - \frac{1}{3}\ln(x + 1) - \ln(x + 2)$.   **63.** $\dfrac{\ln(x + 8)}{\ln 10}$.   **65.** 41.50.   **67.** 22 years.

69. $E = 2.5 \times 10^{11 + 1.5M}$.   **71.** **(a)** 91, 432; **(b)** 8.   **73.** $225\,\ln\frac{9}{4} \approx 182$.

77. $S = 12.4A^{0.26}$.   **79.** **(a)** 3, **(b)** $2 + M_1$.

Exercise 6.3

1. $\dfrac{3}{3x - 4}$.   **3.** $\dfrac{2}{x}$.   **5.** $-\dfrac{2x}{1 - x^2}$.   **7.** $\dfrac{6p^2 + 3}{2p^3 + 3p} = \dfrac{3(2p^2 + 1)}{p(2p^2 + 3)}$.

9. $t\left(\dfrac{1}{t}\right) + (\ln t)(1) = 1 + \ln t$.   **11.** $\dfrac{2\log_3 e}{2x - 1}$.   **13.** $\dfrac{z\left(\dfrac{1}{z}\right) - (\ln z)(1)}{z^2} = \dfrac{1 - \ln z}{z^2}$.

15. $\dfrac{3(2x + 4)}{x^2 + 4x + 5} = \dfrac{6(x + 2)}{x^2 + 4x + 5}$.   **17.** $\dfrac{x}{1 + x^2}$.   **19.** $\dfrac{2}{1 - t^2}$.   **21.** $\dfrac{x}{1 - x^4}$.

23. $\dfrac{4x}{x^2 + 2} + \dfrac{3x^2 + 1}{x^3 + x - 1}$.   **25.** $\dfrac{2(x^2 + 1)}{2x + 1} + 2x\ln(2x + 1)$.   **27.** $\dfrac{3(1 + \ln^2 x)}{x}$.

29. $\dfrac{4\ln^3(ax)}{x}$.   **31.** $\dfrac{x}{2(x - 1)} + \ln\sqrt{x - 1}$.   **33.** $\dfrac{1}{2x\sqrt{4 + \ln x}}$.

35. $3 + 2\ln x$.   **37.** $y = 4x - 12$.

39. Dec. on $(0, 1)$; inc. on $(1, \infty)$; rel. min. when $x = 1$.   **41.** $\dfrac{1}{x(2y + 1)}$.

43. $\dfrac{dr}{dq} = 25\dfrac{(q + 2)\ln(q + 2) - q}{(q + 2)\ln^2(q + 2)}$.   **45.** $\dfrac{1.5E}{\log e}$.

47. **(b)**   **(c)** decreasing at the rate of 0.26.

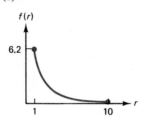

Exercise 6.4

1. $2xe^{x^2+1}$.   **3.** $-5e^{3-5x}$.   **5.** $(6r + 4)e^{3r^2+4r+4} = 2(3r + 2)e^{3r^2+4r+4}$.

7. $x(e^x) + e^x(1) = e^x(x + 1)$.   **9.** $2xe^{-x^2}(1 - x^2)$.   **11.** $\dfrac{e^x - e^{-x}}{2}$.

13. $(6x)4^{3x^2}\ln 4$.   **15.** $\dfrac{2e^{2w}(w - 1)}{w^3}$.   **17.** $\dfrac{e^{1+\sqrt{x}}}{2\sqrt{x}}$.   **19.** $3x^2 - 3^x\ln 3$.

21. $\dfrac{2e^x}{(e^x + 1)^2}$.   **23.** 1.   **25.** $e^{x\ln x}(1 + \ln x)$.

27. $y - e^2 = e^2(x - 2)$ or $y = e^2 x - e^2$.   **29.** e^x.   **31.** $e^z(z^2 + 4z + 2)$.

33. Dec. on $(-\infty, -1)$; inc. on $(-1, \infty)$; rel. min. when $x = -1$.

35. $-\dfrac{e^y}{xe^y + 1}$.   **37.** $dp/dq = -0.015e^{-0.001q}$, $-0.015e^{-0.5}$.   **39.** 50.

41. $dc/dq = 10e^{q/700}$; $10e^{0.5}$; $10e$.   **43.** $100e^{-2}$.

45. Rel. max. when $x = 0$; inf. pt. when   **47.** $-b(10^{A-bM})\ln 10$.
    $x = \pm 1$.

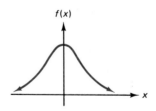

49. **(a)** $dp/dt = 0.8811(0.85)^t \ln(0.85)$; this represents the rate of change of proportion of recalls with respect to length of recall interval; **(b)** -0.10.

51. **(a)** 100; as the altitude increases, practically all people will suffer the bends;
(b) $\dfrac{3,600,000}{e^{0.36h}(1 + 100,000e^{-0.36h})^2}$; **(c)** yes.   **53.** **(a)** 0.

55. **(a)** 1; as time goes on, practically the entire group will be discharged; **(b)** 0.0036.

Exercise 6.5

1. $(x + 1)^2(x - 1)(x^2 + 3)\left[\dfrac{2}{x + 1} + \dfrac{1}{x - 1} + \dfrac{2x}{x^2 + 3}\right]$.

3. $(3x^3 - 1)^2(2x + 5)^3\left[\dfrac{18x^2}{3x^3 - 1} + \dfrac{6}{2x + 5}\right].$

5. $\dfrac{\sqrt{x + 1}\sqrt{x^2 - 2}\sqrt{x + 4}}{2}\left[\dfrac{1}{x + 1} + \dfrac{2x}{x^2 - 2} + \dfrac{1}{x + 4}\right].$

7. $\dfrac{\sqrt{1 - x^2}}{1 - 2x}\left[\dfrac{x}{x^2 - 1} + \dfrac{2}{1 - 2x}\right].$

9. $\dfrac{(2x^2 + 2)^2}{(x + 1)^2(3x + 2)}\left[\dfrac{4x}{x^2 + 1} - \dfrac{2}{x + 1} - \dfrac{3}{3x + 2}\right].$

11. $\dfrac{(8x + 3)^{1/2}(x^2 + 2)^{1/3}}{(1 + 2x)^{1/4}}\left[\dfrac{4}{8x + 3} + \dfrac{2x}{3(x^2 + 2)} - \dfrac{1}{2(1 + 2x)}\right].$

13. $x^{2x+1}\left(\dfrac{2x + 1}{x} + 2\ln x\right).$ **15.** $\dfrac{x^{1/x}(1 - \ln x)}{x^2}.$

17. $2(3x + 1)^{2x}\left[\dfrac{3x}{3x + 1} + \ln(3x + 1)\right].$ **19.** $y = 96x + 36.$

Exercise 6.6

1. $e^{3x} + C.$ **3.** $e^{t^2 + t} + C.$ **5.** $\frac{1}{10}e^{5x^2} + C.$ **7.** $-\frac{1}{6}e^{-2v^3 + 1} + C.$

9. $-\frac{1}{5}e^{-5x} + 2e^x + C.$ **11.** $\frac{1}{3}(e^8 - 1).$ **13.** $\dfrac{4^{7x}}{7\ln 4} + C.$ **15.** $-\frac{1}{7}e^{7/x} + C.$

17. $\ln|x + 5| + C.$ **19.** $\ln|x^3 + x^4| + C.$ **21.** $4\ln|x| + C.$ **23.** $\frac{1}{3}\ln|s^3 + 5| + C.$

25. $-\frac{7}{3}\ln|5 - 3x| + C.$ **27.** $-\frac{3}{4}(z^2 - 6)^{-4} + C.$ **29.** 1. **31.** $\frac{1}{3}\ln|x^3 + 6x| + C.$

33. $\dfrac{3}{2} - \dfrac{1}{e} + \dfrac{1}{2e^2}.$ **35.** $\frac{1}{2}\ln 3.$ **37.** $\frac{1}{2}\ln(x^2 + 1) - \dfrac{1}{6(x^6 + 1)} + C.$

39. $\frac{3}{2}x^2 + x - \ln|x| + C.$ **41.** $\frac{3}{2}\ln(e^{2x} + 1) + C.$ **43.** $\frac{1}{2}(\ln^2 x) + C.$

45. $\ln|\ln(x + 3)| + C.$ **47.** $x^2 - 3x + \frac{2}{3}\ln|3x - 1| + C.$ **49.** $x + \ln|x - 1| + C.$

51. $c = 20\ln|(q + 5)/5| + 2000.$ **53.** $e^2 - 1$ sq. units.

55. $\frac{3}{2} + 2\ln 2 = \frac{3}{2} + \ln 4$ sq units. **57.** $y = \dfrac{1}{2}\ln\dfrac{x^2 + 4}{5} = \ln\sqrt{\dfrac{x^2 + 4}{5}}.$ **59.** \$8639.

61. $\dfrac{1 - e^{-at}}{a} - \dfrac{1 - e^{-bt}}{b}.$ **65.** $\dfrac{i}{k}(1 - e^{-2kR}).$

Review Problems—Chapter 6

1. $\log_3 81 = 4.$ **3.** $\frac{7}{5}.$ **5.** 3. **7.** $\frac{1}{100}.$ **9.** 3. **11.** 3.

13. $2\ln x + \frac{1}{2}\ln(x + 1) - \frac{1}{3}\ln(x^2 + 2).$ **15.** $2x.$ **17.** $y = e^{x^2 + 2}.$

19. $2e^x + e^{x^2}(2x) = 2(e^x + xe^{x^2}).$ **21.** $\dfrac{1}{r^2 + 5r}(2r + 5) = \dfrac{2r + 5}{r(r + 5)}.$

23. $e^{x^2 + 4x + 5}(2x + 4) = 2(x + 2)e^{x^2 + 4x + 5}.$

25. $e^x(2x) + (x^2 + 2)e^x = e^x(x^2 + 2x + 2).$

27. $\dfrac{\sqrt{(x - 6)(x + 5)(9 - x)}}{2}\left[\dfrac{1}{x - 6} + \dfrac{1}{x + 5} - \dfrac{1}{9 - x}\right].$

29. $\dfrac{e^x\left(\dfrac{1}{x}\right) - (\ln x)(e^x)}{e^{2x}} = \dfrac{1 - x\ln x}{xe^x}.$ **31.** $\dfrac{2}{q + 1} + \dfrac{3}{q + 2}.$ **33.** $-7(\ln 10)10^{2 - 7x}.$

35. $\dfrac{4e^{2x+1}(2x-1)}{x^2}$. **37.** $\dfrac{16\log_2 e}{8x+5}$. **39.** $\dfrac{1+2l+3l^2}{1+l+l^2+l^3}$.

41. $(x+1)^{x+1}[1+\ln(x+1)]$. **43.** $2\left(\dfrac{1}{t}\right)+\dfrac{1}{2}\left(\dfrac{1}{1-t}\right)(-1)=\dfrac{5t-4}{2t(t-1)}$.

45. $y\left[\dfrac{3}{2}\left(\dfrac{1}{x^2+2}\right)(2x)+\dfrac{4}{9}\left(\dfrac{1}{x^2+9}\right)(2x)-\dfrac{4}{11}\left(\dfrac{1}{x^3+6x}\right)(3x^2+6)\right]=$

$y\left[\dfrac{3x}{x^2+2}+\dfrac{8x}{9(x^2+9)}-\dfrac{12(x^2+2)}{11(x^3+6x)}\right]$, where y is given in problem.

47. 18. **49.** 2. **51.** $y=2x+2(1-\ln 2)$ or $y=2x+2-\ln 4$. **53.** $\frac{1}{3}$.

55. Inc. $(0,\infty)$; dec. $(-\infty,0)$; rel. min. when $x=0$; conc. up $(-\infty,\infty)$.

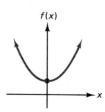

57. $\frac{1}{3}e^{3x+5}+C$. **59.** $\frac{1}{2}\ln|x+1|+C$. **61.** $2\ln|x^3-6x+1|+C$. **63.** $\frac{1}{3}\ln\frac{10}{3}$.

65. $7+e$. **67.** $\frac{1}{2}(e^{2y}+e^{-2y})+C$. **69.** $\ln|x|-\dfrac{2}{x}+C$. **71.** $\dfrac{5^{2x}}{2\ln 5}+C$.

73. $\frac{3}{2}-5\ln 2$. **75.** $6+\ln 3$. **79.** 134,064, 109,762. **83.** 10.

85. $0.008e^{-0.01t}+0.00004e^{-0.0002t}$.

Exercise 7.1

1. $\dfrac{1}{6}\ln\left|\dfrac{x}{6+7x}\right|+C$. **3.** $\dfrac{1}{3}\ln\left|\dfrac{\sqrt{x^2+9}-3}{x}\right|+C$.

5. $\frac{1}{2}[\frac{4}{5}\ln|4+5x|-\frac{2}{3}\ln|2+3x|]+C$. **7.** $\frac{1}{8}(2x-\ln[4+3e^{2x}])+C$.

9. $2\left[\dfrac{1}{1+x}+\ln\left|\dfrac{x}{1+x}\right|\right]+C$. **11.** $1+\ln\frac{4}{3}$.

13. $\frac{1}{2}(x\sqrt{x^2-3}-3\ln|x+\sqrt{x^2-3}|)+C$. **15.** $\frac{1}{144}$.

17. $e^x(x^2-2x+2)+C$. **19.** $2\left(-\dfrac{\sqrt{4x^2+1}}{2x}+\ln|2x+\sqrt{4x^2+1}|\right)+C$.

21. $\dfrac{1}{9}\left(\ln|1+3x|+\dfrac{1}{1+3x}\right)+C$. **23.** $\dfrac{1}{\sqrt{5}}\left(\dfrac{1}{2\sqrt{7}}\ln\left|\dfrac{\sqrt{7}+\sqrt{5}x}{\sqrt{7}-\sqrt{5}x}\right|\right)+C$.

25. $\dfrac{1}{36}\left[\dfrac{(3x)^6\ln(3x)}{6}-\dfrac{(3x)^6}{36}\right]+C=\dfrac{x^6}{36}[6\ln(3x)-1]+C$. **27.** $\dfrac{4(9x-2)(1+3x)^{3/2}}{135}+C$.

29. $\frac{1}{2}\ln|2x+\sqrt{4x^2-13}|+C$. **31.** $\frac{1}{4}[2x^2\ln(2x)-x^2]+C$.

33. $-\dfrac{\sqrt{9-4x^2}}{9x}+C$. **35.** $\frac{1}{2}\ln(x^2+1)+C$. **37.** $\frac{1}{6}(2x^2+1)^{3/2}+C$.

39. $\ln\left|\dfrac{x-3}{x-2}\right|+C$. **41.** $\frac{2}{3}(9\sqrt{3}-10\sqrt{2})$. **43.** $\ln\left|\dfrac{q_n(1-q_0)}{q_0(1-q_n)}\right|$.

45. (a) \$37,599, (b) \$4924. **47.** (a) \$5481, (b) \$535.

Exercise 7.2

1. $\frac{16}{3}$.   **3.** -1.   **5.** 0.   **7.** $\frac{13}{6}$.   **9.** \$12,400.   **11.** \$3321.

Exercise 7.3

1. 0.340, 0.333.   **3.** 1.388, 1.386.   **5.** 0.883.   **7.** 2,361,375.   **9.** 2.967.
11. 0.771.   **13.** $\frac{8}{3}$.

Exercise 7.4

1. $\frac{1}{3}$.   **3.** Div.   **5.** $\dfrac{1}{e}$.   **7.** Div.   **9.** $-\frac{1}{2}$.   **11.** 0.   **13.** (a) 800, (b) $\frac{2}{3}$.

15. $\alpha = e^{-x_c}, \beta = e^{-(1/8)x_c}$.   **17.** $\frac{1}{2}$ sq unit.   **19.** 5000 increase.

Exercise 7.5

1. $y = -\dfrac{1}{x^2 + C}$.   **3.** $y = \frac{1}{3}(x^2 + 1)^{3/2} + C$.   **5.** $y = Ce^x, C > 0$.

7. $y = Cx, C > 0$.   **9.** $y = \sqrt{2x}$.   **11.** $y = \ln\dfrac{x^3 + 3}{3}$.   **13.** $y = \dfrac{4x^2 + 3}{2(x^2 + 1)}$.

15. $N = 20{,}000e^{0.018t}$; $N = 20{,}000(1.2)^{t/10}$; 28,800.   **17.** $2e^{0.946}$ billion.
19. 0.01204; 57.57 sec.   **21.** 2900 years.   **23.** $N = N_0e^{k(t - t_0)}, t \geq t_0$.
25. 12.6 units.   **27.** $q = q_0e^{-kt}$, 0, 13.5%.
29. (a) $A = 400(1 - e^{-t/2})$, (b) 157 grams, (c) 400.

Exercise 7.6

1. 29,400.   **3.** 430,000.   **5.** 1990.   **7.** (b) 375.   **9.** 1:06 A.M.
11. \$62,500.   **13.** $N = M - (M - N_0)e^{-kt}$.

Review Problems—Chapter 7

1. $\dfrac{x^2}{4}[2 \ln(x) - 1] + C$.   **3.** $5 + \frac{9}{4}\ln 3$.   **5.** $\frac{1}{21}(9 \ln|3 + x| - 2 \ln|2 + 3x|) + C$.

7. $\dfrac{1}{2(x + 2)} + \dfrac{1}{4}\ln\left|\dfrac{x}{x + 2}\right| + C$.   **9.** $-\dfrac{\sqrt{9 - 16x^2}}{9x} + C$.   **11.** $\dfrac{3}{2}\ln\left|\dfrac{x - 3}{x + 3}\right| + C$.

13. $\dfrac{e^{7x}}{49}(7x - 1) + C$.   **15.** $\frac{1}{2}\ln|\ln 2x| + C$.   **17.** $x - \frac{3}{2}\ln|3 + 2x| + C$.   **19.** 34.

21. (a) 1.405, (b) 1.388.   **23.** $\frac{1}{18}$.   **25.** Div.   **27.** $y = Ce^{x^3 + x^2}, C > 0$.

29. 144,000.   **31.** 0.0005, 90%.   **33.** $N = \dfrac{450}{1 + 224e^{-1.02t}}$.   **35.** 4:16 P.M.

Exercise 8.1

1. 3.   **3.** 6.   **5.** 6.   **7.** 88.   **9.** 3.   **11.** $2x_0 + 2h - 5y_0 + 4$.

13. $y = -4$.   **15.** $z = 6$.

17.

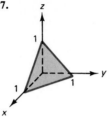

19.

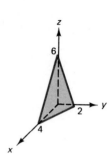

21.

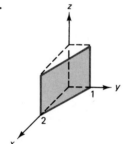

23.

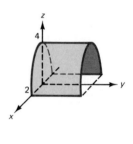

25. Surface is a sphere (top hemisphere is shown).

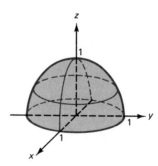

27. 2000.   **29.** $P(r, 5) = \dfrac{5! \, (\frac{1}{4})^r (\frac{3}{4})^{5-r}}{r! \, (5 - r)!}$.

Exercise 8.2

1. $f_x(x, y) = 1$, $f_y(x, y) = -5$.   **3.** $f_x(x, y) = 3$, $f_y(x, y) = 0$.

5. $g_x(x, y) = 5x^4y^4 - 12x^3y^3 + 21x^2 - 3y$, $g_y(x, y) = 4x^5y^3 - 9x^4y^2 + 4y - 3x$.

7. $g_p(p, q) = \dfrac{q}{2\sqrt{pq}}$, $g_q(p, q) = \dfrac{p}{2\sqrt{pq}}$.   **9.** $h_s(s, t) = \dfrac{2s}{t - 3}$, $h_t(s, t) = -\dfrac{s^2 + 4}{(t - 3)^2}$.

11. $u_{q_1}(q_1, q_2) = \dfrac{3}{4q_1}$, $u_{q_2}(q_1, q_2) = \dfrac{1}{4q_2}$.

13. $h_x(x, y) = (x^3 + xy^2 + 3y^3)(x^2 + y^2)^{-3/2}$, $h_y(x, y) = (3x^3 + x^2y + y^3)(x^2 + y^2)^{-3/2}$.

15. $\dfrac{\partial z}{\partial x} = 5ye^{5xy}$, $\dfrac{\partial z}{\partial y} = 5xe^{5xy}$. **17.** $\dfrac{\partial z}{\partial x} = 5\left[\dfrac{2x^2}{x^2 + y} + \ln(x^2 + y)\right]$, $\dfrac{\partial z}{\partial y} = \dfrac{5x}{x^2 + y}$.

19. $f_r(r, s) = \sqrt{r + 2s}\,(3r^2 - 2s) + \dfrac{r^3 - 2rs + s^2}{2\sqrt{r + 2s}}$, $f_s(r, s) = 2(s - r)\sqrt{r + 2s} + \dfrac{r^3 - 2rs + s^2}{\sqrt{r + 2s}}$.

21. $f_r(r, s) = -e^{3-r}\ln(7 - s)$, $f_s(r, s) = \dfrac{-e^{3-r}}{7 - s}$.

23. $g_x(x, y, z) = 6xy + 2y^2z$, $g_y(x, y, z) = 3x^2 + 4xyz$, $g_z(x, y, z) = 2xy^2 + 9z^2$.

25. $g_r(r, s, t) = 2re^{s+t}$, $g_s(r, s, t) = (7s^3 + 21s^2 + r^2)e^{s+t}$, $g_t(r, s, t) = e^{s+t}(r^2 + 7s^3)$.

27. 50. **29.** $\frac{1}{3}$. **31.** 0. **35.** $-\dfrac{ra}{2\left(1 + a\dfrac{n-1}{2}\right)^2}$.

Exercise 8.3

1. 20. **3.** 784.5. **5.** $\dfrac{\partial P}{\partial k} = 1.208648l^{0.192}k^{-0.236}$, $\dfrac{\partial P}{\partial l} = 0.303744l^{-0.808}k^{0.764}$.

7. $\dfrac{\partial q_A}{\partial p_A} = -50$, $\dfrac{\partial q_A}{\partial p_B} = 2$, $\dfrac{\partial q_B}{\partial p_A} = 4$, $\dfrac{\partial q_B}{\partial p_B} = -20$, competitive.

9. $\dfrac{\partial q_A}{\partial p_A} = -\dfrac{100}{p_A^2 p_B^{1/2}}$, $\dfrac{\partial q_A}{\partial p_B} = -\dfrac{50}{p_A p_B^{3/2}}$, $\dfrac{\partial q_B}{\partial p_A} = -\dfrac{500}{3 p_B p_A^{4/3}}$, $\dfrac{\partial q_B}{\partial p_B} = -\dfrac{500}{p_B^2 p_A^{1/3}}$, complementary.

11. $\dfrac{\partial p}{\partial B} = 0.01A^{0.27}B^{-0.99}C^{0.01}D^{0.23}E^{0.09}F^{0.27}$, $\dfrac{\partial p}{\partial C} = 0.01A^{0.27}B^{0.01}C^{-0.99}D^{0.23}E^{0.09}F^{0.27}$.

13. 1120; if a staff manager with an MBA degree had an extra year of work experience before the degree, the manager would receive \$1120 per year in extra compensation.

15. **(a)** -1.015, -0.846; **(b)** one for which $w = w_0$ and $s = s_0$. **17.** **(a)** No, **(b)** 70%.

Exercise 8.4

1. $6xy^2$, $12xy$. **3.** $3xe^{3xy} + 4x^2$, $9xye^{3xy} + 3e^{3xy} + 8x$, $9x(3xy + 2)e^{3xy}$.

5. $(2x + y)(2x^2 + 2xy + y^2 + 1)$, $6x^2 + 8xy + 3y^2 + 1$.

7. $3x^2y + 4xy^2 + y^3$, $3xy^2 + 4x^2y + x^3$, $6xy + 4y^2$, $6xy + 4x^2$.

9. $x(x^2 + y^2)^{-1/2}$, $y^2(x^2 + y^2)^{-3/2}$. **11.** 0. **13.** 1758. **15.** $2e$.

Exercise 8.5

1. $(\frac{14}{3}, -\frac{13}{3})$. **3.** $(2, 5)$, $(2, -6)$, $(-1, 5)$, $(-1, -6)$. **5.** $(50, 150, 350)$.

7. $(-2, \frac{3}{2})$, rel. min. **9.** $(-\frac{1}{4}, \frac{1}{2})$, rel. max. **11.** $(1, 1)$, rel. min.; $(\frac{1}{2}, \frac{1}{4})$, neither.

13. $(0, 0)$, rel. max.; $(4, \frac{1}{2})$, rel. min.; $(0, \frac{1}{2})$, $(4, 0)$, neither. **15.** $(122, 127)$, rel. max.

17. $(-1, -1)$, rel. min. **19.** $l = 24$, $k = 14$. **21.** $p_A = 80$, $p_B = 85$.

23. $q_A = 48$, $q_B = 40$, $p_A = 52$, $p_B = 44$, profit $= 3304$. **25.** $q_A = 3$, $q_B = 2$.

27. 1 ft by 2 ft by 3 ft. **29.** $(\frac{105}{37}, \frac{28}{37})$.

Exercise 8.6

1. $(2, -2)$.   **3.** $(3, \frac{3}{2}, -\frac{3}{2})$.   **5.** $(\frac{4}{3}, -\frac{4}{3}, -\frac{8}{3})$.   **7.** $(6, 3, 2)$.   **9.** $(\frac{2}{3}, \frac{4}{3}, -\frac{4}{3})$.

11. $(3, 3, 6)$.   **13.** Plant 1, 40 units; plant 2, 60 units.   **15.** 74 units (when $l = 8$, $k = 7$).

17. $x = 12$, $y = 8$.   **19.** $x = 10$, $y = 20$, $z = 5$.

Exercise 8.7

1. $\hat{y} = 0.98 + 0.61x$, 3.12.   **3.** $\hat{y} = 0.057 + 1.67x$, 5.90.   **5.** $\hat{q} = 82.6 - 0.641p$.

7. $\hat{y} = 100 + 0.13x$, 105.2.   **9.** $\hat{y} = 8.5 + 2.5x$.

11. **(a)** $\hat{y} = 35.9 - 2.5x$, **(b)** $\hat{y} = 28.4 - 2.5x$.

Exercise 8.8

1. 18.   **3.** $\frac{1}{4}$.   **5.** $\frac{2}{3}$.   **7.** 3.   **9.** 324.   **11.** $-\frac{58}{35}$.   **13.** $\frac{8}{3}$.   **15.** $-\frac{1}{3}$.

17. $\dfrac{e^2}{2} - e + \frac{1}{2}$.   **19.** $-\frac{27}{4}$.   **21.** $\frac{1}{24}$.   **23.** $e^{-4} - e^{-2} - e^{-3} + e^{-1}$.   **25.** $\frac{3}{8}$.

Review Problems—Chapter 8

1.

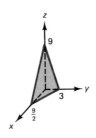

3.

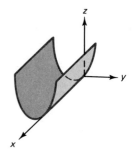

5. $4x + 3y$, $3x + 2y$.   **7.** $\dfrac{y}{(x + y)^2}$, $-\dfrac{x}{(x + y)^2}$.   **9.** $2xze^{x^2yz}(1 + x^2yz)$.

11. $\dfrac{y}{x^2 + y^2}$.   **13.** $2(x + y)$.   **15.** $xze^{yz} \ln z$, $\dfrac{e^{yz}}{z} + ye^{yz} \ln z = e^{yz}\left(\dfrac{1}{z} + y \ln z\right)$.

17. $\partial P/\partial l = 14l^{-0.3}k^{0.3}$, $\partial P/\partial k = 6l^{0.7}\,k^{-0.7}$.   **19.** Competitive products.

21. $(2, 2)$, rel. min.   **23.** 4 ft by 4 ft by 2 ft.   **25.** $(3, 2, 1)$.

27. $\hat{p} = 85.15 - 0.43t$.   **29.** $\dfrac{7}{2}$.   **31.** $\dfrac{1}{210}$.

Exercise 9.1

1. $30°$.   **3.** $90°$.   **5.** $-420°$.   **7.** $225°$.   **9.** 0.   **11.** $\dfrac{\pi}{4}$.   **13.** $-\dfrac{2\pi}{3}$.

15. $\dfrac{7\pi}{4}$.   **17.** Second.   **19.** Fourth.   **21.** Second.   **23.** Quadrantal.

25. $\dfrac{13\pi}{6}, \dfrac{25\pi}{6}; -\dfrac{11\pi}{6}, -\dfrac{23\pi}{6}.$ **27.** $\dfrac{5\pi}{4}.$ **29.** $\dfrac{\pi}{3}.$ **31.** $\dfrac{3\pi}{2}.$ **33.** $\dfrac{3}{5}.$

35. 10π cm.

Exercise 9.2

1. (a) $\dfrac{4}{5}$, (b) $\dfrac{3}{5}$.   **3.** (a) $-\dfrac{3}{5}$, (b) $\dfrac{4}{5}$.   **5.** (a) $-\dfrac{\sqrt{2}}{2}$, (b) $-\dfrac{\sqrt{2}}{2}$.   **7.** $+$.

9. $-$.   **11.** $-$.   **13.** $+$.   **15.** $\sin A = \dfrac{\sqrt{7}}{4}$, $\cos A = \dfrac{3}{4}$, $\sin B = \dfrac{3}{4}$, $\cos B = \dfrac{\sqrt{7}}{4}$.

17. $\dfrac{1}{2}$.   **19.** $\dfrac{\sqrt{2}}{2}$.   **21.** $\dfrac{1}{2}$.   **23.** $a = 3\sqrt{2}$, $b = 3\sqrt{2}$.   **25.** $b = 3\sqrt{3}$, $c = 6$.

27. $a = 1.8$, $b = 3.6$.   **29.** $a = 3.2$, $c = 5.9$.   **31.** $a = 25.5$, $c = 103.2$.

33. (a) 0.0500, (b) 0.0400, (c) 0.0300, (d) 0.0200, (e) 0.0100.   **35.** 339 ft.   **37.** 6 ft.

39. 17.2 m.   **41.** $\sin\dfrac{\pi}{2} = 1$, $\cos\dfrac{\pi}{2} = 0$.   **43.** $\sin\dfrac{3\pi}{2} = -1$, $\cos\dfrac{3\pi}{2} = 0$.

45. $\sin(-3\pi) = 0$, $\cos(-3\pi) = -1$.   **47.** $-a$.   **49.** $-a$.

51. $-\sqrt{0.3455} = -0.5878$.   **53.** $\dfrac{\pi}{4}$.   **55.** $\dfrac{\pi}{6}$.   **57.** $\dfrac{\pi}{3}$.   **59.** $\dfrac{\pi}{6}$.   **61.** $\dfrac{\sqrt{3}}{2}$.

63. $-\dfrac{\sqrt{3}}{2}$.   **65.** $-\dfrac{\sqrt{2}}{2}$.   **67.** $-\dfrac{1}{2}$.   **69.** $-\dfrac{\sqrt{3}}{2}$.   **71.** $\dfrac{1}{2}$.   **73.** $\dfrac{\sqrt{3}}{2}$.

75. $\dfrac{\sqrt{2}}{2}$.

Exercise 9.3

1. $A = 3$, $p = 2\pi$.

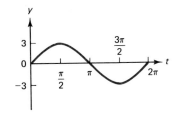

3. $A = 4$, $p = 2\pi$.

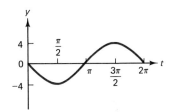

5. $A = \frac{1}{2}$, $p = 2\pi$.

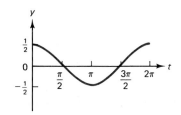

7. $A = 1$, $p = 2\pi$.

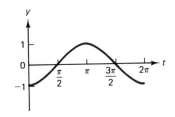

9. $A = 1$, $p = \pi$.

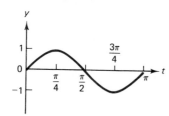

11. $A = 4$, $p = \dfrac{2\pi}{3}$.

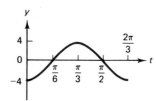

13. $A = 2$, $p = 6$

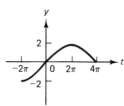

15. Two.

17.

In Problems **19–23**, *answers are in the order:* $\sin t$, $\cos t$, $\tan t$, $\cot t$, $\sec t$, $\csc t$.

19. $\dfrac{4}{5}, \dfrac{3}{5}, \dfrac{4}{3}, \dfrac{3}{4}, \dfrac{5}{3}, \dfrac{5}{4}$.   **21.** $\dfrac{3}{5}, -\dfrac{4}{5}, -\dfrac{3}{4}, -\dfrac{4}{3}, -\dfrac{5}{4}, \dfrac{5}{3}$.

23. $-\dfrac{\sqrt{2}}{2}, -\dfrac{\sqrt{2}}{2}, 1, 1, -\sqrt{2}, -\sqrt{2}$.   **25.** 1.   **27.** $\sqrt{3}$.   **29.** 2.

31. $\tan t = \dfrac{\sqrt{15}}{15}$, $\cot t = \sqrt{15}$, $\sec t = \dfrac{4\sqrt{15}}{15}$, $\csc t = 4$.

33. $\sin t = -\dfrac{3}{5}$, $\cos t = -\dfrac{4}{5}$, $\tan t = \dfrac{3}{4}$, $\cot t = \dfrac{4}{3}$.   **35.** 0.   **37.** 0.

39. Undefined.   **41.** -1.   **43.** -2.   **45.** $\pm\sqrt{5}$.

47. **(a)** 23 days, 28 days, 33 days; **(b)** 0.63, -0.43, -0.10.

Exercise 9.4

1. $\pi \cos \pi t$.   **3.** $(2t + 3) \cos(t^2 + 3t + 5)$.   **5.** $-\dfrac{1}{2} \sin \dfrac{t}{2}$.

7. $2(t + 1) \sec^2(t + 1)^2$.   **9.** $-e^t \csc e^t \cot e^t$.   **11.** $4\pi \sin(2\pi t + 1) \cos(2\pi t + 1)$.

13. $\dfrac{t^2[(-\sin t^2)(2t)] - (\cos t^2)[2t]}{(t^2)^2} = -\dfrac{2(t^2 \sin t^2 + \cos t^2)}{t^3}$.   **15.** $-\dfrac{3 \cos t}{\sin^{5/2} t}$.

17. $t^{1/2}(-\csc^2 t) + (\cot t)\left(\frac{1}{2}t^{-1/2}\right) = \dfrac{\cot t - 2t \csc^2 t}{2\sqrt{t}}.$

19. $6(1 + \sec 2t)^2 \sec 2t \tan 2t.$

21. $(\sin^2 t)[(3 \cos^2 t)(-\sin t)] + (\cos^3 t)[(2 \sin t)(\cos t)] = (\sin t \cos^2 t)(2 \cos^2 t - 3 \sin^2 t).$

23. $\cot t.$ **25.** $\dfrac{1}{\sec t + \tan t}(\sec t \tan t + \sec^2 t) = \sec t.$ **27.** $\dfrac{\pi \sin \pi t}{(1 + \cos \pi t)^2}.$

29. $\dfrac{(1 - \cos t)(\cos t) - (1 + \sin t)(\sin t)}{(1 - \cos t)^2} = \dfrac{\cos t - \sin t - 1}{(1 - \cos t)^2}.$ **33.** $y = \dfrac{1}{2}t - \dfrac{3\sqrt{3} - 5\pi}{6}.$

35. $-2.$ **39.** $\dfrac{70\pi\sqrt{3}}{3}$ mm of mercury per sec.   **41.** 4060, June.   **43.** 4 P.M.

47. $f_s(s, t) = 2(s + t) \cos(s^2 + 2st), f_t(s, t) = 2s \cos(s^2 + 2st).$

49. $f_s(s, t) = \dfrac{2 \tan s \sec^2 s}{\sin^2 t}, f_t(s, t) = -\dfrac{2 \tan^2 s \cos t}{\sin^3 t}.$

Exercise 9.5

1. $-3 \cos t + C.$ **3.** $-\dfrac{\cos 2t}{2} + C.$ **5.** $\dfrac{\sin t^2}{2} + C.$ **7.** $-\dfrac{\cos 3\pi t}{3\pi} + C.$

9. $5 \sin\dfrac{x}{5} + C.$ **11.** $-\dfrac{\cos 6x^2}{12} + C.$ **13.** $\dfrac{\tan 3t}{3} + C.$ **15.** $\dfrac{\sec 5x}{5} + C.$

17. $-2 \cot \sqrt{t} + C.$ **19.** $\dfrac{\sin^5 x}{5} + C.$ **21.** $-\dfrac{2 \cos^{3/2} t}{3} + C.$ **23.** $\dfrac{1}{2}.$

25. $\dfrac{3}{16}.$ **27.** 0.0989.   **29.** $\dfrac{1}{2}$ sq unit.   **31.** $\dfrac{6 - \sqrt{2}}{2}$ sq units.

33. $\ln|\sin x| + C.$ **35.** $2000 + \dfrac{600}{\pi}$ dollars.   **37.** 90.

Exercise 9.6

1. $\dfrac{\pi}{2}.$ **3.** $\dfrac{\pi}{6}.$ **5.** 0.   **7.** $-\dfrac{\pi}{6}.$ **9.** $\dfrac{\pi}{4}.$ **11.** $\dfrac{3\pi}{4}.$ **13.** $\dfrac{1}{3}.$

15. $\dfrac{\sqrt{3}}{2}.$ **17.** 0.   **19.** $\dfrac{\pi}{3}.$ **21.** $\dfrac{3}{\sqrt{1 - 9x^2}}.$ **23.** $\dfrac{3x^2}{1 + x^6}.$

25. $-\dfrac{1}{\sqrt{-t^2 - 4t - 3}}.$ **27.** $\dfrac{9x}{\sqrt{1 - 9x^2}} + 3 \sin^{-1} 3x.$ **29.** $\dfrac{t^2}{1 + t^2} + 2t \tan^{-1} t.$

31. $-\dfrac{e^x}{\sqrt{1 - e^{2x}}}.$ **33.** $\dfrac{2 \sin^{-1} x}{\sqrt{1 - x^2}}.$ **35.** $\dfrac{2}{x\sqrt{x^4 - 1}}.$ **37.** 1.   **39.** $-\dfrac{2x}{(1 + x^2)^2}.$

41. $\dfrac{4\sqrt{2}}{5}.$ **43.** $\dfrac{y}{\sqrt{1 - x^2y^2}}.$ **45.** $\tan^{-1}\dfrac{x}{1000}.$ **47.** 5000π ft/min.

Exercise 9.7

1. $\tan^{-1} t + C.$ **3.** $\dfrac{1}{5} \tan^{-1} 5t + C.$ **5.** $\dfrac{1}{2} \sin^{-1} 2x + C.$ **7.** $\dfrac{1}{4} \sec^{-1} 2t + C.$

9. $\dfrac{\pi}{24}.$ **11.** $\sin^{-1}\dfrac{x}{4} + C.$ **13.** $\dfrac{1}{15} \tan^{-1}\dfrac{5t}{3} + C.$ **15.** $\dfrac{\pi}{12}.$

17. $-\dfrac{1}{4}\sqrt{3 - 4x^2} + C.$ **19.** $\dfrac{1}{2}\sec^{-1} t^2 + C.$ **21.** $\tan^{-1}(t + 2) + C.$

23. $\sin^{-1}\dfrac{t + 1}{3} + C.$ **25.** $\dfrac{\pi}{3}$ sq units.   **27.** $\dfrac{\pi}{16}.$ **29.** $250\pi - 160 + 1000\tan^{-1} 7.$

Review Problems—Chapter 9

1. $150°.$ **3.** $-240°.$ **5.** $3\pi/2.$ **7.** I.   **9.** II.   **11.** $5\pi/3.$

13. $3/5, -4/5, -3/4, -4/3, -5/4, 3/5.$

15. $-5/6, -\sqrt{11}/6, 5\sqrt{11}/11, \sqrt{11}/5, -6\sqrt{11}/11, -6/5.$ **17.** $\sqrt{3}/2.$ **19.** $0.$

21. $\sqrt{3}.$ **23.** $-1/2.$ **25.** $-1.$ **27.** $\sqrt{2}.$ **29.** $\sqrt{3}/3.$ **31.** $-1.$

33. $-2\sqrt{3}/3.$ **35.** $a = 2.8, c = 4.1, B = 0.8.$ **37.** $a = 2.5, b = 9.7, A = 0.3.$

39. $-0.8.$ **41.** $A = 4, p = 2\pi.$

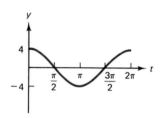

43. $A = 1, p = \dfrac{2\pi}{3}.$ **45.** $A = 2, p = 4\pi.$

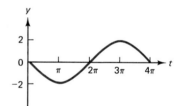

47. $-2\pi\sin(2\pi t + 3).$ **49.** $\dfrac{1}{2}\cos\dfrac{t - 3}{2}.$ **51.** $\dfrac{\sec^2\sqrt{t}}{2\sqrt{t}}.$

53. $\dfrac{1}{(t + 2)^2}\csc\left(\dfrac{t + 1}{t + 2}\right)\cot\left(\dfrac{t + 1}{t + 2}\right).$ **55.** $-3e^{3t}\csc^2 e^{3t}.$

57. $-\dfrac{t\csc\sqrt{t^2 + 4}\cot\sqrt{t^2 + 4}}{\sqrt{t^2 + 4}}.$ **59.** $2t^3\sec^2 t^2 + 2t\tan t^2.$

61. $\dfrac{2 + 3\sec 3t\tan 3t}{2\sqrt{2t + \sec 3t}}.$ **63.** $-\dfrac{2\cos 7t}{7} + C.$ **65.** $-\cot e^t + C.$ **67.** $\dfrac{\sin^7 y}{7} + C.$

69. $\dfrac{(8 - \cos 5t)^5}{25} + C.$ **71.** $1.$ **73.** $-\pi/2.$ **75.** $\pi/6.$ **77.** $5\pi/6.$ **79.** $0.$

81. $0.8.$ **83.** $-\dfrac{7}{\sqrt{1 - 49x^2}}.$ **85.** $\dfrac{e^x}{1 + e^{2x}}.$ **87.** $\dfrac{1}{\sqrt{16 - x^2}}.$ **89.** $0.$

91. $\dfrac{1}{\sqrt{1-x^2}\,\sin^{-1}x}.$ **93.** $-\dfrac{e^{\cos^{-1}x}}{\sqrt{1-x^2}}.$ **95.** $\dfrac{\sin^{-1}5t}{5}+C.$ **97.** $\pi/9.$

99. $\dfrac{1}{4}\sec^{-1}\dfrac{t}{4}+C.$ **101.** $\dfrac{\sqrt{4t^2-15}}{4}+C.$ **103.** $\dfrac{1}{4}\tan^{-1}\!\left(\dfrac{t-2}{4}\right)+C.$

105. 63 ft. **107.** 20.4 m. **109.** $y=-2t+\dfrac{\pi}{2}.$ **111.** (a) $\dfrac{2\pi+3\sqrt{3}}{6}$, (b) 0.

113. 600, March. **115.** $2\sqrt{2}$ sq units.

Exercise A.1

1. $-6.$ **3.** 2. **5.** 11. **7.** $-2.$ **9.** $-63.$ **11.** $-6.$ **13.** $6-x.$

15. $-12x+12y$ (or $12y-12x$). **17.** $-\frac{1}{3}.$ **19.** $-2.$ **21.** 18. **23.** 25.

25. $3x-12.$ **27.** $-x+2.$ **29.** $\frac{8}{11}.$ **31.** $-\dfrac{5x}{7y}.$ **33.** $\dfrac{2}{3x}.$ **35.** 3.

37. $\dfrac{7}{xy}.$ **39.** $\frac{5}{6}.$ **41.** $-\frac{1}{6}.$ **43.** $\dfrac{x-y}{9}.$ **45.** $\frac{1}{24}.$ **47.** $\dfrac{x}{6y}.$

49. Not defined. **51.** Not defined.

Exercise A.2

1. $2^5\,(=32).$ **3.** $w^{12}.$ **5.** $\dfrac{x^8}{y^{17}}.$ **7.** $\dfrac{x^{10}}{y^{50}}.$ **9.** $8x^6y^9.$ **11.** $x^6.$ **13.** $x^{14}.$

15. 5. **17.** $-2.$ **19.** $\frac{1}{2}.$ **21.** 10. **23.** 8. **25.** $\frac{1}{4}.$ **27.** $\frac{1}{32}.$

29. $4\sqrt{2}.$ **31.** $x\sqrt[3]{2}.$ **33.** $4x^2.$ **35.** $3z^2.$ **37.** $\dfrac{9t^2}{4}.$ **39.** $\dfrac{x^3}{y^2z^2}.$ **41.** $\dfrac{2}{x^4}.$

43. $\dfrac{1}{9t^2}.$ **45.** $7^{1/3}s^{2/3}.$ **47.** $x^{1/2}-y^{1/2}.$ **49.** $\dfrac{x^{9/4}z^{3/4}}{y^{1/2}}.$ **51.** $\sqrt[5]{(8x-y)^4}.$

53. $\dfrac{1}{\sqrt[5]{x^4}}.$ **55.** $\dfrac{2}{\sqrt[5]{x^2}}-\dfrac{1}{\sqrt[5]{4x^2}}.$ **57.** $\dfrac{2x^6}{y^3}.$ **59.** $t^{2/3}.$ **61.** $\dfrac{64y^6x^{1/2}}{x^2}.$

63. $xyz.$ **65.** $\frac{1}{9}.$ **67.** $\dfrac{4y^4}{x^2}.$ **69.** $x^2y^{5/2}.$ **71.** $\dfrac{y^{10}}{z^2}.$ **73.** $\dfrac{1}{x^4}.$ **75.** $-\dfrac{4}{s^5}.$

77. $\dfrac{4x^4z^4}{9y^4}.$

Exercise A.3

1. $x^2+9x+20.$ **3.** $x^2+x-6.$ **5.** $10x^2+19x+6.$ **7.** $x^2+6x+9.$

9. $x^2-10x+25.$ **11.** $2y+6\sqrt{2y}+9.$ **13.** $4s^2-1.$

15. $x^3+4x^2-3x-12.$ **17.** $2x^4+2x^3-5x^2-2x+3.$ **19.** $5x^3+5x^2+6x.$

21. $3x^2+2y^2+5xy+2x-8.$ **23.** $z-4.$ **25.** $3x^3+2x-\dfrac{1}{2x^2}.$

27. $x+\dfrac{-1}{x+3}.$ **29.** $3x^2-8x+17+\dfrac{-37}{x+2}.$ **31.** $t+8+\dfrac{64}{t-8}.$

33. $x-2+\dfrac{7}{3x+2}.$

Exercise A.4

1. $2(3x + 2)$.   **3.** $5x(2y + z)$.   **5.** $4bc(2a^3 - 3ab^2d + b^3cd^2)$.

7. $(x - 5)(x + 5)$.   **9.** $(p + 3)(p + 1)$.   **11.** $(4x - 3)(4x + 3)$.

13. $(z + 4)(z + 2)$.   **15.** $(x + 3)^2$.   **17.** $2(x + 4)(x + 2)$.   **19.** $3(x - 1)(x + 1)$.

21. $(6y + 1)(y + 2)$.   **23.** $2s(3s + 4)(2s - 1)$.   **25.** $x^{2/3}y(1 - 2xy)(1 + 2xy)$.

27. $2x(x + 3)(x - 2)$.   **29.** $4(2x + 1)^2$.   **31.** $2(x + 3)^2(x + 1)(x - 1)$.

33. $2(x + 4)(x + 1)$.   **35.** $(x^2 + 4)(x + 2)(x - 2)$.

37. $(y^4 + 1)(y^2 + 1)(y + 1)(y - 1)$.   **39.** $(x^2 + 2)(x + 1)(x - 1)$.

41. $x(x + 1)^2(x - 1)^2$.   **43.** $x(xy - 5)^2$.   **45.** $(x - 2)^2(x + 2)$.

47. $(y - 1)(y + 1)(y^4 + 4)^2$.

Exercise A.5

1. $-\dfrac{y^2}{(y - 3)(y + 2)}$.   **3.** $\dfrac{3 - 2x}{3 + 2x}$.   **5.** $\dfrac{x}{2}$.   **7.** $\dfrac{2}{3}$.   **9.** $\dfrac{2(x + 4)}{(x - 4)(x + 2)}$.

11. $-\dfrac{(2x + 3)(1 + x)}{x + 4}$.   **13.** $x + 2$.   **15.** $\dfrac{5}{3t}$.   **17.** $-\dfrac{1}{p^2 - 1}$.

19. $\dfrac{2x^2 + 3x + 12}{(2x - 1)(x + 3)}$.   **21.** $\dfrac{2x - 3}{(x - 2)(x + 1)(x - 1)}$.   **23.** $\dfrac{35 - 8x}{(x - 1)(x + 5)}$.

25. $\dfrac{x + 1}{3x}$.   **27.** $\dfrac{(x + 2)(6x - 1)}{2x^2(x + 3)}$.   **29.** $\dfrac{x - \sqrt{5}}{x^2 - 5}$.

Exercise A.6

1. $\frac{15}{2}$.   **3.** $\frac{12}{5}$.   **5.** 2.   **7.** 90.   **9.** $-\frac{26}{9}$.   **11.** $-\frac{37}{18}$.   **13.** 3.

15. $\frac{1}{4}$.   **17.** $\frac{3}{2}$.   **19.** $\frac{5}{3}$.   **21.** 3.   **23.** $P = \dfrac{I}{rt}$.   **25.** $x = \dfrac{p + 1}{6}$.

27. $r = \dfrac{S - P}{Pt}$.   **29.** $-2, -1$.   **31.** $4, 3$.   **33.** ± 2.   **35.** $0, 8$.   **37.** $\frac{1}{2}$.

39. $0, \pm 8$.   **41.** $0, 1, -2$.   **43.** $2, -1$.   **45.** $0, 2, -3, 4$.   **47.** $0, \pm 1$.

49. $-5, 3$.   **51.** $\frac{3}{2}$.   **53.** $\dfrac{5 \pm \sqrt{13}}{2}$.   **55.** No real roots.   **57.** $\frac{1}{2}, -\frac{5}{3}$.

59. $4, -\frac{5}{2}$.   **61.** $\dfrac{-2 \pm \sqrt{14}}{2}$.

Exercise A.7

1. $x > 2$.   **3.** $x \le 2$.   **5.** $y > 0$.   **7.** $x \le -\frac{1}{2}$.   **9.** $s < -\frac{2}{5}$.

11. $x \ge -\frac{7}{5}$.   **13.** $x > -\frac{2}{7}$.   **15.** No solution.   **17.** $x < \dfrac{\sqrt{3} - 2}{2}$.

19. $y \le -5$.   **21.** $-\infty < x < \infty$.   **23.** $t > \frac{17}{9}$.   **25.** $x \ge -\frac{14}{3}$.   **27.** $r > 0$.

29. $x \le -2$.   **31.** $y < 0$.

Index